Third Networks and Services

For a listing of recent titles in the
Artech House Communications and Network Engineering Series,
turn to the back of this book.

Third Networks and Services

Mehmet Toy
Hakkı Candan Çankaya

ARTECH
HOUSE
BOSTON | LONDON
artechhouse.com

Library of Congress Cataloging-in-Publication Data
A catalog record for this book is available from the U.S. Library of Congress.

British Library Cataloguing in Publication Data
A catalog record for this book is available from the British Library.

ISBN-13: 978-1-63081-175-4

Cover design by John Gomes

© 2017 Artech House
685 Canton Street
Norwood MA 02062

All rights reserved. Printed and bound in the United States of America. No part of this book may be reproduced or utilized in any form or by any means, electronic or mechanical, including photocopying, recording, or by any information storage and retrieval system, without permission in writing from the publisher.

All terms mentioned in this book that are known to be trademarks or service marks have been appropriately capitalized. Artech House cannot attest to the accuracy of this information. Use of a term in this book should not be regarded as affecting the validity of any trademark or service mark.

10 9 8 7 6 5 4 3 2 1

To the memory of my parents.
To my wife, Fusun, and my sons, Onur, Ozan and Atilla.
—Mehmet Toy

To my father Dr. Cengiz Çankaya and my mother Nurdagül Önder.
To my wife Ebru and my sweet little daughter Selin Alya.
—Hakkı Candan Çankaya

Contents

Foreword by Nan Chen — xvii

Foreword by Michael Lanman — xix

Preface — xxi

Acknowledgments — xxii

CHAPTER 1

Introduction and Overview — 1

1.1 Introduction — 1
1.2 Basic Ethernet — 2
1.3 Synchronization — 2
1.4 Pseudowires — 2
1.5 Protection — 3
1.6 Carrier Ethernet Architectures and Services — 3
1.7 Carrier Ethernet Traffic Management — 4
1.8 Ethernet Operations, Administrations and Maintenance (OAM) — 5
1.9 Circuit Emulation — 6
1.10 Ethernet Local Management Interface — 7
1.11 Provider Backbone Transport — 7
1.12 T-MPLS and MPLS-TP — 9
1.13 Virtual Private LAN Services — 11
1.14 Information Modeling — 13
1.15 The Third Network — 14

CHAPTER 2

Basic Ethernet — 15

2.1 Introduction — 15
2.2 CSMA/CD — 16
2.3 Full Duplex, PAUSE, Autonegotiation — 17
2.4 Repeaters and Hubs — 17
2.5 Bridges — 18
2.6 Switches — 19

2.7	Physical Layer		20
	2.7.1	10-Mbps Ethernet	22
	2.7.2	Fast Ethernet	22
	2.7.3	Gigabit Ethernet	22
	2.7.4	10-Gb, 40-Gb, and 100-Gb Ethernet	23
	2.7.5	LAN PHY	24
	2.7.6	LAN PHY/WAN PHY Sublayers	25
2.8	Temperature Hardening		27
2.9	Standards		27
2.10	Ethernet Frame Types and the EtherType Field		27
	2.10.1	Ethernet Frames	28
	2.10.2	Ethernet Address	29
	2.10.3	Ethernet II or DIX	29
	2.10.4	IEEE 802.3	30
	2.10.5	IEEE 802.2	30
	2.10.6	SNAP	31
2.11	Conclusion		32
	References		32

CHAPTER 3

Synchronization   33

3.1	Introduction		33
3.2	Application Requirements		35
3.3	Synchronization Standards		38
3.4	NTP/SNTP		39
3.5	Precision Time Protocol (IEEE 1588)		43
3.6	Synchronous Ethernet Networks		49
	3.6.1	Clocking Methods for Synchronization	52
	3.6.2	Impact of Packet Network Impairments on Synchronization	55
	3.6.3	Stabilization Period	56
	3.6.4	IWF Synchronization Function	56
	3.6.5	PRC	57
	3.6.6	Operation Modes	58
	3.6.7	Frequency Accuracy of Slave Clock	59
	3.6.8	EEC	59
3.7	Conclusion		60
	References		60

CHAPTER 4

Pseudowires   63

4.1	Introduction		63
4.2	Protocol Layers		63
4.3	Payload Types		64
4.4	Pseudowire Architectures		65
	4.4.1	PWE3 Preprocessing	65
	4.4.2	Payload Convergence Layer	68

		4.4.3	PW Demultiplexer Layer and PSN	70
		4.4.4	Maintenance Reference Model	70
	4.5	Control Plane		71
		4.5.1	PWE3 over an IP PSN	72
		4.5.2	PWE3 over an MPLS PSN	73
	4.6	Multisegment Architecture		73
	4.7	Multisegment Pseudowire Setup Mechanisms		76
		4.7.1	LDP SP-PE TLV	78
	4.8	Resiliency		78
	4.9	Quality of Service and Congestion Control		79
	4.10	Operations and Maintenance		79
	4.11	Security		81
	4.12	Conclusion		82
		References		83

CHAPTER 5

Ethernet Protection 85

	5.1	Introduction		85
	5.2	APS Entities		86
	5.3	Linear Protection		87
		5.3.1	1 + 1 Protection Switching	88
		5.3.2	1:1 Protection Switching	90
		5.3.3	Protection Switching Triggers	91
		5.3.4	APS PDU Format	91
	5.4	Ring Protection		95
		5.4.1	Protection Switching	97
	5.5	Link Aggregation		100
		5.5.1	LAG Objectives	103
		5.5.2	Link Aggregation Operation	103
		5.5.3	LACP	105
		5.5.4	Limitations	106
	5.6	Conclusion		106
		References		106

CHAPTER 6

Carrier Ethernet Architecture and Services 107

	6.1	Introduction		107
	6.2	Architecture		109
		6.2.1	Protocol Stack	110
		6.2.2	ETH Layer Characteristic Information (Ethernet Frame)	114
		6.2.3	ETH Layer Functions	118
		6.2.4	ETH Links	120
	6.3	Interfaces and Types of Connections		122
		6.3.1	UNI	122
		6.3.2	Ethernet Virtual Connection	124
		6.3.3	External Network-Network Interface (ENNI)	126

6.3.4	Operator Virtual Connection	127
6.3.5	VUNI/RUNI	127

6.4 EVC Services and Attributes 130
    6.4.1 Ethernet Service Types 130
    6.4.2 Ethernet Service Definitions 131
    6.4.3 Common Attributes for EVC Services 135
    6.4.4 E-Line Services Attributes and Parameters 140
    6.4.5 E-LAN Services Attributes and Parameters 141
    6.4.6 E-Tree Services Attributes and Parameters 143
6.5 OVC Services and Attributes 144
    6.5.1 Operator Services Attributes 149
6.6 Single CEN and Multi-CEN L2CP (Processing for Single CEN and Multi-CEN) 153
6.7 Conclusion 161
    References 161

CHAPTER 7

Carrier Ethernet Traffic Management 163

7.1 Introduction 163
7.2 Policing 164
7.3 Queuing, Scheduling, and Flow Control 165
7.4 Three-CoS Model 166
7.5 Service-Level Agreements 167
    7.5.1 Frame Delay Performance 167
    7.5.2 Frame Delay Range Performance 169
    7.5.3 Mean Frame Delay Performance 170
    7.5.4 Frame Delay Variation 170
    7.5.5 Frame Loss Ratio 171
7.6 SLSs 172
    7.6.1 Multipoint CoS Performance Objectives 173
    7.6.2 Focused Overload 175
7.7 Application-CoS-Priority Mapping 178
    7.7.1 CoS Identification 178
    7.7.2 PCP and DSCP Mapping 181
7.8 Bandwidth Profile 181
    7.8.1 Bandwidth Profile Algorithms 183
    7.8.2 Models and Use Cases for Bandwidth Profiles 186
7.9 CBS Values, TCP, and Shaping 188
7.10 Conclusion 197
    References 198
    Bibliography 198

CHAPTER 8

Carrier Ethernet Operation, Administration, Management, and Performance 199

8.1 Introduction 199

8.2	Link OAM	202
8.3	Service OAM	205
8.4	Maintenance Entities	207
8.5	Maintenance Points	208
8.6	OAM Addressing and Frame Format	212
8.7	Continuity Check Message (CCM)	215
8.8	Loopback and Reply Messages (LBM and LBR)	220
8.9	Link Trace and Reply Messages (LTM and LTR)	224
8.10	Ethernet Alarm Indication Signal (ETH-AIS)	227
8.11	Ethernet Remote Defect Indication (ETH-RDI)	229
8.12	Ethernet Locked Signal (ETH-LCK)	230
8.13	Performance Measurements	231
8.14	Performance Monitoring	233
8.15	Frame Loss Measurements	234
8.16	Availability	238
8.17	Frame Delay Measurements	242
8.18	Interframe Delay Variation Measurements	243
8.19	Testing	244
	8.19.1 Y.1731 Testing	245
	8.19.2 RFC 2544	246
	8.19.3 Complete Service Testing	249
	8.19.4 Service Activation Testing	249
	8.19.5 SAT Control Protocol and Messages	254
8.20	Performance Monitoring Solutions	263
8.21	Security	266
8.22	OAM Bandwidth	267
8.23	Conclusion	267
	References	267

CHAPTER 9

Circuit Emulation Services 269

9.1	Introduction	269
9.2	Circuit Emulation Functions	271
9.3	Adaptation Function Headers	274
9.4	Synchronization	276
9.5	TDM Application Signaling	277
9.6	CESoETH Defects and Alarms	278
9.7	Performance Monitoring of CESoETH	280
9.8	CESoETH Service Configuration	280
9.9	Conclusion	281
	References	281

CHAPTER 10

Carrier Ethernet Local Management Interface 283

10.1	Introduction	283
10.2	E-LMI Messages	286

10.3	E-LMI Message Elements	288
10.4	E-LMI System Parameters and Procedures	292
	10.4.1 Periodic Polling Process	294
10.5	UNI-C and N Procedures	295
10.6	Conclusion	296
	References	296

CHAPTER 11

Provider Bridges, Provider Backbone Bridges, and Provider Backbone Transport — 297

11.1	Introduction	297
11.2	IEEE 802.1AB	298
	11.2.1 Architecture	299
	11.2.2 Principles of Operation	300
	11.2.3 802.1AB Frame Format	301
11.3	Provider Backbone Bridges	304
	11.3.1 802.1ah Frame Format	305
	11.3.2 PBB Principles of Operation	307
	11.3.3 Provider Backbone Bridge Network	308
11.4	Provider Backbone Transport	312
	11.4.1 Principles of Operation	313
	11.4.2 End-to-End Carrier Ethernet with Multiple PBT Domains and Single Domain	315
	11.4.3 PBB-TE Network	316
	11.4.4 PBT: MPLS Internetworking	318
11.5	Conclusion	318
	References	318

CHAPTER 12

Transport MPLS — 321

12.1	Introduction	321
12.2	Differences from MPLS	323
12.3	Architecture	324
	12.3.1 T-MPLS Interfaces	326
12.4	T-MPLS Frame Structure	328
12.5	T-MPLS Networks	328
	12.5.1 T-MPLS Protection	328
12.6	Conclusion	333
	References	334

CHAPTER 13

MPLS Transport Profile — 335

13.1	Introduction	335
13.2	Frame Format	338
13.3	Architecture	339

		13.3.1	Data Plane	340
		13.3.2	MPLS-TP Router Types	343
		13.3.3	Service Interfaces	343
		13.3.4	IP Transport Service	345
		13.3.5	Generic Associated Channel	348
		13.3.6	Control Plane	349
		13.3.7	Network Management	351
	13.4	OAM		352
		13.4.1	OAM Hierarchy	353
		13.4.2	OAM Functions for Proactive Monitoring	357
		13.4.3	Data Plane Loopback, RDI, AIS	359
		13.4.4	Client Failure Indication, Route Tracing, and Lock Instruct	360
	13.5	Protection Switching		361
	13.6	Security Considerations		361
	13.7	Conclusion		362
		References		362

CHAPTER 14

Virtual Private LAN Services — 365

14.1	Introduction		365
14.2	Data Plane		370
	14.2.1	VPLS Encapsulation	370
	14.2.2	Classification and Forwarding	370
	14.2.3	MAC Address Learning and Aging	371
14.3	LDP-Based VPLS		372
	14.3.1	Flooding, Forwarding, and Address Learning	372
	14.3.2	Tunnel Topology	373
	14.3.3	Discovery	375
	14.3.4	LDP-Based Signaling	376
	14.3.5	Data Forwarding on an Ethernet PW	378
	14.3.6	Hierarchical VPLS	379
14.4	BGP Approach		384
	14.4.1	Auto-Discovery	384
	14.4.2	Signaling	385
	14.4.3	BGP VPLS Operation	386
	14.4.4	Multi-AS VPLS	388
	14.4.5	Hierarchical BGP VPLS	389
14.5	Security		390
14.6	External Network-Network Interface		390
14.7	Conclusion		391
	References		391

CHAPTER 15

Information Modeling — 393

15.1	Introduction	393
15.2	Information Modeling of Carrier Ethernet Services for EMS and NMS	395

	15.2.1	Management of EVC, OVC and ELMI and Discovery of Ethernet	397
	15.2.2	Carrier Ethernet Service OAM Configuration	399
	15.2.3	Carrier Ethernet Service Performance Management	400
	15.2.4	Carrier Ethernet Fault Management	400
	15.2.5	Common Management Objects for Carrier Ethernet Services	402
	15.2.6	Class Diagrams of ITU-T Q.840.1 Carrier Ethernet Management Entities	402
	15.2.7	Carrier Ethernet Management Entities	405
	15.2.8	ENNI and Virtual UNI Related Objects	405
	15.2.9	Fault Management Objects	406
	15.2.10	Performance Monitoring Objects	408
	15.2.11	ENNI and OVC MIBs	408
	15.2.12	SOAM FM MIB [16]	410
	15.2.13	SOAM PM MIB [24]	413
15.3	Service-Level Information Modeling		435
	15.3.1	EVC	438
	15.3.2	OVC	439
	15.3.3	Carrier Ethernet External Interface	440
	15.3.4	UNI	440
	15.3.5	ENNI	441
	15.3.6	ENNI Service	442
	15.3.7	Egress Equivalence Class Identifier	443
15.4	YANG Models for Carrier Ethernet Services		444
	15.4.1	SOAM YANG CFM Module [20]	446
	15.4.2	SOAM FM YANG Module	452
	15.4.3	PM YANG Module [39]	455
15.5	Network Resource Model for Carrier Ethernet Services		459
15.6	Conclusion		461
	References		463

CHAPTER 16

Third Network 465

16.1	Introduction		465
16.2	Life-Cycle Service Orchestration		465
16.3	LSO Management Abstractions and Constructs		468
16.4	Virtualization		470
16.5	Network Virtualization		472
16.6	Virtualized Carrier Ethernet Services		473
	16.6.1	Components of Virtualized Carrier Ethernet Services	474
	16.6.2	Service Chaining for EPL	478
16.7	Cloud Services Architectures		479
	16.7.1	Protocol Stacks and Interfaces	480
	16.7.2	Cloud Services	483
	16.7.3	Network as a Service	484

		16.7.4	Infrastructure as a Service	485
		16.7.5	Security as a Service	489
		16.7.6	Platform as a Service	490
		16.7.7	Software as a Service	490
		16.7.8	Communication as a Service	491
	16.8	Conclusion		492
		References		492

About the Authors 493

Index 495

Foreword by Nan Chen

Four years after MEF, the defining body of carrier Ethernet, introduced Carrier Ethernet 2.0 (CE 2.0) as the gold standard for connectivity, the 15-year-old global market for Carrier Ethernet services has reached $50+ billion, according to IHS Markit. Service providers worldwide count Carrier Ethernet services among the fastest growing services in their wholesale and retail portfolios.

So where do we go next with such a successful market?

The communications industry is in the midst of a paradigm shift from static connectivity services to agile, assured, and orchestrated services that are optimized for the digital economy and the hyperconnected world. MEF refers to these new services as "Third Network" services because they combine the agility of the Internet with the performance and security of CE 2.0 networks.

Emerging Third Network services provide an on-demand experience with user-directed control over service capabilities and cloud connectivity. These services will be delivered over automated, virtualized, and interconnected networks that build upon a CE 2.0 foundation and are powered by new technologies like Lifecycle Service Orchestration (LSO), software-defined networking (SDN), and network functions virtualization (NFV). Third Network services have the potential to be globally certified by MEF like CE 2.0 services are today.

All of the collaborative work that MEF members have put into Carrier Ethernet specifications, operational frameworks, information models, and certification programs for services, equipment, and professionals serve as critical build blocks for creating new orchestrated connectivity services (e.g., wavelengths, CE 2.0, and IP) that can be combined with orchestrated NFV-based services (e.g., virtual security and virtual routers).

MEF is working with leading service and technology providers, open source projects, and other industry organizations to take the huge, high-growth Carrier Ethernet services market to the next level by automating it with LSO. The development of standardized open APIs will allow operators to overcome operational support system (OSS) hurdles that impede their ability to offer agile, on-demand services and to efficiently scale service deployment. LSO will allow service providers not only to orchestrate services on an end-to-end basis within their own networks but also across multiple provider networks. This has true, game-changing potential for service innovation, competition, and revenue generation.

We are entering a new industry phase in which we could see greater divergence between service providers who lead the market and the rest of the pack. Companies that already offer CE 2.0 certified services and are planning to use combinations of LSO, SDN, and NFV to deliver on-demand services with physical and virtual service endpoints, dynamic WAN + cloud solutions, and the like, have the potential to put distance between themselves and those providers that largely have been observing developments up until now.

Communications service providers and their employees will benefit by keeping a close watch over key trends and the quickening pace of network technology and service innovation. Resources like this book, *Third Network Services*, by Mehmet Toy and Hakkı Çankaya have a valuable role to play in helping industry professionals keep up with current and emerging technologies and how they work together to enable network and service transformation.

Nan Chen
Founding President, MEF

Foreword by Michael Lanman

The networking industry is going through revolutionary changes with the introduction of software defined networks (SDN), network function virtualization (NFV), and cloud technologies. Process automation for network services from the order through billing including interactions among operation support systems of service providers is part of this revolution.

After defining feature rich standardized Carrier Ethernet services that have been deployed world-wide, initiation of Third Networks concept with life-cycle service operation (LSO) by MEF should accelerate the automation of these processes.

Third Networks and Services explores theoretical and practical aspects of Carrier Ethernet and related technologies such as virtual private LAN services (VPLS), and information modeling and virtualization of Carrier Ethernet services. The book also explains fundamentals of cloud services and their architectures that are applicable to cloud applications and networking beyond Carrier Ethernet.

Third Networks and Services is a valuable reference for those who are developing and offering not only fixed Layer-2 connectivity services, but also virtualized and cloud-based Layer-3 connectivity services and applications.

Michael Lanman
Senior Vice-President of Enterprise Products
Verizon Communications, Inc.
March 2017

Preface

Networks and Services, which covers Carrier Ethernet and related technologies, was written by Dr. Toy and published by John Wiley in 2012. The book has been used in industry and academia and translated into Turkish by Bahçesehir University in Istanbul, Turkey. The book has received very positive feedback about quality from many colleagues and practitioners in the field. In 2015, Dr. Çankaya, who had been using it as his textbook in his graduate courses, proposed writing the second edition together. This is how *Third Networks and Services* was born.

The term *third network* was first used by Nan Chen in an MEF quarterly meeting. It started as a network of operational systems and turned into life-cycle services operation of not only Carrier Ethernet services, but also Layer-3 connectivity and cloud services.

In this book, we updated most of the Carrier Ethernet-related chapters from *Networks and Services* to include new MEF specifications for interfaces, service OAM, Layer-2 Control Protocol (L2CP), and testing that became the requirements for Carrier Ethernet 2.0 (CE2.0) compliance. Furthermore, we addressed TCP issues, committed burst size (CBS) issues, hierarchical policing, operator virtual connection (OVC) services, Service Activation Testing (SAT) Control Protocol and Protocol Data Unit (PDU), and information modeling associated with Carrier Ethernet. After that, we introduced the Third Network concept, describing life-cycle service orchestration (LSO), virtualization, and cloud architectures and services.

In recent years, with the introduction of software-defined networks (SDN), network function virtualization (NFV), and cloud concepts, the networking industry has been going through a revolution. Automation of the steps involved in LSO has been one of the key focus areas for service providers. We hope that this book will provide insight into fundamentals of fixed, virtualized and cloud-based services, and their life-cycle orchestration.

We thank Michael Lanman and Nan Chen for writing their forewords, our MEF colleagues for their contributions to Carrier Ethernet and LSO, and Stephen Solomon, Kate Skypeck, and Darrell Judd of Artech House for their help with the publication.

Acknowledgments

Hakkı Candan Çankaya would like to acknowledge the excellent work done by Dr. Mehmet Toy for the Networks and Services book. Dr. Çankaya also thanks his wife Ebru for her support and his six-year-old sweet little Selin Alya for her understanding, so that he could work on this book instead of taking her to her favorite neighborhood playground.

CHAPTER 1

Introduction and Overview

1.1 Introduction

The Third Network was initially defined as an agile, assured, and orchestrated network supporting Carrier Ethernet 2.0 and employing software-defined networking (SDN) and network function virtualization (NFV) technologies. The service was the connectivity services that were initially Layer-2 Carrier Ethernet services and later both Layer-2 Carrier Ethernet services and Layer-3 Internet protocol (IP) services, in addition to management services under the life-cycle service orchestration (LSO) umbrella. As Open Cloud Connect (OCC) merged MEF in 2016, the Third Network was expanded to include cloud services. Therefore, the Third Network became an umbrella concept that includes connectivity services, cloud applications, cloud services, and LSO for all.

Ethernet has been the dominant technology for local area networks (LANs) for many decades. With the introduction of circuit concept over Ethernet by IEEE and MEF specifications, Ethernet became a key technology for carriers in metropolitan area networks and wide area networks.

Chapter 2 describes basics of Ethernet such as protocol stack, bridges, switches, and hubs. Chapter 3, 4, and 5 cover the key techniques used in building carrier class Ethernet networks and services, namely, synchronization, pseudowires, and protection.

Chapter 6 begins describing Carrier Ethernet network architectures that incorporate single or multiple operators and services that are currently deployed in the industry and evolving. It is followed by Chapters 7 and 8, describing traffic management and Ethernet Operations, Administrations, and Maintenance (OAM) capabilities of Carrier Ethernet, respectively.

Chapter 9 is devoted to circuit emulation services due to its complexity. For the same reason, Chapter 10 is devoted to Ethernet Local Management Interface (E-LMI).

Addressing scalability of Carrier Ethernet has been questioned at times, despite the availability of S-Tag and C-Tag combination. Provider backbone bridges (PBB) and provider backbone transport (PBT) attempt to resolve this scalability issue. Chapter 11 describes these technologies in details.

Three technologies, namely, Transport multiprotocol lable switching (MPLS), MPLS Transport Profile, and virutal private LAN service (VPLS), can compete or work together with Carrier Ethernet in forming data networks. Chapters 12, 13, and 14 describe them in details.

Chapter 15 covers the information model for the management of the Carrier Ethernet services in general and specifically the YANG models. A common management model is needed for Carrier Ethernet services to ensure that the information in operation support systems (OSSs), SDN orchestrators, SDN controllers, network management systems (NMSs), element management systems (EMSs), and network elements (NEs) from various vendors are logically consistent and allow service providers to readily integrate their capabilities into the management environment.

Finally, Chapter 16 describes the Third Network concept. It describes the LSO and virtualization of Carrier Ethernet services, and provides details of cloud services architectures.

1.2 Basic Ethernet

The Ethernet is physical layer local area network (LAN) technology invented in 1973. Since its invention, the Ethernet has become the most widely used LAN technology because of its speed, low cost, and relative ease of installation. Ethernet PHY supporting rates from 10 Mbps to 100 Gbps are available.

Chapter 2 describes the carrier sense, multiple access, collision detect (CSMA/CD), protocol stack, bridges, switches, and hubs. Toward the end of the chapter, the frame formats defined by IEEE 802.3 and 80.2 are described.

1.3 Synchronization

The time division multiplexing (TDM) networks transporting multiplexed voice and data traffic require highly accurate frequency information to be recoverable from their physical layers. Typically, a TDM circuit service provider will maintain a timing distribution network, providing synchronization traceable to a primary reference clock (PRC) that is compliant with clock quality. A reference timing signal of suitable quality is distributed to the network elements processing the application. One approach is to follow a distributed PRC strategy. An alternative approach is based on a master-slave strategy.

Synchronization in Carrier Ethernet networks is required mainly due to certain applications like mobile backhaul and bandwidth explosion in mobile networks. Frequency, phase, and time synchronizations are needed. Chapter 3 describes two main synchronization techniques, Precision Time Protocol (IEEE 1588 v2) and Synchronous Ethernet networks (SyncE).

1.4 Pseudowires

Pseudowire Emulation Edge-to-Edge (PW3) is a mechanism to emulate telecommunications services across packet-switched networks such as Ethernet, IP, or MPLS.

The emulated services are mostly T1/T3 leased lines, frame relay, Ethernet, and asynchronous transfer mode (ATM) to maximize return on existing assets and minimize operating costs of the service providers. PWs encapsulate cells, bit streams, and protocol data units (PDUs) and transport them across Carrier Ethernet virtual circuits (EVCs), IP, or MPLS tunnels. The transportation of encapsulated data usually require managing the sequence and timing of data units to emulate services such as T1/T3 leased lines and ATM.

Chapter 4 describes various aspects of pseudowires including its architectures, protocol stack, frame forwarding, control plane, resilience, and OAM.

1.5 Protection

Services such as broadcast video, voice over IP, video on demand, and others require 99.999% availability. When failures occur, they are not supposed to be noticed by the subscriber. Automatic protection switching guarantees the availability of resources in the event of failure, and to ensure that switchover is achieved in less than 50 ms. The 50-ms value has been long debated in the industry. In fact, MEF specifications require 500-ms protection switching at user-network interface (UNI) or ENNI. The smaller switching time increases the cost of the equipment. Chapter 5 describes various protection techniques including Linear Protection, Ring Protection, and Link Aggregation Group (LAG)/Link Aggregation Control Protocol (LACP).

1.6 Carrier Ethernet Architectures and Services

Chapter 6 begins describing Carrier Ethernet network architectures and services that are developed by Metro Ethernet Forum and currently deployed in the industry.

Ethernet transport network is a two-layer network consisting of Ethernet media access control (MAC), or (ETH) layer network and Ethernet PHY (ETY) layer network. The ETY layer is the physical layer defined in IEEE 802.3. The ETH layer is the pure packet layer.

The ETH Layer Network is divided into ETH subnetworks that are also called ETH Flow Domain (EFD). An EFD is a set of all ETH flow points transferring information within a given administrative portion of the ETH Layer network. EFDs may be partitioned into sets of nonoverlapping EFDs that are interconnected by ETH links. An IEEE 802.1D bridge represents the smallest instance of an EFD.

The termination of a link is called a flow point pool (FPP). The FPP describes configuration information associated with an interface, such as a user network interface (UNI) or external network-to-network interface (ENNI). An Ethernet frame is exchanged over the ETH layer that consists of Preamble, Start Frame Delimiter (SFD), Destination MAC Address (DA), Source MAC Address (SA), (Optional) 802.1QTag, Ethernet Length/Type (EtherType), User Data, padding if required, frame check sequence (FCS), and extension field, which is required only for a 100-Mbps half-duplex operation. A service frame can be a unicast, multicast, broadcast, and Layer 2 Control Protocol (L2CP) frame.

Two interfaces are defined for Carrier Ethernet, UNI, and ENNI. A connection between UNIs is called Ethernet virtual connection (EVC), while a connection between a UNI and an ENNI is called operator virtual connection (OVC). An EVC or OVC can be point-to-point, point-to-multipoint, or multipoint-to-multipoint. The following services are built on top of these EVCs and OVCs:

- Ethernet line (E-Line) services consisting of Ethernet private line (EPL) and Ethernet virtual private line (EVPL) services;
- Ethernet LAN (E-LAN) services consisting of EPL and EVPL services;
- Ethernet Tree (E-Tree) services consisting of Ethernet private tree (EP-Tree) and Ethernet private virtual tree (EVP-Tree) services;
- Ethernet access services offered over ENNI.

The operator responsible from the end-to-end EVC may order a tunnel between its remote user that is connected to another operator's network and its ENNI gateway. This access of the remote user to the service provider's network is just an access tunnel serving the user. Similarly, there may be multiple operators' networks in between the remote user and the operator that is responsible for the EVC. In this case, the operator networks in the middle provider transit tunnels for the service. Details of these services are described in details in Chapter 6.

1.7 Carrier Ethernet Traffic Management

Traffic management, which is also called packet conditioning, performs queuing, scheduling, and policing frames. The conditioning function includes:

1. Counting bytes and packets that pass and drop;
2. Policing packet flow according to predetermined burst/flow rate (includes both color-aware and color-unaware);
3. Setting color;
4. Setting 1P/DSCP/TOS priority values;
5. Remarking/remapping 1P/DSCP/TOS based on a predefined remarking table;
6. Shaping packets to conformance to predetermined flow rate;
7. Sending packets to a particular queue.

Once frames are classified to a Class of Service (CoS) flow, ingress frames are then policed according to the CoS bandwidth profile assigned to the flow, consisting of: committed information rate (CIR), excess information rate (EIR), committed burst size (CBS), excess burst size (EBS), coupling flag (CF), and color mode (CM).

Frames that are marked "green" by the policer are always queued, and frames that are marked "yellow" are only queued if the fill level of the queue is less than a defined threshold. This ensures that green frames are always forwarded. The Generic Token Bucket Algorithm and various versions have been adopted for this process.

MEF defines three classes as H, M, and L. Their priorities are indicated by priority code points (PCP) or differentiated services code point (DSCP) or type of service (TOS) bytes. Each CoS label has its own performance parameters:

- Class H is intended for real-time applications with tight delay/jitter constraints such as voice over Internet protocol (VoIP).
- Class M is intended for time-critical.
- Class L is intended for non-time-critical data such as e-mail.

Service level agreements (SLAs) between a service provider and subscriber for a given EVC are defined in terms of delay, jitter, loss, and availability performances. SLAs are further defined for various performance tiers (PTs) based on distance.

Chapter 7 also describes shaping, CBS values, and their impact on transmission control protocol (TCP)performance.

1.8 Ethernet Operations, Administrations and Maintenance (OAM)

Ethernet OAM consist of fault management, performance management, testing, and service monitoring capabilities to support Ethernet services described in Chapter 6. Measurements, events/alarms, and provisioning attributes for interfaces (i.e., UNI, VUNI, ENNI), EVC, and OVC are defined.

Ethernet OAM can be categorized as: link layer OAM, service layer OAM, and OAM via Ethernet local management interface (E-LMI).

Link OAM provides mechanisms to monitor link operation and health, and improves fault isolation. The major features covered by this protocol are:

- Discovery of MAC address of the next hop;
- Remote failure indication for link failures;
- Power failure reporting via dying gasp;
- Remote loopback.

Ethernet alarm indication signal (AIS) and remote defect indication (RDI) functions are used to suppress downstream alarms and eliminate alarm storms from a single failure. This is similar to AIS/RDI in legacy services like SONET, TDM, and frame relay.

The service OAM consists of continuity check, loopback, link trace, delay/jitter measurements, loss measurements, and in-service and out-of-service testing protocols to monitor SLAs, identify service level issues, and debug them.

Connectivity fault management (CFM) creates a maintenance hierarchy by defining maintenance domain and maintenance domain levels where maintenance end points (MEPs) determine domain boundaries. Maintenance points that are between these two boundary points, MEPs, are called maintenance intermediate points (MIPs).

Continuity check messages (CCM) are issued periodically by MEPs. They are used for proactive OAM to detect loss of continuity (LOC) among MEPs and discovery of each other in the same domain. MIPs will discover MEPs that are in the same domain using CCMs as well. In addition, CCM can be used for loss measurements and triggering protection switching

The Ethernet loopback function (ETH-LB) is an on-demand OAM function that is used to verify connectivity of a MEP with a peer maintenance point (MP). Loopback is transmitted by a MEP on the request of the administrator to verify connectivity to a particular MP.

Ethernet Link Trace function (ETH-LT) is an on-demand OAM function initiated in a MEP on the request of the administrator to track the path to a destination MEP. They allow the transmitting node to discover connectivity data about the path. The PDU used for ETH-LT request information is called LTM and the PDU used for ETH-LT reply information is called LTR.

Performance measurements can be periodic and on-demand. They are available in 15-minute bins. They can be in-service measurements to monitor health of the network as well as user SLAs and out-of-service measurements before turning up the service for isolating troubles and identifying failed components during failures.

Delay measurement can be performed using the 1DM (one-way delay measurement) or DMM/DMR (delay measurement message/delay measurement reply) PDUs. Loss measurement can be performed by counting service frames using the LMM/LMR (loss measurement message/loss measurement reply) PDUs as well as by counting synthetic frames via the SLM/SLR (synthetic loss message/synthetic loss reply) PDUs.

In addition, Chapter 8 describes availability, testing, and performance monitoring solutions.

1.9 Circuit Emulation

Circuit emulation services (CES) allows Carrier Ethernet networks (CEN) to be able to support legacy equipment of users by supporting legacy interfaces such as TDM. As a results, the users benefit from Carrier Ethernet capabilities without replacing their equipment.

In CES, data streams are converted into frames for transmission over Ethernet. At the destination site, the original bit stream is reconstructed when the headers are removed, payload is concatenated, and the clock is regenerated, while ensuring very low latency. The CEN behaves as a virtual wire.

TDM data is delivered at a constant rate over a dedicated channel. The TDM pseudowire operates in various modes, circuit emulation over packet switched network (CESoPSN), structure agnostic TDM over packet (SAToP), TDMoIP, and high-level data link control emulation over PSN (HDLCoPSN).

In SAToP, TDM is typically unframed. When it is framed or even channelized, the framing and channelization structure are completely disregarded by the transport mechanisms. In such cases, all structural overhead must be transparently transported along with the payload data, and the encapsulation method employed provides no mechanisms for its location or utilization. However, structure-aware TDM transport may explicitly safeguard TDM structure. The unstructured emulation mode is suitable for leased line.

Ethernet CES provides emulation of TDM services, such as N × 64 Kbps, T1, T3, and OC-n, across a metro Ethernet network (MEN), but transfers the data across MEN. From the customer perspective, this TDM service is the same as any other TDM service.

Circuit emulation applications interconnected over a CESoETH service may exchange signaling in addition to TDM data. With structure-agnostic emulation, it is not required to intercept or process circuit emulation signaling. Signaling is embedded in the TDM data stream, and hence it is carried end-to-end across the emulated circuit. With structure-aware emulation, transport of common channel signaling (CCS) may be achieved by carrying the signaling channel with the emulated service such as channel 23 for DS1.

In addition, Chapter 9 describes performance monitoring, provisioning, and fault management of CES over Ethernet.

1.10 Ethernet Local Management Interface

Chapter 10 describes Ethernet local management interface (E-LMI) protocol that operates between the customer edge (CE) device and network edge of the service provider.

The E-LMI protocol includes the following procedures:

- Notification to the CE device of the addition of an EVC;
- Notification to the CE device of the deletion of an EVC;
- Notification to the CE device of the availability (active/partially active) or unavailability (inactive) state of a configured EVC;
- Notification to the CE device of the availability of the remote UNI;
- Communication of UNI and EVC attributes to the CE device.

In order to transfer E-LMI messages between the UNI-C and the UNI-N, a framing or encapsulation mechanism is needed. The E-LMI frame structure is based on the IEEE 802.3 untagged MAC-frame format where the E-LMI messages are encapsulated inside Ethernet frames.

At E-LMI, a STATUS message is sent by the UNI-N to the UNI-C in response to a STATUS ENQUIRY message to indicate the status of EVCs or for the exchange of sequence numbers. STATUS ENQUIRY is sent by the UNI-C to request status or to verify sequence numbers.

The E-LMI procedures are characterized by a set of E-LMI messages that will be exchanged at the UNI. These message exchanges can be asynchronous or periodic. Periodic message exchanges are governed by timers, status counters, and sequence numbers.

1.11 Provider Backbone Transport

Addressing the scalability of Carrier Ethernet is being questioned. Provider backbone bridges (PBB) and provider backbone transport (PBT) described in Chapter 11 attempts to resolve the scalability issue. These extensions to the Ethernet protocols are developed to transform Ethernet to a technology ready for use in MANs/WANs.

Services supported in LAN/MEN such as E-LAN and E-LINE will be supported end-to-end. This results in no changes to the customer's LAN equipment, providing end-to-end usage of the technology, contributing to wider interoperability and low cost. SLAs provide end-to-end performance, based on rate, frame loss, delay, and jitter and enable traffic engineering (TE) to fine-tune the network flows.

Underlying protocols for PBT are as follows:

- IEEE 802.1AB Link Layer Discovery Protocol which is used to discover the network layout and forwarding this information to the control plane or management layer.
- The IEEE 802.1ag protocol to monitor the links and trunks in the PBT layout.
- Expansion of the PBB protocol defined in IEEE 802.1ah
- IEEE 802.1Qay, Provider Backbone Bridges with Traffic Engineering or PBT protocol.

PBB, namely MAC-in-MAC encapsulation, supports complete isolation of individual client-addressing fields as well as isolation from address fields used in the operator backbone. Client provider bridge (PB) frames are encapsulated and forwarded in the backbone network based on new B-DA (backbone destination address), B-SA (backbone source address), and B-VID (backbone VLAN-ID).

Although Q-in-Q supports a tiered hierarchy (i.e., no tag, C-Tag/C-VLAN ID, and S-Tag/S-VLAN ID), the service provider can create 4,094 customer VLANs, which is insufficient for large metropolitan and regional networks. The 802.1ah introduces a new 24-bit tag field (I-SID), service instance identifier, to overcome 12-bit S-VID (S-VLAN ID) defined in PB. This 24-bit tag field is proposed as a solution to the scalability limitations encountered with the 12-bit S-VID defined in provider bridges.

Provider backbone bridges operate the same way as traditional Ethernet bridges. Connectivity fault management (CFM) addresses the end-to-end OAM such as loopback at specific MAC, link trace, and continuity check.

PBB located at the backbone of the PBT network is called backbone core bridge (BCB). The bridge located at the edge of PBT network is called backbone edge bridge (BEB). BCB is an S-VLAN bridge used within the core of a PBBN. Backbone edge bridge (BEB) is a system that encapsulates customer frames for transmission across a provider backbone bridge network (PBBN).

Backbone edge bridge (BEB) has three types: I-type backbone edge bridge (I-BEB), B-type backbone edge bridge (B-BEB), and IB-type backbone edge bridge (IB-BEB).

I-component is responsible for encapsulating frames received from customers, assigning each to a backbone service instance and destination identified by a backbone destination address, backbone source address, and a service instance identifier (I-SID).

B-component is responsible for relaying encapsulated customer frames to and from I-components or other B-components when multiple domains interconnect, either within the same BEB or externally connected, checking that ingress/egress is permitted for frames with that I-SID, translating the I-SID (if necessary), and using it to assign the supporting connection parameters (backbone addresses if necessary and VLAN identifiers) for the PBBN, and relaying the frame to and from the provider network port(s).

PBBN provides multipoint tunnels between provider bridged networks (PBNs) where each B-VLAN carries many S-VLANs.

Traffic-engineered provider backbone bridging (PBT) is intended to bring connection-oriented characteristics and deterministic behavior to Ethernet. It turns off the Ethernet's spanning tree and media-access-control address-flooding and learning characteristics. That lets Ethernet behave more like a traditional carrier transport technology.

The frame format of PBT is the same as the format used for implementing PBB. The difference is in the meaning of the frame's fields. The VID and the B-DA fields together form a 60-bit globally unique identifier.

The control plane is used to manage the forwarding tables of the switches; to create PBT tunnels, all switches need to be controlled from one (PBT) domain. This technique enables circuit-switching on an Ethernet.

Chapter 11 further describes PBT networks and PBT-MPLS interworking.

1.12 T-MPLS and MPLS-TP

Chapters 12, 13, and 14 describe three technologies that are competing or working with Carrier Ethernet in forming data networks, Transport MPLS (T-MPLS), MPLS Transport Profile (MPLS-TP), and VPLS. Although MPLS-TP is supposed to replace T-MPLS, T-MPLS is already deployed. Therefore, Chapter 12 is devoted to it.

T-MPLS offers packet-based alternatives to SONET circuits and promises much greater flexibility in how packet traffic is transported through their metro and core optical networks. In T-MPLS, a new profile for MPLS is created so that MPLS label switched paths (LSPs) and pseudowires can be engineered to behave like TDM circuits or Layer 2 virtual connections.

T-MPLS is intended to be a separate layer network with respect to MPLS. However, T-MPLS will use the same data-link protocol ID (e.g., EtherType), frame format, and forwarding semantics as defined for MPLS frames. Unlike MPLS, it does not support a connectionless mode and is intended to be simpler in scope, less complex in operation, and more easily managed. Layer-3 features have been eliminated and the control plane uses a minimum of IP to lead to low-cost equipment implementations.

As an MPLS subset, T-MPLS abandons the control protocol family that IETF defines for MPLS. It simplifies the data plane, removes unnecessary forwarding processes, and adds ITU-T transport style protection switching and OAM functions (e.g., connectivity verification, alarm suppression, remote defect indication).

The key differences of T-MPLS compared with MPLS include:

- Use of bidirectional LSPs (label-switched paths);
- No PHP (penultimate hop popping) option;

- No ECMP (equal cost multiple path) option.

T-MPLS, similar to MPLS, defines the UNI, which is the interface between a client and service node, and network-network interface (NNI), which is between two service nodes.

In a typical T-MPLS network, a primary LSP and backup LSP are provided. The switching between the primary and secondary LSP tunnels can take place within 50 ms. These T-MPLS tunnels can support both Layer-3 IP/MPLS traffic flows and Layer-2 traffic flows via pseudowires. The T-MPLS protection can be linear or ring.

MPLS-TP is a continuation of T-MPLS. A Joint Working Group (JWT) was formed between the IETF and the ITU-T to achieve mutual alignment of requirements and protocols and come up with another approach. The T-MPLS is renamed to MPLS-TP to produce a converged set of standards for MPLS-TP. The MPLS-TP is a packet-based transport technology based on the MPLS Traffic Engineering (MPLS-TE) and pseudowire (PW) data plane architectures.

The objective is to achieve the transport characteristics of Synchronous Optical Network/Synchronous Digital Hierarchy (SONET/SDH) that are connection-oriented, a high level of availability, quality of service, and extensive OAM capabilities.

With MPLS-TP, network provisioning can be achieved via a centralized NMS and/or a distributed control plane. The Generalized Multi-Protocol Label Switching (GMPLS) can be used as a control plane that provides a common approach for management and control of multilayer transport networks.

Networks are typically operated from a network operation center (NOC) using an NMS that communicates with the network elements. The NMS provides FCAPS management functions (i.e., fault, configuration, accounting, performance, and security management).

For MPLS-TP, NMS can be used for static provisioning while the GMPLS can be used dynamic provisioning of transport paths. The control plane is mainly used to provide restoration functions for improved network survivability in the presence of failures and facilitates end-to-end path provisioning across network or operator domains.

Similar to T-MPLS, MPLS-TP uses a subset of IP/MPLS standards where features that are not required in transport networks such as IP forwarding, penultimate hop popping (PHP), and equal cost multiple path (ECMP) are not supported or made optional. However, MPLS-TP defines extensions to existing IP/MPLS standards and introduces established requirements from transport networks. Among the key new features are comprehensive OAM capable of fast detection, localization, troubleshooting and end-to-end SLA verification, linear and ring protection with sub-50-ms recovery, separation of control and data plane, and fully automated operation without control plane using NMS.

Static and dynamic provisioning models are possible. The static provisioning model is the simplified version commonly known as static MPLS-TP. This version does not implement even the basic MPLS functions, such as Label Distribution Protocol (LDP) and Resource Reservation Protocol-Traffic Extension (RSVP-TE), since the signaling is static. However, it does implement support for GAL (generic associated channel label) and G-ACh (generic associated channel) which is used in supporting OAM functions.

An MPLS-TP label switching router (LSR) is either an MPLS-TP provider edge (PE) router or an MPLS-TP provider (P) router for a given LSP. An MPLS-TP PE router is an MPLS-TP LSR that adapts client traffic and encapsulates it to be transported over an MPLS-TP LSP by pushing a label, or using a pseudowire. An MPLS-TP PE exists at the interface between a pair of layer networks. An MPLS-TP label edge router (LER) is an LSR that exists at the endpoints of an LSP and therefore pushes or pops the LSP label.

An MPLS-TP PE node can support UNI providing the interface between a CE and the MPLS-TP network and NNI providing the interface between two MPLS-TP PEs in different administrative domains.

The details of the MPLS-TP architecture, OAM, and security are described in Chapter 13.

1.13 Virtual Private LAN Services

MPLS facilitates the deployment and management of virtual private networks (VPNs). MPLS-based VPN can be classified as:

- Layer-3 multipoint VPNs or Internet Protocol (IP) VPNs that are often referred to as virtual private routed networks (VPRN);
- Layer-2 point-to-point VPNs, which basically consist of a collection of separate virtual leased lines (VLL) or pseudowires;
- Layer-2 multipoint VPNs, or virtual private LAN services (VPLS).

VPLS is a multipoint service, but unlike IP VPNs, it can transport non-IP traffic and leverages advantages of Ethernet.

Two VPLS solutions are proposed:

- VPLS using BGP that uses BGP for signaling and discovery;
- VPLS using label distribution that uses LDP signaling and basically an extension to the Martini draft.

Both approaches assume tunnel LSPs between PEs. Pseudowires are set up over tunnel LSPs [i.e., virtual connection (VC) LSPs].

In order to establish MPLS LSPs, OSPF-TE and RSVP-TE can be used where OSPF-TE will take bandwidth availability into account when calculating the shortest path, while RSVP-TE allows reservation of bandwidth.

There are two key components of VPLS: PE discovery and signaling. PE discovery can be via provisioning application, BGP, and RADIUS. Signaling can be via targeted LDP and BGP.

In order to offer different classes of service within a VPLS, 802.1p bits in a customer Ethernet frame with a VLAN tag are mapped to EXP bits in the pseudowire and/or tunnel label.

VPLS is a multipoint service; therefore, the entire service provider network appears as a single logical learning bridge for each VPLS. The logical ports of this service provider (SP) bridge are the customer ports as well as the pseudowires on a

VPLS edge (VE). The SP bridge learns MAC addresses at its VEs while a learning bridge learns MAC addresses on its ports. Source MAC addresses of packets with the logical ports on which they arrive are associated in the forwarding information base (FIB) to forward packets.

In LDP-based VPLS, an interface participating in a VPLS must be able to flood, forward, and filter Ethernet frames. Each PE will form remote MAC address to PW associations and associate directly attached MAC addresses to local customer facing ports. Connectivity between PEs can be via MPLS transport tunnels as well as other tunnels over PWs such as GRE, L2TP, and IPSec. The PE runs the LDP signaling protocol and/or routing protocols to set up PWs, setting up transport tunnels to other PEs and delivering traffic over PWs.

A full mesh of LDP sessions is used to establish the mesh of PWs. Once an LDP session has been formed between two PEs, all PWs between these two PEs are signaled over this session. A hierarchical topology can be used in order to minimize the size of the VPLS full mesh when there is a large number of PWs.

Hierarchical VPLS (H-VPLS) was designed to address scalability issues in VPLS. In VPLS, all PE nodes are interconnected in a full mesh to ensure that all destinations can be reached. In H-VPLS, a new type of node is introduced called the multitenant unit (MTU), which aggregates multiple CE connections into a single PE, to reduce the number of PE-to-PE connections.

In BGP approach, VPLS control plane functions mainly auto-discovery and provisioning of PWs are accomplished with a single BGP Update advertisement. In the auto-discovery, each PE discovers other PEs that are part of a given VPLS instance via BGP. When a PE joins or leaves a VPLS instance, only the affected PE's configuration changes while other PEs automatically find out about the change and adapt.

A BGP route target community (or extended communities) is used to identify members of a VPLS. A PE announces usually via I-BGP that it belongs to a specific VPLS instance by annotating its network layer reachability information (NLRI) for that VPLS instance with route target (RT), and acts on this by accepting NLRIs from other PEs that have route target RT.

When a new PE is added by the service provider, a single BGP session is established between the new PE and a route reflector. The new PE then joins a VPLS domain when the VPLS instance is configured on that PE. Once discovery is done, each pair of PEs in a VPLS establishes pseudowires to each other, transmit certain characteristics of the pseudowires that a PE sets up for a given VPLS, and tear down the pseudowires when they are no longer needed. This mechanism is called signaling.

Auto-discovery and signaling functions are typically announced via I-BGP. This assumes that all the sites in a VPLS are connected to PEs in a single autonomous system (AS). However, sites in a VPLS may connect to PEs in different ASs. In this case, I-BGP connection between PEs and PE-to-PE tunnels between the ASs are established.

Hierarchical BGP VPLS is used to scale the VPLS control plane when using BGP.

The advantages of VPLS may be summarized as:

- Complete customer control over their routing resulted in a clear demarcation of functionality between service provider and customer that makes troubleshooting easier.
- Able to add a new site without configuration of the service provider's equipment or the customer equipment at existing sites.
- Minimize MAC address exposure, improving scaling by having one MAC address per site (i.e., one MAC per router) or per service.
- Improve customer separation by having a CE router to block unnecessary broadcast or multicast traffic from customer LANs.
- MPLS core network emulates a flat LAN segment that overcomes distance limitations of Ethernet-switched networks and extends Ethernet broadcast capability across WAN.
- Point-to-multipoint connectivity connects each customer site to many customer sites.
- A single CE-PE link transmits Ethernet packets to multiple remote CE routers.
- Fewer connections required to get full connectivity among customer sites.
- Adding, removing, or relocating a CE router requires configuring only the directly attached PE router. This results in substantial operational expense savings.

1.14 Information Modeling

A common management model is needed for Carrier Ethernet services to ensure that the information in operation support systems (OSSs), SDN orchestrators, SDN controllers, NMSs, EMSs, and network elements from various vendors are logically consistent and allow service providers to readily integrate their capabilities into the management environment.

There are three management information views defined for such an integrated management environment:

- Network element view;
- Overall network view;
- Service view.

The Carrier Ethernet Management Information Model that is defined by MEF facilitates the information flows within the management environment in order to manage and control the Carrier Ethernet services. The management functions that are supported by the information model include configuration management, performance management and monitoring, alarms and testing for the fault management. The information model specifies the interface profiles and data models with entity diagrams that describe the attributes and relations among entities. Facilitating the information flow between various Network Elements and management systems, there is the standardization of functionalities and data representation which are based on NETCONF and YANG. The details of these standards are given in Chapter 15.

1.15 The Third Network

The Third Network was recently defined as an orchestrated network infrastructure that supports the new Carrier Ethernet concept, Carrier Ethernet 2.0. It utilizes virtualization of network functions employing SDN technologies. The new network facilitates agile service activation, service chaining, delivery and support of these services with the required level of performance and security. It also expands to cover cloud-based services and the entire service life cycle under multiple service providers and operators. The detailed ecosystem is framed under a reference architecture called LSO and explained in Chapter 16. The LSO defined three domains: (1) customer domain, (2) service provider domain, and (3) partner (operator) domain. Various interfaces have been defined between and within the domains.

Carrier Ethernet services can be virtualized by dividing the delivery of entire service into separate virtual functional components and attributes including protection, OAM, and synchronization. The provisioning of such virtualized services now requires initiation and orchestration of multiple virtualized functions and set of attributes.

The cloud services architectures are defined by Open Cloud Connect and became the part of the Third Network. These cloud services are grouped under Network as a Service (NaaS), Infrastructure as a Service (IaaS), Platform as a Service (PssS), Software as a Service (SaaS), Communication as a Service (CaaS), and Security as a Service (SECaaS).

CHAPTER 2
Basic Ethernet

2.1 Introduction

The Ethernet is physical layer local area network (LAN) technology invented in 1973. Since its invention, the Ethernet has become the most widely used LAN technology because of its speed, low cost, and relative ease of installation. This is combined with wide computer-market acceptance and the ability to support most of the network protocols.

The Ethernet network system is invented by Robert Metcalfe at Xerox. Metcalfe's Ethernet was modeled after the ALOHA network developed in the 1960s at the University of Hawaii. The ALOHA system used two distinct frequencies in a hub/star configuration, where the hub broadcasts packets to everyone on the outbound channel while clients sending data to the hub on the inbound channel. Data received was immediately re-sent, allowing clients to determine whether or not their data had been received. Any machine noticing corrupted data would wait a short time and then resend the packet. ALOHA was important because it used a shared medium for transmission.

Similarly, Metcalfe's system used a shared medium for transmission, however, detected collisions between simultaneously transmitted frames and included a listening process before frames were transmitted, thereby greatly reducing collisions.

The first Metcalfe system ran at 2.94 Mbps. DEC, Intel, and Xerox (DIX) issued a DIX Ethernet standard for 10-Mbps Ethernet systems by 1980. Also the Institute of Electrical and Electronics Engineers (IEEE) started project 802 to standardize local area networks. In 1985, IEEE published the portion of the standard pertaining to the Ethernet based on the DIX standard: IEEE 802.3 Carrier Sense Multiple Access with Collision Detection (CSMA/CD) Access Method and Physical Layer Specifications. The IEEE standards have been adopted by the American National Standards Institute (ANSI) and by the International Organization of Standards (ISO).

In 1995, 100-Mbps Ethernet over wire or fiber-optic cable that is called Fast Ethernet was standardized by IEEE, IEEE 802.3u. This standard also allowed for equipment that could auto-negotiate the two speeds such that Ethernet device can automatically switch from 10 Mbps to 100 Mbps and vice versa.

2.2 CSMA/CD

Ethernet uses CSMA/CD. The multiple access means that every station is connected to a single copper wire or a set of wires that are connected together to form a single data path. The carrier sense means that before transmitting data, a station checks the wire to see if any other station is already sending data. If the LAN appears to be idle, then the station can begin to send data.

To illustrate how collision takes place, let us consider two Ethernet stations that are 250m apart from each other and send data at a rate of 10 Mbps. Given that light and electricity travel about 1 foot in a nanosecond, after the electric signal for the first bit has traveled about 100 feet down the wire, the same station has begun to send the second bit. If both stations begin transmitting at the same time, then they will be in the middle of the third bit before the signal from each reaches the other station. Therefore, their signals will collide nanoseconds later. When such a collision occurs, the two stations stop transmitting, back off, and try again later after a randomly chosen delay period.

For example, in Figure 2.1, when PC1 begins sending data, the signal passes down the wire and just before it gets to PC2, which hears nothing and thinks that the LAN is idle, begins to transmit its own data. A collision occurs. The second station recognizes this immediately, but the first station will not detect it until the collision signal retraces the first path all the way back through the LAN to its starting point.

Any system based on collision detect must control the time required for the worst round trip through the LAN. For Ethernet, this round trip is limited to 50 microseconds. At a 10-Mbps rate, this is enough time to transmit 500 bits. At 8 bits per byte, this is slightly less than 64 bytes. To make sure that the collision is recognized, the Ethernet requires that a station must continue transmitting until the 50-microsecond period has ended. If the station has less than 64 bytes of data to send, then it must pad the data by adding zeros at the end.

Besides the constant listening, there is an enforced minimum quiet time of 9.6 microseconds between frame transmissions to give other devices a chance. If a collision occurs, retransmission is determined by an algorithm that chooses unique time

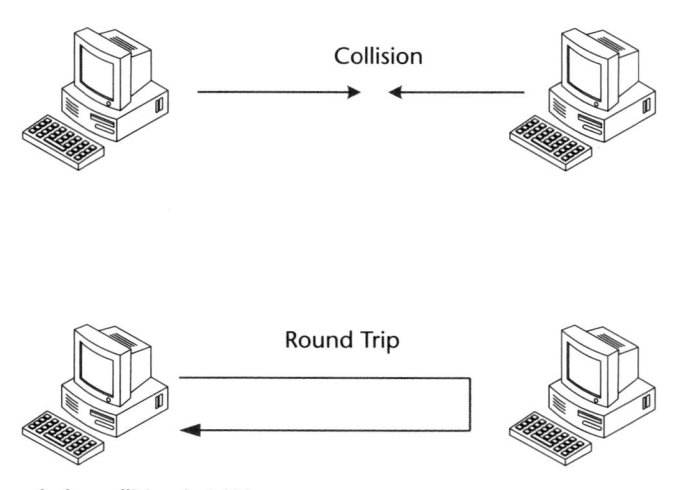

Figure 2.1 Example for collision in LAN.

intervals for resending the frames. The Ethernet interface backs off or waits for the chosen time interval and then retransmits if no activity is detected. The process is repeated until the frame collides up to 16 times and then it is discarded.

The CSMA/CD protocol is designed to allow fair access by all transmission devices to shared network channels. If deferred and retransmitted traffic is less than 5% of the total network traffic, the Ethernet network is considered to be healthy.

The total distance between an Ethernet transmitter and receiver is limited due to timing of the Ethernet signals on the cable as described above. The Ethernet requires a repeater or bridge or switch to go beyond these hard limits. They are discussed in the following sections.

2.3 Full Duplex, PAUSE, Autonegotiation

Reference [1] described a full-duplex Ethernet operation between a pair of stations in which simultaneous transmit and receive are over twisted-pair or fiber-optic cables that support two unidirectional paths. Ethernet devices need support simultaneous transmit and receive functions, in addition to cabling.

Also [1] introduced traffic flow control, called MAC protocol and PAUSE. If traffic gets too heavy, the control protocol can pause the flow of traffic for a brief time period.

When two network interfaces are connected to each other to autonegotiate the best possible shared mode of operation [2], the autonegotiation mechanism detects the speed but not the duplex setting of an Ethernet peer that did not use autonegotiation. An autonegotiating device defaults to half duplex, when the remote does not negotiate.

2.4 Repeaters and Hubs

Repeaters are used when the distance between computers is great enough to cause signal degradation. A repeater provides signal amplifications and retiming that enable the collision domain to extend beyond signal attenuation limitations. It takes the signal from one Ethernet cable and repeats it onto another cable. A repeater has no memory and does not depend on any particular protocol. It duplicates everything, including the collisions. The repeaters, too, must listen and pass their signals only when the line is clear.

If a collision was detected, the repeater can transmit a jam signal onto all ports to ensure collision detection. Repeaters could be used to connect segments so that there were up to five Ethernet segments between any two hosts. Repeaters could detect an improperly terminated link from the continuous collisions and stop forwarding data from it.

Active hubs and repeaters connect LAN segments. Active hubs (see Figure 2.2) are described as the central device in a star topology, and while it connects multiple cable segments at one point, it also repeats the incoming signals before transmitting them on to their destinations. Hubs can be designed to support Simple Network Management Protocol (SNMP) to administer and configure the hub.

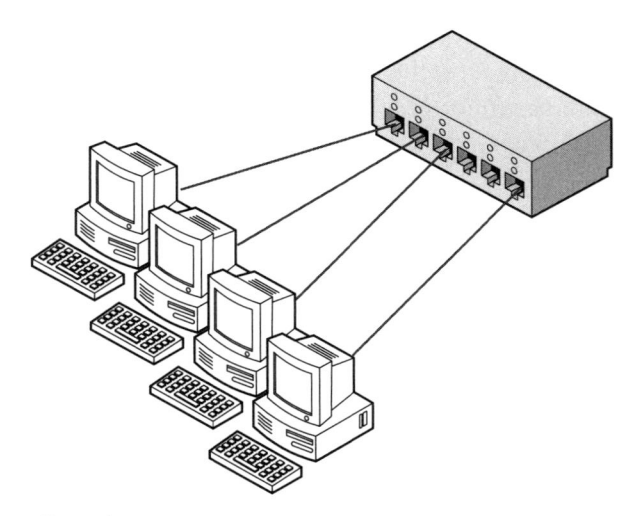

Figure 2.2 Hub configuration.

2.5 Bridges

Bridges transfer MAC-layer packets from one network to another (see Figure 2.3). They serve as gatekeepers between networks. Bridges control traffic by checking source and destination addresses and forwarding only network-specific traffic.

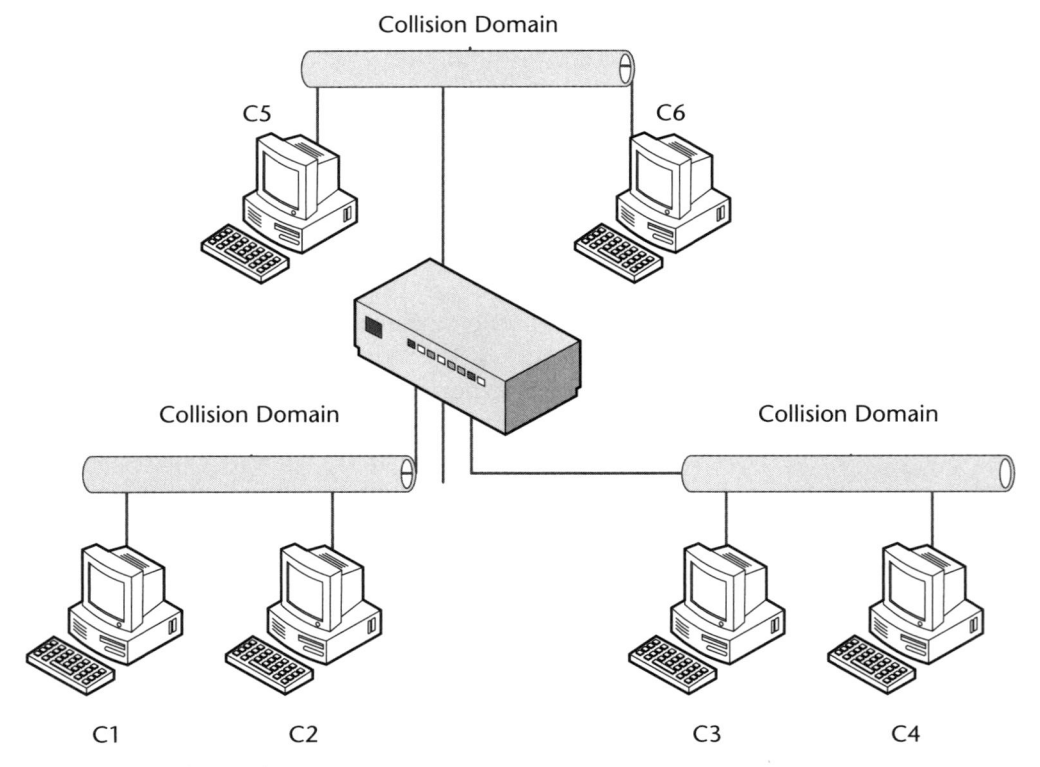

Figure 2.3 Bridge configuration.

Bridges also check for errors and drop traffic that is corrupted, misaligned, and redundant. Bridges help prevent collisions and create separate collision domains by holding and examining entire Ethernet packets before forwarding them on. This allows the network to cover greater distances and add more repeaters onto the network.

Most bridges can learn and store Ethernet addresses by building tables of addresses with information from traffic passing through them. This is a great advantage for users who move from place to place, but it can cause some problems when multiple bridges start network loops.

While repeaters could isolate some aspects of Ethernet segments, such as cable-related issues, they still forwarded all traffic to all Ethernet devices. This created practical limits on how many machines could communicate on an Ethernet network. Also as the entire network was one collision domain and all hosts had to be able to detect collisions anywhere on the network. Segments joined by repeaters had to operate at the same speed, making phased-in upgrades impossible.

To alleviate these problems, bridging was created to communicate at the data link layer while isolating the physical layer. Bridges learn where devices are, by watching MAC addresses. They do not forward packets across segments when they know the destination address is not located in that direction. Broadcast traffic is still forwarded to all network segments.

When handling many ports at the same time, a bridge reads entire Ethernet frame into a buffer, compares the destination address with an internal table of known MAC addresses and make a decision as to whether to drop the packet or forward it to another or all segments.

2.6 Switches

Ethernet switches (Figure 2.4) allow interconnection of bridge ports of different rates, break a network up into multiple networks consisting of a number of smaller collision domains and links. Two types of switches are commonly used:

- Cut-through switches, which read the destination address on the frame header and immediately forward the frame to the switch port attached to the destination MAC address. With this, they improve performance since they only read the header information and transmit rest of the frame without inspection.
- Store and forward switches, which hold the frame until the whole packet is inspected for proper length and CRC integrity, similar to a bridge. This type of switch makes sure the frame is fit to travel before transmitting it. Valid frames are processed for filtering and forwarding and corrupt frames are prevented from forwarding onto an output port.

Unicast bandwidth, multicast bandwidth, and size of MAC address database are the key parameters of the switches. Quality of Service (QoS) procedures partition bandwidth among multiple subscribers and protect unicast and multicast bandwidths. The number of MAC addresses learned from any one customer can be

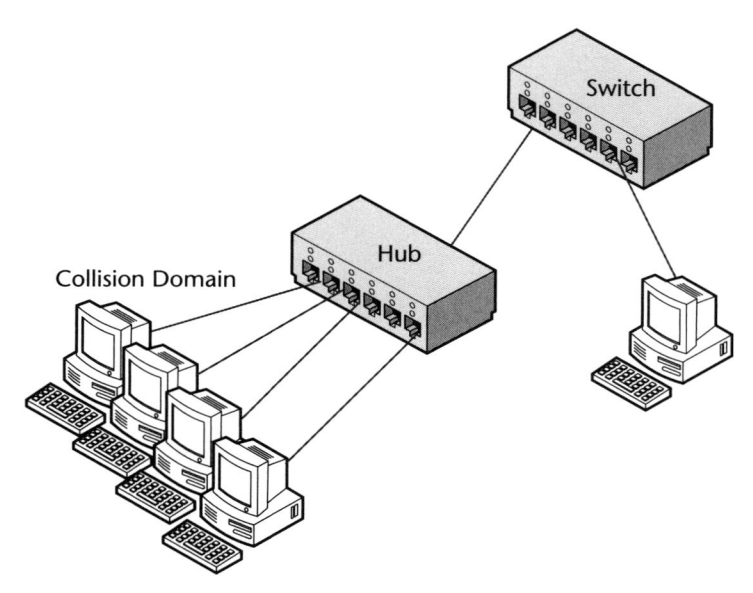

Figure 2.4 Switch configuration.

limited to limit amount of traffic and hide subscribers from malicious or inadvertent attacks against this shared resource.

2.7 Physical Layer

The Ethernet system consists of three basic elements:

- Physical medium carrying Ethernet signals between computers;
- A set of medium access control rules embedded in each Ethernet interface that allow multiple computers to fairly arbitrate access to the shared Ethernet channel;
- An Ethernet frame carrying data over the system.

The Ethernet physical layer evolved over time considerably. The speed ranges from 1 Mbps to 100 Gbps in speed while the physical medium can range from bulky coaxial cable to twisted pair to optical fiber. In general, network protocol stack software will work identically on different medium.

Many Ethernet adapters and switch ports support multiple speeds, using autonegotiation to set the speed and duplex for the best values supported by both connected devices.

The first Ethernet networks, 10BASE5, used thick yellow cable with vampire taps as a shared medium that uses CSMA/CD. Later, 10BASE2 Ethernet used thinner coaxial cable with BNC connectors as the shared CSMA/CD medium. The StarLAN 1BASE5 and 10BASE-T used twisted pairs.

Currently 10BASE-T, 100BASE-TX, and 1000BASE-T utilize twisted pair cables and RJ45 connectors. They run at 10 Mbps, 100 Mbps, and 1 Gbps, respectively.

A data packet on the wire is called a frame. A frame has a preamble and start frame delimiter. After a frame has been sent, transmitters are required to transmit 12 octets of idle characters before transmitting the next frame. This takes 96 ns for rates of 10 Mbps, 100 Mbps, and 1 Gbps, respectively. The maximum net bit rate of 10-Mbps Ethernet is approximately 9.75 Mbps, assuming a continuous stream of a 1,500-byte payload.

Ethernet over fiber is commonly used in structured cabling applications. Fiber has advantages in performance, electrical isolation, and distance, up to tens of kilometers with some versions.

There are two different types of optical fiber: multimode and single-mode. A multimode fiber allows many light propagation paths, while a single-mode fiber allows only one light path.

In multimode fiber, the time it takes for light to travel through a fiber is different for each mode resulting in a spreading of the pulse at the output of the fiber referred to as intermodal dispersion. The difference in the time delay between the modes is called differential mode delay (DMD). Intermodal dispersion limits multimode fiber bandwidth.

A single-mode fiber has higher bandwidth than multimode fiber since it has no intermodal dispersion. This allows for higher data rates over much longer distances than achievable with multimode fiber.

Although single-mode fiber has higher bandwidth, multimode fiber supports high data rates at short distances. Relaxed tolerances on optical coupling requirements allow lower-cost transceivers or lasers. As a result, multimode fiber has dominated in shorter-distance and cost-sensitive LAN applications.

The bandwidth achievable on twisted pair wiring is limited by the quality, distance, and interference associated with each individual pair. Therefore, it is important to utilize lower overhead protocols to transport data for effective utilization of every pair.

Copper access network mainly uses either TDM or ATM inverse multiplexing over ATM (IMA). ATM segments data into 48-byte cells where each segment is encapsulated with a 5-byte header. When carrying Ethernet frames over ATM, there is a partial cell fill at the end of each frame. For example, a 64-byte Ethernet frame must be transported using two 48-byte ATM cells (for a total of 96 bytes used for data) with an additional 10 bytes of overhead for the ATM cell headers. This results in an efficiency of 60% in this example (only 64 out of 106 bytes carry actual data).

IEEE 802.3ah introduced a new encapsulation scheme called 64/65-octet encapsulation that is used on all standard 2BASE-TL and 10PASS-TS Ethernet interfaces. With 64/65-octet encapsulation, at most 5 (7%) of every 69 octets are required for overhead. The efficiency is much higher for frame sizes greater than 64-octets. Each Ethernet frame 2BASE-TL of IEEE 802.3ah also improves efficiency for multipair technologies such as IMA by introducing an aggregation multiplexing and demultiplexing layer into the Ethernet stack that is responsible for taking an Ethernet frame and partitioning it over multiple variable speed links in a manner that best utilizes the speed of each pair. For example, an implementation could partition a frame into variable size fragments, where the size of the fragments depends upon the speed of the link, with the faster links carrying the larger fragments.

The multiple pairs provide protection against facility failures. The pairs in an aggregated group can be scattered across a large system (i.e., distributed bonding).

With distributed bonding, pairs can be aggregated across multiple system components instead of being restricted to a single element or line card. Mid-band Ethernet supports native Ethernet at rates from 2 Mbps to 45 Mbps over bonded copper pairs. The services are exactly the same as fiber-based Carrier Ethernet and conform to the IEEE 802.3ah standard.

Mid-band Ethernet services push the Ethernet service edge well beyond the fiber footprint.

2.7.1 10-Mbps Ethernet

For many years 10BASE2 which is also called ThinNet or Cheapernet was the dominant 10-Mbps Ethernet standards. It is a 50 coaxial cable that connects machines where each machine uses a T-adaptor to connect to its network interface card. It requires terminators at each end.

10BASE-T runs over four wires (i.e., two twisted pairs) on a Category 3 or Category 5 cable. A hub or switch sits in the middle and has a port for each node. This is also the configuration used for 100BASE-T and gigabit Ethernet.

For Ethernet over fiber, 10BASE-F is the standards for 10 Mbps.

2.7.2 Fast Ethernet

100BASE-T is used for any of the three standards for 100-Mbps Ethernet over twisted pair cable, 100BASE-TX, 100BASE-T4, and 100BASE-T2. 100BASE-TX has dominated the market and is synonymous with 100BASE-T.

100BASE-TX is 100-Mbps Ethernet over Category 5 cable that uses two of four pairs. There are two types of it:

- 100BASE-T4: 100-Mbps Ethernet over Category 3 cable as used for 10BASE-T installations. It uses all four pairs in the cable and is limited to half-duplex. It is now obsolete, as Category-5 cables are common.
- 100BASE-T2: 100-Mbps Ethernet over Category-3 cable. Uses only two pairs and supports full-duplex. It is functionally equivalent to 100BASE-TX, but supports old cable. No products supporting this standard were ever manufactured.

However, 100-Mbps Ethernet over fiber is supported by 100BASE-FX.

2.7.3 Gigabit Ethernet

Gigabit Ethernet has the same IEEE 802.3 frame format, full duplex, and flow control methods as 10-Mbps and 100-Mbps Ethernet, only faster. Additionally, it takes advantage of CSMA/CD when in half-duplex mode, supports SNMP and takes advantage of jumbo frames to reduce the frame rate to the end host.

Gigabit Ethernet can be transmitted over Category-5 cable and optical fiber as 1000Base-CX for short distance (i.e., up to 25m) transport over copper, 1000BASE-T over unshielded twisted pair copper cabling (Category 5/5e), 1000Base-SX with 850-nm wavelength over fiber, and 1000Base-LX with 1,300-nm wavelength, which is optimized for longer distances over single-mode fiber.

Gigabit Ethernet types defined in IEEE 802.3 and their descriptions are given in Table 2.1.

2.7.4 10-Gb, 40-Gb, and 100-Gb Ethernet

The 10-Gb Ethernet family of standards encompasses media types for single-mode fiber for long-haul, multimode fiber for distances up to 300m, copper backplane for distances up to 1m and copper twisted pair for distances up to 100m. It was first standardized as IEEE Std. 802.3ae-2002 and later it was included in IEEE Std. 802.3-2008.

10GBASE-SR is designed to support short distances over deployed multimode fiber cabling; it has a range of between 26m and 82m depending on cable type.

10GBASE-LX4 uses wavelength division multiplexing to support ranges of between 240m and 300m over deployed multimode cabling. It also supports 10 km over single-mode fiber.

10GBASE-LR and 10GBASE-ER support 10 km and 40 km, respectively, over single-mode fiber.

10GBASE-T is designed to support copper twisted pair as specified by the IEEE Std. 802.3-2008.

Table 2.1 1-Gbps Interface Types, Related Standards, and Their Descriptions

Type	Description
1000BASE-T	PAM-5 coded signaling, Category 5/Category 5e/Category 6 copper cabling with four twisted pairs
1000BASE-SX	8B10B NRZ-coded signaling, multimode fiber for distances up to 550m
1000BASE-LX	8B10B NRZ-coded signaling, multimode fiber for distance up to 550m or single-mode fiber for distances up to 10 km
1000BASE-LH	It is a long-haul solution and support distances up to 100 km over single-mode fiber
1000BASE-CX	8B10B NRZ-coded signaling, balanced shielded twisted pair over special copper cable, for distances up to 25m
1000BASE-BX10	Bidirectional over single strand of single-mode fiber for distances up to 10 km
1000BASE-LX10	Over a pair of single-mode fibers for distances up to 10 km
1000BASE-PX10-D	Downstream, from head end to tail end, over single-mode fiber using point-to-multipoint topology for distances of at least 10 km
1000BASE-PX10-U	Upstream, from a tail end to the head end, over single-mode fiber using point-to-multipoint topology for distance of at least 10 km
1000BASE-PX20-D	Downstream, from the head end to the tail ends, over single-mode fiber using point-to-multipoint topology for distances of at least 20 km
1000BASE-PX20-U	Upstream, from a tail end to the head end, over single-mode fiber using point-to-multipoint topology for distance of at least 20 km
1000BASE-ZX	Up to 100 km over single-mode fiber
1000BASE-KX	1m over backplane

10GBASE-SW, 10GBASE-LW, and 10GBASE-EW use the WAN PHY, designed to interoperate with OC-192/STM-64 SONET/SDH equipment. They correspond at the physical layer to 10GBASE-SR, 10GBASE-LR, and 10GBASE-ER, respectively, and hence use the same types of fiber and support the same distances.

The 10-Gb Ethernet over fiber interfaces are listed in Table 2.2.

The operation of 10-Gb Ethernet is similar to that of lower-speed Ethernets as well. It maintains the IEEE 802.3 Ethernet frame size and format that preserves Layer-3 and greater protocols. However, 10-Gb Ethernet only operates over point-to-point links in full-duplex mode. Additionally, it uses only multimode and single-mode optical fiber for transporting Ethernet frames. Operation in full-duplex mode eliminates the need for CSMA/CD.

The growth of bandwidth intense applications such as video on demand, high-performance computing, and virtualization drives 40G and 100G Ethernet [3–5] usage in long-haul networks as well as in regional and metro networks; 40G has been developed as an intermediate solution before deploying 100G that can reduce the number of optical wavelengths required. A 1-Tb Ethernet implementation is also being considered.

Tables 2.3 and 2.4 lists 40-Gbps and 100-Gbps interfaces defined by IEEE 802.3.

2.7.5 LAN PHY

The 1IEEE 802.3ae defines two broad physical layer network applications, LAN PHY and wide area network (WAN) PHY.

Table 2.2 10GBASE-X Fiber Interfaces

Interface	PHY	Optics	Description
10GBASE-SR	LAN	850 nm serial	Designed to support short distances over deployed multimode fiber cabling, it has a range of between 26m and 82m depending on cable type; it also supports 300m operation over a new 2,000 MHz.km multimode fiber
10GBASE-LR	LAN	1,310 nm serial	Supports 10 km over single-mode fiber
10GBASE-ER	LAN	1,550 nm serial	Supports 40 km over single-mode fiber
10GBASE-LX4	LAN	4 × 1,310 nm CWDM	Uses wavelength division multiplexing to support ranges of between 240m and 300m over deployed multimode cabling; also supports 10 km over single-mode fiber
10GBASE-SW	WAN	850 nm serial	A variation of 10GBASE-SR using the WAN PHY, designed to interoperate with OC-192/STM-64 SONET/SDH equipment
10GBASE-LW	WAN	1,310 nm serial	A variation of 10GBASE-LR using the WAN PHY, designed to interoperate with OC-192/STM-64 SONET/SDH equipment
10GBASE-EW	WAN	1,550 nm serial	A variation of 10GBASE-ER using the WAN PHY, designed to interoperate with OC-192 / STM-64 SONET/SDH equipment

2.7 Physical Layer

Table 2.3 40GBASE-X Interfaces

Type	Description
40GBASE-SR4	100-m operation over multimode fiber
40GBASE-LR4	10-km operation over single-mode fiber
40GBASE-CR4	10-m operation copper cable assembly
40GBASE-KR4	1-m operation over backplane

Table 2.4 100GBASE-X Interfaces

Name	Description
100GBASE-SR10	100-m operation over multimode fiber
100GBASE-LR4	10-km operation over single-mode fiber
100GBASE-ER4	40-km operation over single-mode fiber
100GBASE-CR10	10-m operation copper cable assembly

The LAN PHY operates at close to the 10-Gb Ethernet rate to maximize throughput over short distances. Two versions of LAN PHY are standardized, serial 10GBASE-R and 4-channel course wave division multiplexing (CWDM) 10GBASE-X.

The 10GBASE-R uses a 64B/66B encoding system that raises the 10-Gb Ethernet line rate from a non-encoded 9.58 Gbps to 10.313 Gbps. However, the 10GBASE-X still uses 8B/10B encoding because all of the 2.5-Gbps CWDM channels that it employs are parallel and run at 3.125 Gbps after encoding.

The MAC to PHY data rate for both LAN PHY versions is 10 Gbps. Encoding is used to reduce clocking and data errors caused by long runs of ones and zeros.

The WAN PHY supports connections to circuit-switched SONET networks. Besides the sublayers added to the LAN PHY, the WAN PHY adds another element called the WAN interface sublayer (WIS). The WIS takes data payload and puts it into a 9.58464-Gbps frame that can be transported at a rate of 9.95328 Gbps. The WIS does not support every SONET feature, but it carries out enough overhead functions, including timing and framing, to make the Ethernet frames recognizable and manageable by the SONET equipment they pass through.

2.7.6 LAN PHY/WAN PHY Sublayers

The physical medium is defined by physical layer while media access layer is defined by ISO data link layer which is divided into two IEEE 802 sublayers, the MAC sublayer and the MAC-client sublayer. The IEEE 802.3 physical layer corresponds to the ISO physical layer.

The MAC sublayer has two primary responsibilities:

- Data encapsulation, including frame assembly before transmission, and frame parsing/error detection during and after reception;
- MAC, including initiation of frame transmission and recovery from transmission failure.

The MAC-client sublayer is the logical link control (LLC) layer defined by IEEE 802.2 which provides the interface between the Ethernet MAC and the upper layers in the protocol stack of the end station. The LLC protocol operates on top of the MAC protocol.

A bridge entity defined by IEEE 802.1 provides LAN-to-LAN interfaces between LANs that use the same protocol (for example, Ethernet to Ethernet) and also between different protocols (for example, Ethernet to token ring).

The protocol header comprises two parts, an LLC protocol header and a Subnetwork Access Protocol (SNAP) header.

The LLC protocol is based on the HDLC link protocol and uses an extended 2-byte address. The first address byte indicates a destination service access point (DSAP) and the second address a source service access point (SSAP). These identify the network protocol entities, which use the link layer service.

A control field is also provided which may support a number of HDLC modes. These include Type 1 (connection-less link protocol), Type 2 (connection-oriented protocol), and Type 3 (connection-less acknowledged protocol).

The SNAP header is used when the LLC protocol carries IP packets and contains the information that would otherwise have been carried in the 2-byte MAC frame type field.

The physical (PHY) layer includes the following sublayers:

- Physical coding sublayer (PCS) that encodes and decodes the data stream between the MAC and PHY layer. There are three categories for the PCS:
    - 10GBASE-R, which is serially encoded (64B/66B) (LAN PHY);
    - 10GBASE-X, which is serially encoded (8B/10B); used for wavelength division multiplexing (WDM) transmissions (LAN PHY);
    - 10GBASE-W, which is serially encoded (64B/66B) and compatible with SONET standards for a 10-Gbps WAN (WAN PHY).
- Physical medium attachment (PMA), which is an optional interface for connection to optical modules. There are two PMA interfaces, 10-Gb Ethernet attachment unit interface (XAUI) and 10-Gb Ethernet 16-bit interface (XSBI). The XAUI is an interface to specialized 10-Gb Ethernet optical modules and system backplanes. The XSBI is a serial optics interface for LAN and WAN PHY applications. Intermediate and long reach optical modules use this interface. It requires more power and more pins than an XAUI.
- Physical medium dependent (PMD), which supports distance objectives of the physical medium dependent sublayer. Four different PMDs are defined to support single and multimode optical fibers:
    - 850 nm serial—MMF, up to 65m;
    - 1,310 nm serial—SMF, up to 10 km;
    - 1,550 nm serial—SMF, up to 40 km;
    - 1,310 nm CWDM—MMF, up to 300m;
    - 1,310 nm CWDM—SMF, 10 km.

2.8 Temperature Hardening

Because of the inherent reach limitations of copper when compared to fiber, equipment must be deployable in controlled environment such as in the central office (CO) and remote terminals (RmTs), and nonenvironmentally controlled basements and wiring terminals.

In harsh environmental conditions, equipment must be environmentally hardened to withstand a diverse range of temperatures from −40°C to 85°C and resist various environmental elements such as dust, rain, snow, sleet, and so forth. For outside plant deployments, equipment must not only be hardened, but must also fit the physical characteristics of the enclosure. In particular, it must adhere to the depth limitations of an RmT, and have only frontal access to all connectors so that it can fit flush against the wall.

To fully leverage Ethernet, copper access equipment needs to support deployments, where the equipment can be located at a considerable distance from the next switch in the network. Network modularity along with long-distance optics, short-distance optics, or Category-5 cabling supports the most flexible deployments.

2.9 Standards

The early development of Ethernet was done by Xerox. Later, it was refined and its second generation is called Ethernet II. It is also called DIX after its corporate sponsors Digital, Intel, and Xerox. As the holder of the trademark, Xerox established and published the standards.

The IEEE formed the 802 committee to look at Ethernet, token ring, fiber optic, and other LAN technology and published:

- 802.3: Hardware standards for Ethernet cards and cables;
- 802.5: Hardware standards for token ring cards and cables;
- 802.2: The new message format for data on any LAN.

The 802.3 standard further refined the electrical connection to the Ethernet. It was immediately adopted by all the hardware vendors.

Standard Ethernet frame sizes are between 64 and 1,518 bytes. Jumbo frames are between 64 and 9,215 bytes. Because larger frames translate to lower frame rates, using jumbo frames on gigabit Ethernet links greatly reduces the number of packets that are received and processed by the end host.

Some of the Ethernet related IEEE standards are given in Table 2.5.

2.10 Ethernet Frame Types and the EtherType Field

There are several types of Ethernet frames:

- The Ethernet Version 2 or Ethernet II frame, which is also called DIX frame (named after DEC, Intel, and Xerox);

Table 2.5 List of IEEE Standards for Ethernet

Name	Description
IEEE 802.3a	10Base-2 (thin Ethernet)
IEEE 802.3c	10 Mbps repeater specifications (clause 9)
IEEE 802.3d	FOIRL (fiber link)
IEEE 802.3i	10Base-T (twisted pair)
IEEE 802.3j	10Base-F (fiber optic)
IEEE 802.3u	100Base-T (Fast Ethernet and autonegotiations)
IEEE 802.3x	Full duplex
IEEE 802.3z	1000Base-X (gigabit Ethernet)
802.3ab	1000Base-T (gigabit Ethernet over twisted pair)
IEEE 802.3ac	VLAN tag (frame size extension to 1,522 bytes)
IEEE 802.3ad	Parallel links (link aggregation)
IEEE 802.3ae	10-Gb Ethernet
IEEE 802.3ah	Ethernet in the first mile
802.3as	Frame expansion
IEEE 802.3at	Power over Ethernet Plus
802.1aj-TPMAR	Architecture
IEEE 802.1ad	QinQ
IEEE 802.3ad	Link aggregation/protection
IEEE 802.1w	Testing
802.1ag	Connectivity fault management
IEEE 1588v2	Synchronization

- Novell's nonstandard variation of IEEE 802.3 (raw 802.3 frame) without an IEEE 802.2 LLC header;
- IEEE 802.2 LLC frame;
- IEEE 802.2 LLC/SNAP frame.

All four Ethernet frames types may optionally contain an IEEE 802.1Q tag to identify the VLAN it belongs to and its IEEE 802.1p priority.

2.10.1 Ethernet Frames

A block of data transmitted on the Ethernet is called a frame. An Ethernet frame has a header (preamble – length), payload (LLC – pad), and a trailer (frame check sequence).

The first 12 bytes of every frame contain the 6-byte destination address and a 6-byte source address. Each Ethernet adapter card comes with a unique factory installed address. Use of this hardware address guarantees a unique identity to each card.

In normal operation, an Ethernet adapter will receive only frames with a destination address that matches its unique address or destination addresses that represent a multicast message. In the promiscuous mode, they receive all frames that appear on the LAN.

2.10.2 Ethernet Address

Each Ethernet network interface card (NIC) has a unique identifier called a MAC address. Each MAC address assigned by the IEEE Registration Authority for the card manufacturers is a 48-bit number, of which the first 24 bits identify the manufacturer, manufacturer ID or organizational unique identifier (OUI). The second half of the address (extension of board ID) is assigned by the manufacturer. The number is usually programmed into the hardware so that it cannot be changed.

At the physical layer, the destination field is preceded by a 7-byte preamble and a 1-byte start of frame delimiter. At the end of the data field is a 4 byte checksum.

There are three common frame formats that will be described in the following sections:

- Ethernet II or DIX;
- IEEE 802.3 and 802.2;
- SNAP.

2.10.3 Ethernet II or DIX

The most common Ethernet Frame format, type II Versions 1.0 and 2.0 of the Digital/Intel/Xerox (DIX) Ethernet specification have a 16-bit subprotocol label field called the EtherType [6]. IEEE 802.3 replaced that with a 16-bit length field, with the MAC header followed by an IEEE 802.2 LLC header. The maximum length of a frame was 1,518 bytes for untagged IEEE 802.3 frames and 1,522 bytes for 802.1Q tagged. The two formats were eventually unified by the convention that values of that field between 64 and 1,522 bytes indicated the use of the new 802.3 Ethernet format with a length field, while values of 1,536 decimal (0600 hexadecimal) and greater indicated the use of the original DIX or Ethernet II frame format with an EtherType subprotocol identifier. This convention allows software to determine whether a frame is an Ethernet II frame or an IEEE 802.3 frame, allowing the coexistence of both standards on the same physical medium.

A 2-byte type code (Figure 2.5) was assigned by Xerox to identify each vendor protocol. To allow collision detect, the 10-Mb Ethernet requires a minimum packet size of 64 bytes. Any shorter message must be padded with zeros. Since short Ethernet frames must be padded with zeros to a length of 64 bytes, each of these higher-level protocols required either a larger minimum message size or an internal length field that can be used to distinguish data from padding.

The preamble is a 64-bit (8-byte) field that contains a synchronization pattern consisting of alternating ones and zeros and ending with two consecutive ones. After synchronization is established, the preamble is used to locate the first bit of the packet. The preamble is generated by the LAN interface card.

In a header, preamble sets bit timing and signals that a frame is being sent (10-Mbps Ethernet). For 100- and 1,000-Mbps Ethernet systems signal constantly and do not need preamble or start frame delimiter fields.

| Pre-Amble (7 Byte) | SFD (1 Byte) | Dest. Addr. (6 Byte) | Src. Addr. (6 Byte) | Type (2 Byte) | Data (46-1500 Byte) | CRC (4 Byte) |

SFD: Start Frame Delimiter (10101011)
Ethernet II Frame (64–1518 Bytes)
Type for Decnet 0x60 0x03
Type for Novell IPX 0x81 0x37
Type for TCP/IP 0x80 0x00
Type for XNS of Xerox 0x60 0x00

Figure 2.5 Frame format of Ethernet II.

2.10.4 IEEE 802.3

The 802.3 format is depicted in Figure 2.6.

To allow collision detect, the 10-Mb Ethernet requires a minimum packet size of 64 bytes. Any shorter message must be padded with zeros. In order for Ethernet to be interchangeable with other types of LANs, it would have to provide a length field to distinguish significant data from padding.

After a frame has been sent transmitters are required to transmit 12 octets of idle characters before transmitting the next frame. For 10M this takes 9,600 ns, 100M 960 ns, and 1,000M 96 ns.

The DIX standard did not need a length field because the vendor protocols that used it (XNS, DECNET, IPX, IP) all had their own length fields. The 802.3 standard replaced the two byte type field with a 2-byte length field.

Xerox had not assigned any important types to have a decimal value below 1,500. Since the maximum size of a packet on Ethernet is 1,500 bytes, there was no conflict or overlap between DIX and 802 standards. Any Ethernet packet with a type/length field less than 1,500 is in 802.3 format with a length while any packet in which the field value is greater than 1,500 must be in DIX format with a type.

2.10.5 IEEE 802.2

The 802.2 variants of Ethernet are not in widespread use on common networks currently. The 802.2 Ethernet is used to support transparent translating bridges between Ethernet and IEEE 802.5 token ring or FDDI networks. The most common framing type used today is Ethernet Version 2, with its EtherType set to 0x0800 for IPv4 and 0x86DD for IPv6.

| Pre-Amble (7 Byte) | SFD (1 Byte) | Dest. Addr. (6 Byte) | Src. Addr. (6 Byte) | Length (2 Byte) | Data (46-1500 Byte) | CRC (4 Byte) |

SFD: Start Frame Delimiter (10101011)
DSAP: Destination Service Access Point
SSAP: Source Service Access Point
SFD: Start Frame Delimiter (10101011)

Figure 2.6 Frame format of IEEE 802.3.

IP traffic can be encapsulated in IEEE 802.2 LLC frames with SNAP. IP Version 6 can also be transmitted over Ethernet using IEEE 802.2 with LLC/SNAP (Figure 2.7).

The IEEE 802.1Q tag, if present, is placed between the source address and the EtherType or length fields. The first two bytes of the tag are the Tag Protocol Identifier (TPID) value of 0x8100. This is located in the same place as the EtherType/length field in untagged frames; therefore, an EtherType value of 0x8100 means the frame is tagged. The true EtherType/length is located after the Q-tag. The TPID is followed by two bytes containing the tag control information (TCI), which is the IEEE 802.1p priority and VLAN ID. The Q-tag is followed by the rest of the frame, using one of the types described above.

The 802.2 header follows the 802.3 header and substitutes type with a new field. The 802.2 header is three bytes long for control packets or the kind of connectionless data sent by all the old DIX protocols. A 4-byte header is defined for connection oriented data, which refers primarily to SNA and NETBEUI. The first 2 bytes identify the SAP. The two SAP fields are set to 0x0404 for SNA and 0xF0F0 for NETBEUI.

2.10.6 SNAP

Some protocols operate directly on top of 802.2 LLC, which provides both datagram and connection-oriented network services. The LLC header includes two additional 8-bit address fields (Figure 2.8), called service access points (SAPs) when both source and destination SAP are set to the value 0xAA, the SNAP service is requested. The SNAP header allows EtherType values to be used with all IEEE 802 protocols, as well as supporting private protocol ID spaces. In IEEE 802.3x-1997, the IEEE Ethernet standard was changed to explicitly allow the use of the 16-bit field after the MAC addresses.

Under SNAP, the 802.2 header appears to be a datagram message (control field 0x03) between SAP ID 0xAA. The first 5 bytes of what 802.2 considers data

Pre-Amble (7 Byte)	SFD (1 Byte)	Dest. Addr. (6 Byte)	Src. Addr. (6 Byte)	Length (2 Byte)	DSAP (0xAA) (1 Byte)	SSAP (0xAA) (1 Byte)	Control (0x03) (1 Byte)	Data (43-1497 Byte)	CRC (4 Byte)

DSAP: Destination Service Access Point
SSAP: Source Service Access Point
SFD: Start Frame Delimiter (10101011)

Figure 2.7 Frame format of IEEE 802.2.

Pre-Amble (7 Byte)	SFD (1 Byte)	Dest. Addr. (6 Byte)	Src. Addr. (6 Byte)	Length (2 Byte)	DSAP (0xAA) (1 Byte)	SSAP (0xAA) (1 Byte)	Control (0x03) (1 Byte)	Org Code (3 Byte)	Type (2 Byte)	Data (43-1492 Byte)	CRC (4 Byte)

DSAP: Destination Service Access Point
SSAP: Source Service Access Point
SFD: Start Frame Delimiter (10101011)

Figure 2.8 Frame format of SNAP.

are actually a subheader ending in the 2-byte DIX type value. Any of the old DIX protocols can convert their existing logic to legal 802 SNAP by simply moving the DIX type field back 8 bytes from its original location.

Ethernet II uses 1 bit to indicate multicast addresses, 802.3 uses 2 bits. On 802.3, the first bit is similar to the multicast bit in that it indicates whether the address is for an individual or for a group, and the second bit indicates whether the address is locally or universally assigned. The second bit is rarely used on Ethernet (CSMA/CD) networks.

Adding IEEE 802.2 LLC information to an 802.3 physical packet format requires three additional fields at the beginning of the data field: a 1-byte destination service access point (DSAP) field, a 1-byte source service access point (SSAP) field, and a 1-byte control field.

2.11 Conclusion

Ethernet is physical layer LAN technology invented in 1973. Since its invention, Ethernet has become the most widely used LAN technology. The 1-Gbps and 10-Gbps LAN interfaces are commonly used today. The demand for 40-Gbps and 100-Gbps Ethernet equipment is expected to grow in the future.

References

[1] IEEE 802.3x, 1997-IEEE Standards for Local and Metropolitan Area Networks: Specification for 802.3 Full Duplex Operation.

[2] IEEE standard 802.3u, 1995-IEEE Local and Metropolitan Area Networks-Supplement - Media Access Control (MAC) Parameters, Physical Layer, Medium Attachment Units and Repeater for 100Mb/s Operation, Type 100BASE-T

[3] Duelk, M., "Consideration for 40 Gigabit Ethernet," IEEE HSSG. http://grouper.ieee.org/groups/802/3/hssg/public/may07/duelk_01_0507.pdf.

[4] Kipp, S., "40 Gigabit Ethernet Answers," IEEE HSSG, May 2007 http://grouper.ieee.org/groups/802/3/hssg/public/may07/kipp_01_0507.pdf.

[5] Metcalfe, B., "Bob Metcalfe on the Terabit Ethernet," http://www.lightreading.com/bob-metcalfe-on-the-terabit-ethernet/v/d-id/703860.

[6] Digital Equipment Corporation, Intel Corporation, Xerox Corporation (September, 1980). The Ethernet: A Local Area Network, ACM SIGCOMM Computer Communications Review, Vol. 11, No. 3, July 1981.

[7] 802.3-2008/Cor. 1-2009 - IEEE Standard for Information Technology-- Local and Metropolitan Area Networks-- Specific Requirements-- Part 3: Carrier Sense Multiple Access with Collision Detection (CSMA/CD) Access Method and Physical Layer Specifications Corrigendum 1: Timing Considerations for PAUSE Operation.

CHAPTER 3

Synchronization

3.1 Introduction

Depending on the application, timing, phase, and/or frequency synchronization are necessary among systems and networks. If one system is ahead of the others, the others are behind that particular one. For a user, switching between these systems would cause time to jump back and forward. This would be very confusing and results in various errors.

Imagine two isolated networks running their own wrong clocks connect to the Internet where some e-mail messages arrive 5 minutes before they were sent and replies arrive 2 minutes before the messages were sent. We all will be very confused.

However, TDM networks optimized for constant rate signals that transport multiplexed voice and data traffic require highly accurate frequency information to be recoverable from their physical layers. Isolated physical clocks, such as piezoelectric crystals, cannot be expected to agree with each other over time. Various effects, such as temperature changes and aging, cause even reliable frequency sources to wander. Frequency discrepancies lead to valid bits being lost and invalid bits inserted, which is called bit slips. In order to avoid bit slips, somewhere in every TDM network there is an extremely accurate primary reference clock (PRC) from which all other TDM clocks in the network directly or indirectly derive their timing. Clocks derived in this manner are said to be traceable to a PRC.

The heart of a synchronous network is a stable and reliable clock source. The PRC is commonly used to describe a very accurate clock. Global Positioning System (GPS) receivers are typically used to drive the reference clock where GPS is a satellite-based sourced clocking system. Fixed, Earth-mounted reference station signals are added in the calculations of the GPS reception to increase the accuracy/clock quality in differential GPS (DGPS) systems.

Some countries such as Germany, Japan, and France also maintain a radio-based distribution with atomic clock sources with 2-MW power and 3,000-km radius reach, providing high-quality clocking and low variances.

The primary clock signal has to be distributed downwards to the edges of the network to sync up all necessary devices. Special-purpose T1 or E1 circuits are typically used to deliver clocking to the relevant boxes. At the tail end, a cellular base station controller could be the main device enabling cell sites to align with other cell sites within a network, allowing users to tax the network by roaming at speeds of up to 200 km/hr (130 mph). High-speed roaming and the handoff of calls between cell towers truly tax the infrastructure and require the utmost in clock synchronization. One major requirement for synchronous networks is their ability to keep clocks in sync once they lose their connection from the master source(s), called holdover time. Holdover time depends solely on the accuracy of the built-in oscillator. The longer the devices can run autonomously, the more expensive those components become. A holdover time of 24 hours is essential, while more than a week is commercially infeasible for deployments within every leaf node in the network.

The Ethernet is the low-cost technology of the choice for enterprise and residential networks, and expected to be used in WAN as well, with the introduction of Carrier Ethernet capabilities. Migration from TDM networks to Ethernet-based packet technology introduces new challenges. Packet switching was originally introduced to handle asynchronous data. However, applications such as the transport of TDM service and the distribution of synchronization over packet networks require the strict synchronization requirements of those applications. An acceptable level of quality such as limited slip rate must be maintained.

Synchronization in TDM networks is well understood and implemented. Typically, a TDM circuit service provider will maintain a timing distribution network, providing synchronization traceable to a PRC that is compliant with clock quality [1].

In some private network applications involving circuit emulation, it may be sufficient to distribute a non-PRC quality level (QL) common clock toward CES interworking function (IWF). However, the use of synchronization timing below PRC QL could result in internetworking difficulties between different network domains.

In order to achieve synchronization, network operators have distributes a reference timing signal of suitable quality to the network elements processing the application. One approach is to follow a distributed PRC strategy. An alternative approach is based on a master-slave strategy.

A different performance can be requested in case the packet network is part of an access network or is the underlying layer of the core network. The distribution of a synchronization reference over a portion of a core network may be requested to comply with strict jitter and wander requirements. However, in the access network, requirements may be relaxed to allow a distribution of a timing reference signal with a performance level lower than PRC quality to support the timing requirements of the end node.

The nodes involved in packet-oriented transmission technology such as ATM nodes do not require any synchronization for the implementation of the packet-switching function. In fact, at any entrance point of a packet switch, an individual device provides packet timing adaptation of the incoming signal to the internal timing. For example, in ATM networks, idle cell stuffing is used to deal with frequency differences. Transmission links do not need to synchronize with each other.

However, synchronization functions in packet networks, especially on the boundary of the packet networks, are dependent on the services carried over the network. As the packet network supports TDM-based applications, it needs to provide correct timing at the traffic interfaces. For TDM-based services, the

IWF may require network-synchronous operation in order to provide acceptable performance.

The transport of TDM signals through packet networks requires that the signals at the output of the packet network comply with TDM timing requirements in order for TDM equipment to interwork. This is independent of the application whether it is voice or data.

The adaptation of TDM signals into the packet network is CES [2–4]. Jitter and wander limits at traffic and/or synchronization interfaces, long-term frequency accuracy that can influence the slip performance, and total delay, which is critical for real-time services such as the voice service, must be satisfied.

The PDH timing requirements for traffic interfaces are mainly related to jitter, wander, and slip performance where timing information is only present at the physical layer. These values are specified in [5] for the network based on 2,048-Kbps hierarchy and in [6] for the network based on 1,544-Kbps hierarchy. In addition, [7] specified the applicable slip rate objectives.

Time distribution relates to the transfer of time rather than frequency. Frequency is a relative measure but is generally assumed to be measured relative to a frequency standard. Time differs from frequency in that it represents an absolute, monotonically increasing value that can be generally traced to the rotation of the earth (e.g., year, day, hour, minute, second). Mechanisms to distribution time are significantly different than what is used to distribute frequency.

Timestamps may be used in some network applications to support generation of frequency. The notion of time carried by these timestamps compared with the time generated by the local oscillator can be used to recover a frequency reference for the local oscillator. Differential methods can also be used to recover timing from packets. In this case, the timestamp needs only to be a relative and can be used as an estimate of phase. Because phase and frequency are related, it is possible to use this relative information to recreate a frequency reference. This is known as differential timing [8]. As an example, synchronous residual time stamp (SRTS) [9] is a well-known method standardized for use in ATM AAL1 that allows relative phase to be signaled as a timestamp, to be sent across a packet network to be used to recreate the frequency of a PDH signal.

3.2 Application Requirements

While accurate clocking is necessary for running services like cell towers with mobile applications, each network operator has its own clocking mechanism with varying levels of accuracy. Accuracy is highly dependent on the resources available to the operator.

An American National Standards Institute (ANSI) standard entitled "Synchronization Interface Standards for Digital Networks" (ANSI/T1.101) [10] defined the stratum levels and minimum performance requirements for digital network synchronization. The requirements for the stratum levels are shown in Table 3.1, which provides a comparison and summary of the drift and slip rates for the strata clock systems. Stratum 0 is defined as a completely autonomous source of timing, which has no other input, other than perhaps a yearly calibration. The usual source of Stratum 0 timing is an atomic standard (Caesium Beam or Hydrogen Maser).

Table 3.1 Accuracy Requirements for Stratum Clocking

Stratum	Accuracy/ Adjust Range	Pull-In-Range	Stability	Time to First Frame Slip
1	1×10^{-11}	NA	NA	72 days
2	1.6×10^{-8}	Must be capable of synchronizing to clock with accuracy of $\pm 1.6 \times 10^{-8}$	1×10^{-10}/day	7 days
3E	1.0×10^{-6}	Must be capable of synchronizing to clock with accuracy of $\pm 4.6 \times 10^{-6}$	1×10^{-8}/day	3.5 hours
3	4.6×10^{-6}	Must be capable of synchronizing to clock with accuracy of $\pm 4.6 \times 10^{-6}$	3.7×10^{-7}/day	6 minutes (255 in 24 hours)
4E	32×10^{-6}	Must be capable of synchronizing to clock with accuracy of $\pm 32 \times 10^{-6}$	Same as accuracy	Not yet specified
4	32×10^{-6}	Must be capable of synchronizing to clock with accuracy of $\pm 32 \times 10^{-6}$	Same as accuracy	NA

The minimum adjustable range and maximum drift is defined as a fractional frequency offset (f/f) of 1×10^{-15} or less.

Clock QLs are defined by the industry standards organizations to maintain clock quality in the network for time sensitive services need synchronization, to avoid overflow or underflow of slip buffers, bit errors, and other adverse effects. Reference [5] provided criteria for controlled slip rate.

Synchronization in Carrier Ethernet networks is required mainly due to mobile backhaul and bandwidth explosion in mobile networks, and replacement of the synchronous physical layer with the Ethernet. The need for transport and recovery of clocking information has two main aspects:

- Time of day (ToD) for correct logging, access control, time stamping, and so forth;
- Clock phase, where high precision is needed to sync up mobile cells or to drive TDM circuits.

Three types of synchronization are needed: frequency, phase, and time. Tables 3.2 and 3.3 list some of the applications and required synchronization types.

References [6] for E circuits and [11] for T circuits set limits on the magnitude of jitter and wander at network interfaces. The wander may not exceed given values anywhere in the network. Thus, a circuit emulation link, for example, may consume only part of the wander budget:

- GSM, WCDMA, and CDMA2000 require frequency accuracy of 0.05 ppm at air interface.
- CDMA2000 requires time synchronization at the ± 3-μs level (± 10 μs worst case).
- WCDMA TDD mode requires 2.5-μs time accuracy between neighboring base stations [i.e., ± 1.25 μs of Coordinated Universal Time (UTC)].

These requirements are too difficult to achieve without good transparent clocks or boundary clocks (BC) in each intermediate node. Some cellular operators do

3.2 Application Requirements

Table 3.2 Synchronization Applications

Synchronization Type	Application	Required or Targeted Quality
Frequency	TDM support (circuit emulation service)	PRC traceability
	Third-Generation Partnership Project (3GPP2) base stations [including Long Term Evolution (LTE)]	Frequency assignment shall be better than ± 0.05 ppm for frequency division duplex (FDD), and micro to femtocells ± 0.25 ppm.
	IEEE 802.16 (WiMAX)	Unsynchronized orthogonal frequency division multiple access (OFDMA): frequency accuracy shall be better than ±2 ppm.
	Digital Video Broadcasting Terrestrial/Handheld (DVB-T/H)	Frequency accuracy depends on radio frequency, down to a few parts per billion.
Time Phase (relative time) Time of day (ToD; wall-clock, absolute time)	3GPP2 code division multiple access (CDMA)	Time alignment error should be less than 3 microseconds for phased and shall be less than 10 microseconds for ToD
	3GPP Universal Multiple Telecommunications Service (UMTS) time-division duplex (TDD)	Intercell synchronization accuracy must be better than ±2.5 microseconds between base stations (or < ±1.25 microseconds from common source).
	DVB-T/H single-frequency network (SFN)	All transmitters within a single-frequency network must broadcast an identical signal to within 1-microsecond accuracy.
	3GPP LTE Multi-MEDI broadcast over a single frequency network (MBSFN)	Cell synchronization accuracy should be better than or equal to 3 microseconds for SFN support.
	802.16D/e TDD	Requirements depend on: mode, modulation, application, implementation, and option used; it likely would have to be better than 5 microseconds (TBC); ≤±1/16 × cyclic prefix.
	SLA monitoring and correlation of logs	Y.1731 [12] delay and jitter measurements.

have control over the transport network so they could use IEEE1588-compliant switches for achieving time synchronization.

The short-term goal is to improve precision to <1 ms. The target is a few orders of magnitude below average delay (i.e., ~10–100 microseconds).

For correlation, the finer the time-stamping, the faster the correlation.

In order to emulate TDM service, two things have to interwork:

- The emulated service has to transport the phase information or has to follow a certain clock discipline.
- The payload has to be transparently carried through the network. The lowest layer must set the line speed and structure of the circuit. A select number of timeslots are then used for signaling, payload, and clocking information.

Typical speeds for TDM-circuit emulations start at 64-Kbps ISDN-B channels and can go up to OC-3/STM-1 levels running at 155 Mbps. In order to recover potential jitter that might occur during transmission, it is common to use buffering to compensate. Yet buffering will always add latency to the overall service

Table 3.3 Time-Phase Requirements

Application	Time/Phase Synchronization Accuracy
CDMA2000	±3 ms with respect to UTC (during normal conditions), ±10 ms of UTC (when the time sync reference is disconnected)
WCDMA (TDD mode)	2.5-ms phase difference between base stations
TD-SCDMA (TDD mode)	3-ms phase difference between base stations
LTE (TDD)	3-ms time difference between base stations (small cell), 10 ms time difference between base stations (large cell).
Multicast broadcast single frequency network (MBSFN)	<±1 ms with respect to a common time reference (continuous timescale)
WiMAX (TDD mode)	Depend on several parameters. As an example ±0.5 ms and ±5 ms have been mentioned for a couple of typical cases.
IP network delay monitoring	Depends on the level of quality that shall be monitored. As an example, ±100 ms with respect to a common time reference (e.g., UTC) may be required. ±1 ms has also been mentioned.
Billing and alarms	±100 ms with respect to a common time reference (e.g., UTC)

performance. Values over 100 ms are quite long and might interfere with VoIP services. Since the payload type may be unknown, a fully transparent service must forward every bit.

Payload-aware emulations are far more effective on bandwidth utilization, but they are also more costly on component complexity and software management. In order to recover clock information, three typical methods are used:

- Internal clocking by using its own onboard oscillator where the device clocks a line directly from its own independent source.
- External clocking that takes the clock information from an external connection and distributes it among all connected services belonging to this phase system. Typically, a primary reference input such as GPS or building integrated timing supplies/sync supply units (BITS/SSU) also known as Stand-alone Synchronization Equipment (SASE) is used as a grandmaster clock.
- Loop-timed/line-derived clocking where the clock source is taken from a selected TDM circuit.

Separate clock disciplines can be handled within one device, potentially supporting diverse applications that do not have the same clock source.

Table 3.4 describes timing methods for mobile backhaul as an example.

3.3 Synchronization Standards

In order to provide clocking across Ethernet, the following options are available:

- Synchronous Ethernet Layer-1 embedded timing information on the physical layer requiring full hardware support to link up. This approach provides only phase, not ToD.
- IEEE 1588v2 (PTPv2)-Layer-2 embedded operation, administration and maintenance (OAM) frames with highest priority to ship clock/phase and

Table 3.4 Timing Technologies for Circuit Emulation Services (CES)

Technology	Frequency Accuracy	Frequency (MTIE/TDEV)	Phase/Time (Time of Day)
Internal XO	Yes	No	No
GPS	Yes	Yes	Yes
BITS/SSU (SASE)	Yes	Yes	Yes
Synchronous Ethernet (Physical Layer)	Yes	Yes	No
IEEE 1588v2 (Protocol Layer)	Yes	Yes	Yes
NTP v4 (Protocol Layer)	Yes	Yes	Yes
T1/E1 (Leased Line)	Yes	Yes	No

MTIE/TDEV: Maximum time interval error/time deviation.

ToD information across the packet network. Hardware NIC (network interface card) support is advised for higher accuracy, but the packets remain standard Ethernet frames.
- IETF RFC4330 (NTPv4)-Layer-3 ToD information with high precision, but slow sync-up times and no phase lock support to aid TDM-emulation services.

Most of the standards relevant to the options above are from three different standards organizations, IETF [13–16], ITU-T [8, 17–23], and IEEE [24, 25].

Emulation of a native service is very different than transporting of the payload of a TDM line carrying a known protocol type. The difference is described as content-aware or payload-agnostic, where any type of payload can be carried along with the source clock information. The main difference is related to the amount of bandwidth consumed when transporting the emulated service. If a payload-aware service detects no activity within the line, little to no bandwidth is being used to transport the service. Whereas in case of payload-agnostic mode, every data bit from the TDM port has to be transported and reconstituted across the network. After reconstruction, the TDM service, a clock recovery process called phase locked loop (PLL) ensures that a stable clock will be delivered despite packet delay variances (jitter) or other interfering factors.

There are two main standards currently being used to run pseudowire services on most platforms: circuit emulation over PSN (CESoPSN) [26] and structure-agnostic TDM over packet (SAToP) [27]. With these two standards, it is possible to emulate a clock-sensitive service across a variable packet core transport.

3.4 NTP/SNTP

Network Time Protocol (NTP) is an Internet protocol used to synchronize the clocks of computers to a time reference. A full implementation of the NTP protocol is too complicated for many systems. A simplified version of the protocol, namely Simple Network Time Protocol (SNTP), had been defined. SNTP is basically also NTP, but lacks some internal algorithms that are not needed for all types of servers.

NTP needs a reference clock that defines the true time to operate. All clocks are set towards that true time. NTP uses UTC as reference time.

NTP is a fault-tolerant protocol that will automatically selects the best of several available time sources to synchronize to. Multiple candidates can be combined to minimize the accumulated error.

NTP is highly scalable. A synchronization network may consist of several reference clocks. Each node of such a network can exchange time information either bidirectional or unidirectional. Propagating time from one node to another forms a hierarchical graph with reference clocks at the top.

Having available several time sources, NTP can select the best candidates to build its estimate of the current time. The protocol is highly accurate, using a resolution of less than a nanosecond (about 2^{-32} seconds). Even when a network connection is temporarily unavailable, NTP can use measurements from the past to estimate current time and error.

NTP implementations are supported by most of the popular UNIX and Windows operating systems. Among those are: AIX, FreeBSD, HP-UX, Irix, Linux, NetBSD, SCO UNIX, OpenBSD, OSF/1, Solaris, System V.4, Windows Vista, Windows NT 4.0, Windows 2000, Windows XP, and Windows .NET Server 2003.

According to a survey of the NTP network [8], there were at least 175,000 hosts running NTP in the Internet. Among these there were over 300 valid stratum-1 servers. In addition there were over 20,000 servers at stratum 2, and over 80,000 servers at stratum 3 (Figure 3.1).

Currently, there are version 3 and version 4 implementations of NTP available. NTPv4 [28] algorithms can deal with high delay variations a bit better than the

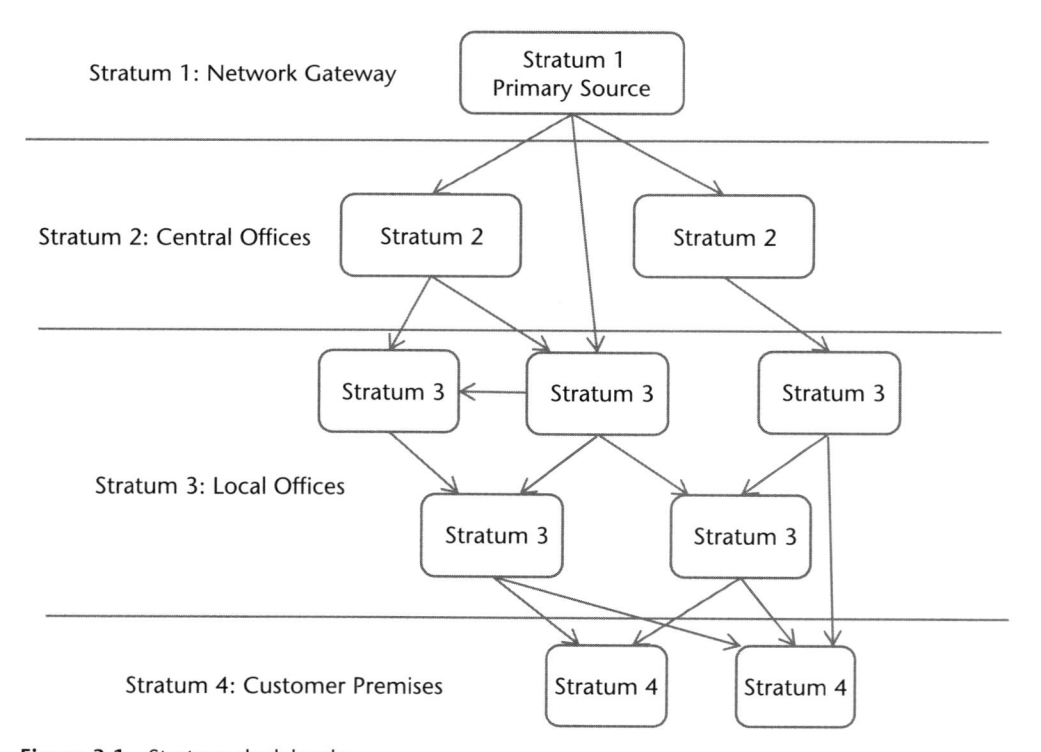

Figure 3.1 Stratum clock levels.

version 3. However, NTPv4 [28] uses floating point operations, whereas NTPv3 [29] used integer arithmetic. This might be an issue for older systems without a floating point unit.

The new features of version 4 compared to version 3 are:

- Use of floating-point arithmetic instead of fixed-point arithmetic;
- Redesigned clock discipline algorithm that improves accuracy, handling of network jitter, and polling intervals;
- Support for the nanokernel kernel implementation that provides nanosecond precision as well as improved algorithms;
- Public-key cryptography known as autokey that avoids having common secret keys;
- Automatic server discovery (manycast mode);
- Fast synchronization at startup and after network failures (burst mode);
- New and revised drivers for reference clocks;
- Support for new platforms and operating systems.

However, SNTP Version 4 [30] accommodates Internet Protocol Version 6 (IPv6) and Open Systems Interconnection (OSI) addressing. Certain optional extensions to the basic Version 3 model, including an anycast mode and an authentication scheme are designed specifically for multicast and anycast modes

NTP and SNTP services are defined on UDP/IP-Port: 123. SNTP can talk to NTP servers. NTP is more complex than SNTP and can handle multiple sources. SNTP accuracy today is typically between 1 ms and 50 ms. NTPv4 claims to be close to n 10 μs, depending on network and hardware. GPS receivers provide typically <10 ns–200 ns, Stratum-1.

With (S)NTP, the transmission time stamp is being transmitted in the synchronization packet itself, compared to PTP (IEEE1588) where it is transmitted in a following packet. In this way measurement of transmission, reception and transmission of measured time stamps can be decoupled.

Systems use NTP to synchronize clocks to the ToD master (NTP time server). Clock sync accuracy is affected by frame jitter in time sync request/response messages (Figures 3.2 and 3.3):

Round-trip delay = T4 – T1
One-way delay = ½ round-trip or 1/2 (T4 – T1)
Current time = T3 + ½(T4 – T1) + (T5 – T4)

Slave clock accuracy depends on packet delay and jitter. Packet delay can be accounted for in computation, but jitter affects tracking accuracy. In one-way delay measurement, jitter affect is additive. For example, if there is 1-ms jitter between slave 1 and the master, and 1 ms between slave 2 and the master, we may end up with 2-ms jitter between slave 1 and slave 2. In order to minimize jitter impact on the accuracy, various proprietary algorithms are developed by various vendors.

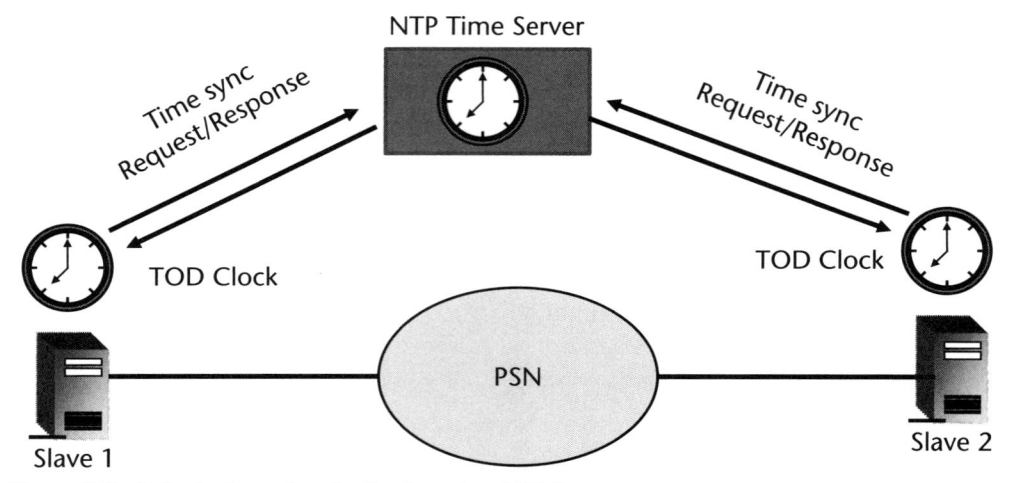

Figure 3.2 ToD clock synchronization based an NTP time server.

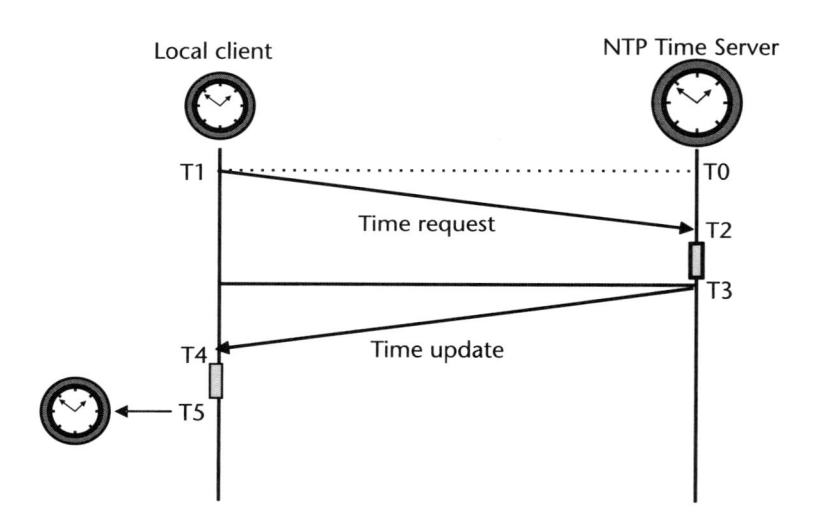

Round-trip delay = T4−T1
One-way delay = ½ round-trip or 1/2 (T4−T1)
Current time = T3 + ½(T4−T1) + (T5−T4)

Figure 3.3 Current time calculation based on an NTP server.

Real-Time Protocol (RTP) v1 [31] and RTPv2 [32] are among them. Real-Time Transport Control Protocol (RTCP) [32] provides feedback on the quality of the data distribution and carries a persistent transport-level identifier for an RTP source. All participants send RTCP packets; therefore, the rate must be controlled in order for RTP to scale up to a large number of participants. By having each participant send its control packets to all the others, each can independently observe the number of participants. The number of participants is used to calculate the rate at which the packets are sent.

3.5 Precision Time Protocol (IEEE 1588)

In recent years, an increasing number of systems utilize a more real-time clocks and distributed architecture, all of which are synchronized to each other within the system. These clocks are used to manage distributed file systems, backup and recovery systems, and many other similar activities. These computers typically interact via LANs and the Internet. In this environment, the most widely used technique for synchronizing the clocks is the Network Time Protocol (NTP) or the related SNTP.

IEEE 1588 is designed to fill a niche not well served by either of the two dominant protocols, NTP and GPS. IEEE 1588 is designed for local systems requiring very high accuracies beyond those attainable using NTP. It is also designed for applications that cannot bear the cost of a GPS receiver at each node, or for which GPS signals are inaccessible.

IEEE-1588 enables submicrosecond synchronization of clocks by having a master clock send multicast synchronization message frames containing timestamps. All IEEE-1588 receivers correct their local time based on the received timestamp and an estimation of the one-way delay from transmitter to receiver.

IEEE 1588 is a protocol designed to synchronize real-time clocks in the nodes of a distributed system that communicate using a network. IEEE 1588 defines a protocol enabling precise synchronization of clocks applicable to systems communicating by local area networks supporting multicast messaging including but not limited to Ethernet. The protocol will enable heterogeneous systems that include clocks of various inherent precision, resolution, and stability to synchronize.

The Precision Time Protocol (PTP) is a high-precision time protocol for synchronization used in measurement and control systems residing on a local area network. Accuracy in the submicrosecond range may be achieved with low-cost implementations.

PTP was originally defined in the IEEE 1588-2002 [33]. In 2008 a revised standard, IEEE 1588-2008 [25], was released which is also known as PTPv2. The PTPv2 improves accuracy, precision and robustness but is not backwards compatible with the original 2002 version [34].

PTP is based on IP multicasting and can be used on any network that supports multicasting. It can be scaled for a large number of PTP nodes. Precision is typically in a range of 100 ns to 100 μs depending on real-time capabilities of an end system.

The master cyclically transmits a unique synchronization (SYNC) message to the related slave clocks at defined intervals, by default every 2 seconds.

The master clock measures the exact time of transmission. The slave clocks measure the exact times of reception. The master then sends in a second message, the follow-up message, the exact time of transmission of the corresponding sync message to the slave clocks.

On reception of the sync message and, for increased accuracy, on reception of the corresponding follow-up message, the slave clock calculates the correction (offset) in relation to the master clock taking into account the reception time stamp of the sync message. The slave clock must then be corrected by this offset. If there were to be no delay over the transmission path, both clocks would now be synchronous.

For highly accurate synchronization, the time of transmission and reception of PTP messages can be determined as precisely and as closely as possible to the hardware.

The data format for PTP over Ethernet is depicted in Figure 3.4.

Although PTP, which is 1588 v2, can be implemented over any packet network, the major focus has been on the development of PTP over UDP/IPv4. The protocol stack is depicted in Figure 3.5.

The protocol will support system-wide synchronization accuracy in the submicrosecond range with minimal network and local clock computing resources. The default behavior of the protocol will allow simple systems to be installed and operated without an administrator's involvement.

The IEEE 1588 organizes the clocks into a master-slave hierarchy (Figure 3.6) based on observing the clock property information contained in multicastSync messages. Each slave synchronizes to its master (based on Sync, Delay_Req, Follow_Up, and Delay_Resp messages exchanged between master and its slave) (Figure 3.7).

The time base in a IEEE 1588 system is the time base of the grandmaster clock such as a Stratum-1 source. The grandmaster clock time base is implementation-dependent. All other clocks synchronize to the grand master. If the grandmaster clock maintains a UTC time base, the 1588 protocol distributes the appropriate leap-second information to the slaves. Figure 3.8 depicts the use of the 1588 clocking in the CES application.

In the synchronization process, master clock sends Sync and Follow_up messages (Figure 3.9). The Sync message contains an estimate of the sending time (~t1).

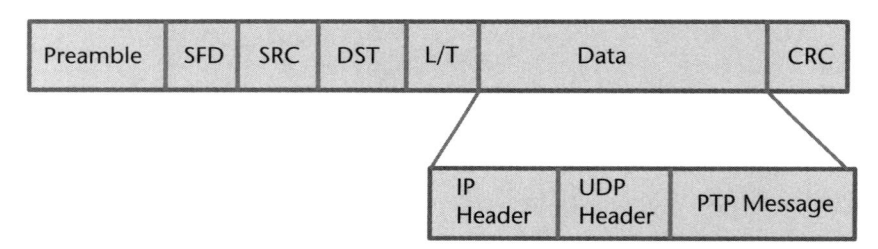

Figure 3.4 Data format for PTP over Ethernet.

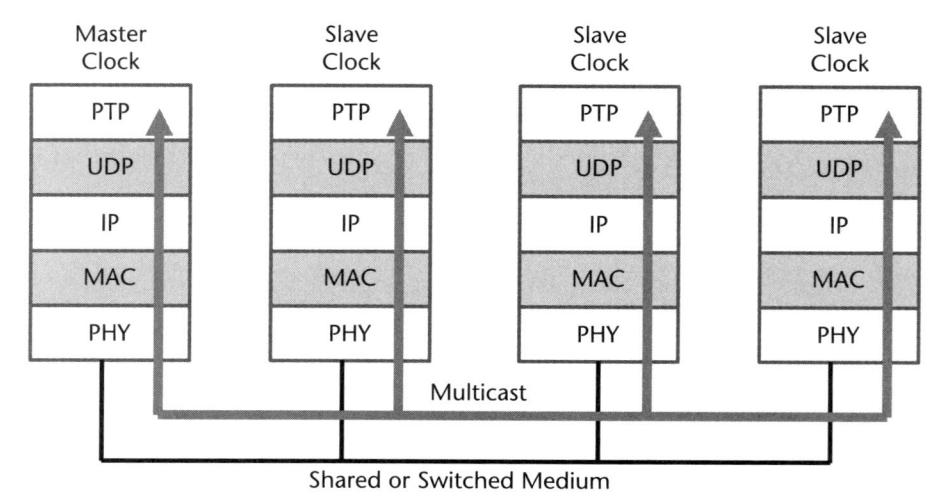

Figure 3.5 IEEE 1588 protocol stack.

3.5 Precision Time Protocol (IEEE 1588)

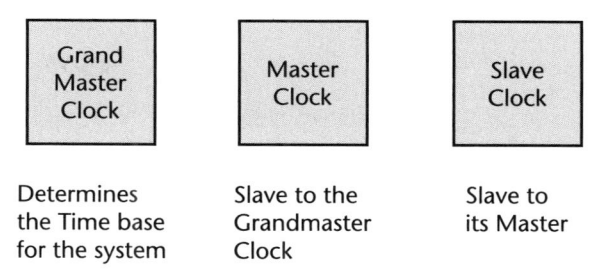

Figure 3.6 Clock hierarchy.

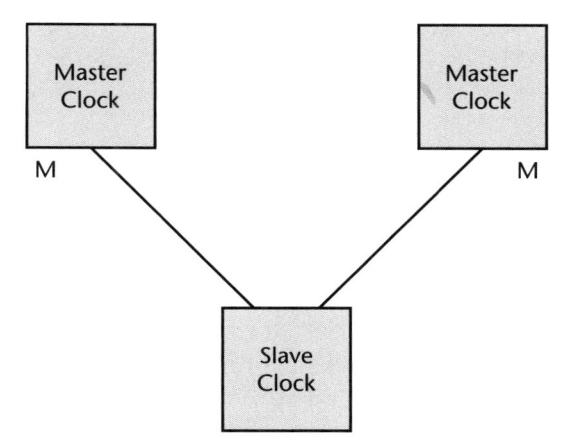

Figure 3.7 Slave receiving inputs from multiple masters.

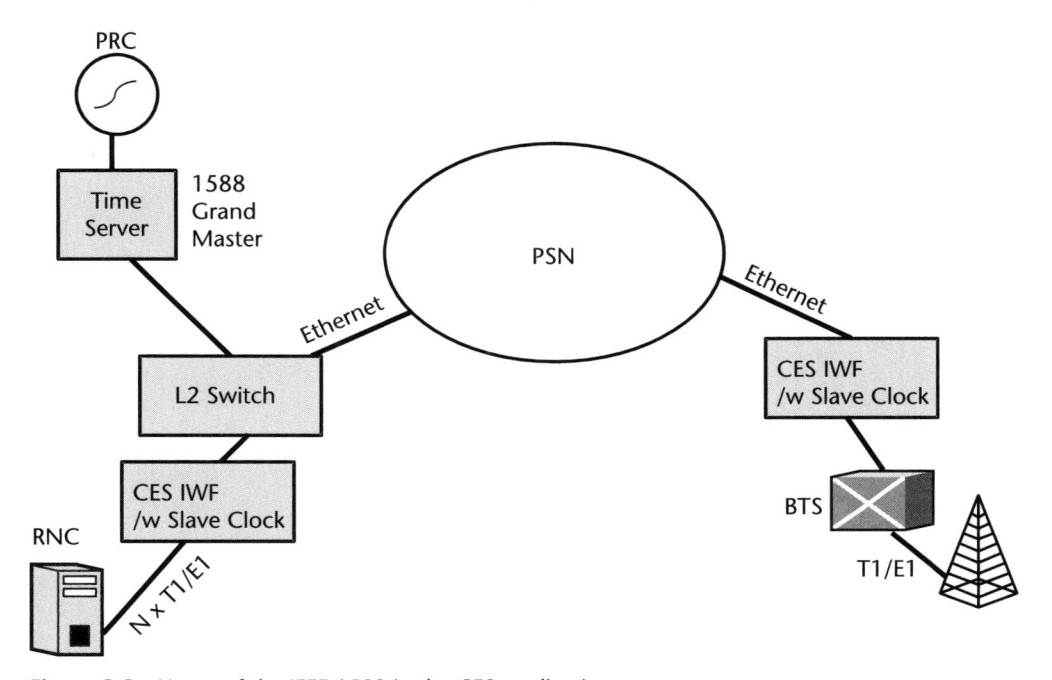

Figure 3.8 Usage of the IEEE 1588 in the CES application.

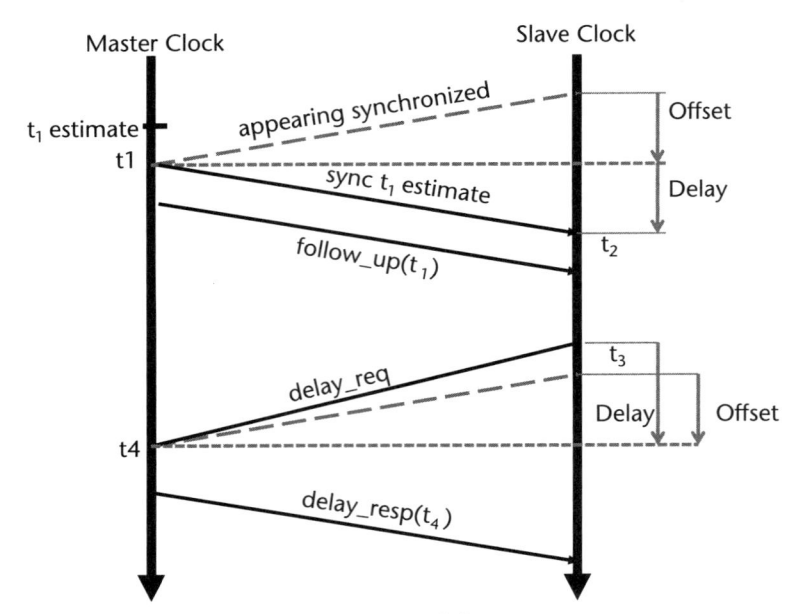

Figure 3.9 Message exchanges between master and slave.

When received by a slave clock the receipt time is noted. Follow_Up message is always associated with the preceding Sync message and contain the "precise sending time= (t1)"as measured as close as possible to the physical layer of the network. When received by a slave clock, the "precise sending time" is used in computations rather than the estimated sending time contained in the Sync message.

However, the slave clock sends Delay_Req messages. The slave measures and records the sending time (t3). When Delay_Req is received by the master clock, the receipt time is noted (t4). In response, master sends Delay_Resp message containing the receipt time of the associated Delay_Req message (t4). When received by a slave clock, the receipt time is noted and used in conjunction with the sending time of the associated Delay_Req message as part of the latency calculation.

Synchronization computation (in the slave clock):

offset = receipt time − precise sending time − one way delay (for a Sync message)
one way delay = {MS Delay + SM Delay}/2 (assumes symmetric delay)
MS Delay = receipt time − precise sending time (for a Sync message)
SM Delay = Delay_Req receipt time − precise sending time (of a Delay_Req message)

From this offset the slave corrects its local clock.

In order to synchronize a pair of clocks using 1588, a Sync message is sent from master to slave (Figure 3.9). The apparent time difference between the two clocks is measured:

MS_Time_Difference = slave's receipt time −master's sending time = t2−t1
MS_Time_Difference = offset + MS Delay

3.5 Precision Time Protocol (IEEE 1588)

where MS delay is the time for Sync message to travel from master to slave, which is a combination of propagation and queuing delay.

After that, a Delay_Req message from slave is sent to master and the apparent time difference between the two clocks is measured:

SM_Time_Difference = master's receipt time − slave's sending time = $t_4 - t_3$
SM_Time_Difference = −offset + SM Delay

where SM Delay is the time for Delay_Req message to travel from slave to master, which is a combination of propagation and queuing delay.

From the above, we have the following two equations:

MS_Time_Difference = offset + MS Delay (3.1)
SM_Time_Difference = −offset + SM Delay (3.2)

From (3.1) and (3.2),

offset = {(MS_Time_Difference − SM_Time_Difference) − (MS Delay − SM Delay)}/2
MS Delay + SM Delay = {MS_Time_Difference + SM_Time_Difference}

If we assume

MS Delay = SM Delay = One_Way_Delay

then,

offset = {MS_Time_Difference − SM_Time_Difference}/2
one_way_delay = {MS_Time_Difference + SM_Time_Difference}/2
For MS_Time_Difference = 10 minutes and SM_Time_Difference = −5 minutes, we can have offset = {10 − (−5)}/2 = 7.5 minutes and one_way_delay = {10 + (−5)}/2 = 2.5 minutes.

IEEE 1588 v2 employs a two-way methodology, where packets are sent back and forth from the clock master to the clock slaves. This overcomes high-amplitude, ultra-low-frequency wander that defeats other methods such as adaptive clock recovery techniques. Also, the standard is virtually independent of the physical media and can flow over low-speed, twisted-pair, high-speed optical fiber, wireless, or even satellite links without requiring equipment design modifications.

It is not limited to TDM circuit emulation like the in-band solutions, but it can support CES better than adaptive clocking by distributing a precise network clock to every interworking function node in the system. It also can be used for any pure, packet-based network, providing synchronization for future backhaul networks to be deployed by mobile operators.

The standard can distribute time/phase, frequency, or both. Telecom operators can use it to sell a synchronization service to customers (residential, wireless operators, and so forth). It is resilient because a failed network node can be routed around. It also is resilient because the synchronization can come from one or more grandmaster clock nodes.

IEEE 1588 v2 packets fully comply with Ethernet and IP standards and are backward compatible with all existing Ethernet and IP routing and switching equipment. There is no requirement for intermediate switches or routers to be IEEE 1588 v2 aware. They see these timing packets as normal packet data.

The protocol calls for synchronization packets with time stamps to be sent from master clocks to all slave clocks and for individual slave clocks to send time-stamped packets to the master. The clock grandmaster maintains a time base locked to a primary reference clock and establishes a separate synchronization session with each of the slaves it serves. The master and slave exchange timing packets according to the syntax of the IEEE 1588 v2 protocol.

This process provides the timing clock recovery algorithm with the time stamps it needs to precisely recreate the master time base. From this time base, the synchronization signals used by the network equipment are synthesized. The timing clock recovery algorithm filters most noise, packet queuing delay, and propagation delay created by the transport network.

Accuracy of time stamping is influenced by protocol stack and execution environment. Messages generated or received by the PTP application are delayed within sending and receiving clocks. Delay components are cumulative and consist of constant and varying parts, called transit delay and delay variation (jitter).

Almost all of these effects can be bypassed if time stamps are taken as near as possible to the physical layer with hardware assistance.

To cope with the heavily fluctuating delay of storing and forwarding network elements such as switches/routers, two approaches are envisaged:

- Boundary clocks (BCs) are used (Figure 3.10). BCs forward frames as switches are used to, but they also contain a clock. PTP frames are processed by the BC in order to synchronize the internal clock and to synchronize other clocks connected to the BC. This leads to a hierarchical topology of clocks. A BC's synchronization performance is independent of the network load.
- Drift and offset are estimated by statistical methods. The methods require a long history of measured data. This requires reasonably stable oscillators and results in slow reaction to topological or environmental changes.

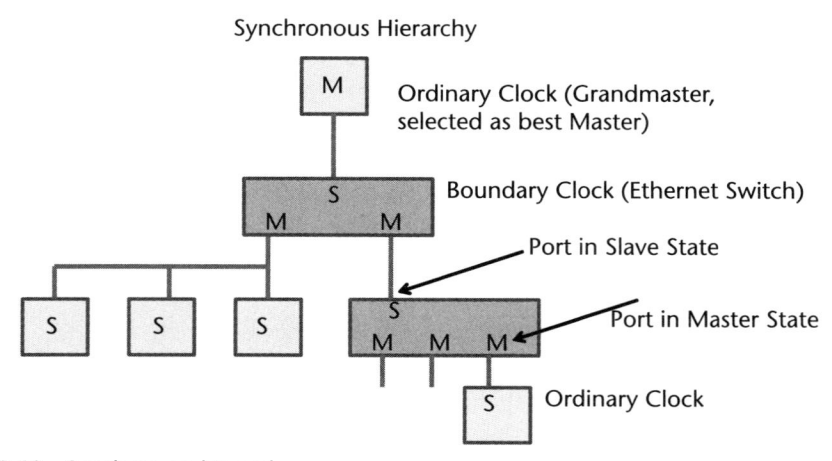

Figure 3.10 Synchronous hierarchy.

In a network using IEEE 1588, PTP configures and segments the network automatically. Each node employs a grand master clock (GMC) algorithm, also known as a "best master clock" algorithm, to determine the most accurate clock in the segment. Each node advertises its clock properties and features to other nodes using Sync messages. Once the GMC is selected, all nodes synchronize to it by adjusting their clocks accordingly. The GMC algorithm runs periodically, providing resynchronization as nodes are added or removed. Boundary clocks and transparent clocks both provide accurate distribution of the PTP across network devices such as switches, routers, and repeaters. A transparent clock forwards PTP event messages (but does not act as a master or slave) and provides correction for the delay time across the device. A boundary clock synchronizes other IEEE 1588 clocks across the subnets defined by a router, a switch, or other devices that could block the transmission of all IEEE 1588 messages. A boundary clock also eliminates the jitter typically generated by network devices due to internal buffering. In addition, a boundary clock typically acts as a master for all connected subnets, and appropriately retransmits 1588 management messages.

IEEE 1588 boundary clocks or transparent switches reduce the effect of jitter in Ethernet-based IEEE 1588 networks. A switch acting as a boundary clock runs the PTP and is synchronized to an attached master clock. The boundary clock in turn acts as a master clock to all attached slaves.

Even with hardware assistance, some fluctuations can still be observed due to quantization effects resulted from time stamp resolution and jitter in the data path and oscillator instabilities. Stochastic fluctuations may be removed by filtering and averaging algorithms. Long-term averaging requires a reasonable oscillator stability. If a topology change occurs such as a fast reconfiguration in ring configuration, then filtering and averaging slow the convergence. If reconfiguration can be reliably detected, filtering and averaging should be bypassed to accelerate convergence.

Issues with IEEE 1588 may be expressed as:

- IEEE 1588 only allows the values of sync interval to be 1, 2, 8, 16, and 64 seconds. It is difficult to maintain performance in a loaded network with sync packet rate of 1 pps and an inexpensive oscillator.
- IEEE 1588 relies on a symmetric network.
- IEEE 1588 does not have provision for redundancy support.
- IEEE 1588 relies on boundary clocks topology. Boundary clocks are not available in legacy telecom networks.
- IEEE-1588 only supports multicast.
- Long PTP messages consume too much bandwidth.

3.6 Synchronous Ethernet Networks

There are two methods for clocking distribution:

- Plesiosynchronous and network synchronous methods, called synchronous Ethernet;
- Packet-based methods such as IEEE 1588 (PTP) as described in previous sections.

Synchronous Ethernet is defined by the ITU a means of using the Ethernet to transfer timing (frequency) via the Ethernet PHY layer. This is a general case of layer-1 timing and was introduced in [8].

Synchronous Ethernet networks (SyncE) use the PHY clock transmissions and generates the clock signal from a bit stream similar to traditional SONET/SDH/PDH PLLs. Each node in the packet network recovers the clock (Figure 3.11).

The IEEE 802.3 standards require that the line rate of Ethernet operate within a specific rate (±100 ppm) relative to an absolute reference. According to the synchronous Ethernet, this rate must be traceable to an external reference. As a result, Ethernet devices requiring frequency recovery via synchronous Ethernet need to support a synchronization status message. Synchronous Ethernet ports nominally operate within a frequency tolerance range of ±4.6 ppm. However, in order to operate with nonsynchronous interfaces, synchronous Ethernet receivers must also operate at ±100 ppm in order to maintain data continuity.

Ethernet networks are free-running (±100 ppm). However, in case of synchronous Ethernet, a master-slave synchronization architecture at the physical layer is used to provide reference timing signal distribution over packet networks from the backbone level to the access level. This method can be used to provide timing recovery at the IWFs for constant bit rate (CBR) services transported over packet networks. It could also be used to provide a reference timing signal down to edge access equipment in a pure Ethernet network supporting synchronous Ethernet.

Within existing Ethernet technology, the service is effectively asynchronous. In synchronous Ethernet, existing Ethernet services will continue to be mapped into and out of the Ethernet physical layer at the appropriate rates.

Hierarchical timing distribution is recommended for synchronous Ethernet networks. Timing should not be passed from a synchronous Ethernet in free-run/holdover mode to a higher-quality clock because the higher-quality clock should not follow the synchronous Ethernet signal during fault conditions.

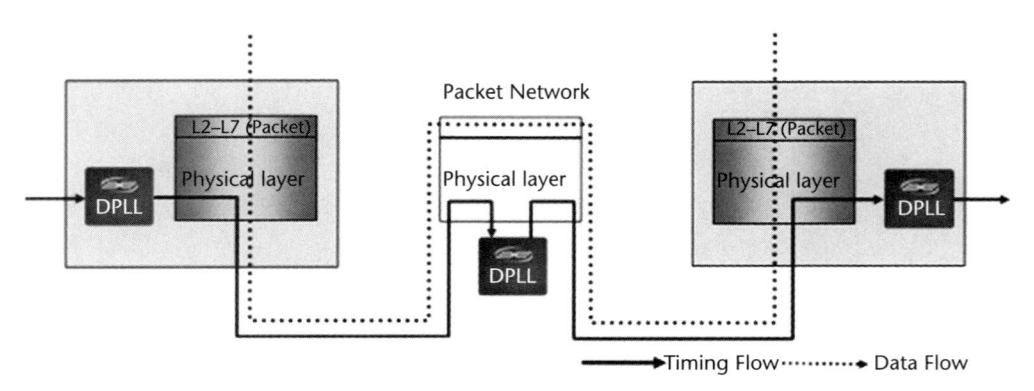

Figure 3.11 Timing through the Ethernet PHY.

3.6 Synchronous Ethernet Networks

In general, a reference timing signal traceable to a PRC is injected into the Ethernet switch using an external clock port to deliver a physical layer clock from the Ethernet switch to the IWF (Figure 3.12). This signal is extracted and processed via a synchronization function before injecting timing onto the Ethernet bit stream. The synchronization function provides filtering and may require holdover.

Clearly, there may be a number of Ethernet switches between the element where the reference timing signal is injected and the IWF. In such cases, the synchronization function within the Ethernet switch must be able to recover synchronization line timing from the incoming bit stream.

The network clock is the clock used to discipline the synchronization function within the Ethernet switch. The clock injected into the synchronization function will be synchronous (i.e., locked to the network clock).

The packet-based methods rely on timing information carried by the packets (e.g., sending dedicated time stamp messages as shown in Figure 3.13). Methods using two-way transfer of timing information are also possible such as NTP or similar protocols. In some cases, this is the only alternative to a PRC-distributed approach.

Timing recovery for constant bit rate services such as TDM circuit emulation service require that the timing of the signal is similar on both ends of the packet network and is handled by the IWFs. The service clock is preserved in such a way that the incoming service clock frequency is replicated as the outgoing service clock frequency.

Functional blocks defined for synchronization-related functions include clock functions, time distribution functions, clock selection functions and IWF are necessary to implement circuit emulation. These functional blocks such as ITU-T G.8262 [17] clocking may be embedded into network equipment or into other equipment such as a SSU/SASE.

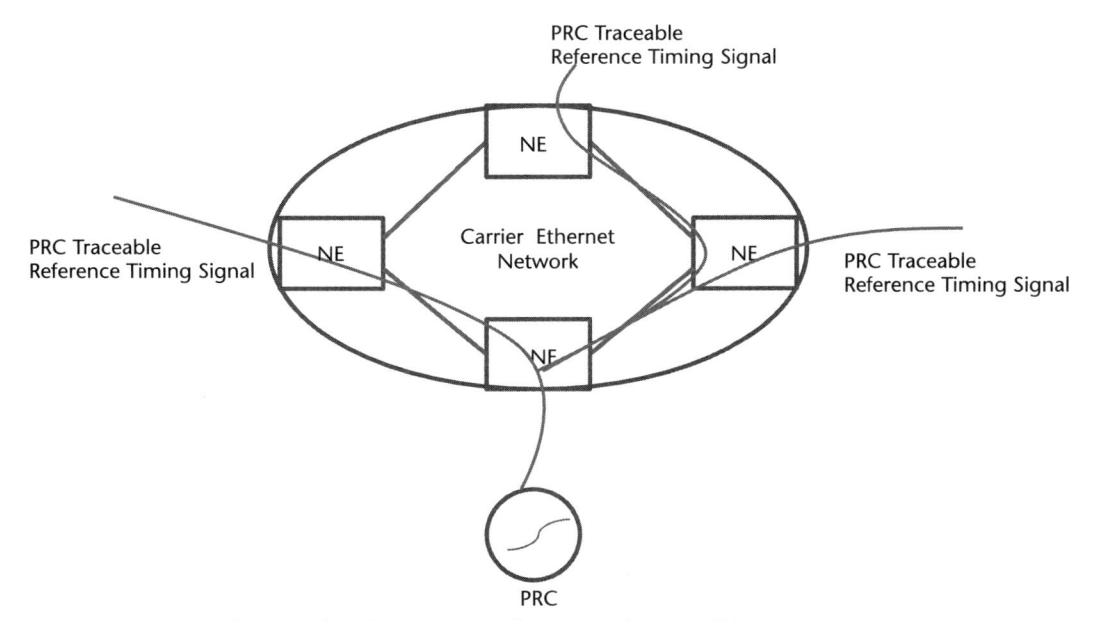

Figure 3.12 Master-slave synchronization network over synchronous Ethernet.

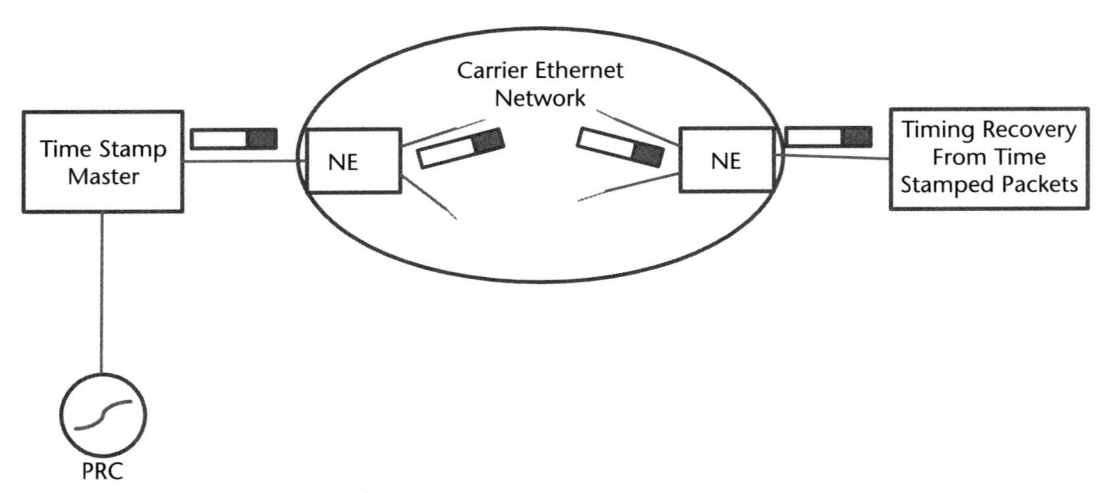

Figure 3.13 Timing distribution via time stamps in packet systems.

Clocks within the network have been categorized based on performance, and a master/slave synchronization scheme is employed. As in GPS, in some cases, the satellite is used for the distribution of frequency.

Four clocking methods are used for synchronization of packet networks [8]:

- Network: synchronous operation;
- Differential methods;
- Adaptive methods;
- Reference clock available at the TDM end systems.

3.6.1 Clocking Methods for Synchronization

Network-synchronous operation method refers to the fully network-synchronous operation by using a PRC-traceable network derived clock or a local PRC such as GPS as the service clock. This method does not preserve the service timing.

According to the differential methods, the difference between the service clock and the reference clock is encoded and transmitted across the packet network (Figure 3.14). The service clock is recovered on the far end of the packet network making use of a common reference clock (i.e., service timing is preserved). The synchronous residual time stamp (SRTS) method [9] is an example of this family of methods.

Differential methods may work with IWF reference clocks that are not PRC-traceable.

In the adaptive methods, the timing can be recovered based on the interarrival time of the packets or on the fill level of the jitter buffer. It should be highlighted that the method preserves the service timing (Figure 3.15).

When the reference clock is available at the TDM end systems, there is no need to recover the timing since both the end systems have direct access to the timing reference and will retime the signal leaving the IWF.

3.6 Synchronous Ethernet Networks

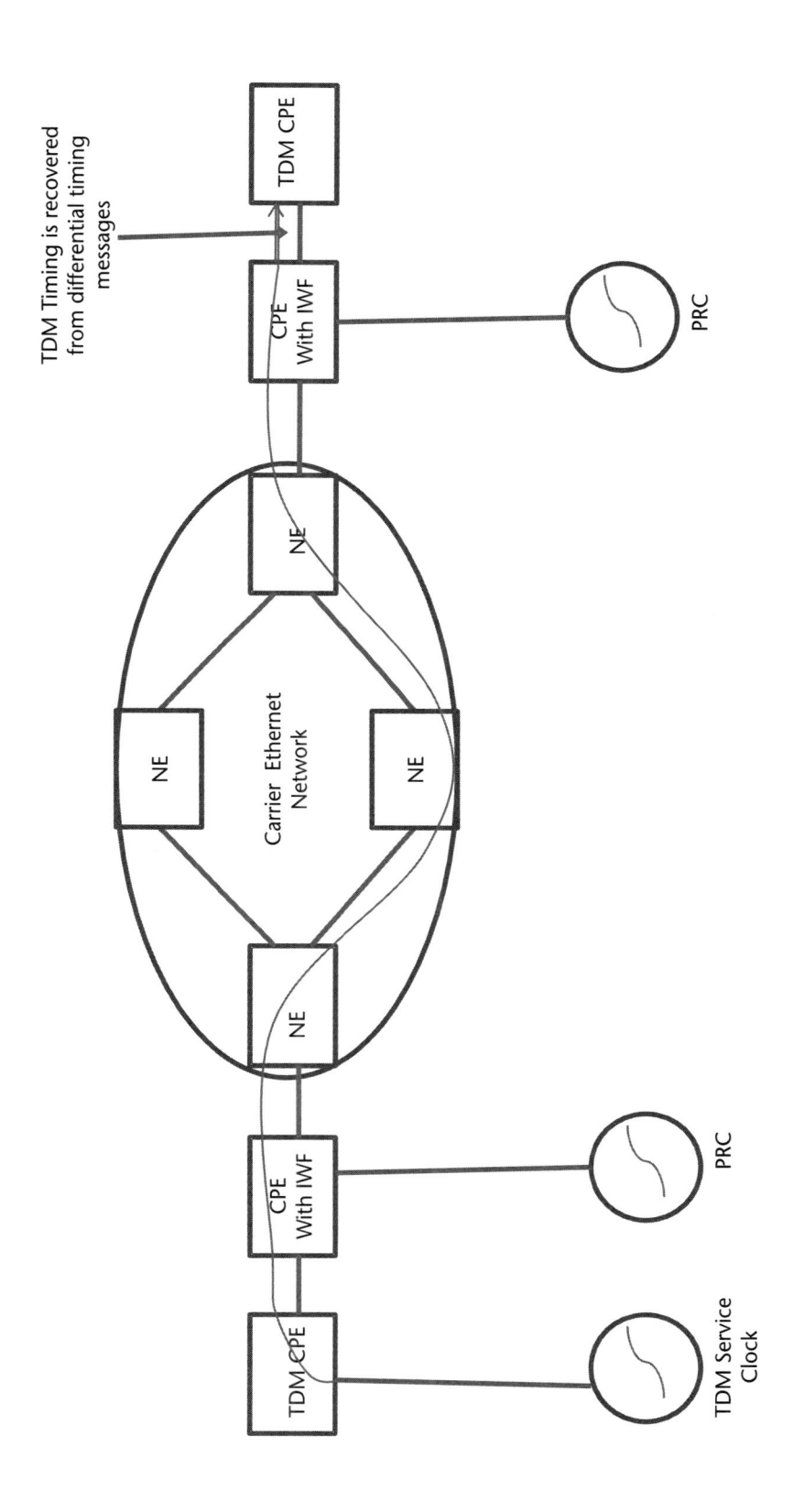

Figure 3.14 Example of timing recovery operation based on differential methods.

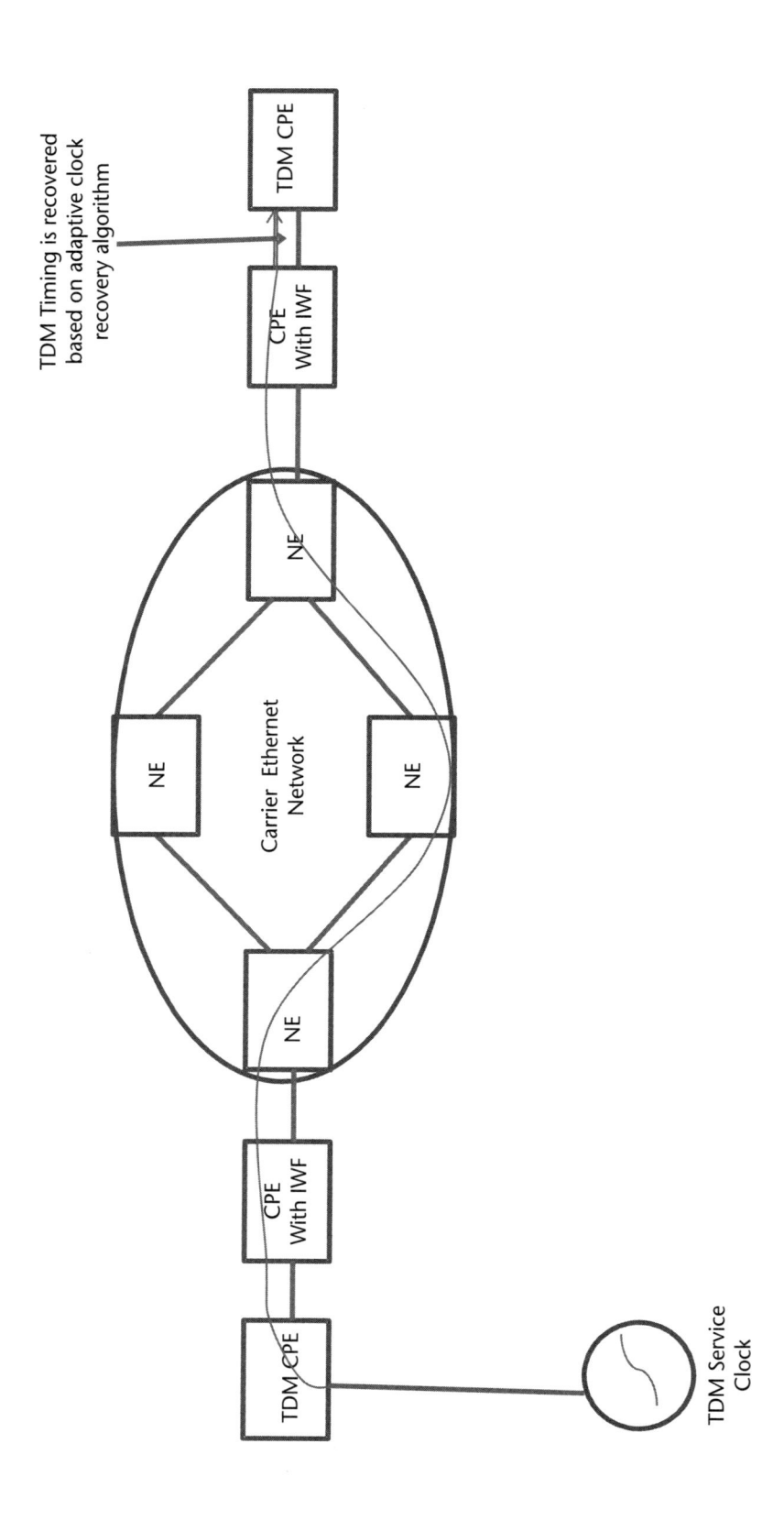

Figure 3.15 Adaptive method.

3.6.2 Impact of Packet Network Impairments on Synchronization

Fundamentally, network synchronization is required in Layer-1 networks to manage buffers. PDH, SDH, and OTN networks have Layer-1 buffers to accommodate changes in rates. These buffers are managed by mechanisms such as stuff bytes and pointers, together with system clocks to accommodate different clocking domains.

In packet networks, the data is delivered over the network in packets or frames, rather than being carried as a continuous stream of constant bit rate. Packets may be statistically multiplexed. Within a single switch, multiple packet streams may have to converge onto a single output buffer, resulting a buffer contention. Packets can be delayed or dropped.

As packets may traverse different routes, a stream of packets from ingress to egress may exhibit significant packet delay variation. Additionally, packets may be misordered resulting in additional buffering. Therefore, large buffers are required to perform packet-level processing.

The following performance parameters affect timing distribution and clock recovery:

- Packet transfer delay and delay variation, in differential methods, should not affect clock recovery performance when a network reference clock is available at both ends and differential methods are used.

In adaptive methods, adaptive recovery of the service clock from a packet stream containing constant bit rate data is usually achieved by some computing function of the arrival rate or arrival times of the packets at the destination node.

If the delay through the packet network is constant, the frequency of arrival of packets at the destination node is not affected by the network. There may be a phase lag in the recovered clock due to the delay through the network, but there should be no frequency or phase wander. If the delay varies, it may be perceived by a clock recovery process as a change in phase or frequency of the original service clock.

- Packet loss ratio is when TDM circuits carried over packet networks may be extremely vulnerable to bit errors caused by packet loss. A single bit error in the packet leads to the whole packet being discarded, yielding a burst of consecutive bit errors in the recovered TDM stream. Hence, even moderate levels of packet loss (from the viewpoint of conventional packet network) may cause unavailability of a TDM circuit.
- Packet severe loss block outcomes [35, 36] define a severe loss block outcome as occurring when, for a block of packets observed at an ingress interface during time interval T, the ratio of lost packets to total packets exceeds a threshold. Similar effects are expected in Ethernet networks. During these impairments, the timing recovery mechanism has to handle the total loss of packets.
- Packet error ratio was when, in order to meet the network requirements, limiting the jitter and wander production of the synchronous Ethernet solution in a wide area network environment will be necessary.

The synchronization function within the synchronous Ethernet switch should be based upon the performance characteristics of an embedded clock. Such a clock will ensure proper network operation occurs when the clock is synchronized from another similar synchronous Ethernet clock or a higher-quality clock. For consistency with existing synchronization networks, the embedded clock may be based upon [34].

3.6.3 Stabilization Period

The stabilization period is an important parameter during the start-up phase and when switching between timing references. When equipment operates in a holdover mode for hours, the phase error when selecting a new clock reference is largely dominated by the phase error caused by the frequency error of the clock in holdover. When the adaptive method is used, the requirement on the stabilization period may depend on the actual phase noise in the packet network. A large packet delay variation may require a long period before the clock can lock to the timing reference.

The filter implementation and the characteristics of the internal oscillator are important as well. In fact, depending on the holdover characteristics, longer time could be accepted when switching from a reference to a second reference, as a good holdover can allow longer locking periods.

The stabilization period size is still under study. For testing purposes, a period of 900 seconds is proposed for adaptive methods.

3.6.4 IWF Synchronization Function

The IWF provides the necessary adaptations between TDM and packet streams. The possible supported timing options for the Tx clock are:

- Timing from recovered source clock carried by the TDM input (loop-timing or line-timing);
- Timing from the network clock (the network clock can be derived either from the physical layer of the traffic links from the packet network or through an external physical timing interface, e.g., 2,048 kHz);
- Timing from a free-running clock (it shall provide an accuracy according to relevant TDM/CBR service interface, e.g., 2,048 kbps shall comply with ITU-T Rec. G.703, ±50 ppm);
- Differential methods;
- Adaptive timing (including clock recovery using dedicated time stamps).

Depending on the services to be provided, a suitable subset of the listed timing options shall be supported. Slip control in the TDM Tx direction is necessary to control possible over/underflow in the playout buffer. Slips shall be performed on n × 125 μs frames. When a TDM transmitter and/or receiver clocks are in holdover or are traceable to clocks in holdover, and a synchronous clock recovery technique (differential method or network-synchronous operation) is used, slips (most likely uncontrolled) will occur.

When selecting a new timing source, the output wander may temporarily exceed the output wander limit. However, the output wander must be within the output wander limit by the end of a period called the stabilization period.

Another characteristic that is relevant for the IWF is the latency. The latency requirements are normally defined on the network level specifying the total latency in the end-to-end connection.

Network-synchronous operation maybe used when a signal traceable to PRC is available at the IWFs and it is not required to preserve the service clock. Differential methods maybe used when a PRC traceable reference is available at the IWF. With this method, it is possible to preserve the service clock. Adaptive methods may be used when the delay variation in the network can be controlled. With this method, it is possible to preserve the service clock.

3.6.5 PRC

A typical synchronous Ethernet architecture will have a PRC located in one of three locations dependent on the overall architecture:

- Core-located: The PRC will be located at the core node, centrally located with some form of distribution to the IWF.
- Access-located: The PRC will be located at some point further back within the network, typically at the multiservice access point. This architecture might use more PRC nodes compared to the core-located architecture.
- IWF-located: The PRC will be located geographically with the IWF and there will be a direct synchronization connection to the IWF, therefore, there will be one PRC per IWF.

Synchronization status messaging (SSM) provides a mechanism for downstream Ethernet switches to determine the traceability of the synchronization distribution scheme back to the PRC or highest-quality clock that is available. The synchronization function processes the SSM. Synchronization status messages for Ethernet implement the SSM channel using an IEEE 802.3 Organizational-Specific Slow Protocol (OSSP).

For upstream network failure conditions, the synchronization function takes appropriate action based on the SSM, presets priorities, and selects an alternate synchronization feed from the network or an external interface.

SSM messages represent the quality level of the system clocks located in the various network elements. Quality level refers to the holdover performance of a clock. A "heartbeat" message is used to provide a continuous indication of the clock quality level. A message period of one second meets the message rate requirements of IEEE 802.3 slow protocols. To minimize the effects of wander that may occur during holdover, an event type message with a new SSM quality level is generated.

To protect against possible failure, the lack of the messages is considered to be a failure condition.

An ESMC information PDU, containing the current QL used by the system clock selection algorithm, is generated once per second.

For synchronous Ethernet, the slow protocol used for the transmission of the SSM code relies on the use of a heartbeat timer. ESMC information PDUs are sent periodically at a rate of one PDU per second. Lack of reception of an ESMC information PDU within a 5-second period results in the QL being set to do-not-use (DNU).

Synchronous Ethernet equipment will require a reference source selection mechanism to provide traceability to upstream elements and ultimately the PRC with respect to frequency. The selection mechanism controls the physical timing flows within the equipment. The selection mechanism must be able to select an appropriate external reference source, traffic reference source, and internal clock (i.e., local oscillator).

3.6.6 Operation Modes

The operation mode can be asynchronous or synchronous. Ethernet equipment that is not synchronous Ethernet-capable work in an asynchronous mode, where each input interface gets its timing from its input signal, which is within a frequency range of ±100 ppm (±20 ppm for 10G WAN). The output interfaces might each have a free-running oscillator generating timing within a frequency rate of ±100 ppm (±20 ppm for 10G WAN).

Synchronous Ethernet equipments are equipped with a system clock. Synchronous Ethernet interfaces are able to extract the received clock and pass it to a system clock. This equipment clock may work in several modes, quality level (QL), or priority modes [37–38]. Each interface of a synchronous Ethernet equipment might be configured to work in either nonsynchronous or synchronous operation mode.

A synchronous Ethernet interface configured in asynchronous mode is an interface that, for the receive side, does not pass the recovered clock to the system clock and is therefore not a candidate reference to the synchronization selection process. It does not process the Ethernet synchronization messaging channel (ESMC) that may be present and therefore cannot extract the QL value.

On the transmit side, its output frequency might be synchronized to the embedded Ethernet equipment clock (EEC), but this remains unknown to the receive interface at the other termination of the link. Since an asynchronous interface does not generate an ESMC and therefore does not transmit a QL, this interface does not participate to the synchronization network. It is functionally identical to an asynchronous interface.

A synchronous Ethernet interface can be configured in synchronous operation mode.

Its receive side is able to extract the frequency of its input signal and passes it to a system clock that can be an EEC or better quality clock. It processes the ESMC and extracts the QL values that are required for a frequency reference.

The transmit part of the interface is locked to the output timing of the system clock and generates the ESMC to transport a QL. The 1G copper Ethernet interfaces perform link auto-negotiation to determine the master and slave clocks for the link.

3.6.7 Frequency Accuracy of Slave Clock

For Option 1, under free-running conditions, the EEC output frequency accuracy should not be greater than 4.6 ppm with regard to a reference traceable to a [ITU-T G.811] clock [1]. One month and 1 year for the time interval for this accuracy have been proposed.

For Option 2, under prolonged holdover conditions, the output frequency accuracy of the different types of node clocks should not exceed 4.6 ppm with regard to a reference traceable to a primary reference clock, over a time period of 1 year (Table 3.5).

3.6.8 EEC

Reference [8] outlined minimum requirements for timing devices used in synchronous Ethernet for synchronizing the equipment. It supports clock distribution based on network-synchronous line-code methods as in synchronous Ethernet. It is called embedded Ethernet equipment clock (EEC).

EEC has Option 1 and Option 2. The first option, referred to as EEC Option 1, applies to synchronous Ethernet equipment that is designed to interwork with networks optimized for the 2,048-Kbps hierarchy. These networks allow the worst-case synchronization reference as specified in [39–40]. The second option, referred to as EEC Option 2, applies to synchronous Ethernet equipment that is designed to interwork with networks optimized for the 1,544-Kbps hierarchy [37].

A synchronous Ethernet equipment slave clock needs to support all of the requirements specific to one option and not mix requirements between EEC Options 1 and 2.

The noise generation of an EEC represents the amount of phase noise produced at the output when there is an ideal input reference signal or the clock is in holdover state. The performance level for a suitable reference is at least 10 times more stable than the output level.

The clock ability to limit this noise is described by its frequency stability. The maximum time interval error (MTIE) and time deviation (TDEV) characterize noise generation performance. MTIE and TDEV are measured through an equivalent 10-Hz, first-order, lowpass measurement filter, at a maximum sampling time of 1/30 seconds. The minimum measurement period for TDEV is twelve times the integration period.

Wander generation bounds for locked mode for both options are given in [17].

Table 3.5 Option 1 and Option 2 Functionalities

Functionality	Option 1	Option 2
Minimum pull-in range	±4.6 ppm (independent from internal oscillator frequency offset)	±4.6 ppm (independent from internal oscillator frequency offset)
Hold-in range	Not required	±4.6 ppm (independent from internal oscillator frequency offset)
Pull-out range	±4.6 ppm	NA

When a clock is not locked to a synchronization reference, the random noise components are negligible compared to deterministic effects like initial frequency offset.

However, jitter generation have different limits for different interface rates. The peak-to-peak amplitude measured over a 60-second interval should not exceed 0.5 for 1-Gbps interface with measuring filter bandwidth of 2.5 kHz to 10 MHz and 10-Gbps interface with measuring filter bandwidth of 20 kHz to 80 MHz [17].

Bounds for noise tolerance, noise transfer, transit response, and holdover performance of slave clocks should be also satisfied [17]. In order to determine the EEC noise tolerance, the worst-case network limit is used. The tolerance of an EEC indicates the minimum phase noise level at the input of the clock to be accommodated while maintaining the clock within performance limits, without causing any alarms or switch reference or going into a holdover mode.

The noise transfer characteristics determine EEC properties related to the transfer of excursions of the input phase relative to the carrier phase.

Transient response and holdover performance defines the ability to withstand disturbances and avoid transmission defects or failures. The short-term phase transient response and long-term phase transient response (holdover) requirements are given in [17].

3.7 Conclusion

Depending on the application, timing, phase, and/or frequency synchronization is necessary among systems and networks. For example, TDM networks require highly accurate frequency information to be recoverable from their physical layers. In every TDM network there is at an extremely accurate PRC from which all other TDM clocks in the network derive their timing.

Clocking distribution is performed either via synchronous Ethernet or packet-based methods such as IEEE 1588 (PTP).

The 1588 v2 is picking up momentum in the industry to provide time, phase, and frequency synchronization due to the fact that it does not require changing existing hardware.

References

[1] ITU G.811, Timing Characteristics of Primary Reference Clocks, September 1997.
[2] MEF 18, Abstract Test Suite for Circuit Emulation Services over Ethernet, 2007.
[3] MEF 3, Circuit Emulation Service Definitions, Framework and Requirements in Metro Ethernet Networks, 2004.
[4] MEF 8, Implementation Agreement for the Emulation of PDH Circuits over Metro Ethernet Networks. 2004.
[5] ITU-T G.823, The control of Jitter and Wander Within Digital Networks Which Are Based on the 2048 kbit/s Hierarchy, 2000.
[6] ITU-T G.824, The Control of Jitter and Wander within Digital Networks Which Are Based on the 1544 kbit/s Hierarchy, 2000.

[7] ITU-T G.822, Controlled Slip Rate Objectives on an International Digital Connection, 1993.

[8] ITU G.8261, Timing and Synchronization Aspects in Packet Networks, May 2006.

[9] ITU-T I.363.1, B-ISDN ATM Adaptation Layer Specification: Type 1 AAL, 1996.

[10] ANSI/T1.101-1987, Synchronization Interface Standard for Digital Networks, 1987.

[11] Lewis, J., "Hybrid Mode Synchronous Ethernet and IEEE 1588 in Wireless TDD Applications," Voice & Timing Soultions for a New Global Network, 2010.

[12] ITU-T Y.1731, OAM Functions and Mechanisms for Ethernet Based Networks, 2008.

[13] RFC 5905, Network Time Protocol Version 4: Protocol and Algorithms Specification, June 2010.

[14] RFC 3985, Pseudowire Emulation Edge-to-Edge (PWE3) Architecture, 2005.

[15] RFC 4197, Requirements for Edge-to-Edge Emulation of Time Division Multiplexed (TDM) Circuits over Packet Switching Networks, 2005.

[16] RFC 4553, Structure-Agnostic Time Division Multiplexing (TDM) over Packet (SAToP), 2006.

[17] ITU G.8262, ITU-T G.8262, Timing Characteristics of Synchronous Embedded Ethernet Equipment Clock (EEC), July 2010.

[18] ITU-T G.8260, Definitions and Terminology for Synchronization in Packet Networks, 2010.

[19] ITU-T G. 8271, Basic of Time and Phase Synchronization, 2011.

[20] ITU-T G.8261.1, Network Requirements for Frequency Synchronization, 2011.

[21] ITU-T G.8264, Distribution of Timing Information Through Packet Networks, 2008.

[22] ITU-T G.8265, Architecture and Requirements for Packet Based Frequency Delivery, 2010.

[23] ITU-T G.8265.1, Precision Time Protocol Telecom Profile for Frequency Synchronization, 2010.

[24] IEEE 1588v2 Standard for Local and Metropolitan Area Networks - Timing and Synchronization for Time-Sensitive Applications in Bridged Local Area Networks - 802.1as (Draft 7.6), November 2010.

[25] IEEE 1588-2008, Standard for a Precision Clock Synchronization Protocol for Networked Measurement and Control Systems, 2008.

[26] RFC 5086, Structure-Aware Time Division Multiplexed (TDM) Circuit Emulation Service over Packet Switched Network (CESoPSN), 2007.

[27] RFC 4553, Structure-Agnostic Time Division Multiplexing (TDM) over Packet (SAToP), June 2006.

[28] RFC 5905, Network Time Protocol Version 4: Protocol and Algorithms Specification, June 2010.

[29] RFC 1305, Network Time Protocol (Version 3) Specification, Implementation and Analysis, March 1992.

[30] RFC 2030, Simple Network Time Protocol (SNTP) Version 4 for IPv4, IPv6 and OSI, October 1996

[31] RFC 1889, RTP: A Transport Protocol for Real-Time Applications, January 1996.

[32] RFC 3550, RTP: A Transport Protocol for Real-Time Applications, July 2003.

[33] IEEE 1588-2002, "Standard for a Precision Clock Synchronization Protocol for Networked Measurement and Control Systems," 2002.

[34] National Institute of Standards and Technology (NIST), "IEEE 1588 Systems," http://www.nist.gov/intelligent-systesm-division/ieee-1588-systems.

[35] ITU-T Y.1540, Internet Protocol Data Communication Service – IP Packet Transfer and Availability Performance Parameters, 2003.

[36] ITU-T Y.1561, Performance and Availability Parameters for MPLS Networks, 2005.

[37] ITU-T G.813, Timing Characteristics of SDH Equipment Slave Clocks (SEC), 2003.

[38] ITU-T G.781, Synchronization Layer Functions, 1999.

[39] ITU-T G.803, Architecture of Transport Networks Based on SDH, 2003.

[40] ITU-T G.812, Timing Requirements of Slave Clocks Suitable for Use as Node Clocks in Synchronization Networks, 2006.

CHAPTER 4
Pseudowires

4.1 Introduction

Pseudowire emulation edge-to-edge (PW3) is a mechanism to emulate telecommunications services across packet-switched networks such as Ethernet, IP, or MPLS. The emulated services are mostly T1/T3 leased lines, frame relay, Ethernet and ATM to maximize return on existing assets and minimize operating costs of the service providers (SPs).

Pseudowires (PWs) encapsulate cells, bit streams, and PDUs and transport them across EVCs and IP or MPLS tunnels. The transportation of encapsulated data usually require managing the sequence and timing of data units to emulate services such as T1/T3 leased lines and ATM.

For a customer equipment, a PW appears to be a dedicated circuit for the emulated service. This provides simplicity to implement various services over PSNs.

4.2 Protocol Layers

The logical protocol layer structure is needed to support a PW is shown in Figure 4.1.

The payload is transported over the encapsulation layer. The encapsulation layer carries any information, not already present within the payload itself including sequencing, and the interface to the PW demultiplexer layer.

The PW demultiplexer layer provides the ability to deliver multiple PWs over a single PSN tunnel.

The PSN convergence layer provides the enhancements needed to make the PSN conform to a consistent interface to the PW, making the PW independent of the PSN type.

A detailed protocol stack reference model for PWs is given in Figure 4.2 [1].

The PW provides the CE with an emulated physical or virtual connection to its peer at the far end. Native service PDUs from the CE are passed through an encapsulation layer at the sending PE and then sent over the PSN. The receiving PE removes the encapsulation and restores the payload to its native format for transmission to the destination CE.

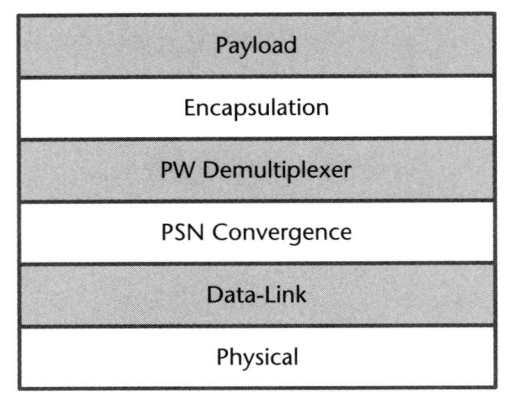

Figure 4.1 Logical protocol layering model.

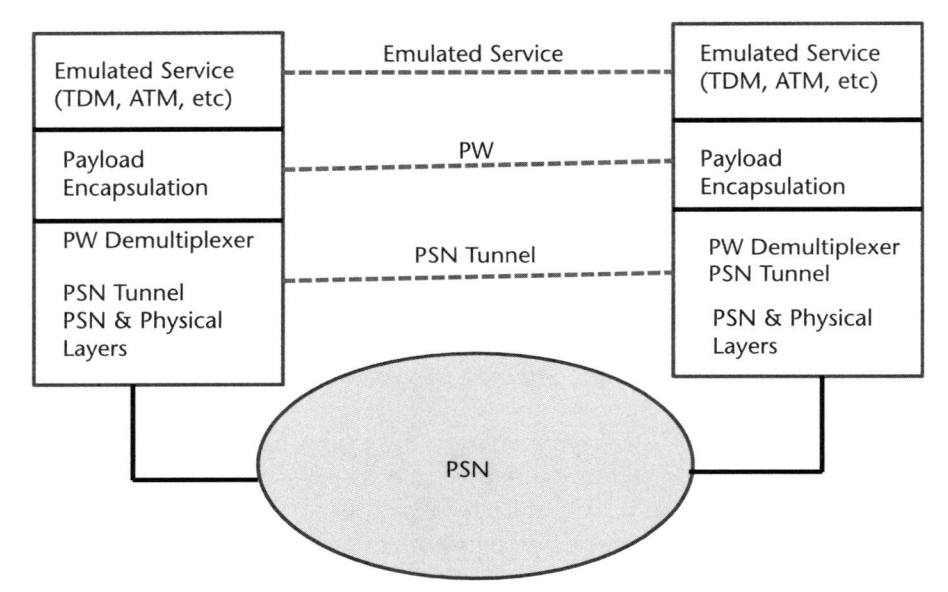

Figure 4.2 PWE3 protocol stack reference model.

4.3 Payload Types

The native data units (payloads) can be packet, cell, bit stream, and structured bit stream. PW services for these payloads are given in Table 4.1.

A packet payload is a variable-size data unit delivered to the PE via an attachment circuit (AC) and relayed across the PW as a single unit or in multiple segments where the combined size of packet payload and its PWE3 and PSN headers is larger than the PSN path MTU. A packet payload may need sequencing and real-time support.

In some applications such as frame relay, the packet payload is selected based on an identifier, part of the forwarder function. For example, frame relay PDUs will be selected based on the frame delay data-link connection identifier (DLCI).

Table 4.1 Payload Types and Corresponding PW Services

Payload Type	PW Service
Structured Bit Stream	SONET/SDH (e.g., SPE, VT, NxDS0).
Bit Stream	Unstructured E1, T1, E3, T3
Packet	Ethernet, HDLC framing, frame relay, ATM AAL5 PDU
Cell	ATM

A cell payload is created by capturing, transporting, and replaying groups of octets presented on the wire in a fixed-size format. For example, the cell payloads for ATM and MPEG transport stream packets (DVB) are 53 octets and 188 octets, respectively.

To reduce per-PSN packet overhead, multiple cells may be concatenated into a single payload. The cell payload service will need sequence numbering and may also need time synchronization.

A bit stream payload is created by capturing, transporting, and replaying the bit pattern on the emulated wire. For example, E1 and T1 send all-ones to indicate failure that can be detected without any knowledge of the structure of the bit stream. Sequencing and time synchronization are needed.

A structured bit stream payload is created by using some knowledge of the underlying structure of the bit stream to capture, transport, and replay the bit pattern on the emulated wire. Sequencing and time synchronization are needed.

Some parts of the original bit stream may be stripped in the PSN-bound direction. For example, in structured SONET, the section and line overhead may be stripped. The stripped information may appear in the encapsulation layer to facilitate the reconstitution.

4.4 Pseudowire Architectures

A PW is a connection between two PE equipment connecting two ACs. An AC can be a frame relay DLCI, an ATM VCI/VPI, an Ethernet port, a VLAN, an MPLS LSP, and so forth. Figure 4.3 illustrates the network reference model for point-to-point PWs [2].

The two PEs (Provider Edge 1 and Provider Edge 2) have to provide one or more PWs on behalf of their client CEs (CE 1 and CE 2) to enable the client CEs to communicate over the PSN. A PSN tunnel is established to provide a data path for the PW. The PW traffic is invisible to the core network, and the core network is transparent to the CEs. Native data units (bits, cells, or packets) arrive via the AC, are encapsulated in a PW-PDU, and are carried across the underlying network via the PSN tunnel. The PEs perform the necessary encapsulation and decapsulation of PW-PDUs and handle any other functions required by the PW service, such as sequencing or timing.

4.4.1 PWE3 Preprocessing

Applications such as frame relay have to perform operations on the native data units received from the CE (including both payload and signaling traffic) before

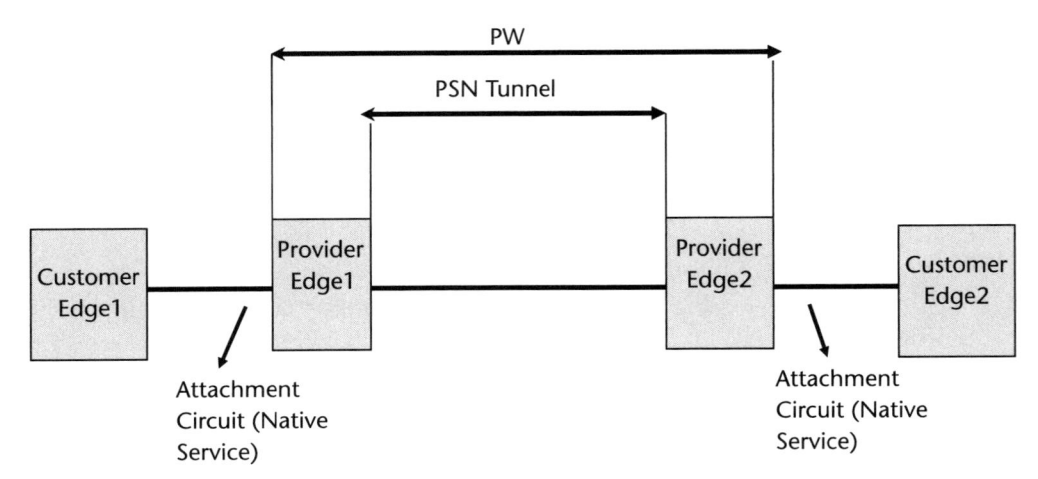

Figure 4.3 PWE3 network reference model.

they are transmitted across the PW by the PE. Examples include Ethernet bridging, SONET cross-connect, translation of locally significant identifiers such as virtual circuit identifier (VCI) virtual path identifier (VPI), or translation to another service type. These operations could be carried out in an external equipment or within the PE where processed data is then presented to the PW via a virtual interface within the PE.

Both PEs or one of the PEs can support preprocessing (PREP) functionality. Figure 4.4 shows the interworking of one PE with preprocessing, and a second without this functionality. The functional interface between PREP and the PW is that represented by a physical interface carrying the service. This reference point effectively defines the necessary interworking specification.

Figure 4.5 illustrates how the protocol stack reference model is extended to include the provision of preprocessing.

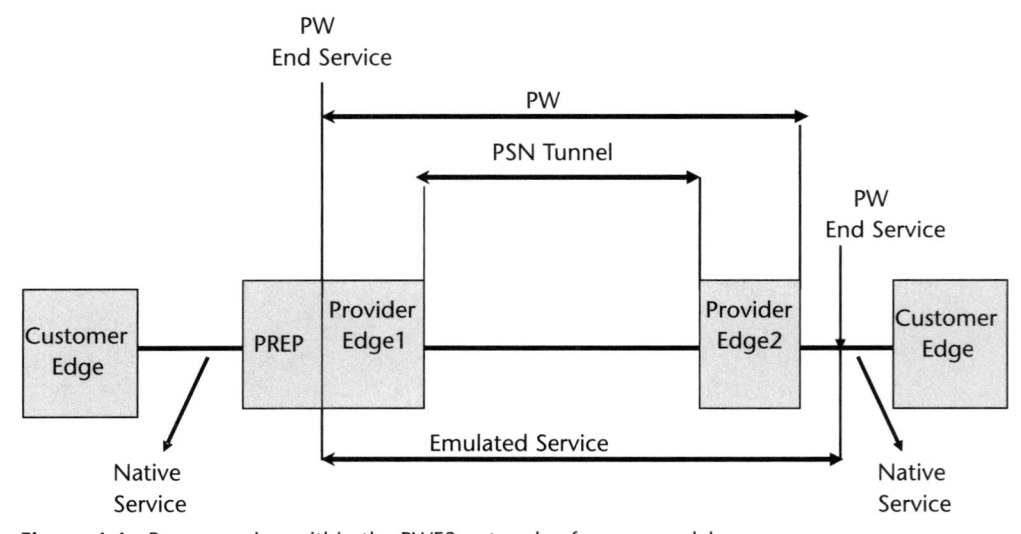

Figure 4.4 Preprocessing within the PWE3 network reference model.

4.4 Pseudowire Architectures

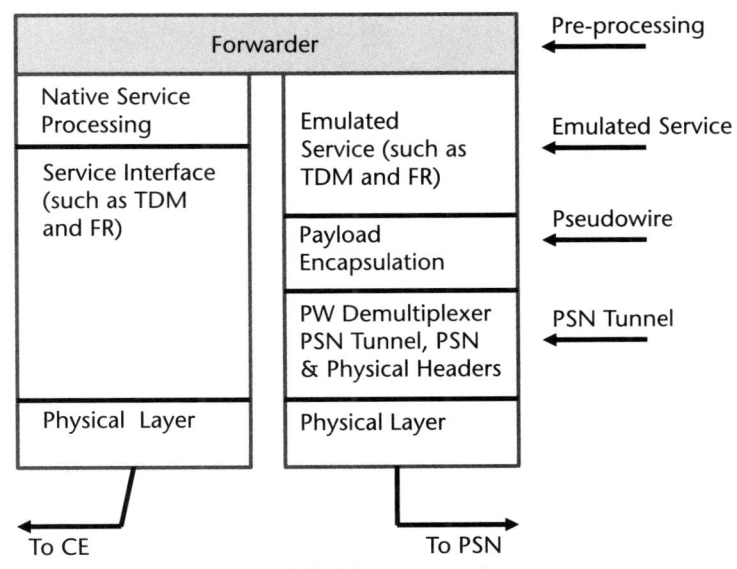

Figure 4.5 Protocol stack reference model with preprocessing.

The required preprocessing can be divided into two components as forwarder (FWRD) and native service processing (NSP). Forwarders can be single input and single output as well as multiple inputs and multiple outputs as depicted in Figures 4.6 and 4.7.

Figure 4.6 shows a simple forwarder (i.e., single input and single output) that performs some type of filtering operation. Figure 4.7 shows a forwarding between multiple ACs and multiple PWs where payloads are extracted from one or more ACs and directed to one or more PWs.

Some applications required some form of data or address translation, or some other operation requiring knowledge of the semantics of the payload. This is the function of the native service processor. The use of the NSP approach simplifies the design of the PW by restricting a PW to homogeneous operation.

Figure 4.8 illustrates the relationship between NSP, forwarder, and PWs in a PE. The NSP function may apply any transformation operation on the payloads as they pass between the physical interface to the CE and the virtual interface to the forwarder. These transformation operations will be limited to those that have been implemented in the data path, and that are enabled by the PE configuration. A PE device may contain more than one forwarder.

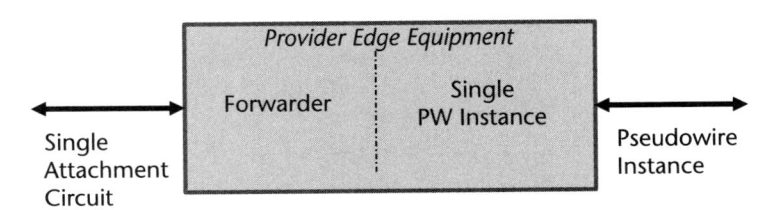

Figure 4.6 Simple point-to-point service

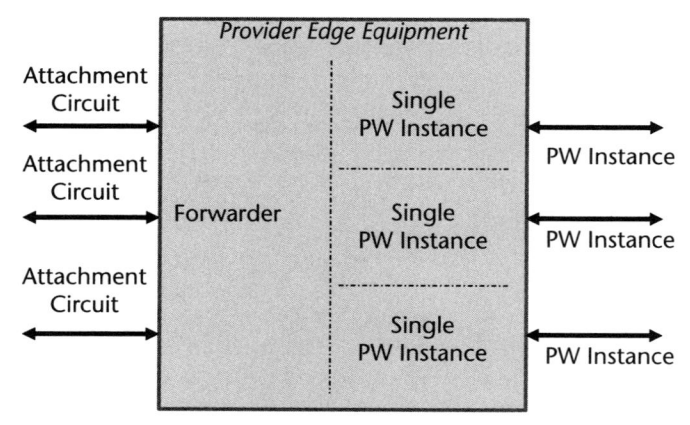

Figure 4.7 Multiple AC to multiple PW forwarding.

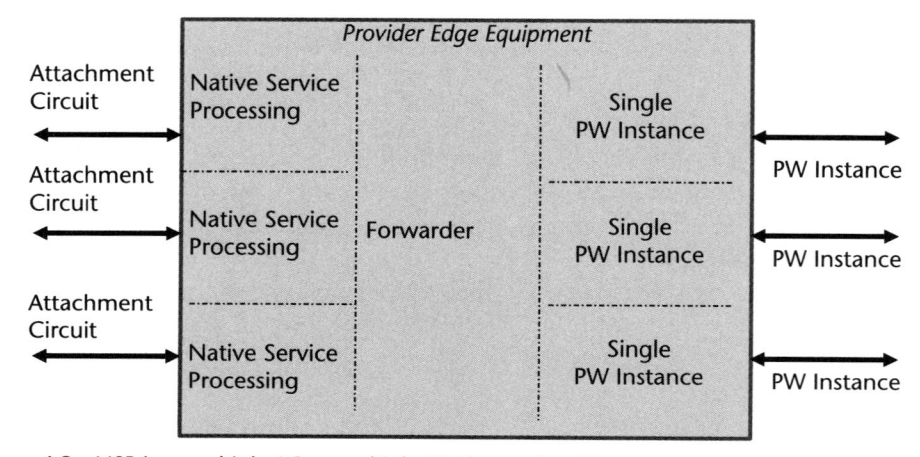

Figure 4.8 NSP in a multiple AC to multiple PW forwarding PE.

4.4.2 Payload Convergence Layer

The primary task of the payload convergence layer is the encapsulation of the native data units containing an L2 or L1 header (payload) in PW-PDUs. The payload convergence header carries the additional information. This information may be used to replay the native data units at the CE-bound physical interface.

From its underlying PW demultiplexer and PSN layers, the PW encapsulation layer and its associated signaling require a reliable control channel for signaling line events, status indications, and CE-CE events that are translated and sent between PEs.

In addition, the encapsulation layer may require the following depending on applications:

1. A high-priority indicated via DSCP or EXP bit or a bit in the tunnel header, unreliable, sequenced channel. A typical use is for CE-to-CE signaling.
2. A sequenced channel for data traffic that is sensitive to packet reordering.
3. An unsequenced channel for data traffic insensitive to packet order.

4.4 Pseudowire Architectures

The PW encapsulation layer provides the necessary infrastructure to adapt the specific payload type being transported over the PW to the PW demultiplexer layer that is used to carry the PW over the PSN. It consists of payload convergence, timing and sequencing sublayers. The sublayering and its context with the protocol stack are shown in Figure 4.9.

The payload convergence sublayer provides the provision of per-packet signaling and other out-of-band information while the timing and sequencing layers provide timing and sequencing services to the payload convergence layer.

The sequencing function provides frame ordering, frame duplication detection, and frame loss detection that are key to the emulation of the invariant properties of a physical wire.

The size of the sequence-number space depends on the speed of the emulated service, and on the maximum time of the transient conditions in the PSN. A sequence number space greater than 2^{16} may be needed to prevent the sequence number space from wrapping during the transient.

When packets carrying the PW-PDUs traverse a PSN, they may arrive out of order at the destination PE. For some services, the frames (control frames, data frames, or both) must be delivered in order. Providing a sequence number in the sequence sublayer header for each packet is one possible approach. Alternatively, it is possible to drop misordered PW PDUs instead of trying to sort PW PDUs into the correct order.

For some native services, the receiving PE has to play out the native traffic as it was received at the sending PE. The timing information either sent between the two PEs or in some cases received from an external reference.

Some applications such as CES require clock recovery and timed payload delivery. Clock recovery is the extraction of output transmission bit timing information from the delivered packet stream. Timed delivery is the delivery of noncontiguous PW PDUs to the PW output interface with a constant phase relative to the input interface. The timing of the delivery may be relative to a clock derived from the packet stream received over the PSN clock recovery or to an external clock.

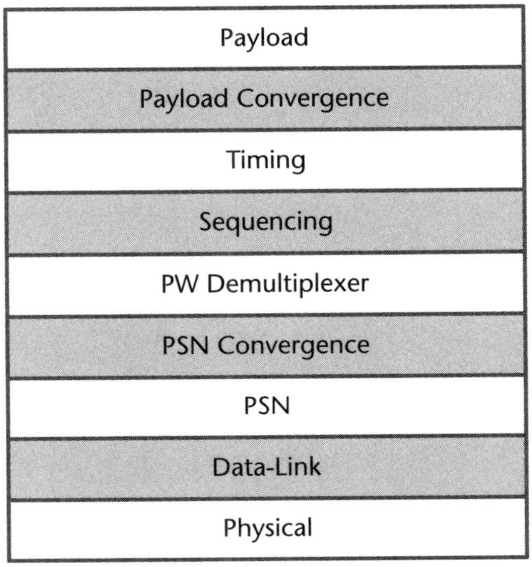

Payload
Payload Convergence
Timing
Sequencing
PW Demultiplexer
PSN Convergence
PSN
Data-Link
Physical

Figure 4.9 PWE3 encapsulation layer in context.

There will be cases where the combined size of the payload and its associated PWE3 and PSN headers may exceed the PSN path MTU. Then fragmentation and reassembly have to be performed for the packet to be delivered.

A PE implementation may not to support fragmentation and will drop packets that exceed the PSN MTU.

4.4.3 PW Demultiplexer Layer and PSN

The purpose of the PW demultiplexer layer is to allow multiple PWs to be carried in a single tunnel. This minimizes complexity and conserves resources. Some types of native service are capable of grouping multiple circuits into a trunk (e.g., multiple ATM VCs in a vitrual path (VP), multiple Ethernet VLANs on physical media, or multiple DS0 services within a T1 or E1). A PW may interconnect two end trunks. That trunk would have a single multiplexing identifier.

The demultiplexer layer provides three main functions: multiplexing, fragmentation, and identifying PDU length and delivering PDU.

If the PSN provides a fragmentation and reassembly service of adequate performance, it may be used to obtain an effective MTU that is large enough to transport the PW PDUs.

PDU delivery to the egress PE is the function of the PSN layer. If the underlying PSN does not provide all the information necessary to determine the length of a PW-PDU, the encapsulation layer must provide it.

It is a common practice to use an error detection mechanism such as a CRC or similar mechanism to ensure end-to-end integrity of frames. The PW service-specific mechanisms must define whether the packet's checksum shall be preserved across the PW or be removed from PE-bound PDUs and then be recalculated for insertion in CE-bound data.

For protocols such as ATM and frame relay (FR), the checksum is restricted to a single link (e.g., FR DLCI or ATM VPI/VCI).

For congestion consideration, if the traffic carried over the PW is known to be TCP friendly, packet discard in the PSN will trigger the necessary reduction in offered load, and no additional congestion avoidance action is necessary.

If the PW is operating over a PSN that provides enhanced delivery, the PEs should monitor packet loss to ensure that the requested service is actually being delivered. If it is not, then the PE should assume that the PSN is providing a best-effort service and should use the best-effort service congestion avoidance measures described next.

4.4.4 Maintenance Reference Model

Signaling between CEs and PE is used to maintain the PW components as depicted in Figure 4.10 [1].

The CE (end-to-end) signaling is between the CEs. This signaling could be frame relay PVC status signaling, ATM SVC signaling, TDM CAS (channel-associated signaling), and so forth.

The PW/PE maintenance is used between the PEs (or NSPs) to set up, maintain, and tear down PWs, including any required coordination of parameters.

4.5 Control Plane 71

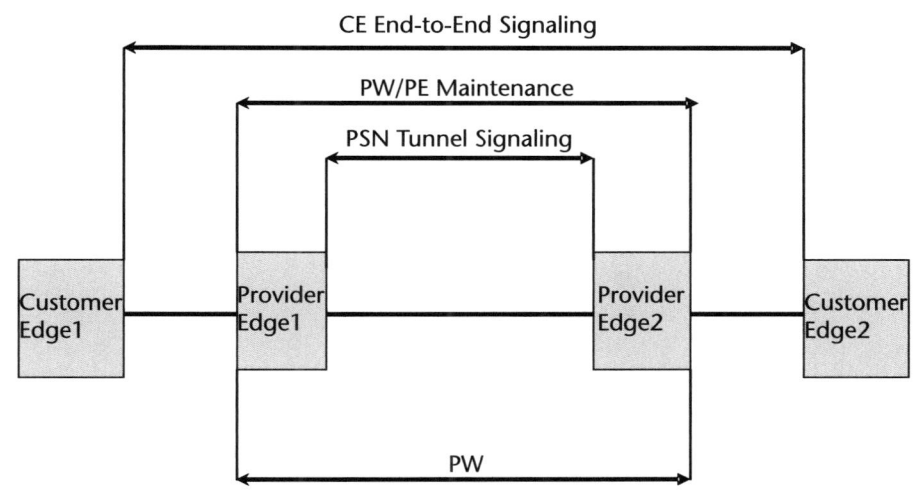

```
                    CE End-to-End Signaling
                   ┌─────────────────────────┐
                        PW/PE Maintenance
                      ┌───────────────────┐
                        PSN Tunnel Signaling
                         ┌─────────────┐
  ┌────────┐         ┌────────┐     ┌────────┐         ┌────────┐
  │Customer│─────────│Provider│     │Provider│─────────│Customer│
  │Edge1   │         │Edge1   │     │Edge2   │         │Edge2   │
  └────────┘         └────────┘     └────────┘         └────────┘
                         └──────────PW──────────┘
```

Figure 4.10 PWE3 maintenance reference model.

The PSN tunnel signaling controls the PW multiplexing and some elements of the underlying PSN. Examples are L2TP control protocol, MPLS LDP, and RSVP-TE.

4.5 Control Plane

PWE3 control plane services include setting up and tearing down a PW, monitoring it, and dealing with various fault conditions. The control plane messages such as Ethernet flow control and TDM tone signaling maybe exchanged in-band while the messages such as the signaling VC of an ATM VP and TDM CCS signaling maybe exchanged out-of-band. The control plane messages should be transported by using either a higher priority or a reliable channel provided by the PW demultiplexer layer.

The control plane services are:

- Set up or tear down of PWs: A PW must be set up before an emulated service can be established and must be torn down when an emulated service is no longer needed. Setup or teardown of a PW can be triggered by an operator command, from the management plane of a PE, by signaling setup or teardown of an AC (e.g., an ATM SVC), or by an auto-discovery mechanism.
- Status monitoring: Some native services such as ATM have mechanisms for status monitoring. For these services, the corresponding emulated services must specify how to perform status monitoring. For status changes including PW up and down, a notification should be sent to the management system. When the physical link (or subnetwork) between a CE and a PE fails, all the emulated services that go through that link (or subnetwork) will fail. Then it is desirable that a single notification message be used to notify failure of the whole group of emulated services.

- Misconnection and payload type mismatch: Misconnection can breach the integrity of the system and the payload mismatch can disrupt the customer network. The tunneling mechanism and its associated control protocol can be used to deal with mismatch issues. For example, a PW-TYPE identifier is exchanged during the PW setup that is used to verify the compatibility of the ACs.
- Packet loss, corruption, and out-of-order delivery: Packet loss, corruption, and out-of-order delivery on the PSN path between the PEs may occur. For some payload types, these errors can be mapped either to a bit error burst or to loss of carrier on the PW. If a native service has some mechanism to deal with bit error, the corresponding PWE3 service should provide a similar mechanism.
- Keep-alive: If a native service has a keep-alive mechanism, the corresponding emulated service must provide a mechanism to propagate it across the PW.

In the following sections, two examples are given.

4.5.1 PWE3 over an IP PSN

PWE3 over an IP PSN protocol structure is depicted in Figure 4.11.

Timing and sequencing are provided by real-time transport protocol (RTP) [3]. The encapsulation layer may also carry a sequence number. In that case, sequencing should be provided either by the PW encapsulation or RTP.

PW demultiplexing is provided by the PW label that can be an MPLS label, an L2TP session ID, or a UDP port number. When PWs are carried over IP, the PSN convergence layer will not be needed.

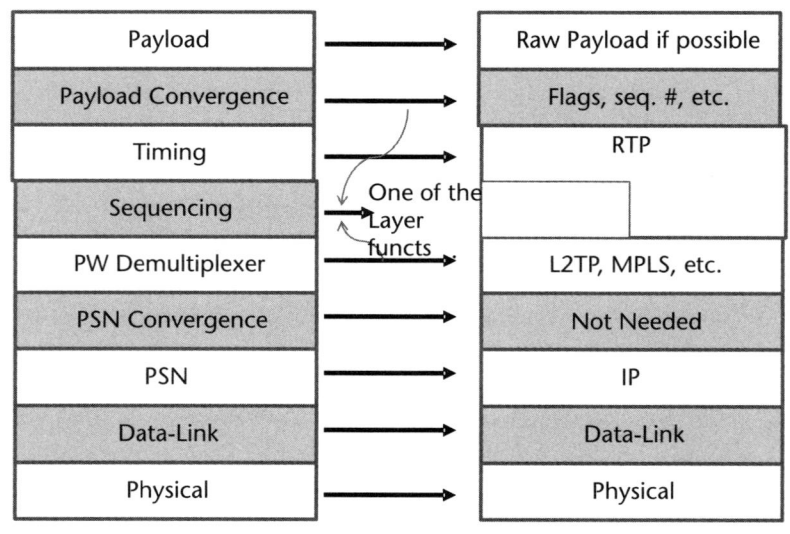

Payload	→	Raw Payload if possible
Payload Convergence	→	Flags, seq. #, etc.
Timing	→	RTP
Sequencing	One of the Layer functs	
PW Demultiplexer	→	L2TP, MPLS, etc.
PSN Convergence	→	Not Needed
PSN	→	IP
Data-Link	→	Data-Link
Physical	→	Physical

Figure 4.11 PWE3 over an IP PSN.

4.5.2 PWE3 over an MPLS PSN

The protocol layering for PWE3 over an MPLS PSN is given in Figure 4.12. An inner MPLS label is used to provide the PW demultiplexing function. A control word is used to carry most of the information needed by the PWE3 encapsulation layer and the PSN convergence layer in a compact format. The flags in the control word provide the necessary payload convergence. A sequence field provides support for both in-order payload delivery and a PSN fragmentation service within the PSN convergence layer (supported by a fragmentation control method).

The Ethernet pads all frames to a minimum size of 64 bytes. The MPLS header does not include a length indicator. Therefore, to allow PWE3 to be carried in MPLS to pass correctly over an Ethernet datalink, a length correction field is needed in the control word. As with an IP PSN, where appropriate, timing is provided by RTP [3].

4.6 Multisegment Architecture

PWs may span multiple domains of one or more networks. Figure 4.13 shows a multisegment case where terminating PE1 (T-PE1) and terminating PE2 (T-PE2) provide PWE3 service to CE1 and CE2. One PSN tunnel extends from T-PE1 to S-PE1across PSN1, and a second PSN tunnel extends from S-PE to T-PE2 across PSN2 [4].

PWs are used to connect the ACs attached to T-PE1 to the corresponding ACs attached to T-PE2. Each PW on PSN tunnel 1 is switched to a PW in the tunnel across PSN2 Sat S-PE1 to complete the multisegment PW (MS-PW) between T-PE1 and T-PE2. S-PE1 is the PW switching point called as the PW switching provider edge (S-PE). PW1 and PW2 are segments of the same MS-PW while PW2 and PW4

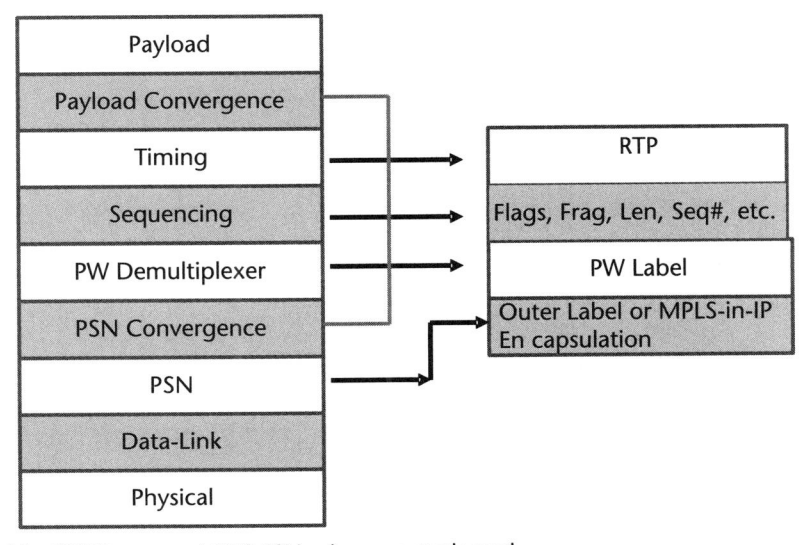

Figure 4.12 PWE3 over an MPLS PSN using a control word.

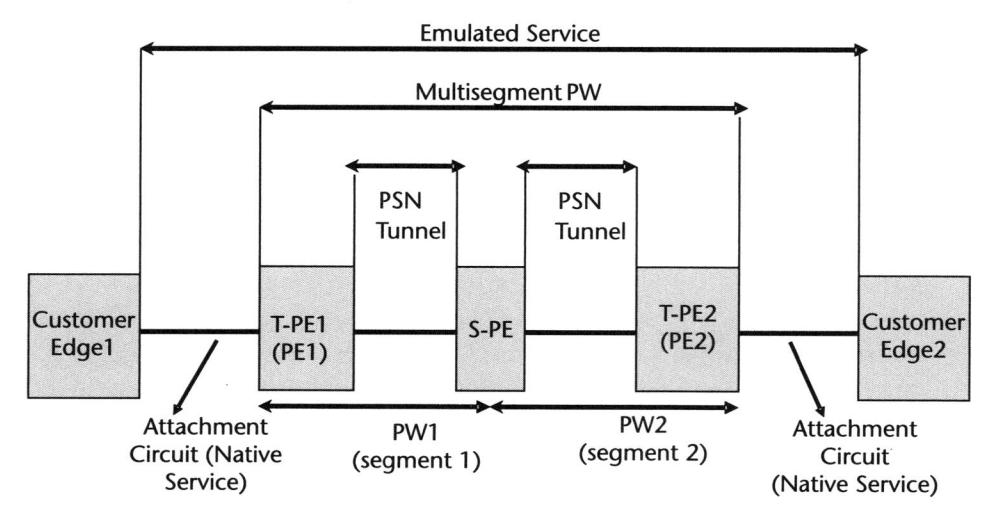

Figure 4.13 PW switching reference model [4].

are segments of another PW. PW segments of the same MS-PW (e.g., PW1 and PW2) may be of the same PW type or different types, and PSN tunnels (e.g., PSN Tunnel 1 and PSN Tunnel 2) can be the same or different technology.

There are two methods for switching a PW between two PW domains. In the first method (Figure 4.14), the two separate control plane domains terminate on different PEs [5].

In Figure 4.15, PWs in two separate PSNs are stitched together using native service attachment circuits. PE2 and PE3 only run the control plane for the PSN to which they are directly attached.

In Figure 4.15, SPE1 runs two separate control planes: one toward TPE1, and one toward TPE2. The PW switching point (S-PE) is configured to connect PW Segment 1 and PW Segment 2 together to complete the multisegment PW between

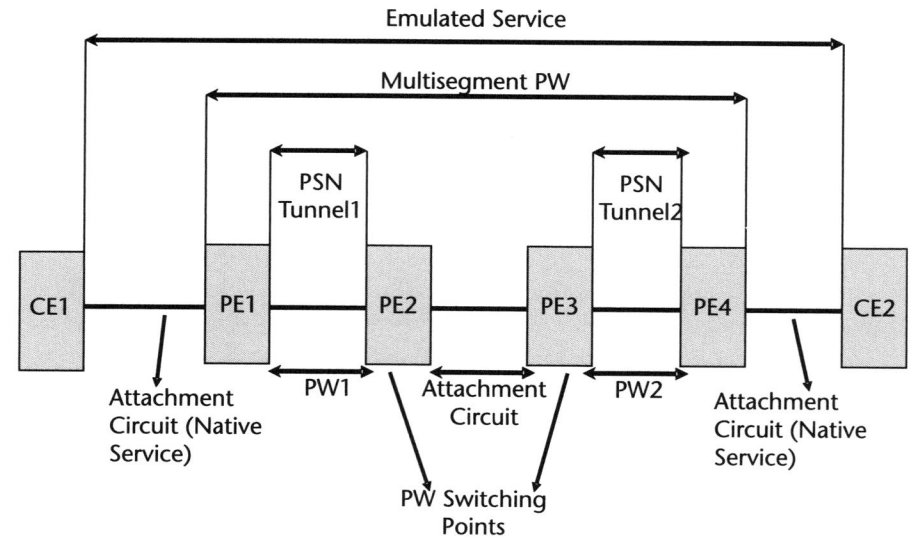

Figure 4.14 PW switching using AC reference model.

4.6 Multisegment Architecture

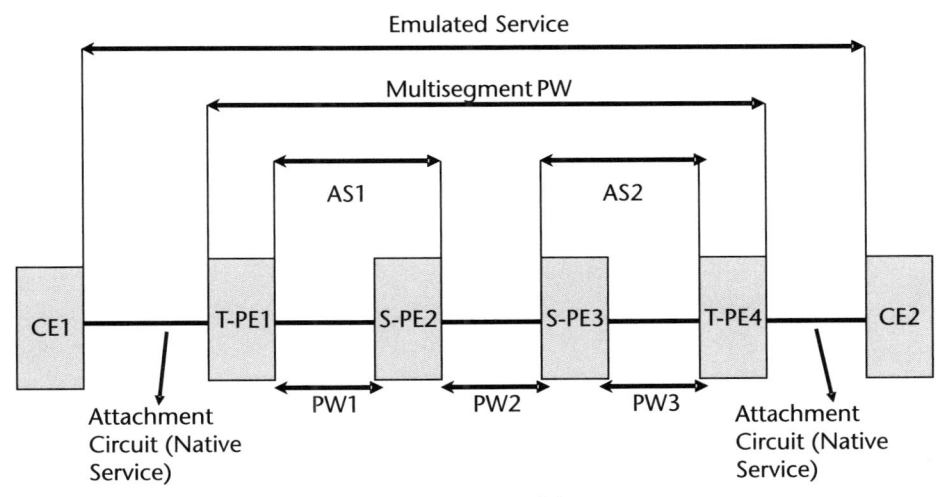

Figure 4.15 PW switching interprovider reference model.

TPE1 and TPE2. PW Segment 1 and PW Segment 2 must be of the same PW type. However, PSN Tunnel 1 and PSN Tunnel 2 do not have to be the same technology because the PEs can adapt the PDU encapsulation between the different PSN technologies.

The PWs in each PSN are established independently where each PSN is treated as a separate PW domain. For example, in Figure 4.15, PW1 can be setup between PE1 and PE2 using the LDP targeted session while PW2 is setup between PE2 and PE3 at the same time. However, the ACs are configured as the same PW type.

An S-PE switches an MS-PW from one segment to another based on the PW identifiers such as PW label of MPLS PWs. In Figure 4.15, the domains that PSN Tunnel 1 and PSN Tunnel 2 traverse could be IGP areas in the same IGP network.

A PW may transit more than one S-PEs along its path as shown in Figure 4.15 where S-PE2 at the border of one AS1 (Autonomous System 1) and S-PE3 at the border of AS2. The MS-PW between T-PE1 and T-PE4 is composed of three segments, PW1 segment in AS1, PW2 segment between two border routers S-PE2 and S-PE3 acting as the switching PEs, and PWE3 segment in AS2. AS1 and AS2 could belong to the same provider or to two different providers.

PWE3 defines the signaling and encapsulation techniques for establishing Single Segment SS-PWs between a pair of terminating PEs (T-PEs), and in the vast majority of cases, this will be sufficient. MS-PWs may be useful in the following situations:

- A PW extends from a T-PE in one provider domain to a T-PE in another provider domain.
- It may not be feasible to establish a direct PW control channel between the T-PEs, residing in different provider networks, to set up and maintain PWs.
- PWE3 signaling protocols and PSN types may differ in different provider networks. The terminating PEs may be connected to networks employing different PW signaling and/or PSN protocols. In this case, it is not possible to use an SS-PW. An MS-PW with the appropriate signaling protocol interworking

performed at the PW switching points can enable PW connectivity between the terminating PEs in this scenario.

- In deploying PWs edge to edge in large service provider networks, the PWs will be tunneled over PSN TE-tunnels with bandwidth constraints. A single-segment PW architecture would require that a full mesh of PSN TE-tunnels be provisioned to allow PWs to be established between all PEs. In this environment, the network is either partitioned into a number of smaller PWE3 domains or consisting of a sparse mesh of PSN TE-tunnels and PW signaling adjacencies.
- Service providers wish to extend PW technology to access and metro networks in order to reduce maintenance complexity and operational costs. For example, in hierarchical IP/MPLS networks, access networks connected to a backbone use PWs up to the edge of the backbone where they can be terminated or switched onto a PW segment crossing the backbone. The use of PWE3 switching between the access and backbone networks can potentially reduce the PWE3 control channels and routing information processed by the access network T-PEs.

4.7 Multisegment Pseudowire Setup Mechanisms

An MS-PW can traverse multiple service provider administrative domains. Furthermore, it can traverse multiple autonomous systems within the same administrative domain or different administrative domains. As a result, PWs and PW control channels such as targeted LDP may cross autonomous system (AS) boundaries.

A multisegment PW can be configured statically where the switching points (S-PEs) and T-PEs are manually provisioned for each segment. The static configuration of MPLS labels for MPLS-PW segments and the cross-connection of them are useful when an MS-PW crosses provider boundaries and two providers do not want to run any PW signaling protocol between them.

A multisegment PW can be also configured using signaling mechanisms via either predetermined routes or signaled dynamic routes. In predetermined route configuration, the PW is established along an administratively determined route using an end-to-end signaling protocol with automated stitching at the S-PEs. In signaled dynamic route configuration, the PW is established along a dynamically determined route using an end-to-end signaling protocol with automated stitching at the S-PEs. The number of S-PEs traversed is only limited by the TTL field of PW MPLS label which is set by the originating PE. In establishing an MPLS PW via signaling, LDP with FEC 128 (i.e., PWid FEC Element) or LDP with FEC 129 (i.e., Generalized PWid FEC Element) can be used. In establishing Layer 2 Tunneling Protocol version 3 (L2TPv3) PW, L2TPv3 is used.

When the MS-PW segments are dynamically signaled, the signaled MS-PW segments can be on the path of a statically configured MS-PW, signaled/statically routed MS-PW, or signaled/dynamically routed MS-PW. Segments are dynamically rerouted around failure points when segments are set up using the dynamic setup method.

4.7 Multisegment Pseudowire Setup Mechanisms

For the MPLS PW setup, there would be four PW switching alternatives:

- Switching between two static control planes;
- Switching between static and dynamic LDP control planes;
- Switching between two LDP control planes using the same FEC type;
- Switching between LDP using FEC 128 and LDP using the generalized FEC 129.

For static control plane switching, the S-PE is configured to direct the MPLS packets from one PW into the other. There is no control protocol involved. It is possible to have one of the control planes is a simple static PW configuration and the other control plane is either a dynamic LDP FEC 128 or generalized PW FEC. In this case, the static control plane is considered similar to an AC.

In switching between LDP using FEC 128 and LDP using the generalized FEC 129, the PE using the generalized FEC 129 can be active or passive. A PE that assumes the active role will send the LDP PW setup message, while a passive role PE will simply reply to an incoming LDP PW setup message. The S-PE will always remain passive until a PWid FEC 128 LDP message is received, which will cause the corresponding generalized PW FEC LDP message to be formed and sent. If a generalized FEC PW LDP message is received while the switching point PE is in a passive role, the corresponding PW FEC 128 LDP message will be formed and sent.

Control plane switching between MPLS-PW and L2TPv3-PW can be static or dynamic as well. The switching alternatives are:

- Switching between static MPLS PW and static L2TPv3 PW where there is no control protocol involved. The S-PE maps MPLS PW Label to L2TPv3 Session ID as well MPLS tunnel label to PE destination IP address.
- Switching between a static MPLS PW and a dynamic L2TPv3 PW where the static control plane is considered identical to an AC.
- Switching between a static L2TPv3 PW and a dynamic LDP/MPLS PW where the static control plane is considered identical to an AC.
- Switching between a dynamic LDP/MPLS PW and a dynamic L2TPv3 PW where the switching point assumes an initial passive role and does not initiate an LDP/MPLS or L2TPv3 PW until it has received a label mapping or incoming-call-request from one side of the node. MPLS PWs are made up of two unidirectional label-switched paths (LSPs) bonded together by FEC identifiers. L2TPv3 PWs are bidirectional in nature and setup via message exchanges.

In dynamic route selection for an MS-PW, S-PEs and T-PEs discover S-PEs on the path to a destination T-PE. After that, the S-PEs along the MS-PW are automatically selected.

4.7.1 LDP SP-PE TLV

The edge-to-edge PW might traverse several switching points, in separate administrative domains as described above. For management and troubleshooting reasons, it is useful to record information about the switching points at the S-PEs that the PW traverses. This is accomplished by using a PW switching point PE TLV (SP-PE TLV) [5].

The SP-PE TLV may appear once for each switching point traversed, and cannot be of length zero. The SP-PE TLV is appended to the PW FEC at each S-PE, and the order of the SP-PE TLVs in the LDP message must be preserved. The SP-PE TLV is necessary to support some of the virtual circuit connectivity verification (VCCV) functions for MS-PWs. The SP-PE TLV is encoded as in Figure 4.16.

The SP-PE TLV format is shown in Figure 4.16. The SP-PE TLV length field specifies the total length of all the following SP-PE TLV fields in octets. Sub-TLV type field encodes how the value field is to be interpreted. The SP-PE TLV contains sub-TLVs to describe various characteristics of the S-PE traversed. The length field specifies the length of the value field in octets. The value field is the octet string of length octets that encodes information to be interpreted as specified by the type field.

4.8 Resiliency

A PW segment, a contiguous set of PW segments, as well as the end-to-end path can be protected. The protection and primary paths for the protected segment(s) share the same respective segments end points. A protection path for a PW segment, sequence of segments, or end-to-end path is signaled.

Traffic is switched from a primary PW to secondary PW when an element on the path of a primary MS-PW fails. The primary and backup paths may be statically configured, statically specified for signaling, or dynamically selected via dynamic routing depending on the MS-PW establishment mechanism. Backup and primary paths should have the ability to traverse separate S-PEs. For example, a backup PW can be configured with a different T-PE from the primary PW.

The protection mechanism can automatically revert to a primary PW from a backup PW once the primary path is recovered from failures.

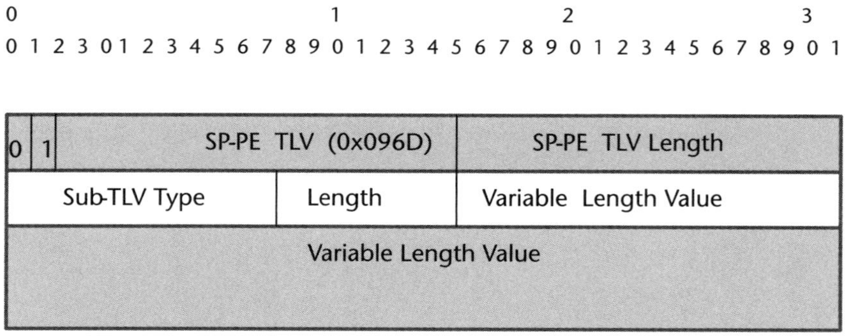

Figure 4.16 SP-PE TLV.

4.9 Quality of Service and Congestion Control

PWs are intended to support emulated services with strict packet/frame loss, delay, and jitter requirements satisfied by reserving sufficient network resources such as bandwidth and buffer and by providing appropriate scheduling priority and drop precedence throughout the network.

Path provisioning is frequently performed through quality-of-service (QoS) reservation protocols or network management protocols. QoS provisioning for MS-PWs that may transmit across network domains under the control of multiple entities is much more difficult than that for SS-PWs that remain within a single administrative domain.

When the T-PE attempts to signal an MS-PW, signaling identifies the CoS associated with an MS-PW, carries the traffic parameters used by the admission control for an MS-PW per CoS, and separates traffic parameter values to be specified for the forward and reverse directions of the PW.

The signaling protocol prioritizes the PW setup and support maintenance operation among PWs.

For SS-PWs, a traffic engineered PSN tunnel (i.e., MPLS-TE) may be used to ensure that sufficient resources are reserved in the P-routers to provide QoS to PWs on the tunnel. In this case, T-PEs will provide admission control of PWs onto the PSN tunnel and accounting for reserved and available bandwidth on the tunnel.

For MS-PWs, each S-PE maps a PW segment to a PSN tunnel where S-PEs and T-PEs automatically binds a PW segment to a PSN tunnel based on CoS and bandwidth requirements. S-PEs and T-PEs associate a CoS marking, such as EXP field value of MPLS PWs, with PW PDUs to specify packet treatment.

Different administrative domains may use different CoS values to imply the same CoS treatment. S-PEs at administrative domain boundaries translate from one CoS value to another as a PW PDU crosses from one domain to the next.

The CoS and bandwidth of the MS-PW are configurable at T-PEs and S-PEs. Each domain individually implements a method to control congestion. This can be by QoS reservation, or other congestion control method.

Each PSN carrying the PW may be subject to congestion. Each PW segment will handle any congestion independently of the other MS-PW segments.

4.10 Operations and Maintenance

The PE reports the status of the interface and tabulates statistics for PW that help monitor the state of the network and measure service-level agreements (SLAs) for the PW. Typical counters are as follows:

- Counts of PW-PDUs sent and received, with and without errors;
- Counts of sequenced PW-PDUs lost;
- Counts of service PDUs sent and received over the PSN, with and without errors;
- Service-specific interface counts;
- One-way delay and delay variation.

End-to-end connectivity and the exact functional path can be identified by connection verification and traceroute mechanisms available at PEs. Connection verification and other alarm mechanisms can alert the operator that a PW has lost its remote connection.

OAM mechanisms defined in ITU-T I.610 can be used for the attachment circuits to detect defects in the network, localize, diagnose, and communicate PW defect states on the PW attachment circuit.

Defect states for SS-PWs between AC and PWs are propagated across a PWE3 network following the failure and recovery from faults. For MS-PWs, a common PW OAM mechanism agreed by all PE routers along the MS PW is supported end to end. Failure of a segment is notified to other segments of an MS-PW. At the S-PE, defects on a PSN tunnel are propagated to all PWs that utilize that particular PSN tunnel.

The S-PE can behave as a segment end point and pass T-PE to T-PE PW OAM messages transparently. Both MS-PWs and SS-PWs can measure round-trip delay, one-way delay, jitter, and packet loss ratio.

Single-segment PWs (SS-PW) and MS-PW capabilities are signaled using the virtual circuit connectivity verification (VCCV) parameter included in the interface parameter field of the PWid FEC TLV or the interface parameter sub-TLV of the Generalized PWid FEC TLV [6]. When a switching point exists between PE nodes, it is required to be able to continue operating VCCV end-to-end across a switching point and to provide the ability to trace the path of the MS-PW over any number of segments.

When an MS-PW includes SS-PWs that use the L2TPv3, the MPLS PW OAM is terminated at the S-PE connecting the L2TPv3 and MPLS segments. Status information received in a particular PW segment can be used to generate the appropriate status messages on the following PW segment. In the case of L2TPV3, the status bits in the circuit status can be mapped directly to the PW status bits defined in [2].

VCCV messages are specific to the MPLS data plane and cannot be used for an L2TPv3 PW segment. VCCV messages from L2TPv3 PW segments must be translated to those for MPLS PW segments and vice versa.

As stated above, the S-PE performs a standard MPLS label swap operation on the MPLS PW label where the PW label TTL is decreased at every S-PE. Once the PW label TTL reaches the value of 0, the packet is sent to the control plane to be processed. Hence, by controlling the PW TTL value of the PW label, it is possible to select exactly which S-PE will respond to the VCCV packet.

In the PW switching with attachment circuits (Figure 4.14), PW status messages indicating PW or attachment circuit faults is mapped to fault indications or OAM messages on the connecting AC.

In the PW control plane switching (Figure 4.17), the status of the PWs is forwarded unchanged from one PW to the other by the control plane switching function.

Communication of the fault status of one of the locally attached PW segments at an S-PE may be needed. For LDP, this can be accomplished by sending an LDP notification message in Figure 4.17.

This message is then relayed by each S-PE unchanged. The T-PE decodes the status message and the included SP-PE TLV to detect exactly where the fault occurred.

4.11 Security 81

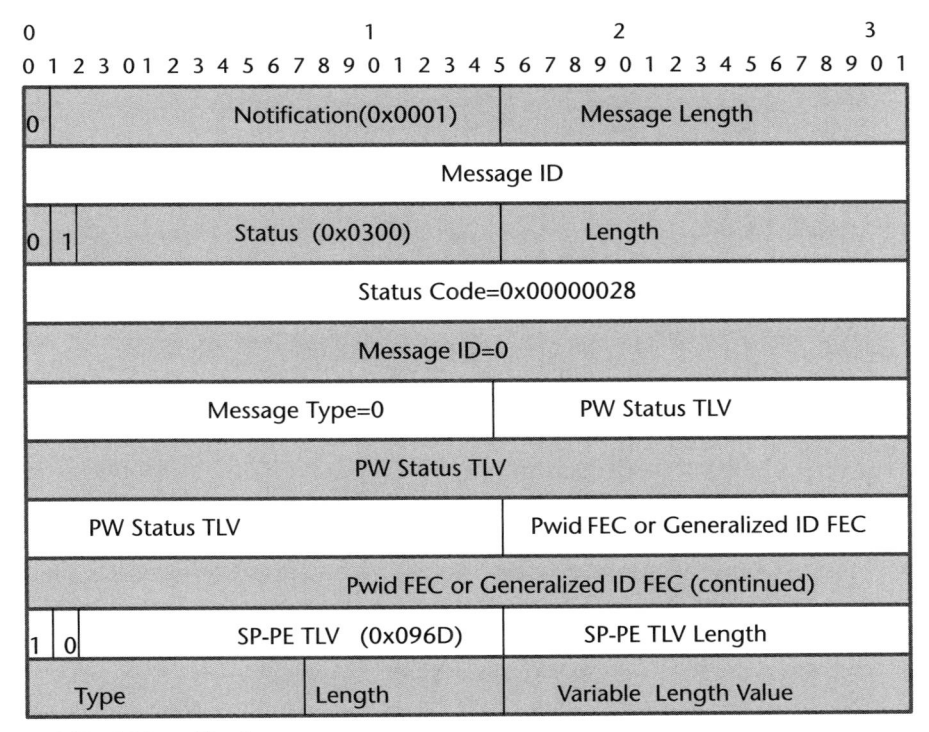

```
 0                   1                   2                   3
 0 1 2 3 0 1 2 3 4 5 6 7 8 9 0 1 2 3 4 5 6 7 8 9 0 1 2 3 4 5 6 7 8 9 0 1
```

0	Notification(0x0001)	Message Length
Message ID		
0	1	Status (0x0300)	Length
Status Code=0x00000028		
Message ID=0		
Message Type=0	PW Status TLV	
PW Status TLV		
PW Status TLV	Pwid FEC or Generalized ID FEC	
Pwid FEC or Generalized ID FEC (continued)		
1	0	SP-PE TLV (0x096D)	SP-PE TLV Length
Type	Length	Variable Length Value

Figure 4.17 LDP notification message.

At the T-PE, if there is no SP-PE TLV included in the LDP status notification, then the status message can be assumed to have originated at the remote T-PE.

4.11 Security

PWE3 provides no means of protecting the integrity, confidentiality, or delivery of the native data units. The relatively weak security mechanisms represent a greater vulnerability in an emulated Ethernet connected via a PW.

Controlling PSN access to the PW tunnel end point may protect against PW Demultiplexer and PSN tunnel services disruption. By restricting PW Tunnel end point access to legitimate remote PE sources of traffic, the PE may reject traffic that would interfere with the PW demultiplexing and PSN tunnel services. Security protocols such as IPSec may be used by the PW demultiplexer layer in order to provide authentication and data integrity of the data between the PW demultiplexer end points.

Security needs to be provided for data-plane and control plane [6]. For dataplane security, packets of an MS-PW traveling to a PE or an AC should be delivered to the intended recipients. Packets from outside an MS-PW entering the MS-PW should be consistent with the policies of the MS-PW.

MS-PWs that cross service provider (SP) domain boundaries may connect one T-PE in a SP domain to a T-PE in another provider domain. They may also transit other provider domains even if the two T-PEs are under the control of one SP.

When there is one or more PDUs are falsely inserted into an MS-PW at any of the originating, terminating, or transit domains as a result of a malicious attack or fault in the S-PE, there should be a mechanism for the end-to-end authenticity of MS-PW PDUs.

For control-plane security, an MS-PW connects two attachment circuits. It is important to make sure that PW connections are not arbitrarily accepted from anywhere, or else a local attachment circuit might get connected to an arbitrary remote attachment circuit.

Directly interconnecting the S-PEs using a physically secure link, and enabling signaling and routing authentication between the S-PEs, eliminates the possibility of receiving an MS-PW signaling message or packet from an untrusted peer.

S-PEs in different provider networks may reside at each end of a physically secure link, or be interconnected by a limited number of trusted PSN tunnels, each S-PE will have a trust relationship with a limited number of S-PEs in other ASs.

Static manual configuration of MS-PWs at S-PEs and T-PEs provides a greater degree of security. If an identification of both ends of an MS-PW is configured and carried in the signaling message, an S-PE can verify the signaling message against the configuration.

An incoming MS-PW request/reply is not accepted unless its IP source address is known to be the source of an eligible peer which 0 is an S-PE or a T-PE with which the originating S-PE or T-PE has a trust relationship.

The set of eligible peers could be preconfigured (either as a list of IP addresses or as a list of address/mask combinations) or automatically generated from the local PW configuration information.

The S-PE and T-PE drop the unaccepted signaling messages in the data path to avoid a denial-of-service (DoS) attack on the control plane. S-PEs that connect one provider domain to another provider domain usually rate-limit signaling traffic in order to prevent DoS attacks on the control plane.

Even if a connection request appears to come from an eligible peer, its source address can be spoofed. Source address filtering at the border routers of that network could eliminate the possibility of source address spoofing.

A PW connects two attachment circuits. In order to prevent a local attachment circuit get connected to an arbitrary remote attachment circuit, LDP connections are not arbitrarily accepted from anywhere. An incoming session request must not be accepted unless its IP source address is known to be the source of an eligible peer.

Even if a connection request appears to come from an eligible peer, its source address can be spoofed. A mean of preventing source address spoofing must be in place such as source address filtering at the border routers that could eliminate the possibility of source address spoofing.

The LDP MD5 authentication can be used to provide integrity and authentication for the LDP messages and protect against source address spoofing.

4.12 Conclusion

PWE3 is a mechanism to emulate telecommunications services across PSNs. The emulated services are mostly T1/T3 leased lines, frame relay, Ethernet and ATM to maximize return on existing assets and minimize operating costs of the service providers. Therefore, payloads can be packet, cell, bit stream, and structured bit stream.

A PW is a connection between two PE equipment connecting two ACs. It can span one or multiple networks. The PWE3 control plane services provide capabilities for setting up and tearing down a PW, monitoring it and dealing with various fault conditions.

The PE reports the status of the interface and tabulates statistics for PW that help to monitor the state of the network and measure SLAs for the PW. The end-to-end connectivity and the exact functional path can be identified by connection verification and trace route mechanisms available at PEs.

PWE3 provides no means of protecting the integrity, confidentiality, or delivery of the native data units. Security needs to be provided for data-plane and control plane. For data-plane security, packets of an MS-PW traveling to a PE or an AC should be delivered to intended recipients. Packets from outside an MS-PW entering the MS-PW should be consistent with the policies of the MS-PW.

References

[1] RFC 3985, Pseudo Wire Emulation Edge-to-Edge (PWE3) Architecture, 2005.
[2] RFC 3916, Requirements for Pseudo-Wire Emulation Edge-to-Edge (PWE3), September 2004.
[3] RFC 3550, RTP: A Transport Protocol for Real-Time Applications, July 2003.
[4] RFC 5254, Requirements for Multi-Segment Pseudowire Emulation Edge-to-Edge (PWE3), October 2008
[5] RFC 6073, Segmented Pseudowire, January 2011.
[6] RFC 4447, Pseudowire Setup and Maintenance Using the Label Distribution Protocol (LDP), April 2006.

CHAPTER 5

Ethernet Protection

5.1 Introduction

Services such as broadcast video, voice over IP, and video on demand require five-nine availability that means a high-level survivability against failures. When failures do occur, they are not supposed to be noticed by the subscriber. The main purpose of automatic protection switching is to guarantee the availability of resources in the event of failure and to ensure that switchover is achieved in less than 50 ms so that the network will not be affected, although 50 ms is currently being debated in the industry. In fact, [1] mandated 500-ms protection switching at UNI or ENNI. The smaller switching time increases the cost of the equipment. The authors' view is that at higher speeds such as 1G and 10G, protection switching time should be even smaller than 50 ms in order to reduce frame loss so that the service can stay within frame loss ratio (FLR) limits. The protection switching time is the minimum possible time for the sum of persistent (i.e., nontransient) failure detection, speed of light propagation, signaling protocol time, and regaining sync alignment.

To a degree, both the Spanning Tree Protocol (STP) and Rapid Spanning Tree Protocol (RSTP) have been proven to be effective for preventing loops and assuring that backup paths are available. However, both protocols are slow to respond to network failures. They are slow on the order of 30 seconds or more. Automatic protection switching (APS), Ethernet protection switched rings, and link aggregation groups (LAGs) are among the proposed solutions.

Maintenance operations can benefit from network flexibility and from the ability to transmit traffic over either working entity or the protection entity. Therefore, the APS mechanism needs to support traffic to be manually switched over from one trunk to the other.

In a communication network, traffic protection can be implemented in many different ways for various network topologies such as star, loop, and mesh. Loop-based networks are attractive for enabling redundancy in a network consisting of a relatively large number of nodes. A redundant path for each node is provided by just one additional link that closes two adjacent branches to form a loop. However, loops are not allowed in Ethernet-based transport networks, as Ethernet frames would circulate forever. The loop has to be broken at some point, so that Ethernet transport is enabled.

Ethernet automatic protection switching (EAPS) is an exemplary solution for Layer-2 loop protection, which is comparable to solutions such as Ethernet protection switched rings (EPSR) and Ethernet ring protection (ERP) [2]. The terms loop and ring are synonyms here.

The EAPS ring consists of a master node and one or more transit nodes. The two ring ports of the master node are configured as the primary port and the secondary port. The master node blocks logically the secondary port except for a control VLAN. The master node sends periodic health check packets from the primary port through the control VLAN towards the secondary port. When a fault occurs in the ring, the master detects this either by missing health check packets or by special fault detection packets generated by one of the transit nodes.

EAPS and similar solutions are on/off-type mechanisms without adaptation to available link capacities in the ring. Therefore, they do not allow any load balancing. In the spanning tree, loops resulting from redundant paths are broken by use of the STP algorithm. The STP breaks loops by disabling Ethernet switch ports so that the remaining active links build up a tree topology. In a failure case, when an active link breaks, STP calculates a new tree, taking then the appropriate so far disabled links into use. The RSTP converges faster. However, both STP and RSTP are on/off-type mechanisms without adaptation to available link capacities in the ring, therefore not allowing any load balancing.

Resilient Packet Ring (RPR; IEEE 802.17), which is independent of the underlying physical layer, provides a fast protection switching (less than 50 ms). The RPR concept is based on two counterrotating rings designed to transport Ethernet frames efficiently. There are no dedicated protection resources. Both rings transport traffic using shortest paths.

In this chapter, we describe linear protection, ring-based protection, and LAGs.

5.2 APS Entities

APS includes detection of failures [signal failure (SF) or signal degrade (SD)] on a working channel, switching traffic transmission to a protection channel, selecting traffic reception from the protection channel, and (optionally) reverting back to the working channel once failure is repaired. The entities are (see Figure 5.1):

- Working entity, which is used when no failure exists;
- Protection entity, which is used when a failure exists;
- Head end, which is the entity transmitting data to working/protection channel;
- Tail end, which is the entity receiving data from the working/protection channel;
- Bridge: function at the head end that connects traffic (including extra traffic) to the working and protection channels;
- Selector: function at the tail end that extracts traffic (perhaps extra traffic) from the working or protection channel;

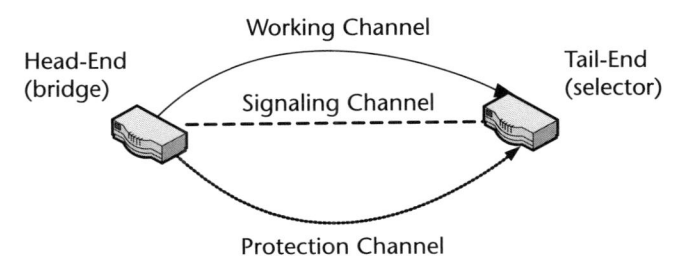

Figure 5.1 APS entities.

- APS signaling channel: channel used to communicate between the head end and the tail end for APS purposes;
- Trail termination: function responsible for failure detection including injection and extraction of operations administration management (OAM).

Protection switching is usually triggered by a failure, although the operator may manually force a protection switch. A failure is declared when a fault condition persists long enough for the ability to perform the required function. Failures are SF or SD (of various types) and may be detected by a physical layer, indicated by signaling (e.g., AIS), and detected by OAM mechanisms. When there is no SF or SD, the state is called no request (NR). EAPS is a linear protection scheme designed to protect VLAN-based Ethernet networks.

With EAPS, a protected domain is configured with two paths: a working path and a protection path. Both working and protection paths can be monitored using an OAM protocol like connectivity fault management (CFM). Normally, traffic is carried on the working path and the protection path is disabled. If the working path fails, APS switches the traffic to the protection path, and the protection path becomes the active path.

The EAPS can be unidirectional and bidirectional. The unidirectional switching utilizes fully independent selectors at each end of the protected domain. Bidirectional switching attempts to configure the two end points with the same bridge and selector settings, even for a unidirectional failure. Unidirectional switching can protect two unidirectional failures in opposite directions on different entities.

Protection switching can be unidirectional as well as bidirectional. With unidirectional switching, the selectors at each end are fully independent. Therefore, two ends must be coordinated (i.e., APS communications). Because there is no signaling, it could be faster than bidirectional switching and easy to implement. However, the bidirectional switching management can be easier since directions traverse same network elements.

5.3 Linear Protection

APS uses two modes of operation: linear 1 + 1 protection switching architecture and linear 1:1 protection switching architecture. The linear 1 + 1 protection switching architecture operates with either unidirectional or bidirectional switching. The linear 1:1 protection switching architecture operates with bidirectional switching.

The linear protection architecture is defined in [3] where protection switching occurs at the two distinct end points of a point-to-point Ethernet entity such as EVC. Between these end points, there will be both working and protection transport entities. In the linear 1 + 1 protection switching architecture, the normal traffic is copied and fed to both working and protection paths with a permanent bridge at the source of the protected domain. The traffic on the working and protection transport entities is transmitted simultaneously to the sink of the protected domain, where a selection between the working and protection transport entities is made.

In the linear 1:1 protection switching architecture, the normal traffic is transported on either the working path or on the protection path using a selector bridge at the source of the protection domain. The selector at the sink of the protected domain selects the entity that carries the normal traffic.

Protection switching can be revertive or nonrevertive. In revertive operation, normal traffic signal is restored to the working unit after the condition(s) causing a switch has cleared. In the case of clearing a command such as forced switch, this happens immediately. In the case of clearing of a defect, this generally happens after the expiry of a wait-to-restore (WTR) timer to avoid chattering of selectors in the case of intermittent defects.

In nonrevertive operation, normal traffic signal is allowed to remain on the protection transport entity even after a switch reason has cleared. This is generally accomplished by replacing the previous switch request with a do not revert (DNR) request.

Revertive mechanisms may be preferable when the working channel has better performance [free bandwidth, bit error rate (BER), delay], when there are frequent switches (easier to manage), and when there is extra traffic. However, nonrevertive also has the advantages of only one service disruption due to protection switching and it may be simpler to implement. Protection switching can be triggered either by an administrator or signal failures. Details are discussed in the following section.

5.3.1 1 + 1 Protection Switching

The 1 + 1 protection is a simplest and fastest form of protection, but wastes the capacity of the protecting entity. The head-end bridge always sends data on both working and protecting entities while the tail end selects the transport entity to use.

The 1 + 1 protection is often provisioned as nonrevertive, to avoid a second interruption to the normal traffic signal. However, there are reasons to be revertive as well. The primary path could be a favorable path for traffic in terms of delay, capacity, and so forth.

The 1 + 1 bidirectional linear protection switching architecture is shown in Figure 5.2 [3]. The protected Ethernet entity (i.e., EVC, OVC, Link) connection is permanently bridged to both the working transport entity and the protection transport entity. The Ethernet frames are received only from the working entity.

When a fault such as loss of signal occurs on the working entity, the traffic is switched from the working entity to the protecting entity (Figure 5.3) [3]. Both directions are switched even when a unidirectional defect occurs. APS coordination protocol is necessary to achieve coordination.

The 1 + 1 unidirectional linear protection switching architecture is shown in Figure 5.4 [3]. The protected Ethernet connection is permanently bridged to both

5.3 Linear Protection

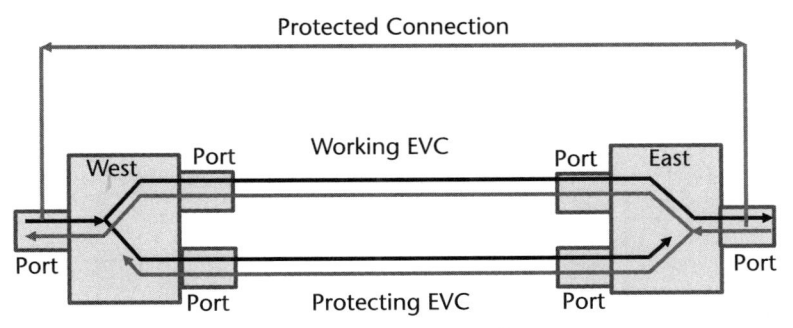

Figure 5.2 The 1 + 1 bidirectional protection switching architecture.

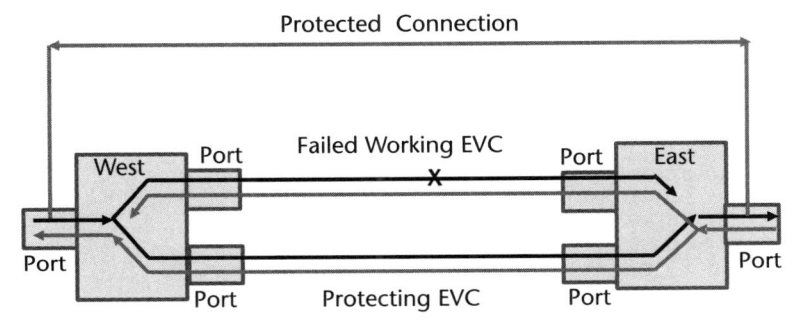

Figure 5.3 The 1 + 1 bidirectional protection switching for failed EVC.

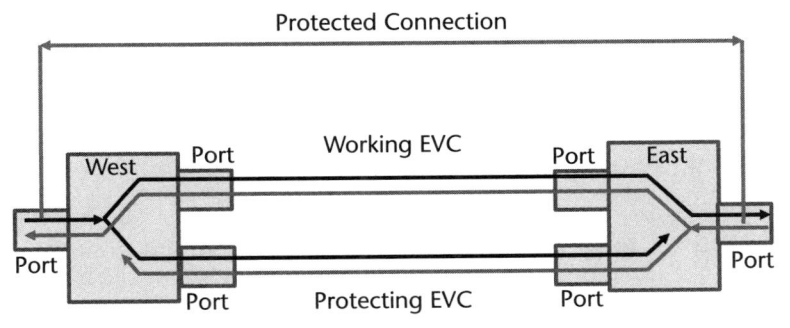

Figure 5.4 The 1 + 1 unidirectional protection switching architecture.

the working transport entity and the protection transport entity. The traffic is received only from the working entity for both directions.

When a fault such as loss of signal occurs on the working entity in the west-to-east direction, each direction is switched independently. Selectors at the sink of the protected domain operate only based on the local information; therefore, APS coordination protocol is not necessary (Figure 5.5) [3]. Traffic that is not affected from this failure continues to be received via the working entity.

When there are failures on both working and protecting entities (Figure 5.6) [3], both entities are switched. Unidirectional protection switching can protect this type of dual defect scenarios while bidirectional protection switching cannot. This

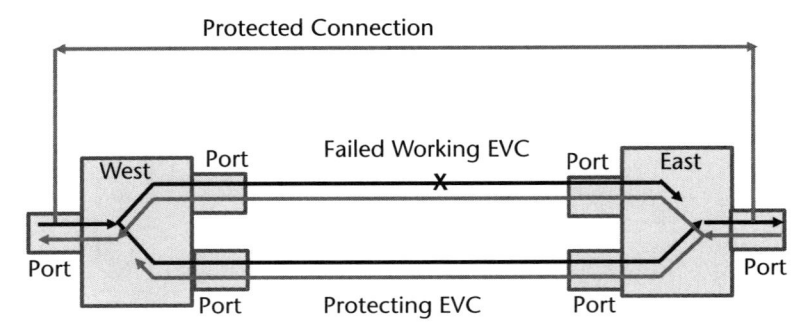

Figure 5.5 The 1 + 1 unidirectional protection switching for failed EVC in the west-to-east direction.

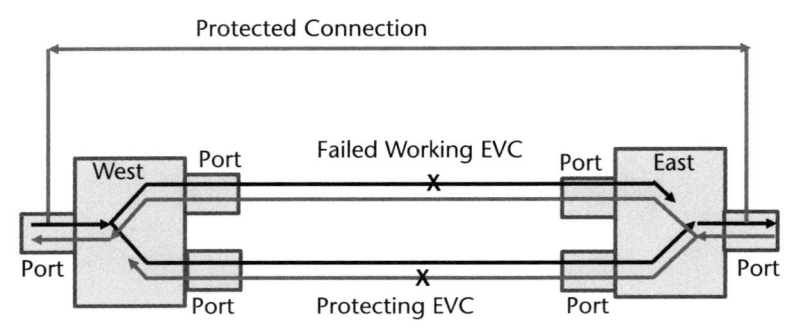

Figure 5.6 The 1 + 1 unidirectional protection switching for failed EVC in both directions.

is possible if each of the transport entities has a healthy path that does not use the faulty units to carry the traffic.

5.3.2 1:1 Protection Switching

In the 1:1 protection scheme, the head-end bridge usually sends data on the working channel. When failure is detected, it starts sending frames over the protecting entity and the tail end selects the protecting entity. The protecting channel is used for traffic other than the one is being protected, when it is not used for protecting the working entity. This is the main difference between 1:1 and 1 + 1 protection schemes where the capacity of the protecting entity is not wasted. The failure is detected by the tail end via the CCM; therefore, the APS signaling is necessary.

The 1:1 protection is usually revertive, but it is possible to define the protocol to permit a nonrevertive operation for 1:1 protection. In general, the choice of revertive or nonrevertive will be the same at both ends of the protection group. However, a mismatch of this parameter does not prevent interworking.

Figure 5.7 [3] illustrates the 1:1 linear protection switching architecture, where the normal traffic is being transmitted over the working entity and the tail-end protection switching process determines the specific output for Ethernet Connection A is transferred.

When there is a failure on working entity (Figure 5.8), at the head end, the traffic for connection A is forwarded to the protection transport entity, and at the tail end, the traffic for connection A is received from the protection transport entity.

5.3 Linear Protection

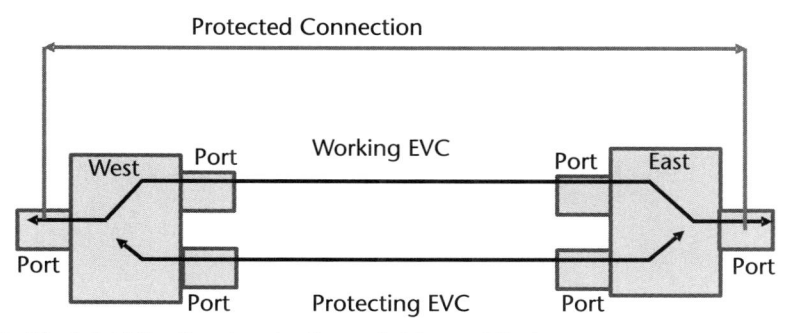

Figure 5.7 The 1:1 bidirectional protection switching architecture.

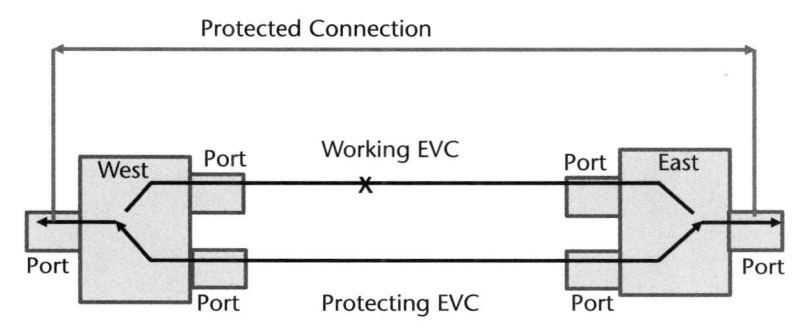

Figure 5.8 The 1:1 bidirectional protection switching for failed EVC.

5.3.3 Protection Switching Triggers

Protection witching can be initiated by a network administrator or network failures. A network operator can trigger protection switching in forced switch or manual switch modes if it has a higher priority than any other local request or the tail-end request. Similarly, a declared SF or SD on a working entity can switch traffic to the protecting entity if the protecting entity is healthy and the detected SF condition has a higher priority than any other local request or the far-end request. In the bidirectional 1 + 1 and 1:1 architectures, the received APS protocol requests to switch have a higher priority than any other local request.

In order to monitor both working and protecting entities, MEPs at the head end and tail end must be active and transmitting CCMs. SF condition can be due to loss of connectivity (LOC) as a result of not receiving three CCMs within a configured time interval or other failure reasons discussed in Chapter 7.

The protection switching process also requires APS communication in order to coordinate its switching behavior with the tail end of the protected domain if the protection switching architecture is not 1 + 1 unidirectional protection switching. APS PDU is transmitted and received between the same MEP pair on the protection transport entity where CCM is transmitted for the monitoring (Figure 5.9) [3].

5.3.4 APS PDU Format

Protection switching control is achieved via Ethernet OAM PDU as defined in [4].

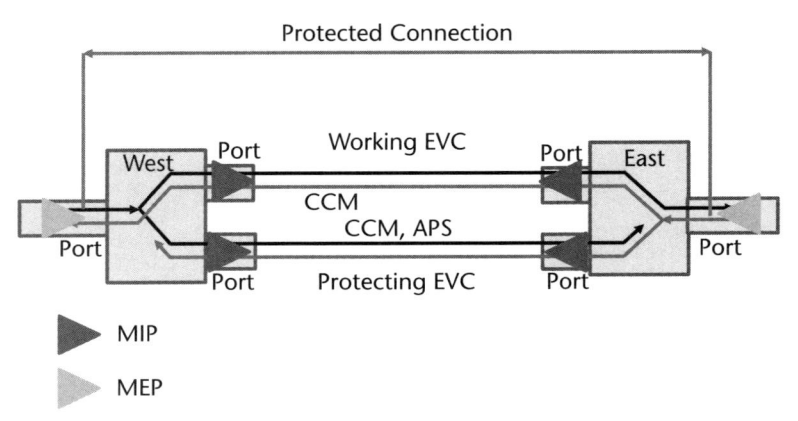

Figure 5.9 MEPs in a 1 + 1 bidirectional EVC protection switching architecture.

Proactive OAM frames for Ethernet APS Function (ETH-APS) need to be counted; 4 octets in the APS PDU are used to carry APS-specific information (Figures 5.10 and 5.11) [3]. The TLV offset field is required to be set to 0x04.

Table 5.1 describes code points and values for APS-specific information.

5.3.4.1 Revertive and Nonrevertive Modes

In the revertive mode of operation, the flow reverts to the working facility as soon as the failure has been corrected. In conditions in which working traffic is being received via the protection entity, if local protection switching requests have been previously active and now become inactive, a local wait-to-restore (WTR) state is entered. Because this state now represents the highest priority local request, it is indicated on the transmitted request/state (Table 5.1) [3] information and maintains the switch. This state normally times out and becomes a no request (NR) state after the wait-to-restore timer has expired. The wait-to-restore timer is deactivated earlier if any local request of higher priority preempts this state.

In nonrevertive mode of operation, the protection facility is treated as the working facility. In conditions in which working traffic is being transmitted via the protection entity, if local protection switching requests have been previously active and now become inactive, a local do not revert state (Table 5.1) is reached. Because

Figure 5.10 APS PDU format.

Figure 5.11 APS-specific information format.

Table 5.1 Code Points and Field Values for APS-Specific Information

Request/state		1111	Lockout of protection (LO)	Priority
		1110	Signal fail for protection (SF-P)	Highest
		1101	Forced switch (FS)	
		1011	Signal fail for working (SF)	
		1001	Signal degrade (SD)	
		0111	Manual switch (MS)	
		0101	Wait to restore (WTR)	
		0100	Exercise (EXER)	
		0010	Reverse request (RR)	
		0001	Do not revert (DNR)	
		0000	No request (NR)	lowest
		Others	Reserved for future international standardization	
Protection type	A	0	No APS channel	
		1	APS channel	
	B	0	1 + 1 (permanent bridge)	
		1	1:1 (no permanent bridge)	
	D	0	Unidirectional switching	
		1	Bidirectional switching	
	R	0	Nonrevertive operation	
		1	Revertive operation	
Requested signal		0	Null signal	
		1	Normal traffic signal	
		2–255	(Reserved for future use)	
Bridged signal		0	Null signal	
		1	Normal traffic signal	
		2–255	(Reserved for future use)	

this state now represents the highest priority local request, it is indicated on the transmitted request/state information and maintains the switch, thus preventing reversion back to the released bridge/selector position in nonrevertive mode under NR conditions.

5.3.4.2 Transmission and Acceptance of APS

Traffic units that carry APS PDU are called APS frames. The APS frames are transported via the protection transport entity only, being inserted by the head end of the protected domain and extracted by the tail end of the protected domain.

A new APS frame must be transmitted immediately when a change in the transmitted status occurs.

The first three APS frames should be transmitted as fast as possible after the status change of the protection end point so that fast protection switching is possible even if one or two APS frames are lost or corrupted. For the fast protection switching in 50 ms, the interval of the first three APS frames is desirable to be 3.3 ms, which is the same interval as CCM frames for fast defect detection. APS frames after the first three frames should be transmitted with the interval of 5 seconds.

5.3.4.3 Protection Types

The valid protection types are:

- 000x: 1 + 1 unidirectional, no APS communication;
- 100x: 1 + 1 unidirectional with APS communication;
- 101x: 1 + 1 bidirectional with APS communication;
- 111x: 1:1 bidirectional with APS communication.

The default value, all zeros, matches the only type of protection that can operate without 1+1 unidirectional APS.

If the B bit mismatches, the selector is released as 1:1 and 1 + 1 are incompatible. This will result in a defect. If the A bit mismatches, the side expecting APS will fall back to 1 + 1 unidirectional switching without APS communication. If the D bit mismatches, the bidirectional side will fall back to unidirectional switching. If the R bit mismatches, one side will clear switches to WTR and the other will clear to DNR. The two sides will interwork and the traffic is protected.

Requested signal indicates the signal that the near end requests be carried over the protection path. For NR, this is the null signal when the far end is not bridging normal traffic signal to the protection entity. For LO, this is the null signal. For exercise, this is the null signal when exercise replaces NR or the normal traffic signal in the case where exercise replaces DNR. For all other requests, this will be the normal traffic signal requested to be carried over the protection transport entity.

Bridged signal indicates the signal that is bridged onto the protection path. For 1 + 1 protection, this reflects the permanent bridge. For 1:1 protection, this will indicate what is actually bridged to the protection entity, either the null signal, or normal traffic signal.

5.3.4.4 Timers

There are three types of timers: hold-off timer, wait-to-restore timer, and guard timer. The hold-off timer coordinates timing of protection switches at multiple layers or across cascaded protected domains, allowing either a server layer protection switch to have a chance to fix the problem before switching at a client layer or an upstream protected domain to switch before a downstream domain.

Each protection group should have a provisionable hold-off timer. The suggested range of the hold-off timer is 0 to 10 seconds in steps of 100 ms with an accuracy of ±5 ms.

When a new defect or more severe defect occurs, this event will not be reported immediately to protection switching, if the provisioned hold-off timer is set to a nonzero value: the hold-off timer will be started. When the hold-off timer expires, it will be checked whether a defect still exists on the trail that started the timer. If it does, that defect will be reported to protection switching.

A WTR timer prevents frequent operation of the protection switch due to an intermittent defect, and a failed working transport entity must become fault-free. After the failed entity becomes fault-free for the WTR period, a normal traffic can use it.

After the failed working transport entity meets this criterion, a fixed period of time shall elapse before a normal traffic signal uses it again. This period, called the wait-to-restore (WTR) period, may be configured by the operator in 1-minute steps between 5 and 12 minutes; the default value is 5 minutes. An SF (or SD if applicable) condition will override the WTR.

In the revertive mode of operation, when the failed working transport entity is no longer in SF (or SD if applicable) condition, a local WTR state will be activated. The WTR timer deactivates earlier when any request of higher priority preempts this state.

Ring automatic protection switching (R-APS) messages are continuously transmitted, copied, and forwarded at every ring node around the ring. This can result in a message corresponding to an old request, which is no longer relevant, being received by ring nodes. The reception of messages with outdated information could result in erroneous interpretation of the existing requests in the ring and lead to erroneous protection switching decisions. The guard timer is used to prevent ring nodes from receiving outdated R-APS messages. During the duration of the guard timer, all received R-APS messages are ignored by the ring protection control process. This allows that old messages still circulating on the ring may be ignored; as a result, a node will be unaware of new or existing ring requests transmitted from other nodes.

The period of the guard timer may be configured by the operator in 10-ms steps between 10 ms and 2 seconds, with a default value of 500 ms. This time should be greater than the maximum expected forwarding delay for which one R-APS message circles around the ring.

5.4 Ring Protection

Ethernet protection switched rings (EPSR) run over standards Ethernet interfaces and, similar to STP, provides a polling mechanism to detect ring-based faults and failover accordingly. Unlike STP, EPSR uses a fault detection scheme that alerts the ring that a break has occurred and indicates that it must take action instead of making a calculation. When a fault is detected, the ring automatically sends traffic over a protected reverse path and can converge in less than 50 ms.

EPSR operates over standard Ethernet ports in a domain, and can be used with either the STP or the RSTP. The EPSR domain is a collection of VLANs of user frames, the control VLAN, and the associated switch ports. The controlling node for an EPSR domain is called the master node (Figure 5.12) [5], which is responsible for status polling, collecting error messages, and controlling the flow of traffic in an EPSR domain. All other nodes in an EPSR domain are transit nodes. Transit nodes generate failure notices and receive control messages from the master. An APS protocol is used to coordinate the protection actions over the ring.

The primary port of the master node determines the direction of traffic flow. This port is always operational. The secondary port of the master node remains active, but blocks all protected VLANs from operating until ring failover. All control messages are sent and received over a control VLAN. This is the only VLAN that is never blocked. Because traffic flow is based on the master node, its placement is

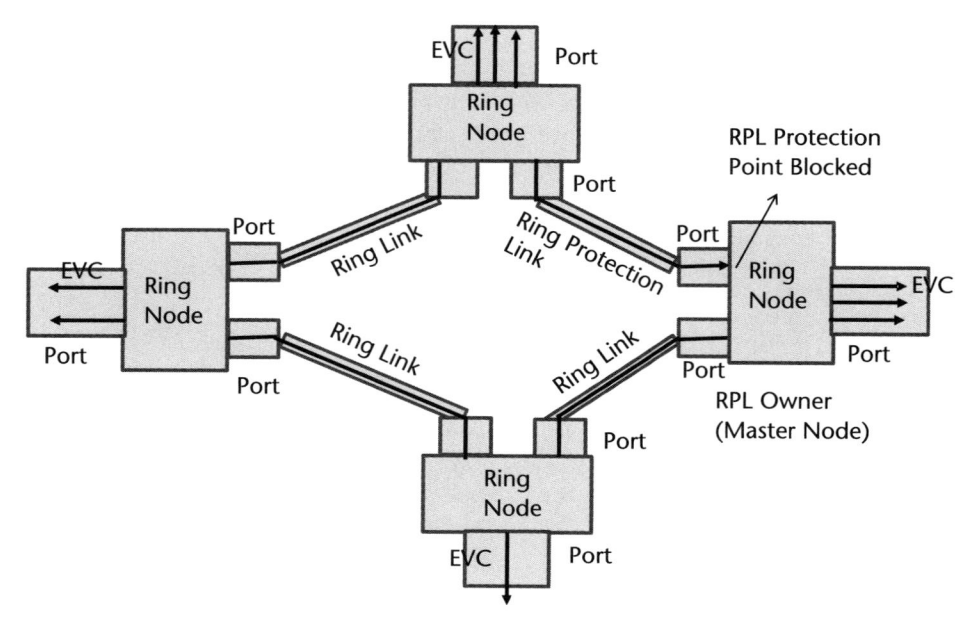

Figure 5.12 Ethernet ring protection switching architecture.

critical. Most scenarios dictate that the least-used node should be designated as the master node, thereby assuring optimum spatial reuse of the bandwidth.

Each ring node is connected to adjacent nodes participating in the same ring, using two independent links. The link between two adjacent nodes is called a ring link. A port on a ring link is called a ring port. The minimum number of nodes on a ring is two.

The ring protection switching architecture is based on the principle of loop avoidance and the utilization of learning, forwarding, and addressing table mechanisms, which is part of the Ethernet flow forwarding function. Loop avoidance in the ring is achieved by guaranteeing that, at any time, traffic may flow on all but one of the ring links. This particular link is called the ring protection link (RPL), and under normal conditions this link is blocked. One designated node, the RPL owner, is responsible to block traffic over the RPL. Under a ring failure condition, the RPL owner is responsible to unblock the RPL, allowing the RPL to be used for traffic. Under a normal operation, the master node's secondary port is blocked for all protected VLANs; only the control VLAN remains unblocked on the secondary port. The master node periodically sends out health-check messages through its primary port at times specified by the user. These health-check messages are then received on the master node's secondary port.

Multiple EPSR domains can operate over the same physical ring (Figure 5.13) [5]. For example, while one EPSR domain operates with normal traffic flow through the westbound interface, a second EPSR domain could operate through the eastbound interface. This feature further enhances the spatial reuse capability of EPSR that allows the operator to accurately control the flow of traffic to maximize the available bandwidth of high-speed links in both directions.

The Ethernet rings could support a multi-ring/ladder network that consists of conjoined Ethernet rings (Figure 5.14).

5.4 Ring Protection

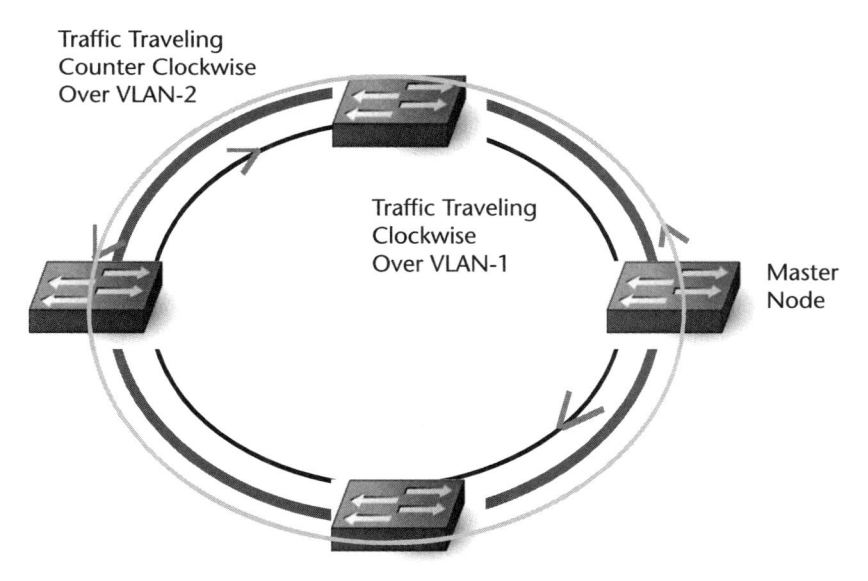

Figure 5.13 Multiple ESR domains.

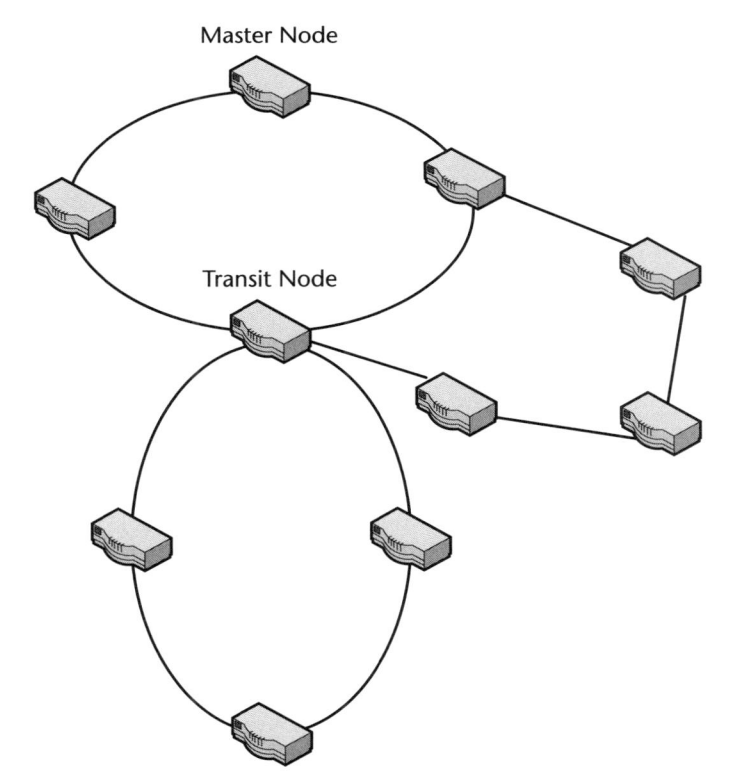

Figure 5.14 Conjoined Ethernet rings.

5.4.1 Protection Switching

The ring can enter into ring-fault operation by either the master node failing to receive two consecutive health-check messages on its secondary port or a transit node sending an EPSR LINK-DOWN control message to the master node. When the ring enters into such a state, the master node unblocks its secondary port and flushes its forwarding database (FDB). It also sends a RING-DOWNFLUSH-FDB control message to all transit nodes instructing them to flush their FDBs as well. Then, by normal bridge learning, all paths and communication are restored (Figure 5.15).

When the master node either starts to receive its health check messages on its secondary port, or when the failed transit node sends a LINK_UP message, the master node will then restore the ring to its original topology. The master node accomplishes this by reblocking its secondary port, flushing its FDB, and sending a control message to all transit nodes instructing them to flush their FDBs. The transit node whose port has just come back up is responsible for preventing a network loop while the master's secondary port is still in operation. It accomplishes this by blocking the protected VLANs on the restored port and placing them in a preforwarding state. When the ring restoration control message arrives from the master node, the transit node then flushes its FDB and unblocks the preforwarding port.

When the failure (i.e., defect condition such as signal failure) condition is detected, the nodes adjacent to failed ring link initiates protection switching mechanism.

In a revertive operation, after the clearing of an SF condition on a ring link with a defect, the position of the blocked port of the ring link is maintained until the WTR timer expires. An RPL owner will initiate reversion when the WTR timer expires, prior to any other higher priority event or command.

In an Ethernet ring, without congestion, with less than 1,200 km of ring fiber circumference, and fewer than 16 nodes, the switching time for a failure on a ring link is expected to be less than 50 ms.

Protection switching is performed at all ring nodes as defined in G.8032 [5]. An APS protocol coordinates ring protection actions around the ring.

Protection switching is triggered when:

- SF is declared on one of the ring links, and the detected SF condition has a higher priority than any other local request or far-end request.
- The received APS protocol requests to switch and it has a higher priority than any other local request.

However, manual switching is not defined in [5].

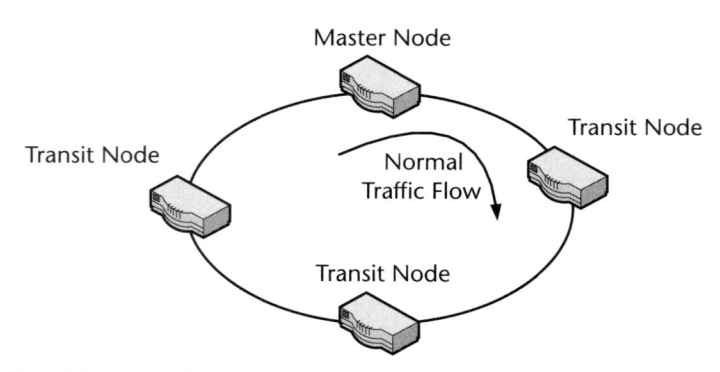

Figure 5.15 Ring failure operation.

5.4 Ring Protection

In revertive operation, the traffic channel is restored to the working transport entity, that is, blocked on the RPL, after the condition causing a switch has cleared. In the case of clearing of a defect, the traffic channel reverts after the expiry of a WTR timer, which is used to avoid toggling protection states in the case of intermittent defects.

In nonrevertive operation, the traffic channel is allowed to use the RPL, if it is not failed, after a switch condition has cleared. Because in Ethernet ring protection the working transport entity resources may be more optimized, in some cases it is desirable to revert to the normal path once all ring links are available. This is performed at the expense of an additional traffic interruption. In some cases, there may be no advantage to revert to the normal working transport entities immediately. In this case, a second traffic interruption is avoided by not reverting the protection switching.

Figure 5.16 illustrates a situation where a protection switch has occurred due to a signal-fail condition on one ring link. In this case traffic channel is blocked bidirectionally on the ports where the failure is detected and bidirectionally unblocked at the RPL connection point.

In a revertive operation, when the failure is recovered from, the traffic channel will resume use of the recovered ring link only after the traffic channel has been blocked on the RPL.

The ring links of each node may be monitored by individually exchanging continuity check messages (CCM) defined in [4] on the maintenance entity group end points (MEP) shown in Figure 5.17.

In Figure 5.17, MEPs on each ring port are used for monitoring the ring link.

If an MEP detects a defect, which contributes to an SF defect condition, this will inform the ERP control process that a failure condition has been detected.

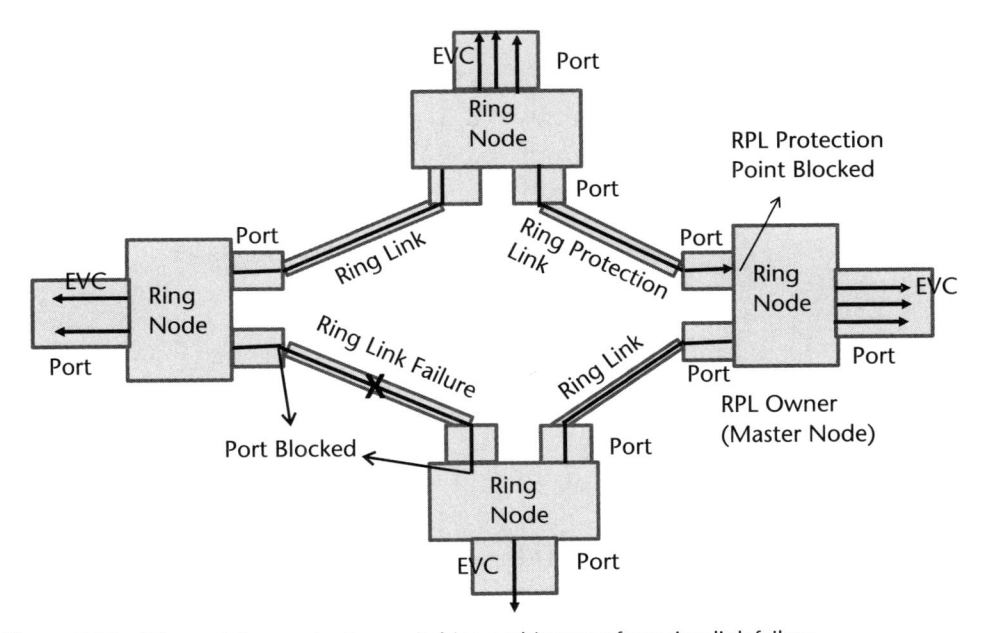

Figure 5.16 Ethernet ring protection switching architecture for a ring link failure.

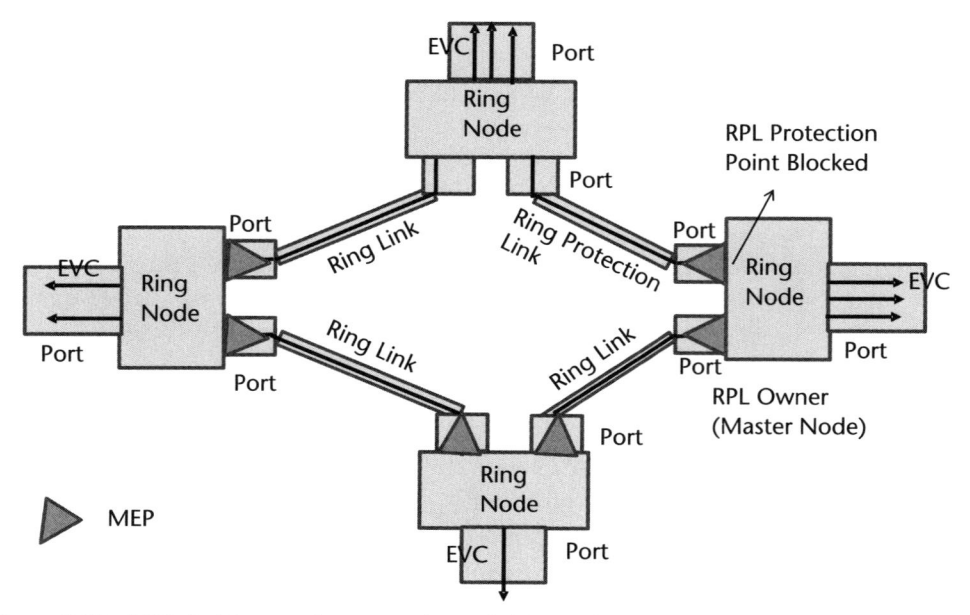

Figure 5.17 MEPs in Ethernet ring protection switching architecture.

An ERP control function uses the ETH_CI_SSF information, forwarded from the ETHx/ETH-m_A_Sk, to assert the signal-fail condition of the ring link.

The ring protection mechanism requires the APS protocol to coordinate the switching behavior among all ring nodes. The ring APS protocol communication is performed using R-APS PDUs. R-APS PDUs are transmitted and received at an ERP control process.

Blocking traffic from one or more VLANs is achieved by VLAN ID filtering on a port. This results in blocking the transmission and reception of traffic on one ring port.

R-APS channel VLAN traffic forwarding is always blocked at the same ports where traffic channel is blocked. This only prevents R-APS messages received at one ring port from being forwarded to the other ring port. However, this does not prevent R-APS messages, locally generated at the ERP control process, from being transmitted over both ring ports, and also allows R-APS messages received at each port to be delivered to the ERP control process.

A filtering database flush consists of removing the learned MAC addresses of the ring ports from the node's filtering database.

Ring protection is based on loop avoidance. This is achieved by guaranteeing that at any time traffic may flow on all but one of the ring links.

5.5 Link Aggregation

IEEE 802.3ad-based link aggregation [6, 7] provides a method for aggregating two or more parallel physical links together to form LAGs such that a MAC client can treat the LAG as if it were a single link. The link aggregation is able to increase

5.5 Link Aggregation

the capacity and availability of the communications channel between devices using existing fast Ethernet and gigabit Ethernet technology. Two or more gigabit Ethernet connections are combined in order to increase the bandwidth capability and to create resilient and redundant links. As a result, LAG is a capability that has been adopted and leveraged in Carrier Ethernet networks. The LAG can be used at the external network-to-network interface (ENNI) [8] when there are multiple links between two Carrier Ethernet networks (CENs) (Figure 5.18). Similarly, there may be multiple links between a CEN device at UNI-N and a CE device at UNI-C (Figure 5.18) [9, 10]. These multiple links are used to enhance the resiliency of the EVC service that is received by the end customer at UNI-C. Obviously, options of delivering resiliency for the service depend on the number of links. If only one, then the resiliency service attribute [9, 10] is recorded as "none." If the number of links is two, then these links can be "2-Link Active-standby," "All-Active," or "Other" and the resiliency service attribute indicates one of these options. If there are three or more links, then the resiliency service attribute has either "All-Active" or "Other" assigned to it. When the UNI resiliency service attribute is "2-Link Active-Standby," only one link functions in the active mode and the other functions in the standby mode. In case of a failure over the active link, the standby link takes over. When the UNI resiliency service attribute is set to "All-Active," then all links may be used to deliver the services. The service frames can be distributed among the links based on the service being delivered. Each service can be given a priority list of the links to be used in case the most prior link does fail. The list may include all links in a sequential order and the frames are dropped only after all links in the list have failed. Obviously, the longer the list, more resilient the service becomes. To assign the priority list, each service is first given a conversation ID based on C-tags; then conversation IDs are mapped to a specific list of links (represented by the link IDs) where the sequence represents the priority. The link IDs have meanings only locally where the LAG is defined. When "All-Active" is in practice, the mapping of the link priority list to the services being delivered is negotiated by the service provider and the subscriber. The collective setup of these mappings for services

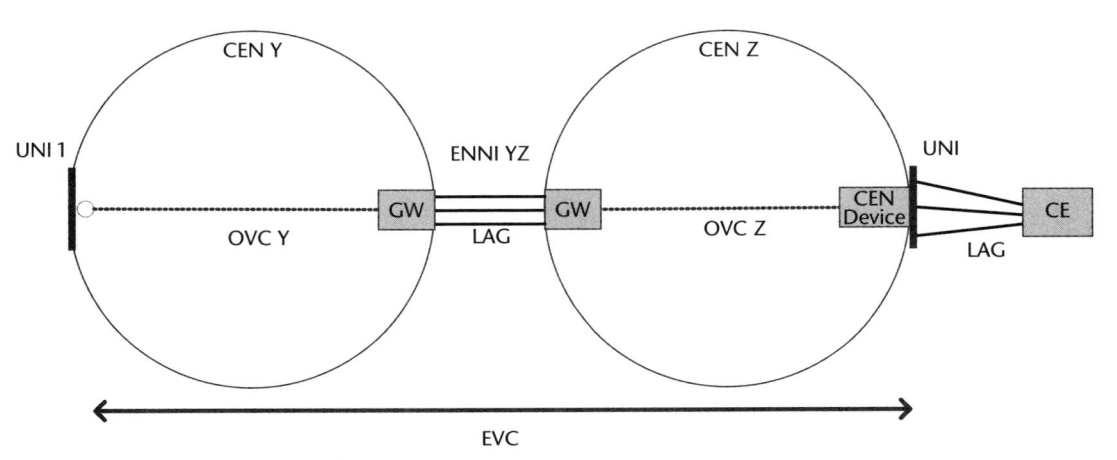

Figure 5.18 ENNI and UNI with LAG.

being delivered presents an opportunity for load balancing the multiple links and improved resiliency of the service considering failures.

Link aggregation also provides load balancing where the processing and communications activity is distributed across several links in a trunk so that no single link is overwhelmed.

The link aggregation is an optional sublayer between a MAC client and the MAC as depicted in Figure 5.19. It is dynamic and provides more functionality through the Link Aggregation Control Protocol (LACP). LACP dynamically detects whether links can be aggregated into a link aggregation group and does so when links become available. IEEE 802.3ad was designed for point-to-point link aggregation only.

Once a port is a member of a LAG, it will always remain a LAG member even if there is just a single link active in the LAG. This has the benefit of improving resiliency.

LACP supports the automatic creation of LAGs by exchanging LACP packets between LAN ports. LACP frames (i.e., LACPDUs) are sent down all links that have the protocol enabled. If it finds a device on the other end of the link that also has LACP enabled, it will also independently send frames along the same links enabling the two units to detect multiple links between themselves and then combine them into a single logical link. LACP packets are exchanged only between ports in passive and active modes. In active mode, it will always send frames along the configured links. In passive mode, however, the port participates in the control if its partner is active.

In addition to automatic link aggregation, LACP maintains the LAG and therefore detects link layer failures. LACP packets are exchanged end-to-end; thus, if a link in the core were to fail and the local port(s) do not register the failure, LACP will time out and remove the port from the LAG. The default LACP settings, with the long timers, will remove the port from the LAG in 90 seconds. If short timers were used, the port can be removed in 3 seconds.

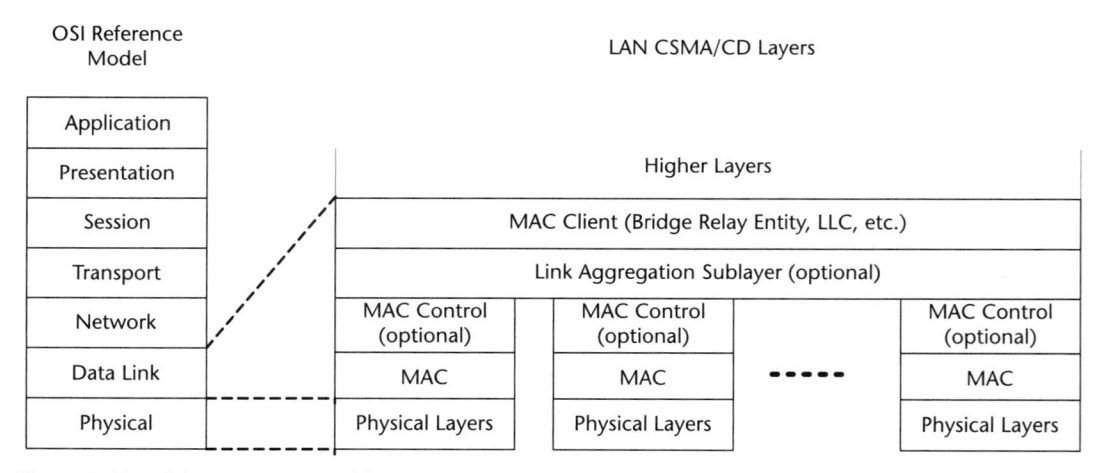

Figure 5.19 Link aggregation sublayer.

5.5.1 LAG Objectives

The Standard 802.3ad [6] lists the following main goals for LAG:

- Increased bandwidth by combining multiple links into one logical link;
- Increased availability by not causing failure from the perspective of MAC client when a single link within LAG fails or is replaced;
- Linearly incremental bandwidth by increasing bandwidth in unit multiples (i.e., depending on physical layer options such as 100 Mbps or 1 GbE) as opposed to the order of magnitude;
- Load sharing by distributing MAC client traffic across multiple links;
- Automatic configuration of LAGs and allocating individual links to those groups;
- Rapid configuration and reconfiguration in the event of changes in physical connectivity; link aggregation quickly converges to a new configuration, typically on the order of 1 second or less;
- Deterministic behavior;
- Low risk of duplication or misordering of frames;
- Support of existing IEEE 802.3 MAC clients by not changing existing higher-layer protocols or application to use link aggregation;
- Backwards compatibility with aggregation-unaware devices;
- Accommodation of differing capabilities and constraints by accommodating devices with differing hardware and software constraints on link aggregation;
- No change to the IEEE 802.3 frame format;
- Network management support: the standard specifies appropriate management objects for configuration, monitoring, and control of link aggregation.

Link aggregation does not support multipoint aggregations, dissimilar MACs, half-duplex operation, and operation across multiple data rates. Dissimilar MACs such as gigabit Ethernet and FDDI are not supported. However, dissimilar PHYs such as copper and fiber are supported. All ports in a link aggregation group must be operating in full-duplex mode. Also all links in an LAG must operate at the same data rate.

5.5.2 Link Aggregation Operation

Link aggregation function consists of frame distribution, frame collection, aggregator parser/multiplexer, aggregation control, and control parser/multiplexer (Figure 5.20).

The frame distribution function submits frames from MAC client to the appropriate port, based on a frame distribution algorithm. In reverse, frame collection block passes frames received from the various ports to the MAC client. The combination of distribution and collection is called the aggregator. The frame distribution may include an explicit marker protocol with marker generator/receiver that

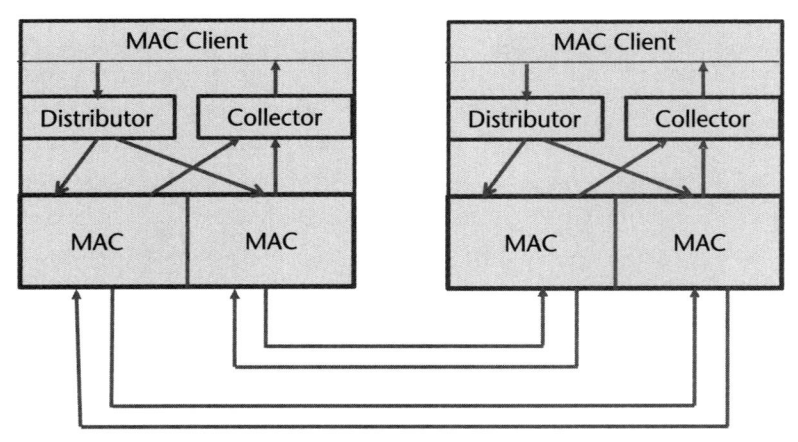

Figure 5.20 Link aggregation functional diagram.

searches for a marker identifying the last frame of a conversation. On transmit, the aggregator multiplexer passes frame transmission requests from the distributor and marker generator/responder to the appropriate port. On receive, the aggregator parser distinguishes marker request, marker response, and MAC client PDUs and passes them to appropriate marker responder, marker receiver, and collector, respectively. The aggregation control incorporates LACP. On transmit, the control parser/multiplexer passes frame transmission requests from the aggregator and control entities to the appropriate port. On the receive, the control parser/multiplexer distinguishes LACPDUs from other frames and passes PACPDUs and other frames to appropriate entities and ports.

Each network interface controller is assigned a unique MAC address. In most cases, this address is used as source and destination address during the transmission of packets. Aggregated links appear as a single logical network interface with one virtual MAC address. The MAC address of one of the interfaces belonging to the aggregated link provides the virtual address of the logical link. All ports in a link aggregation group must be in the same VLAN or VLANs. Ports in a LAG can be distributed over different modules.

LACP-enabled ports with the same key must have the same VLAN membership. On LACP-disabled ports with the same key, VLAN membership can be different. This usually happens when VLANs are added or deleted from these ports, but before LACP is reenabled on these ports, VLAN membership must be the same for ports with the same key.

In order to change the VLANs membership on an LAG or the ports membership within the VLANs, LACP needs to be disabled on the ports. Once the changes are completed, LACP can be enabled again on all appropriate port members.

In aggregated links, the link on which to transmit a given frame must be selected. Sending one long frame may take longer than sending several short ones; therefore, the short frames may be received earlier than one long frame. The order has to be restored at the receiver side. Thus, an agreement has been made: all frames belonging to one conversation (or EVC) must be transmitted through the same physical link. This guarantees correct ordering at the receiving end station. For this reason, no sequencing information may be added to the frames.

Traffic belonging to separate EVCs can be sent through various links in a random order. An EVC may need to be transferred to another link because the originally mapped link is out of service (failed or configured out of the aggregation) or a new link has become available, relieving the existing ones. This can be realized either by means of a delay time that the distributor must somehow determine or through an explicit marker protocol that searches for a marker identifying the last frame of a conversation. The distributor inserts a marker message behind the last frame of a conversation. After the collector receives this marker message, it sends a response to the distributor, which then knows that all frames of the conversation have been delivered. After that, the distributor can send frames of these types of conversations via a new link without delay.

5.5.3 LACP

LACP uses the following parameters:

- LACP priority, which is configured at the system level and at the port level:
  - Port priority, which can be configured automatically or through the CLI. LACP uses the port priority to decide which ports should be put in standby mode when there is a hardware limitation that prevents all compatible ports from aggregating. LACP also uses the port priority with the port number to form the port identifier.
  - System priority, which is configured on each device running LACP. The system priority can be configured automatically or through the CLI. LACP uses the system priority with the device MAC address to form the system ID and also during negotiation with other systems.
- LACP keys, which are used to determine which ports are eligible to be aggregated into a LAG where keys do not need to match between two LACP peers. LACP automatically configures an administrative key value on each port configured to use LACP. A port's ability to aggregate with other ports is determined by LACP timers, which determine the failover times. The default timer settings are: timeout: 3; fast-periodic-time: 1,000 (ms); and slow-periodic-time: 30,000 (ms).

The user can choose to use either the fast or slow timer, which is set on the port level. By default, the long timer can be used. Hence, a link is determined ineligible to be aggregated if it does not receive an LACPDU for a period of: timeout × slow-periodic-time = 3 × 30 seconds = 90 seconds.

Link aggregation is compatible with the Spanning Tree Protocol (STP/RSTP/MSTP). LAGs must be in the same STP group(s). The operation of the LACP is only affected by the physical link state or its LACP peer status. When a link goes up and down, the LACP is notified. The STP forwarding state does not affect the operation of LACP module. LACPDU can be sent even if the port is in STP blocking state.

5.5.4 Limitations

In 802.3ad, all physical ports in the link aggregation group must reside on the same logical device, which in most scenarios will leave a single point of failure when the physical switch to which both links are connected goes offline.

In most implementations, all the ports used in an aggregation consist of the same physical type, such as all copper ports (10/100/1000BASE-T), all multimode fiber ports, or all single-mode fiber ports. However, all the IEEE standard requires is that each link be full duplex and all of them have an identical speed.

Some equipment support static configuration of link aggregation. Link aggregation between similarly statically configured equipment will work, but will fail between a statically configured equipment and an equipment configured for LACP.

5.6 Conclusion

Protection switching is a key capability for the high availability of Carrier Ethernet services. In this chapter, we have described APS, ring protection, and LAG methods. The 50-ms protection switching time mentioned here has been challenged and 500 ms is recommended. Currently, LAG is the mostly deployed protection scheme, especially at ENNI and access. As applications requiring high availability transported over Carrier Ethernet services, more APS and ring protection deployments can be expected.

References

[1] MEF 32, Requirements for Service Protection Across External Interfaces, July 2011.
[2] RFC 3619, Extreme Networks' Ethernet Automatic Protection Switching (EAPS) Version 1, October 2003.
[3] ITU-T G.8031, Ethernet Linear Protection Switching, 2006.
[4] ITU-T Y.1731, OAM Functions and Mechanisms for Ethernet Based Networks, 2008.
[5] ITU-T G.8032, Ethernet Ring Protection Switching, 2012.
[6] IEEE Std 802.3ad-2000, Amendment to Carrier Sense Multiple Access with Collision Detection (CSMA/CD) Access Method and Physical Layer Specifications-Aggregation of Multiple Link Segments.
[7] IEEE Std. 802.1AX-2014, "IEEE Standard for Local and Metropolitan Area Networks – Link Aggregation," December 2014.
[8] MEF 26, External Network-Network Interface (ENNI), Phase 2, January 2012.
[9] MEF 10.3, Ethernet Services Attributes Phase 3, October 2013.
[10] MEEF 10.3.2, Amendment to MEF 10.3 – UNI Resiliency Enhancement, October 2015.

CHAPTER 6

Carrier Ethernet Architecture and Services

6.1 Introduction

Ethernet has been a dominant technology in enterprise networks, allowing bandwidth partitioning, user segregation, and traffic prioritization, introduced by IEEE 802.1P and 1Q. In 1998, IEEE 802.1d combined VLAN and prioritization capabilities and introduced another service provider tag to scale the addressing structure. These initial capabilities formed the base for Carrier Ethernet.

In 2001, the Metro Ethernet Forum was formed to define Ethernet-based carrier services. MEF defined Carrier Ethernet for business users as "a ubiquitous, standardized, carrier-class Service and Network defined by five attributes (Standardized Services, QoS, Service Management, Reliability, Scalability) that distinguish it from familiar LAN based Ethernet" [1]. Similarly, it is defined for service providers as: "A set of certified network elements that connect to transport Carrier Ethernet services for all users, locally & worldwide; Carrier Ethernet services that are carried over physical Ethernet networks and other legacy transport technologies."

The Ethernet has been used as a LAN connectivity technology in enterprise. With the introduction of Carrier Ethernet, the Ethernet is being used as a wide area service. It is an easy to use, widely available, and well-understood technology. The Ethernet simplifies network operations [operations, administration, maintenance, and provisioning] and is cost-effective due to widespread usage of the Ethernet interface. Bandwidth can be added in increments. A single user interface can connect to multiple services such as Internet, VPN, voice, and video services.

The Ethernet was developed to connect computers on corporate networks. Since its inception, it has become the predominant technology used in corporate networks. Now carriers, which typically provide phone and Internet access services, are putting it to use in their metropolitan area networks. As a result, Ethernet has been replacing wide area technologies such as frame relay, ATM, and dedicated leased line.

With the Ethernet, the same protocol is being used for LAN and MAN. Therefore, there is no protocol conversion between LAN and WAN. People trained to provision and maintain WAN can be used to provision and maintain LAN. This results in lower equipment and operational cost. Due to these benefits, Carrier Ethernet services and technologies claim about a market share of approximately $80 billion [2]. Among many others, there are some important features that resonates more with the end customers [3]. These include extensive service coverage within and/or between metro areas, availability of end-to-end SLAs with performance visibility, variety of available bandwidth, and fast service turn-up.

For Carrier Ethernet to be successful, it has to display the same properties of current WAN technologies. The networks must scale to support the hundred thousands of customers to adequately address metropolitan and regional distances. The network availability needs to be 99.999%, with a range of protection mechanisms capable of sub-50-ms recovery. Service management systems must be available to provision new services, manage service-level agreements (SLAs) and troubleshoot the network under fault conditions. Standardized services must be offered to support existing applications, and allow service providers to extend their geographic reach through interoperability agreements with competitors. QoS and traffic engineering schemes are needed to prioritize traffic streams and ensure that service-level parameters are agreed.

The Ethernet is scalable from 10 Mbps to 100 Gbps with finer granularity as low as 56 Kbps and 1 Mbps. Carrier Ethernet allows incremental bandwidth assignment. Most carriers provide services that start at 10 Mbps and can scale up to 100 Gbps. These are packet-based services far faster and more powerful than traditional frame relay and ATM services. With enhanced OAM, end-to-end monitoring and trouble isolation are possible. Link protection, equipment, and ring protection mechanisms [4] have been developed to provide necessary reliability for Ethernet networks.

With QoS and synchronization [5, 6] capabilities, applications with strict performance requirements can be supported.

Carrier requirements for access include fault detection and isolation, performance monitoring and statistics, and protection failover. A core network Ethernet mesh located in strategic point of presence (PoPs) will provide for the redundancy and performance required. The service is not required to be a fully protected service except at the core.

The Ethernet becomes connection-oriented by disabling unpredictable functions such as MAC learning, Spanning Tree Protocol, and broadcast storms. Furthermore, this performance-grade Ethernet enhances the scalability and minimizes the cost and complexity of aggregation and transport. Connection-oriented Ethernet, combined with a transport-style intelligent control plane and service-level management, simplifies the operations of the network and allows the Ethernet to be managed as easily as circuits, using the same procedures and systems. There are multiple organizations that work on the standardization and best practices of Carrier Ethernet technologies and services. Among those organizations are ITU, IEEE, and MEF. MEF has been developing specifications for Carrier Ethernet architecture, service, management, test, and measurement. Management specs cover OAM requirements, models and definitions. Architecture specs define an architectural reference model and set of common linguistic tools for the technical teams. The architecture is layered. Each layer has been decomposed into the adaptation, connection, and termination elements. Service specs define Carrier Ethernet services models, definitions, and service parameters and attributes. Test and measurement define test methodologies and test suites that enable conformance to MEF services as defined in the MEF specifications.

In this chapter, we review Carrier Ethernet fundamentals in terms of architectures and services that can be used in service deployments on single Carrier Ethernet network as well as multiple operators' Carrier Ethernet networks.

6.2 Architecture

The Ethernet transport network is a two-layer network [7] consisting of Ethernet MAC (ETH) layer network and Ethernet PHY (ETY) layer network. The ETY layer is the physical layer defined in IEEE 802.3. This layer can be called as the section layer as well. The ETH layer is the pure packet layer that is also called the path layer.

Ethernet transport network consists of a set of access groups of the same type where the access groups demarcate the point of access into the ETH layer network. The information transferred between the access groups is the characteristic information. The association of two or more access points creates a connectionless transport entity called as the Ethernet connectionless trail (ECT).

The ETH layer network depicted in Figure 6.1 is divided into ETH subnetworks that are also called ETH flow domain (EFD). An EFD is a set of all ETH (termination) flow points transferring information within a given administrative portion of the ETH layer network. EFDs may be partitioned into sets of nonoverlapping EFDs recursively that are interconnected by ETH links. An IEEE 802.1D bridge represents the smallest instance of an EFD.

A link is a topological component that describes a fixed topological relationship between subnetworks, along with the capacity supported by an underlying server layer network domain (LND) trail.

The termination of a link is called a flow point pool (FPP). The FPP describes configuration information associated with an interface, such as a UNI or ENNI. The FPP is associated with the trail termination of the underlying server trail used to perform adaptation and transport of the characteristic information of the client LND.

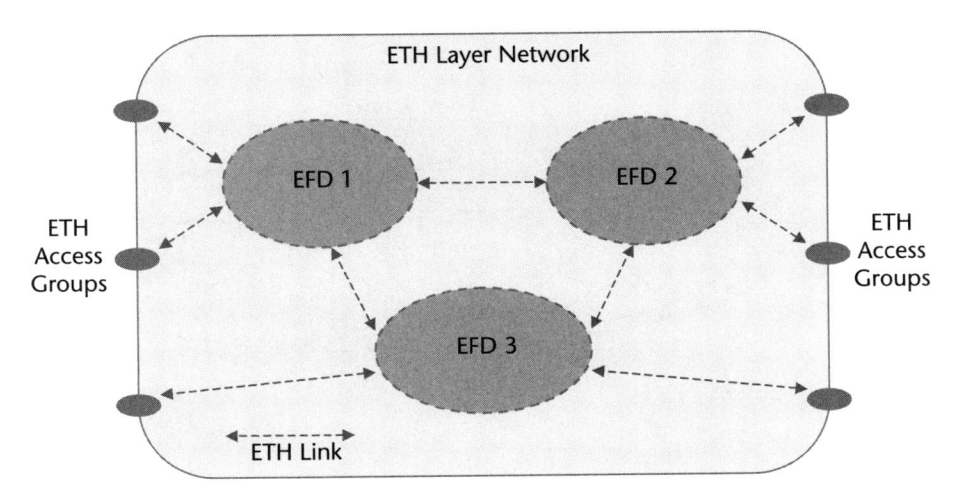

Figure 6.1 ETH layer topological components.

A subset of flow points can form a FPP link. The FPP link represents available capacity between a pair of flow domains, or a flow domain and an access group, or a pair of access groups. EFD may be partitioned into smaller flow domains interconnected by FPP links (Figure 6.2).

A group of collocated flow termination functions that are connected to the same flow domain or flow point pool link is called an access group.

The ETH layer is responsible for the instantiation of Ethernet MAC oriented connectivity services and the delivery of Ethernet PDUs presented across internal and external interfaces. All service-aware aspects associated with Ethernet MAC flows, including operations, administration, maintenance, and provisioning capabilities required to support Ethernet connectivity services are functions of the ETH layer.

An LND represents an administration's view of the layer network responsible for transporting a specific type of characteristic information such as IP, MPLS, and SONET. Layer networks may use transport resources in other layer networks. In Figure 6.3, the ETH LND uses the resources of the MPLS LND.

6.2.1 Protocol Stack

The protocol stack for a Carrier Ethernet Network with an ETH layer can be represented as in Figure 6.4. Data plane, control plane, and management plane cross all these layers. Packets of application layer are represented in the form of Ethernet service layer PDUs at ETH service. In turn, Ethernet service PDUs are converted into transport layer frames or packets, depending on whether transport protocol is Ethernet, or SONET/SDH or MPLS, and so forth.

The data plane defines Ethernet frames, tagging, and traffic management. The control plane defines signaling and control. The management plane defines provisioning, device discovery, protection, and OAM.

Figure 6.5 depicts the Ethernet services layer in a network encompassing multiple service providers.

The ETH layer network provides the transport of adapted information through an ETH connectionless trail between ETH access points [8]. An example of the ETH layer network and transport components are shown in Figure 6.6.

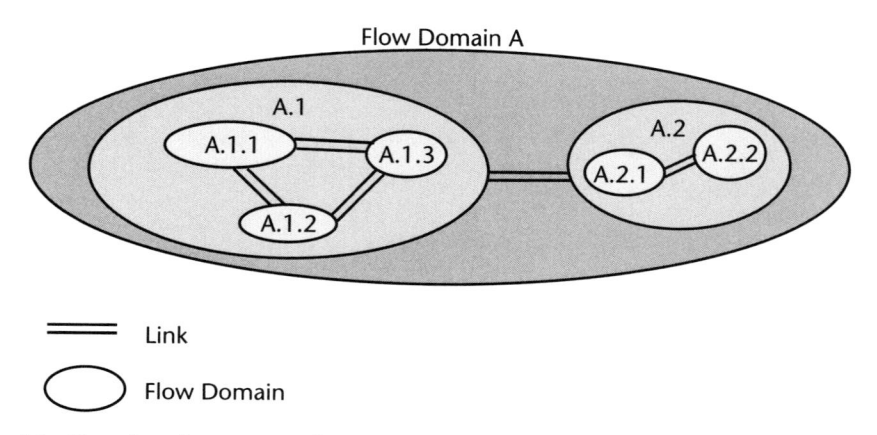

Figure 6.2 Flow domains connected via flow point pool links.

6.2 Architecture

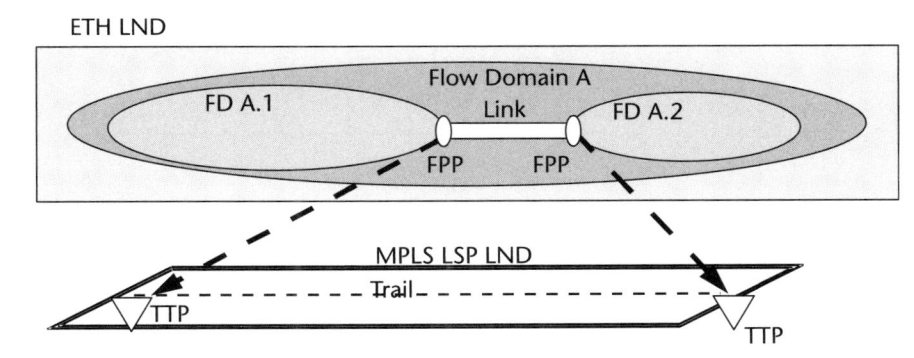

Figure 6.3 FPP, link, and trail.

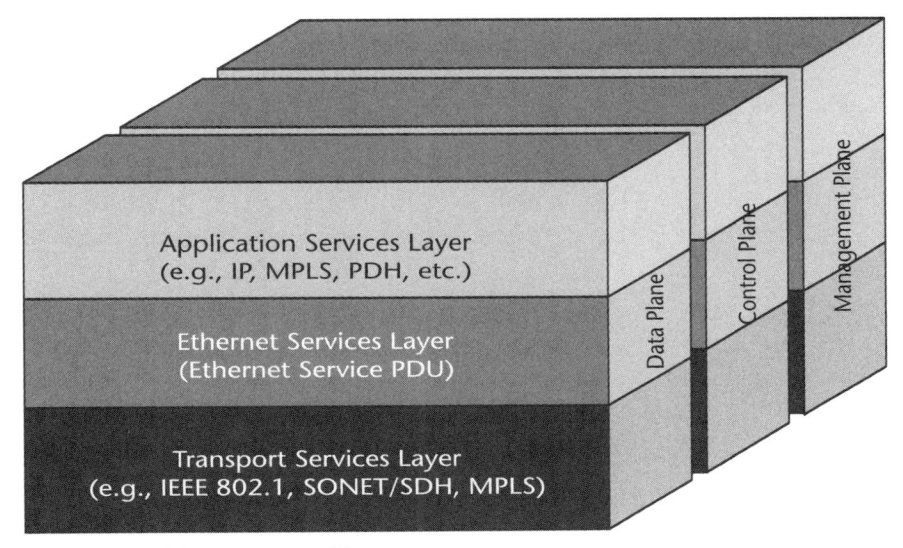

Figure 6.4 Carrier Ethernet protocol layers.

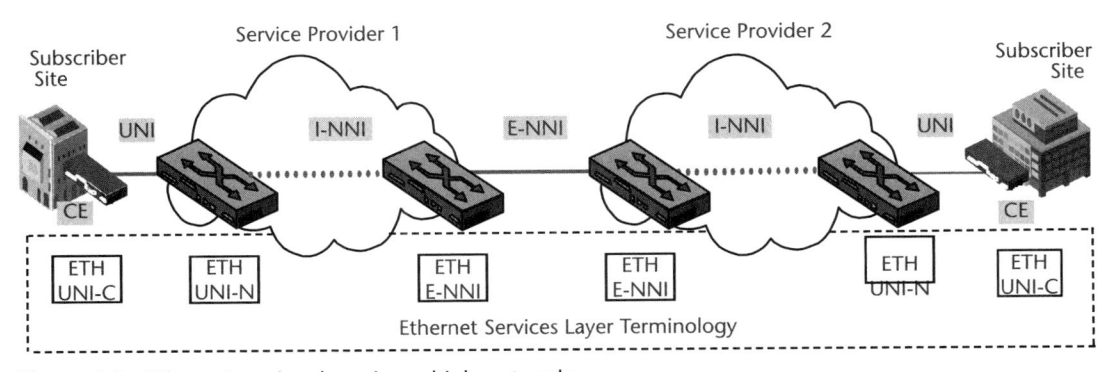

Figure 6.5 Ethernet services layer in multiple networks.

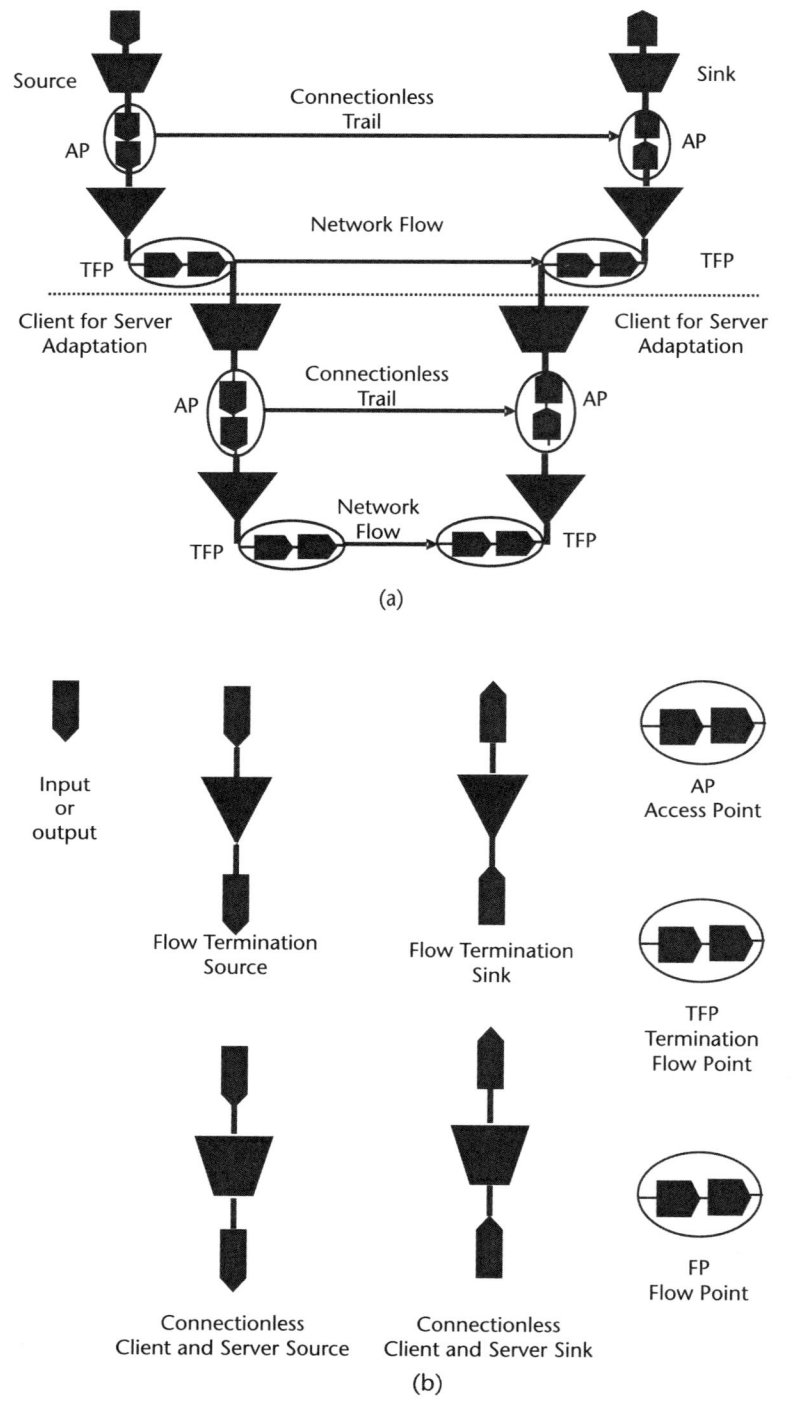

Figure 6.6 (a) Client-server relationship between a connectionless client layer network and connectionless server layer network. (b) Conventions.

The transport entities provide transparent information transfer between layer network reference points. Two basic monitored entities of transport are flows and trails. Flows may be decomposed into network flows, flow domain flows, and link flows.

A flow is an aggregation of one or more traffic units with an element of common routing. It is unidirectional and can contain another flow. Flows can be multiplexed together in the same layer network.

A link flow represents the fixed relationship between the ends of the link. It is delimited by flow points and transfers information transparently across an FPP link. A link flow represents a pair of adaptation functions and a trail in the server layer network.

A grouping of traffic units that are transferred transparently across a flow domain is called a flow domain flow. The ports associated with flow points at the boundary of the flow domain delimit it. Flow domain flows are constructed from a concatenation of flow domain flows and link flows.

A network flow is a grouping of traffic units that are transferred transparently across a layer network. It is delimited by the termination flow points (TFPs). In general, network flows are constructed from a concatenation of flow domain flows and link flows. The TFP is formed by binding the port of a flow termination to either a flow domain port or a port on an FPP link.

A connectionless trail represents the transfer of monitored adapted characteristic information of the client layer network between access points. It is delimited by two access points, one at each end of the connectionless trail. It represents the association between a source and destination on a per traffic unit or datagram basis.

Adaptation and flow termination functions are used in describing the architecture of connectionless layer networks. Adaptation source adapts the client layer network characteristic information into a form suitable for transport over a trail in a connection-oriented server layer network or connectionless trail in a connectionless server layer network in the server layer network. The adaptation source function input to output relation is a many-to-one relationship or a one-to-many.

Adaptation sink converts the server layer network trail in a connection-oriented server layer network or connectionless trail information in a connectionless server layer network information into the characteristic information of the client layer network. Labeling, scheduling, buffering, queuing, multiplexing, traffic dropping, segmentation, and reassembly are examples of an adaptation function.

Flow termination is accomplished by flow termination source and flow termination sink. Flow termination source accepts adapted characteristic information from a client layer network at its input, adds information to allow the connectionless trail to be monitored, and presents the characteristic information of the layer network at its output(s). A flow termination function adds or extracts information to monitor a connectionless trail as well as it provides no overhead to monitor the connectionless trail. Flow termination sink accepts the characteristic information of the layer network at its input, removes the information related to connectionless trail monitoring, and presents the remaining information at its output.

Using these transport components, it is possible to completely describe the logical topology of a connectionless layer network. An example of the ETH layer network containing the transport processing functions, transport entities, topological components, and reference points are shown in Figure 6.7.

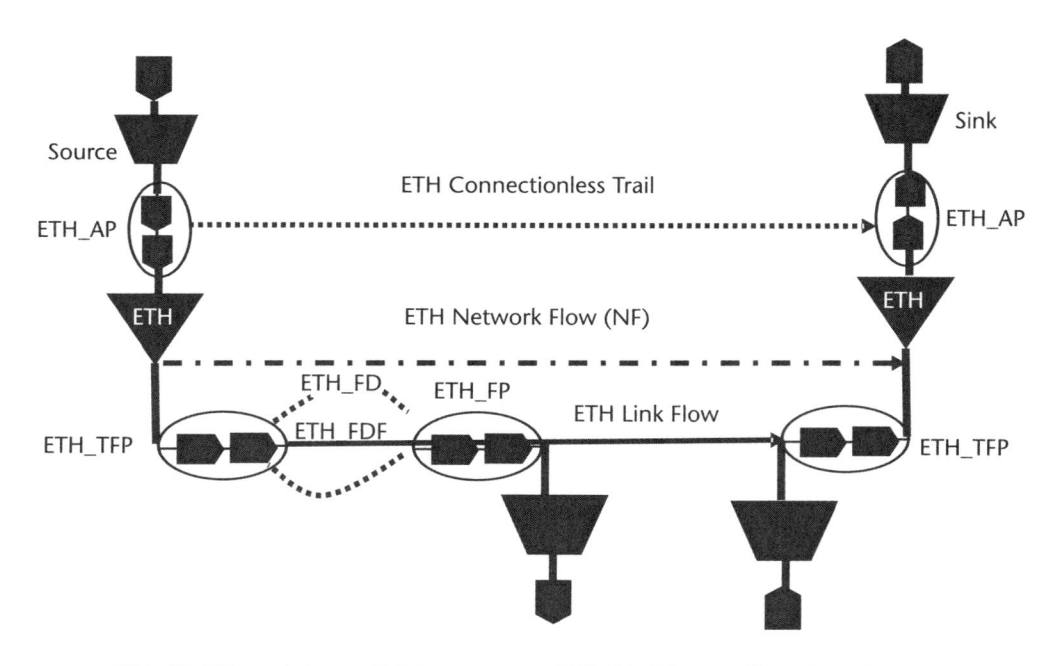

Figure 6.7 Ethernet layer network for unicast flow.

ETH_AP: Ethernet Access Point
ETH_FDF: Ethernet Flow Domain Flow
ETH_FD: Ethernet Flow Domain
ETH_TFP: Ethernet Termination Flow Point

6.2.2 ETH Layer Characteristic Information (Ethernet Frame)

The characteristic information, which is the Ethernet Frame, is exchanged over ETH layer [9]. It consists of a preamble, start frame delimiter (SFD), destination MAC address (DA), source MAC address (SA), (optional) 802.1QTag, Ethernet length/type (EtherType), user data, padding if required, frame check sequence (FCS), and extension field, which is required only for a 100-Mbps half-duplex operation (Figure 6.8).

Seven bytes of preamble allow the physical layer to reach its steady-state synchronization with the received frame's timing.

The length of padding for n-byte-long user data is max [0, minFrameSize − (8 × n + 2 × addressSize + 48)] bits [9].

The maximum possible size of the data field is maxUntaggedFrameSize − (2 addressSize + 48)/8 octets, where minFrameSize is 64 octets [9].

A cyclic redundancy check (CRC) is used by the transmit and receive algorithms to generate a CRC value for the FCS field. The FCS field contains a 4-octet (32-bit) CRC value. This value is computed as a function of the contents of the source address, destination address, length, data, and pad (that is, all fields except the preamble, SFD, FCS, and extension).

The optional 4-octet 802.1QTag is composed of a 2-octet 802.1QTagType and the tag control information (TCI). TCI contains 3 bits of CoS/priority information, the single-bit canonical format indicator (CFI) and the 12-bit VLAN identifier (VLAN ID). The VLAN ID and CoS/priority are optional information elements of the frame. Figure 6.8 illustrates the ETH_CI for common IEEE 802.3-2002/2005 compliant frame formats.

6.2 Architecture

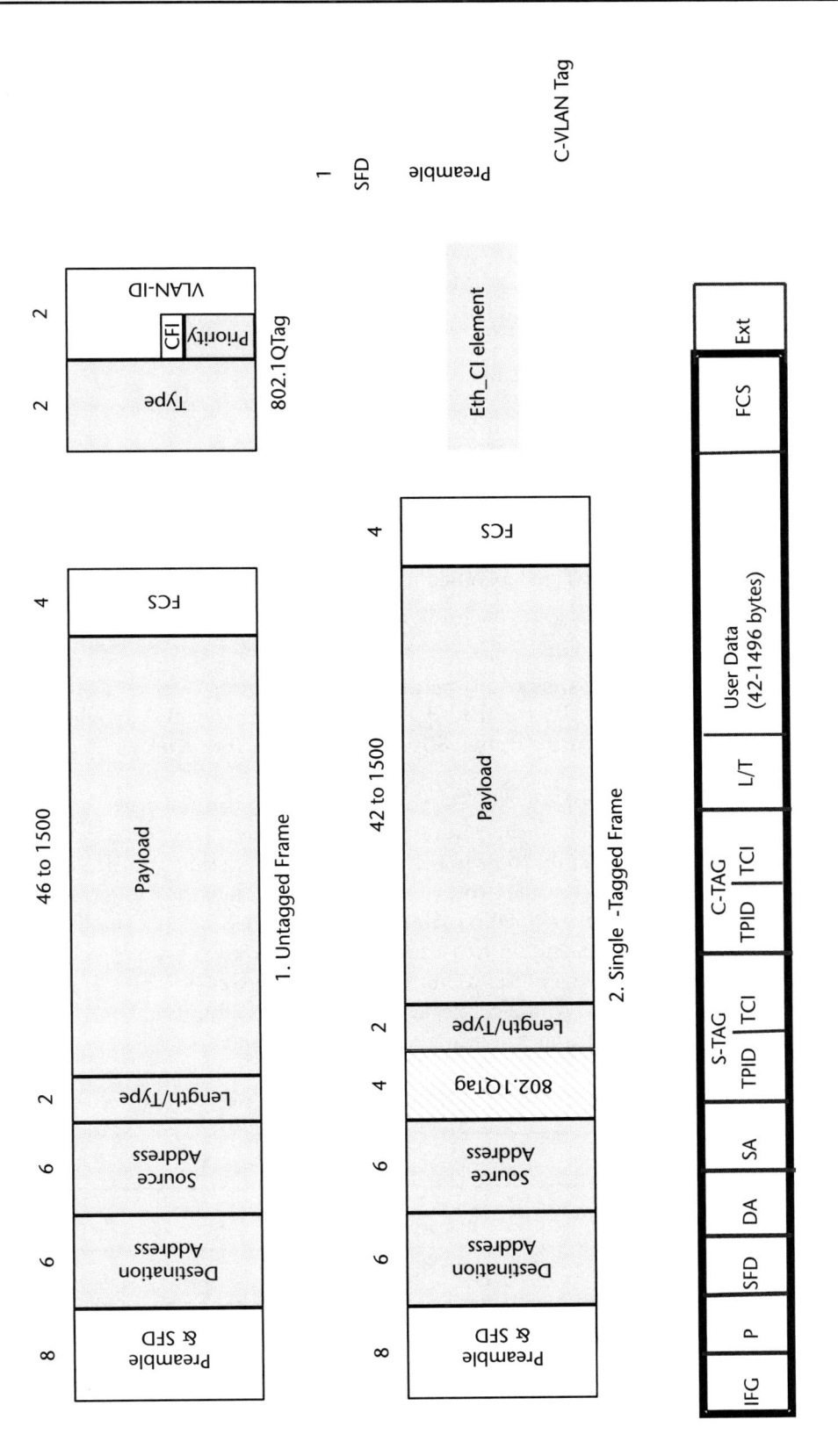

Figure 6.8 ETH layer characteristic information (ETH_CI)

CFI is always set to zero for Ethernet switches. CFI is used for compatibility reasons between Ethernet type network and token ring type network. If a frame received at an Ethernet port with a CFI is set to 1, then that frame should not be forwarded.

The user data conveys the APP layer PDU, either in a raw format or as a logical link control (LLC) encapsulated PDU where the LLC provides a link service access point (LSAP) for access to higher layers and additional LSAPs for each implemented link layer protocol such as Link Layer Discovery Protocol (LLDP) and STP.

ETH layer PDUs are the data frames used to exchange the ETH_CI across standardized ETH layer interfaces and associated reference points (i.e., UNIs and ENNIs). The Ethernet service frame is the ETH layer PDU exchanged across the UNI:

- IFG: Interframe gap, 12 bytes;
- TCI (Tag control information=VID+CFI/DE+PCP)- 2 bytes;
- S-Tag TPID: Service tag identifier, 2 bytes (0x88-A8);
- C-Tag TPID: Customer tag identifier, 2 bytes (0x81-00);
- L/T (Length/Type): Length of frame or data type, 2 bytes (L/T field of a tagged MAC frame always uses the type interpretation and contains the 802.1Q Protocol Type: a constant equal to 0x81-00);
- FCS (Frame check sequence)-4 bytes;
- Ext: Extension field;
- P/SFD (Preamble/start of frame delimiter): Alternate ones and zeros for the preamble, 11010101 for the SFD. This allows for receiver synchronization and marks the start of frame, 8 Bytes (P-7 bytes, SFD-1 byte).

With introduction of the 802.3z standard for Gigabit Ethernet in 1998, an extension field was added to the end of the Ethernet frame to ensure that it would be long enough for collisions to propagate to all stations in the network. The extension field is appended as needed to bring the minimum length of the transmission up to 512 bytes (as measured from the destination address field through the extension field). It is required only in half-duplex mode, as the collision protocol is not used in full-duplex mode. Nondata bits, referred to as extension bits, are transmitted in the extension field so the carrier is extended for the minimum required time. Figure 6.9 illustrates a frame with an extension field appended.

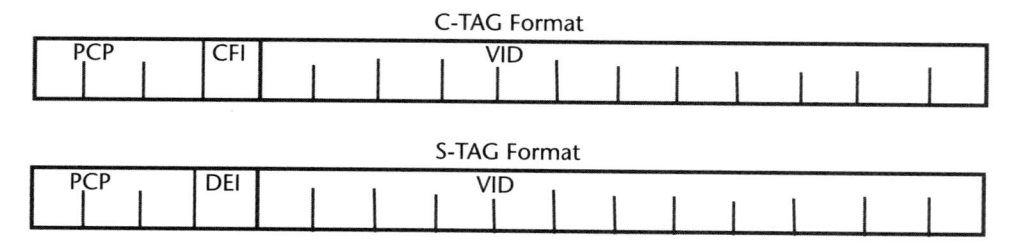

PCP: Priority Code Point, 3 bits
CFI: Canonical Format Indicator, 1 bit
VID: VLAN Identifier, 12 bits (0-4094)
DEI: Drop Eligibility Bit

Figure 6.9 C-tag and S-tag.

CoS/priority information associated with an ETH layer PDU can be conveyed explicitly in the 3-bit priority field.

The IEEE 802.1Q standard adds four additional bytes to the standard IEEE 802.3 Ethernet frame and is referred to as the VLAN tag. Two bytes are used for the tag protocol identifier (TPID) and the other 2 bytes are used for tag control identifier (TCI). The TCI field is divided into PCP, CFI, and VID.

- Tag Protocol Identifier (TPID) is a 16-bit field set to a value of 0x8100 in order to identify the frame as an IEEE 802.1Q-tagged frame. This field is located at the same position as the EtherType/length field in untagged frames and is thus used to distinguish the frame from untagged frames.
- Priority code point (PCP) of TCI is a 3-bit field that refers to the IEEE 802.1p priority. It indicates the frame priority level. Values are from 0 (best effort) to 7 (highest); 1 represents the lowest priority. These values can be used to prioritize different classes of traffic (voice, video, data).
- Canonical format indicator (CFI of TCI) is a 1-bit field. If the value of this field is 1, the MAC address is in noncanonical format. If the value is 0, the MAC address is in canonical format. It is always set to zero for Ethernet switches. CFI is used for compatibility between Ethernet and token ring networks. If a frame received at an Ethernet port has a CFI set to 1, then that frame should not be bridged to an untagged port.
- VLAN identifier (VID) of TCI is a 12-bit field specifying the VLAN to which the frame belongs. The hexadecimal values of 0x000 and 0xFFF are reserved. All other values may be used as VLAN identifiers, allowing up to 4,094 VLANs. The reserved value 0x000 indicates that the frame does not belong to any VLAN; in this case, the 802.1Q tag specifies only a priority and is referred to as a priority tag. On bridges, VLAN 1 (the default VLAN ID) is often reserved for a management VLAN; this is vendor-specific.

IEEE 802.1ad introduces a service tag, S-tag, next to SA. The S-tag is used to identify the service. The subscriber's VLAN tag (C-VLAN Tag) remains intact and is not altered by the service provider anywhere within the provider's network. Tag Protocol Identifier (TPID) of 0x88a8 is defined for S-tag.

Q-in-Q networks do not provide any separation between the provider and subscribers' MAC addresses. Therefore, provider switches must learn all MAC addresses in the network, regardless of whether they belong to the service provider or to the subscriber.

Most Ethernet control protocols [i.e., bridged protocol data units (BPDUs)] used by subscribers' networks must not interact with the provider's networking equipment. For example, Spanning Tree Protocol (STP) instances in the subscriber network must not interact with STP instances used in the provider network. In the subscriber's STP BPDUs need to be tunneled through the provider's network. BPDUs are identified by their destination MAC address.

The service frame is just a regular Ethernet frame beginning with the first bit of the destination address through the last bit of the FCS. A service frame that contains an IEEE 802.1Q tag can be up to 1,522 bytes long and a service frame that does not contain an IEEE 802.1Q tag can be up to 1,518 bytes.

A service frame can be a unicast, multicast, broadcast, and L2CP frame. Unicast service frame is a service frame that has a unicast destination MAC address. Multicast service frame is a service frame that has a multicast destination MAC address. Broadcast service frame is a service frame with the broadcast destination MAC address. Layer-2 control protocol service frame is a frame whose destination MAC address is one of the addresses listed in Table 6.1. Some Layer-2 control protocols share the same destination MAC address and are identified by additional fields such as the EtherType and a protocol identifier. Therefore, disposition of service frames carrying Layer-2 control protocols may be different for different protocols that use the same destination MAC address.

6.2.3 ETH Layer Functions

The ETH layer functions may be categorized as:

- ETH conditioning functions;
- ETH EVC adaptation function (EEAF);
- ETH EVC termination function (EETF);
- ETH connection function (ECF)
- APP to ETH adaptation function (EAF)
- ETH flow termination function (EFTF)
- ETH to TRAN adaptation function (TAF)

Service frames are Ethernet frames that are exchanged between the Carrier Ethernet network and the customer edge across the UNI. A service frame sent from the Carrier Ethernet network to the customer edge at a UNI is called an egress service frame. A service frame sent from the customer edge to the CEN at a UNI is called an ingress service frame.

The ETH conditioning functions are frame classification, filtering, metering, marking, policing, and shaping that can be applied to flows. There are three types of ETH conditioning functions:

- The ETH flow conditioning function (EFCF) is responsible for the conditioning of the subscriber flow into and out of a subscriber EFD. The ingress and egress process on flows to the CEN at the UNI-C is called EFCF.
- The ETH subscriber conditioning function (ESCF) is for the conditioning the subscriber flows toward the CEN and from the CEN. At the ingress toward the service provider EFD (CEN), per user contract with service provider (i.e.,

Table 6.1 MAC Addresses for L2 Control Protocols

MAC Addresses5	Description
01-80-C2-00-00-00 through 01-80-C2-00-00-0F	Bridge block of protocols
01-80-C2-00-00-20 through 01-80-C2-00-00-2F	GARP block of protocols
01-80-C2-00-00-10	All-bridges protocol

SLA), classification, filtering, CoS instance identification, policing, marking, and shaping are performed. At the egress, the same functions except policing are performed.

- The ETH provider conditioning function (EPCF) is responsible for conditioning flow(s) between two CENs at the ENNI.

The ETH EVC adaptation function (EEAF) is responsible for the adaptation of service frames into and out of EVCs. The EEAF source:

- Maps conditioned ingress service frames into their corresponding EVC PDUs;
- Adapts the subscriber CoS ID into service provider CoS indication per CoS instance;
- Multiplexes ingress service frames into their corresponding EVC;
- Buffers and schedules ingress service frames according to a scheduling algorithm as per CoS instance.

However, the EEAF sink demultiplexes egress service frames from various EVCs into their corresponding service flow instances and adapts service provider CoS information into the Subscriber CE-VLAN-CoS information, if applicable.

The ETH EVC termination function (EETF) is responsible for the creation and termination of EVC trails. The EETF source multiplexes management (e.g., OAM), control, and data plane PDUs and relays adapted ETH layer PDU towards the service provider EFD. The sink receives the adapted EVC PDU from the service provider EFD and demultiplexes management (e.g., OAM), control and data plane PDUs.

The ETH connection function (ECF) facilitates the creation of point-to-point or multipoint connections.

The ETH adaptation function (EAF) is responsible for the adaptation of the APP layer PDUs to the ETH layer. It forms LLC PDUs, allocates EtherType per client application and/or LLC type (if LLC is present), pads to minimum transmission unit size, and multiplexes adapted client PDUs towards ETH flow termination function (EFTF), demultiplexes adapted client PDUs from EFTF, processes EtherType and perform decapsulation, and extract LLC PDU (if LLC is present) and relays it to client process (as per EtherType).

The EFTF is responsible for the creation and termination of ETH network flows and supports the protocol interface between APP layer and ETH layer. The EFTF source prepares the Ethernet service frame, formats the ETH layer PDU, and relays the ETH layer PDU toward the target EFD. However, the EFTF sink receives the ETH layer PDU from EFD, extracts user data, and relays adapted client PDU to EAF.

The ETH to TAF is responsible for the adaptation of the ETH layer PDUs to its serving TRAN layer such as Ethernet, SONET/SDH, ATM, FR, and MPLS. The TAF source buffers and schedules frames, allocate VLAN ID field value, pads payload to meet minimum transmission unit size, generate service frame FCS, encapsulate/encodes (e.g., adaptation) ETH_CI according to TRAN layer.

The TAF source also multiplexes EVC PDUs into ETH link, adapts rate into the TRANs layer, and inserts adapted ETH Layer data stream into payload of TRAN layer signal.

The sink TAF performs FCS verification of Ethernet MAC frame and Ethernet MAC frame filtering of subscriber frames not intended to be forwarded across the UNI, extracts adapted ETH_CI from payload of the TRAN layer signal, and de-multiplexes encapsulated EVC PDUs.

6.2.4 ETH Links

There are two types of links, access links and trunk links. The link between the port in the subscriber's CE supporting the UNI-C and the port in the service provider's PE supporting the UNI-N is an example to an access link. However, the link interconnects ports between service provider NEs are an example to a trunk link.

For any access link at the UNI-C, there is a one-to-one relationship between the ETH access link and its underlying TRAN link.

Figure 6.10 illustrates the relationships between the UNI functional elements and the associated ETH access link, using the Ethernet layer functions.

When any UNI-N is indirectly attached to an UNI-C via a transport multiplexing function (TMF), the ETH access link can be multiplexed by the TRAN layer. In this case there will be a many-to-one relationship between the ETH access link and its underlying TRAN link.

Figure 6.11 demonstrates the relationships between the UNI components and the associated ETH access link.

The UNI-C is always connected to the UNI reference point by an IEEE 802.3 PHY. The UNI-N implements the service provider side of the UNI functions. The

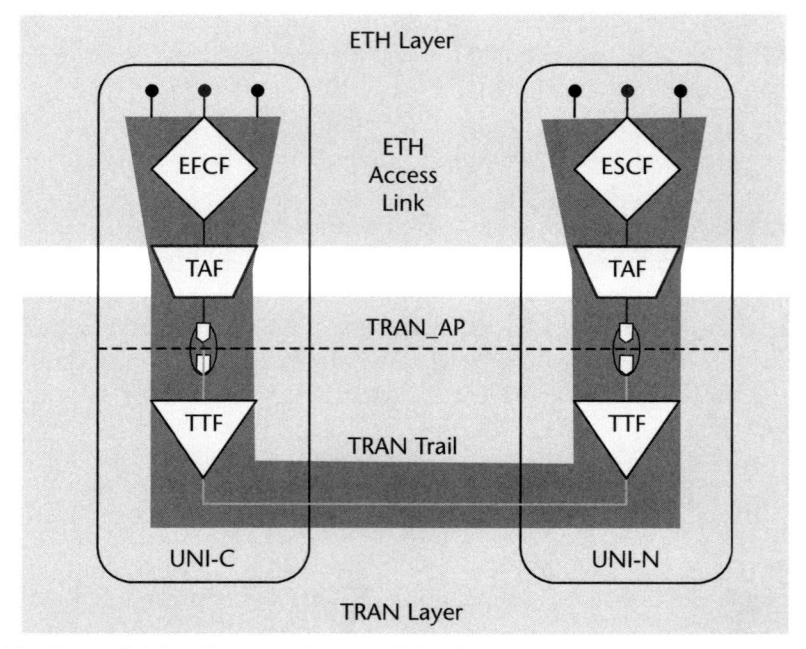

Figure 6.10 Access link for direct attachment of UNI-C to an UNI-N.

6.2 Architecture

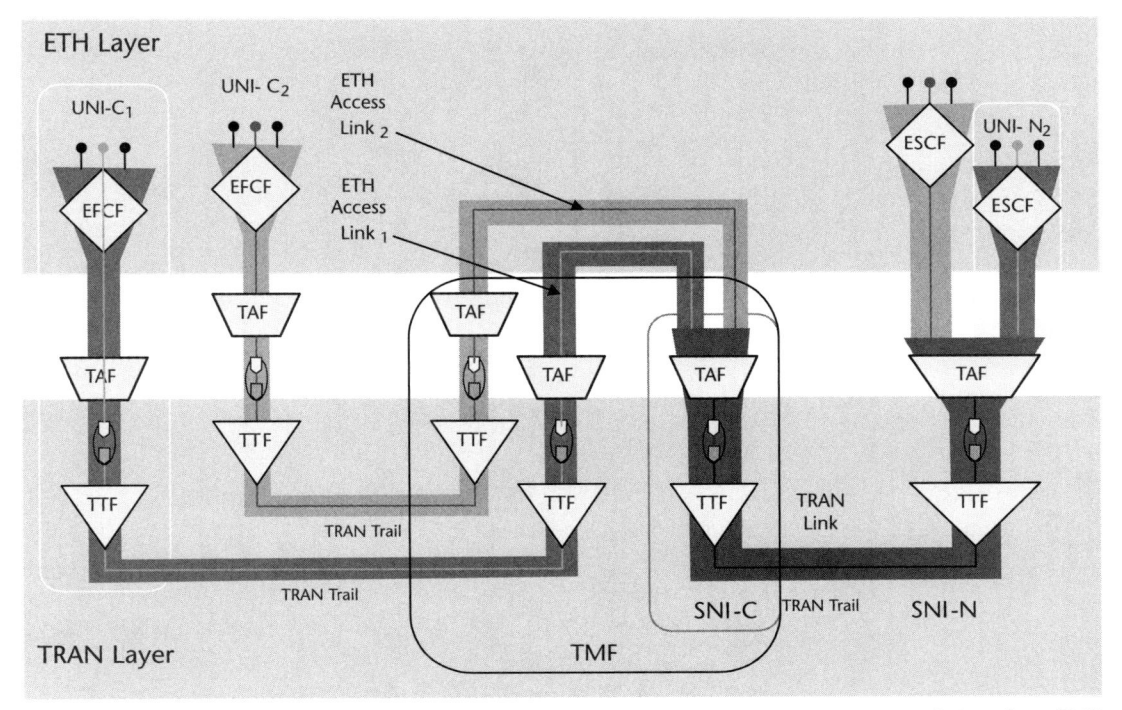

Figure 6.11 Access links for indirect attachment of an UNI-C to an UNI-N via a service node interface (SNI) from a TMF.

reference point between the access network and the PE equipment is called the service node interface (SNI) (see Figure 6.12).

In an ETH trunk link, multiple EVCs may ride on. An ETH trunk link is instantiated via TAF function and underlying TRAN on an internal network to network interface (INNI), and an ETH provider conditioning and TAF functions along with an underlying TRAN trail on an ENNI.

The ETH trunk link implemented via an INNI includes a TRAN layer specific TAF and an underlying TRAN trail.

Figure 6.13 illustrates the relationships between the INNI and the associated ETH trunk link. Figure 6.14 illustrates the relationships between the ENNI functional elements and the associated ETH trunk link.

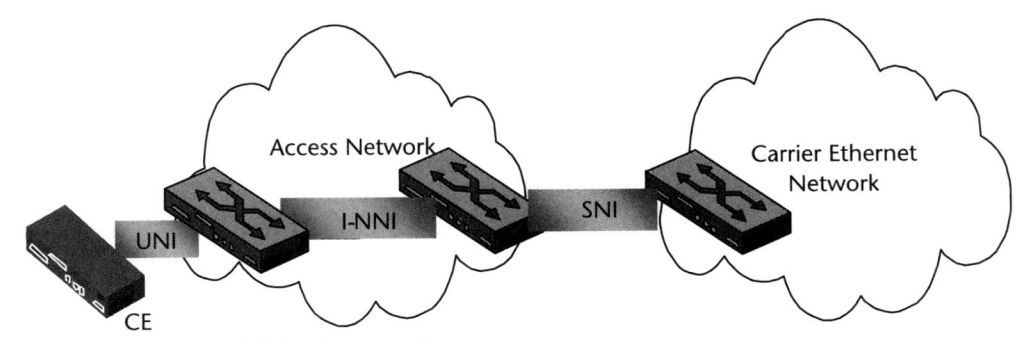

Figure 6.12 UNI and SNI reference points.

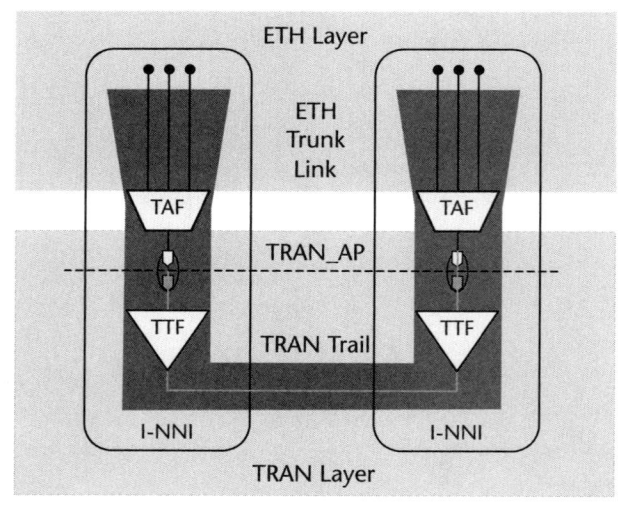

Figure 6.13 An ETH trunk link between two I-NNIs.

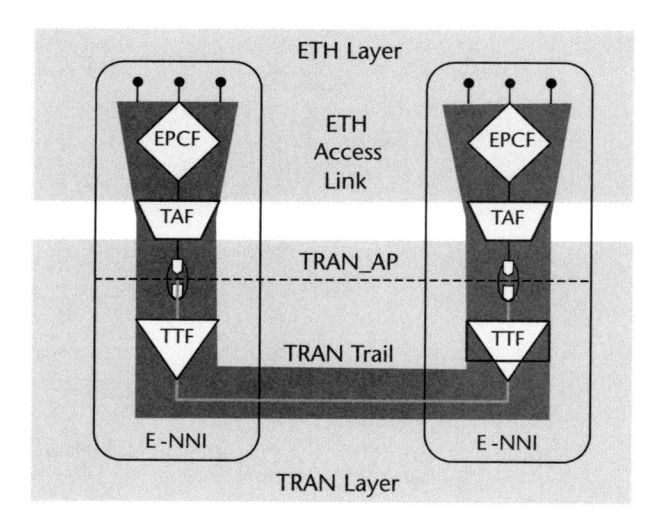

Figure 6.14 An ETH access link among E-NNIs.

6.3 Interfaces and Types of Connections

This section describes the interface between user and network, the interface between networks, and types of connections between interfaces.

6.3.1 UNI

The standard user interface to a CEN is called user-to-network interface (UNI) [1]. UNI is dedicated physical demarcation point between the responsibility of the service provider and the responsibility of a subscriber.

UNI functions are distributed between CPE and CEN, as UNI-C and UNI-N, respectively

6.3 Interfaces and Types of Connections

UNI-C executes the processes of the customer side while UNI-N executes the processes of the network side. CPE may not support UNI-C functions and use a network interface device (NID) to support the demarcation point between the CPE and the CEN, as depicted in Figure 6.15, which is the most common implementation.

The UNI consists of a data plane, a control plane and a management plane. The data plane defines Ethernet frames, tagging, and traffic management. The management plane defines provisioning, device discovery, protection, and OAM. The control plane is expected to define signaling and control, which has not yet been addressed in the standards yet.

Ethernet demarcation is analogous to a CSU/DSU in frame relay. It provides separation between carrier WAN and enterprise LAN and enables testing and monitoring of both LAN and WAN.

The physical medium for UNI can be copper, coax, or fiber. The operating speeds can be 1 to 10 Mbps, 100 Mbps, 1 Gbps or 10 Gbps where the 10-Gbps rate is only supported over fiber.

Service frames are Ethernet frames that are exchanged between the CEN and the customer edge across the UNI. A service frame sent from the CEN to the customer edge at a UNI is called an egress service frame. A service frame sent from the customer edge to the CEN at a UNI is called an ingress service frame.

MEF defines UNI on a port or an Ethernet virtual circuit (EVC) basis with CIR/EIR (committed/excess information rate), and CBS/EBS (committed/excess burst size).

Traffic is classified/prioritized based on TOS, DSCP, and 802.1P.

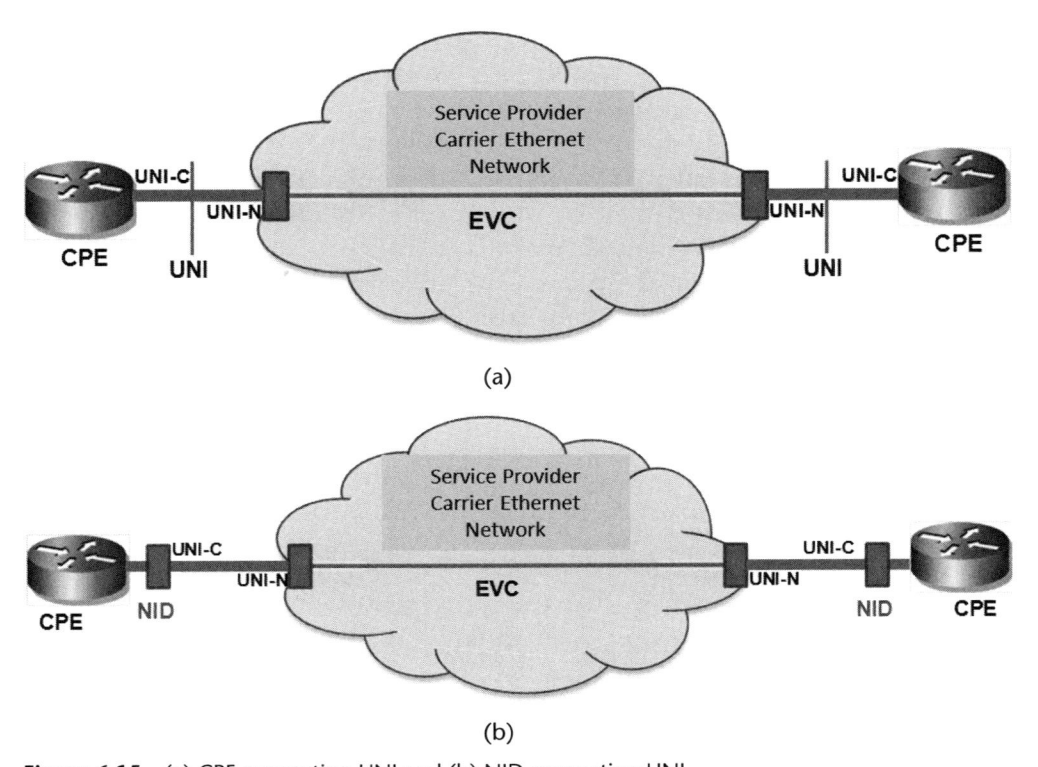

Figure 6.15 (a) CPE supporting UNI and (b) NID supporting UNI.

MEF 11 [10] introduced 3 types of UNI, UNI Type 1, Type 2, and Type 3. In UNI Type 1, which is defined in MEF 13 [11], and the service provider and customer manually configure the UNI-N and UNI-C for services. MEF 13 divides UNI Type I into two categories:

- Type 1.1: Nonmultiplexed UNI for services like EPL;
- Type 1.2: Multiplexed UNI for services like EVPL.

UNI Type 2, which is defined in MEF 20 [12], supports an automated implementation model allowing UNI-C to retrieve EVC status and configuration information from UNI-N, enhanced UNI attributes, and additional fault management and protection functionalities. UNI Type 2 is divided as UNI Type 2.1 and Type 2.2. The Type 2.1 includes service OAM, enhanced UNI attributes such as bandwidth profile per egress UNI and L2CP handling mandatory features, and link OAM, port protection via Link Aggregation Protocol and E-LMI (as defined in MEF 16) optional features. In Type 2.2, the optional features of Type 2.1 will become the mandatory features of UNI.

UNI Type 3, which has not yet been defined, may support UNI-C to request, signal, and negotiate EVCs and its associated service attributes to the UNI-N.

6.3.2 Ethernet Virtual Connection

An EVC as defined in MEF 10.3 [13] is a logical representation of an Ethernet service between two or more UNIs and establishes a communication relationship between UNIs. An UNI may contain one or more EVCs (Figure 6.16).

While connecting two or more UNIs, EVC prevents data transfer between subscriber sites that are not part of the same UNI. The service frame cannot be delivered back to the originating UNI either.

The service frames cannot leak into or out of an EVC. This capability, which is similar to a frame relay or ATM permanent virtual circuit (PVC), allows UNIs constructing their own Layer-2 private networks. At a UNI, multiple EVCs can be constructed to form multiple private networks connections.

Point-to-point EVC (Figure 6.15) supports communication between only two UNIs. All ingress service frames at one UNI, with the possible exception of Layer-2 control protocol messages are typically delivered to the other UNI.

Multipoint-to-multipoint EVC supports any-to-any communication between two or more UNIs. This EVC creates a service that behaves like a switched Ethernet environment, and is an essential component of E-LAN services. Additional UNIs can be added to the multipoint-to-multipoint EVC.

In a multipoint-to-multipoint EVC, a single broadcast or multicast ingress service frame, that is determined from the destination MAC address, at a given UNI is replicated in the MEN and one copy would be delivered to each of the other UNIs in the EVC.

Broadcast service frames, which are defined by all 1s destination MAC addresses and multicast service frames with multicast destination MAC addresses, are replicated and delivered to all other UNIs in the EVC.

6.3 Interfaces and Types of Connections 125

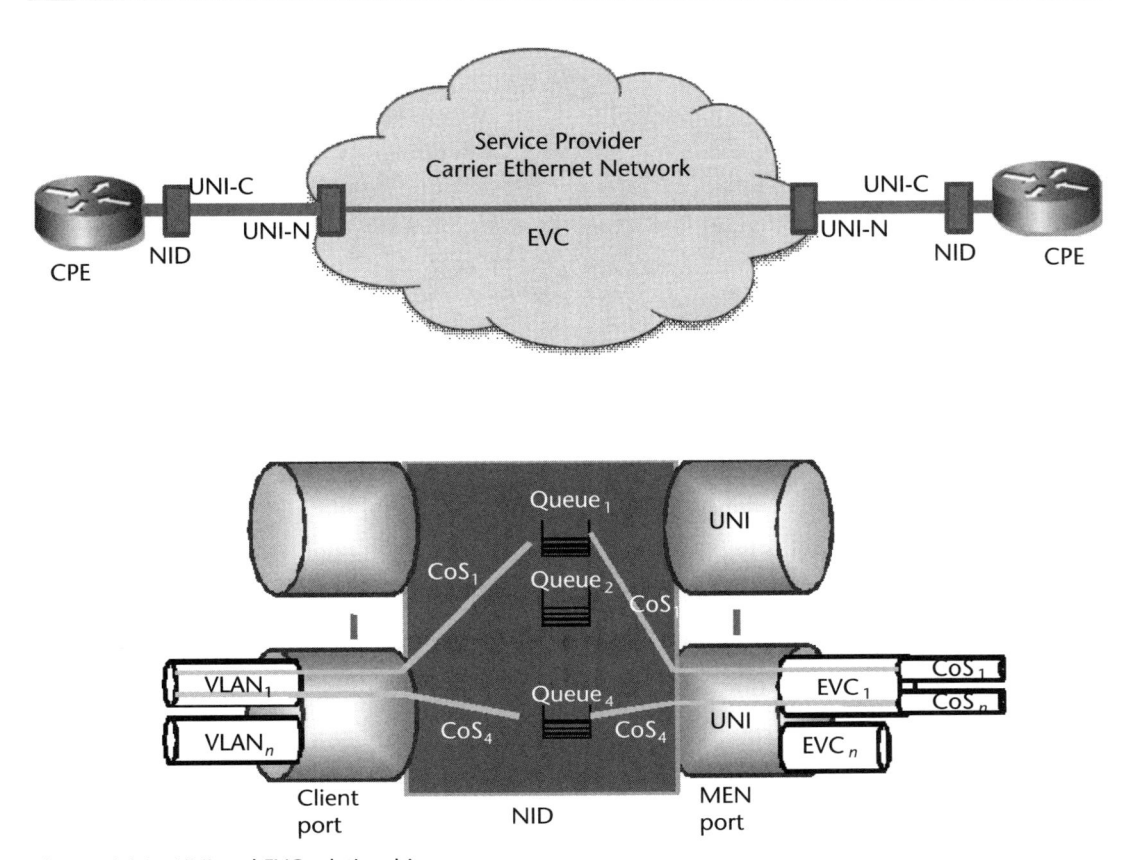

Figure 6.16 UNI and EVC relationship.

Unicast service frames with unicast destination MAC addresses are managed in one of two ways:

- They can be replicated and delivered to all other UNIs in the EVC. This makes the EVC behave like a shared-media Ethernet.
- The CEN can learn which MAC addresses are "behind" which UNIs by observing the source MAC addresses in service frames and deliver a service frame to only the appropriate UNI when it learns the destination MAC address. When it has not yet learned the destination MAC address, it replicates the service frame and delivers it to all other UNIs in the EVC. In this case, the CEN behaves like a MAC learning bridge.

Point-to-multipoint EVC, also called rooted-multipoint EVC, supports communication between two or more UNIs, but does not support any-to-any communication. UNIs are designated as root or leaf. Transmissions from the root are delivered to the leaves, and transmissions from the leaves are delivered to the root(s). No communication can occur between the leaves or between the roots. This EVC can be used to create a hub-and-spoke connection arrangement without needing to configure multiple point-to-point EVCs.

In a given UNI, subscriber flows may be mapped into one or more EVCs. When multiple subscriber flows mapped into a single EVC, there could be multiple bandwidth profiles associated with the EVC where there is a bandwidth profile per class of service (CoS) instance. The bandwidth profile defines bandwidth and performance parameters, such as delay, jitter, and availability per CoS instance.

Similarly, multiple CE-VLAN IDs can be mapped to one EVC where each CE-VLAN ID may belong to a different service (i.e., service multiplexing). This configuration is called a bundling map. When there is a bundling map, the EVC must have the CE-VLAN ID preservation, which allows the subscriber to use the same VLAN values at all sites.

6.3.3 External Network-Network Interface (ENNI)

Most of the domestic and international communications pass through more than one service provider/operator network. This is expected to be true for Carrier Ethernet services as well. An EVC between two UNIs may have to travel multiple operator CENs. One operator/service provider may not be able to support all UNIs of a subscriber. As a result, multiple CENs need to interface each other. This interface between two CENs is called the ENNI, depicted in Figure 6.17.

An ENNI can be implemented with one or more physical links. However, when there is no protection mechanism such as LAG among multiple links connecting two operator CENs, each link represents a distinct ENNI.

Similar to UNI, ENNI consists of a data plane, a control plane and a management plane. The data plane defines Ethernet frames, tagging, and traffic management. The management plane defines provisioning, device discovery, protection, and OAM. The control plane is expected to define signaling and control, which is not addressed in the standards yet.

The physical medium for ENNI can be copper, coax or fiber. However, the fiber interface between ENNI gateways is more common. The operating speeds can be 1 Gbps or 10 Gbps.

Similar to UNI, ENNI has a bandwidth profile. The parameters of the bandwidth profile are CIR in bits per second, CBS in bytes that is greater than or equal to the largest MTU size allowed for the ENNI frames, EIR in bits per second, EBS in bytes that are greater than or equal to the largest MTU size allowed for the

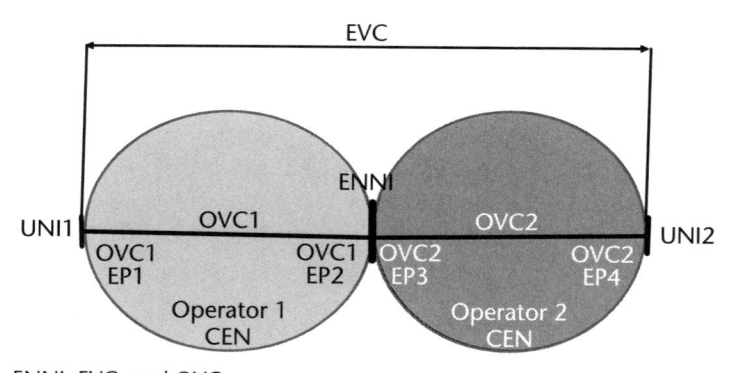

Figure 6.17 ENNI, EVC, and OVC.

6.3 Interfaces and Types of Connections 127

ENNI frames, coupling flag (CF) and color mode (CM). The CF has value 0 or 1 and CM has the value of color-blind or color-aware.

6.3.4 Operator Virtual Connection

When the EVC travels multiple operator CENs, the EVC is realized by concatenating pieces of other connections in each operator network, which is called operator virtual connections (OVCs). For example, in Figure 6.17, EVC is constructed by concatenating OVC 1 in CEN 1 with OVC 2 in CEN 2.

Each OVC endpoint is associated with either a UNI or an ENNI and at least one OVC endpoint associated by an OVC at an ENNI (see Figure 6.18).

Hairpin switching occurs when an OVC associates two or more OVC endpoints at a given ENNI. An ingress S-tagged ENNI frame at a given ENNI results in an egress S-tagged ENNI frame with a different S-VLAN ID value at the given ENNI.

An OVC endpoint (EP) represents the OVC termination point at an external interface such as ENNI or UNI. The endpoint map specifies the relationship between S-tagged ENNI frame and OVC endpoint within an operator CEN. An ingress S-tagged ENNI frame that is not mapped to an existing endpoint will be discarded.

Table 6.2 illustrates an example for endpoint map where S-VLAN IDs are mapped to EP Identifier which is a string administered by the operator.

Multiple S-VLAN IDs can be bundled and mapped to an OVC endpoint. In that case, the OVC endpoint should be configured to preserve S-VLAN IDs. S-VLAN ID preservation cannot be supported in hairpin switching configuration.

When there is bundling, frames originated by more than one subscriber may be carried by the OVC resulting duplicate MAC addresses. To avoid possible problems due to this duplication, MAC address learning may be disabled on the OVC.

6.3.5 VUNI/RUNI

Ordering and maintaining EVCs crossing an ENNI can be cumbersome due to coordinations between entities of two or more carriers. In order to somewhat simplify this process, the operator responsible from the end-to-end EVC may order a tunnel between its remote user that is connected to another operator's network and its ENNI gateway, that can accommodate multiple OVCs, instead of ordering the

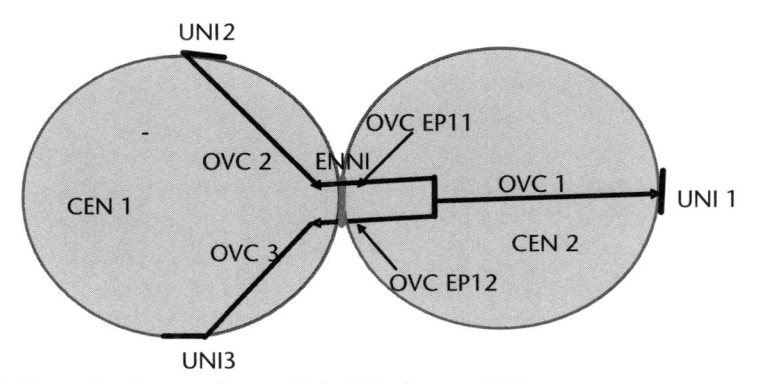

Figure 6.18 Example of supporting multiple OEPs for one OVC.

Table 6.2 Endpoint Map Example

S-VLAN ID Value	End Point Identifier	End Point Type
100	comcast-twc-enni_1_10	OVC End Point
200	comcast-twc-enni_2_20	OVC End Point

OVCs one by one. This access of the remote user to the service provider's network is called UNI tunnel access (UTA). The remote user end of the tunnel is called remote UNI (RUNI) while the service provider end of this tunnel is called virtual UNI (VUNI) (Figure 6.19) [14].

The VUNI has service attributes similar to those of a UNI, and is paired with a remote UNI. Its main function is to specify the processing rules applicable to ENNI frames present in the VUNI provider domain and associate them with a given UTA instance.

In Figure 6.20, the CE at UNI Y participates in EVCs 1, 2, and 3. These EVCs have the service provider agreed bandwidth profile attributes and CoS markings. At the remote UNI, service frames of EVC 1, 2, and 3 are exchanged with the CE. Such frames may be C-tagged, priority tagged, or untagged. The remote UNI is instantiated by the operator as a UNI where the network operator maps all service frames to the single OVC endpoint supporting the UTA OVC.

A single bandwidth profile and CoS may be applied at this remote UNI OVC endpoint. At the UTA OVC endpoint at the network operator's side of the ENNI, an S-VLAN ID is used to map ENNI frames to the OVC endpoint supporting the UTA, and applies a UTA specific single bandwidth profile and CoS.

Given there is a bandwidth profile per UTA OVC, it will become a challenge how to satisfy bandwidth requirements per EVC basis. This issue somewhat solvable when we have CIR = 0 and EIR > 0 and CIR > 0 and EIR = 0. But it is very difficult to solve when CIR > 0 and EIR > 0.

In the VUNI provider's network, the relationship between the UTA OVC and the VUNI is realized by an S-VLAN ID present at the ENNI, whose value is negotiated between the VUNI provider and the network operator. At the ENNI, when receiving an ENNI Frame, the VUNI provider maps (using the endpoint map) a

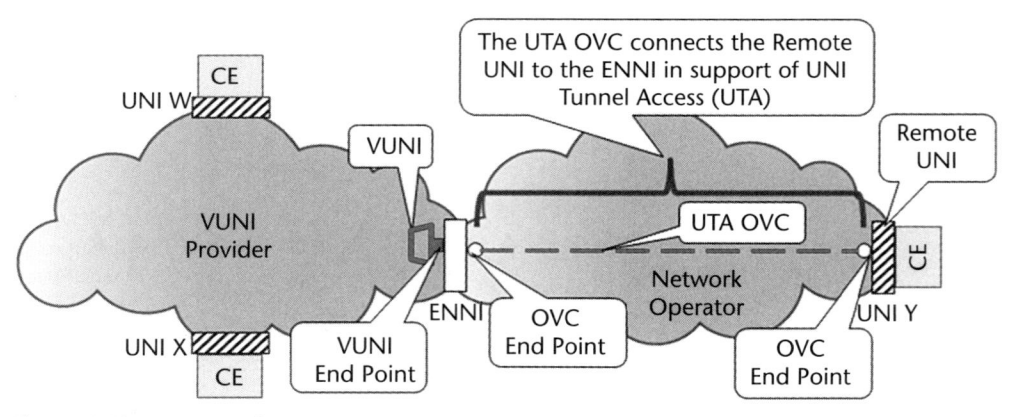

Figure 6.19 UNI tunnel access.

6.3 Interfaces and Types of Connections

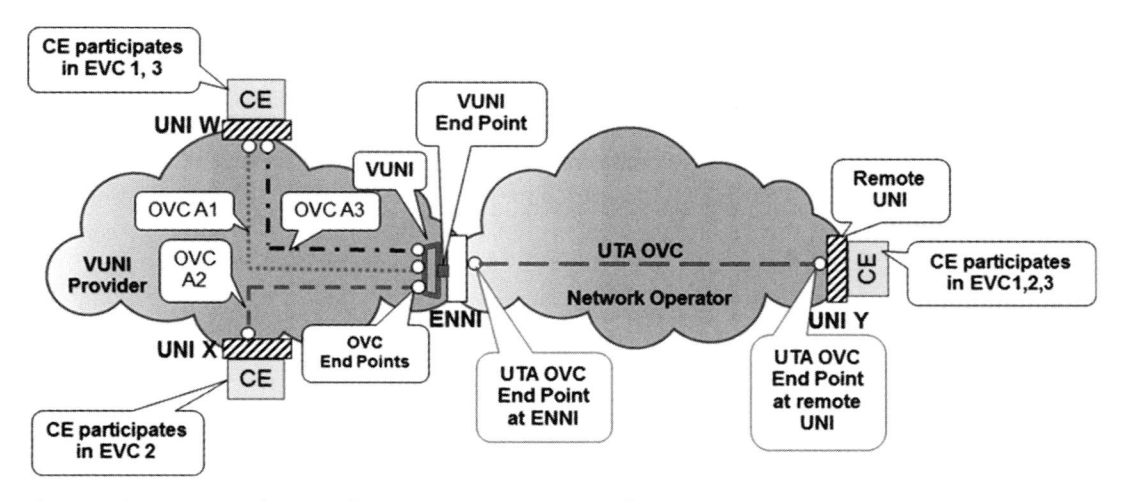

Figure 6.20 EVCs implemented using VUNI, UTA OVC, and remote UNI.

single S-VLAN ID to a VUNI endpoint associated with a VUNI. The VUNI then maps frames based on their CE-VLAN ID to the appropriate OVC endpoint for OVCs A1, A2, and A3. In the reverse direction, the VUNI multiplexes frames from OVCs A1, A2, and A3 into a tunnel denoted by a unique S-VLAN ID, which is associated with the network operator's UTA OVC. Note that A1, A2, and A3 have nonoverlapping CE-VLAN IDs at the VUNI.

Multiple VUNIs can be associated with a single ENNI. Figure 6.21 shows how the UTA and a VUNI may be employed to instantiate the services across ENNI.

At the ENNI between CEN B and CEN C, Figure 6.21 shows the mapping of frames with an S-VID of 2023 to a VUNI endpoint representing VUNI A.

In summary, UTA is composed of a UTA OVC component in the operator CEN and associated VUNI in the provider CEN and remote UNI in the operator CEN components. Requirements for each component will be described in the following sections.

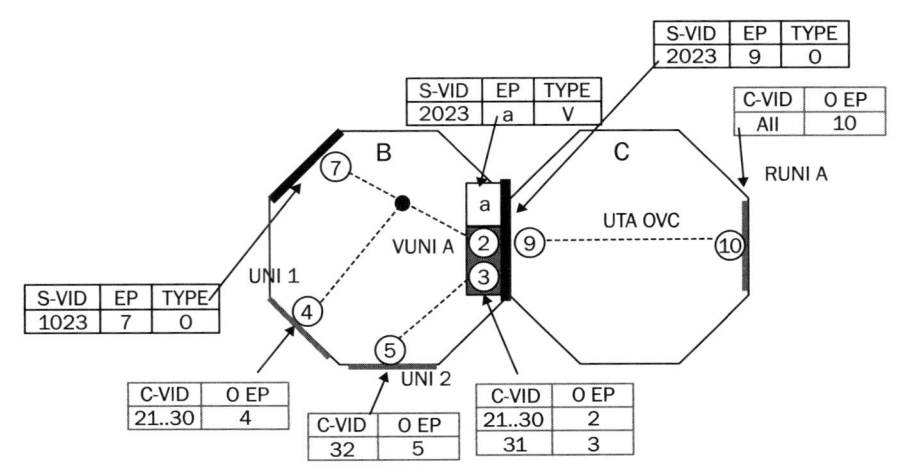

Figure 6.21 Example multiple EVCs supported by UTA and VUNI.

6.4 EVC Services and Attributes

Carrier Ethernet services are defined in terms of what is seen by customer edge (CE) and independent from technology inside CEN with a set of attributes including QoS, standardized services, scalability, reliability, and service management. These attributes are reflected in UNI and EVC parameters.

The physical demarcation point between a subscriber and service provider is called UNI. A demarcation point between two service providers is called ENNI. If it is between service provider internal networks, it is called INNI.

The CE and CEN exchange service frames across the UNI. The frame transmitted toward the service provider is called an ingress service frame and the frame transmitted toward the subscriber is called an egress service frame.

EVC is the logical representation of an Ethernet connection that is defined between two or more UNIs. EVCs can be point-to-point, point-to-multipoint (or rooted multipoint), or multipoint-to-multipoint as depicted in Figure 6.22.

There are three service types that are built over these EVCs, namely: (1) E-Line, (2) E-LAN, and (3) E-Tree. There are two Ethernet services that are defined for each one of the three Ethernet service types. These are port-based and VLAN-based Ethernet services. Port-based services use all-to-one bundling at the UNI and are called private services. The VLAN-based services use CE-VLAN ID to identify the corresponding EVC as multiple EVCs share the same UNI interface. These types of services are called virtual private services in the Carrier Ethernet terminology; see this classification in Table 6.3.

6.4.1 Ethernet Service Types

The Ethernet line (E-Line) service type uses point-to-point EVC and can deliver a variety of Ethernet services. From a simple scale, it can just deliver a symmetrical best-effort traffic between two UNIs of the same bandwidth, or it may be used

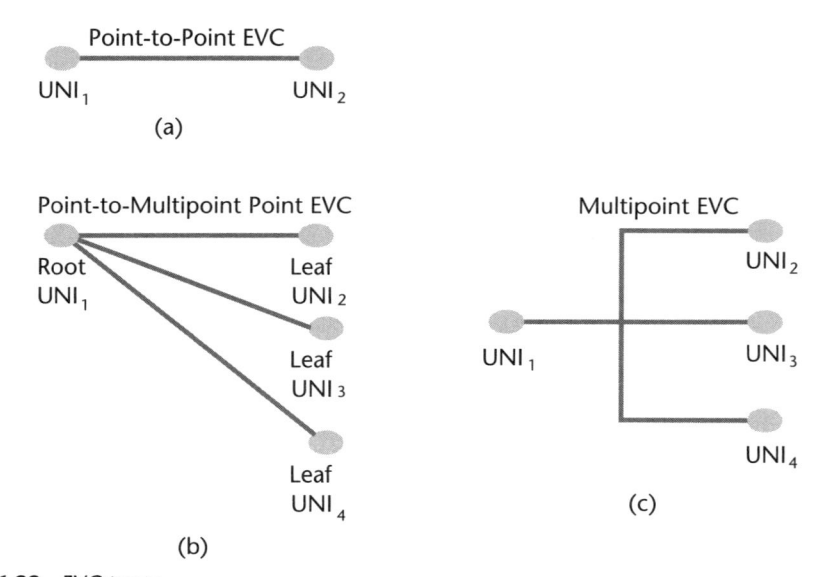

Figure 6.22 EVC types.

6.4 EVC Services and Attributes

Table 6.3 Carrier Ethernet Services

Service Type	Port-Based (All-to-One Bundling)	VLAN-Based (Service Multiplexed)
E-Line over point-to-point EVC	Ethernet private line (EPL)	Ethernet virtual private line (EVPL)
E-LAN over multipoint-to-multipoint EVC	Ethernet private LAN (EP-LAN)	Ethernet virtual private LAN (EVP-LAN)
E-Tree over rooted Multi-point EVC	Ethernet private tree (EP-Tree)	Ethernet virtual private tree (EVP-Tree)

between two UNIs of different line rates with some specific bandwidth profiles in a given CoS. All these attributes are defined in attribute lists for UNIs and the EVC.

The Ethernet LAN (E-LAN) service type uses multipoint-to-multipoint EVC and can deliver a variety of Ethernet services. In a more advanced form, an E-LAN service can be configured to carry a specific level of performance objectives.

The Ethernet tree (E-Tree) service type uses rooted-multipoint EVC and also delivers a variety of Ethernet services. As a simple service, it can deliver traffic from a single root to all leaves with no performance expectations, while leaves cannot communicate with each other. This can be an applicable service for a video application where multicast and/or broadcast are needed. In a more complicated version, this service type may deliver traffic from a multiple of root UNIs, instead of just one. Roots can communicate with each other; however, the leaves can only communicate with its corresponding root.

6.4.2 Ethernet Service Definitions

An Ethernet service can be defined by setting up appropriate attribute values for an Ethernet service type. There are two Ethernet services defined for each service type: (1) private service, and (2) virtual private service.

The Ethernet private line (EPL) service can be defined by setting appropriate attribute values for an E-Line service type. The service is between two UNI ports where there is no multiplexing at UNI and provides transparency for service frames (Figure 6.23). The EPL is used to replace a TDM, FR and ATM private lines. There is a single Ethernet virtual connection (EVC) per UNI. For cases where EVC speed is less than the UNI speed, the CE is expected to shape traffic.

New in MEF 6.2 [15], an EPL service can have multiple envelopes specified for a UNI and can have multiple bandwidth profiles with in an envelope, if the attribute, called token share attribute, is enabled.

The EVPL service can be defined by setting another set of attribute values for an E-Line service type similar to EPL service.

An EVPL (Figure 6.24) can be used to create services similar to the EPL. As a result of EVC multiplexing based on CE-VLAN-ID, an EVPL does not provide as much transparency of service frames as with an EPL, since some of service frames may be sent to one EVC while other service frames may be sent to other EVCs. EVPL can also treat some L2CP service frames differently based on their destination addresses as detailed in multi-CEN L2CP [16]. New in MEF 6.2 [15], an EVPL service can have multiple envelopes specified for a UNI and can have multiple bandwidth profiles within an envelope, if the token share attribute is enabled.

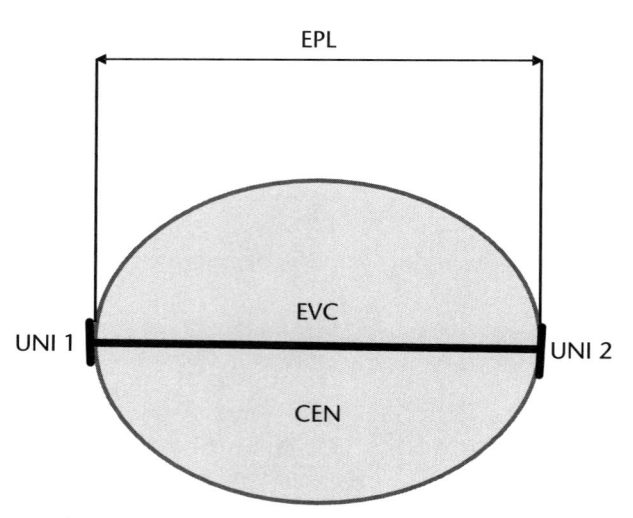

Figure 6.23 EPL example.

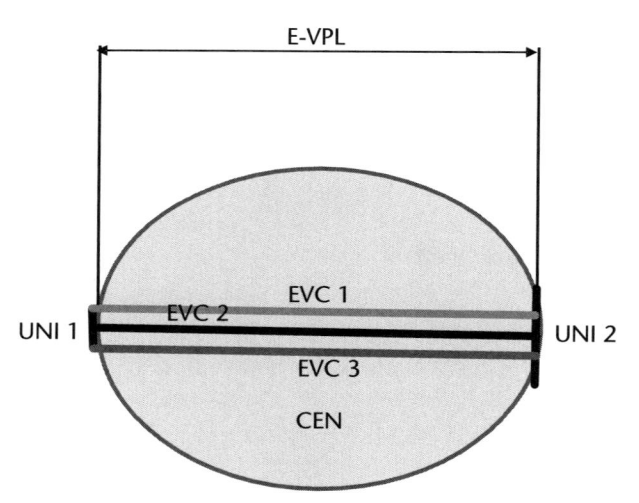

Figure 6.24 EVPL service.

The Ethernet private LAN (EP-LAN) service can be defined by setting appropriate attribute values for an E-LAN service type, which is used to create multipoint L2 VPNs and transparent LAN service with a specific performance assurance as well as no-performance assurance. This service sets a good foundation for IPTV and multicast service delivery networks. In the EP-LAN service (Figure 6.25), CE-VLAN tag is preserved and L2 control protocols can be tunneled. Therefore, the customer can configure the VLANs across the network without needing the service provider. This is an additional flexibility offered to the customer by reducing the coordination with the service provider. This service can also preserve the CE-VLAN CoS information and be able to define multiple CoS names and bandwidth profile flows. Updating MEF 6.1, MEF6.2 allows one or more envelopes to be specified at a UNI, each of which can encompass multiple bandwidth profile flows based on CoS name, if the token share attribute is enabled.

6.4 EVC Services and Attributes

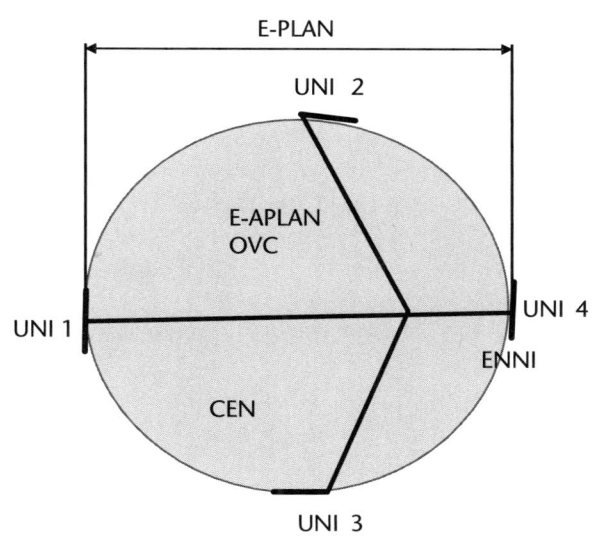

Figure 6.25 EP-LAN service.

Like the EP-LAN service, the Ethernet virtual private LAN (EVP-LAN) service is also defined by setting appropriate attribute values for an E-LAN service type. However, in this service type, the customer can subscribe to an additional service or services from the same UNI that is originally hosting the Ethernet LAN service. To accomplish this capability, the EVCs can be multiplexed on the same UNI by bundling and create EVP-LAN services (Figure 6.26). Note that E-LAN and E-Line service types can be multiplexed in the same UNI where some of the EVCs can be part of the E-LAN while the others can be part of E-Line services. Furthermore, UNIs can be added and removed from E-LAN service without disturbing the users on the E-LAN. Similar to EP-LAN, MEF6.2 also allows for EVP-LAN to have one or more envelopes to be specified at a UNI each of which can encompass multiple bandwidth profile flows based on CoS name, if the token share attribute is enabled.

The Ethernet private tree (EP-Tree) service can be defined by setting appropriate attribute values for an E-Tree service type, in which services can be distributed from a few centralized sites to the rest of the sites in a point-to-multipoint fashion that resembles a LAN service type. These centralized sites are known as root sites

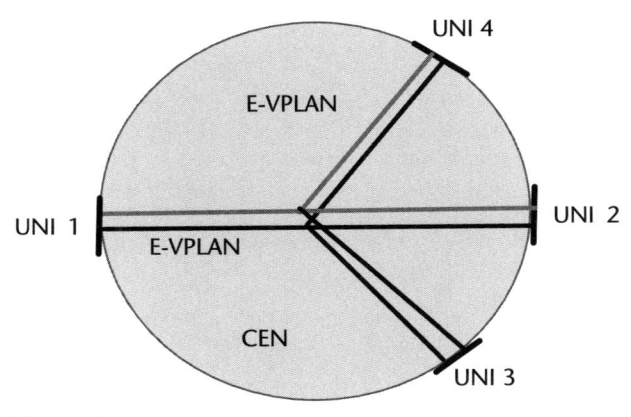

Figure 6.26 E-VPLAN service.

and the rest are leaf sites. The EP-Tree service that is depicted in Figure 6.27 is a subset of an EP-LAN service where traffic from any leaf site UNI can be sent/received to/from a root site UNI but never being forwarded to another leaf site UNI. The service is used for applications requiring point-to-multipoint topology such as video on demand, Internet access, triple-play backhaul, mobile cell site backhaul, and franchising applications. EP-Tree service has the same benefits that EP-LAN has. The EP-Tree service can support all to-one bundling with CE-VLAN tag preservation. Unlike MEF 6.1, MEF 6.2 allows the EP-Tree service to have one or more envelopes to be specified at a UNI, each of which can encompass multiple bandwidth profile flows based on CoS name, if the token share attribute is enabled.

Like EP-Tree Service, the Ethernet virtual private tree (EVP-Tree) service is also defined by setting appropriate attribute values for an E-Tree service type. However, in this virtualized service, the customer can subscribe to an additional point-to-point or multipoint to multipoint service or services, from the same UNI that is originally hosting the Ethernet Tree service. See Figure 6.28 for two EVP-Tree services sharing the same UNIs in an EVP-LAN fashion with the same dedicated root site UNI. This virtualized service would also require the all-to-one bundling to be disabled. Unlike MEF 6.1, MEF 6.2 allows EVP-Tree service to have one or more envelopes to be specified at a UNI, each of which can encompass multiple bandwidth profile flows based on CoS name, if the token share attribute is enabled.

Some of these services can support applications with strict performance requirements such as voice and best effort performance requirements such as internet access (Table 6.4).

Ethernet service delivery is independent of underlying technologies. At the access, Carrier Ethernet service may ride over TDM, SONET, Ethernet, and PON technologies. At the backbone, Carrier Ethernet service may ride over IP/MPLS network.

Table 6.5 compares Carrier Ethernet-based VPN service and IP-based VPN services in access networks.

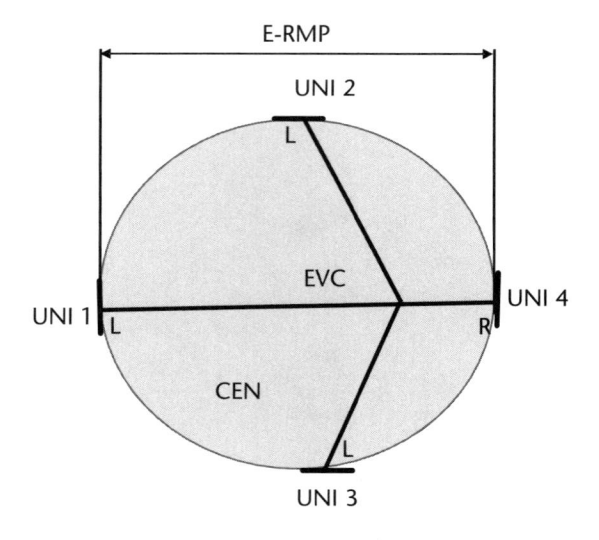

Figure 6.27 EP-Tree service.

6.4 EVC Services and Attributes

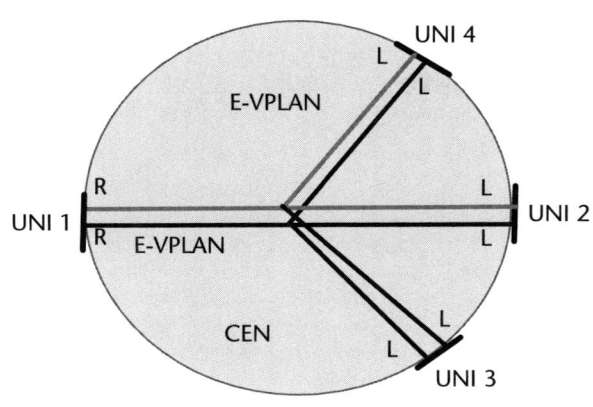

Figure 6.28 EVP-Tree service.

Table 6.4 Possible Applications for Carrier Ethernet Services

EPL Applications	EVPL Applications	ELAN Applications
Healthcare, financial, pharmaceuticals	Local/regional government, education, healthcare, financial	Local/regional government, primary and secondary education
Medical imaging, data center connectivity, business continuity, CAD engineering	VoIP, video, point-to-point intra-metro connectivity, classes of service corporate networking, photo imaging, videoconferencing	Collaboration, multisite connectivity, multipoint Ethernet, administration

Table 6.5 Comparison of Ethernet and IP-Based VPNs

Service Attribute	Ethernet Service	IP Service
Customer handoff	Ethernet UNI	Ethernet port (or PDH circuit)
Service identification	VLAN ID/EVC	IP address
CoS identification	PCP	DSCP/ToS
Packet/frame routing/forwarding	MAC address (E-LAN), VLAN ID (E-Line)	IP address
Fault management	Link Trace, Continuity Check (Layer 2 Ping), Loopbacks	Traceroute, ICMP Ping
Performance management	Frame delay, frame delay variation, frame loss ratio, service availability	Packet delay, packet delay variation, packet loss

In the following sections, we will describe the common attributes for all service types and specific attributes and parameter values for each service.

6.4.3 Common Attributes for EVC Services

The common attributes and parameter values for all services are grouped into three categories: (1) common UNI attributes, (2) common EVC attributes, and (3) common EVC per UNI attributes.

6.4.3.1 Common UNI Attributes

The following is a list of the UNI service attributes that each Ethernet service has. The values of some attributes may change depending on the service type.

- UNI ID: An arbitrary text string as specified in MEF 10.3 [13] is assigned by the service provider and it uniquely defines the interface within the CEN. A service provider usually subscribes to a naming convention that may physically define the location of the interface including the name of the data center, device, slot, and port.
- Physical layer: This layer lists the physical layers that the UNI may use for each of its physical link.
- Synchronous mode: This mode lists either enabled or disabled for each physical link to receive bit clock reference at the physical layer. If this attribute is enabled, then the subscriber device can receive clock synchronization from the network.
- Number of links: This specifies the number of physical links that the UNI implements. This value should be at least 1. If protection is implemented by applying UNI resiliency, then one can use value of 2 for this attribute.
- UNI resiliency: The values for this attribute can be none, 2-link aggregation, or other. If there is only one physical link implemented on the UNI, the value of this attribute should be none. If the UNI has two physical links implemented, then the value should represent the fact by taking the value of 2-link aggregation. If the UNI implements more than 2 physical links, the value should be other.
- Service frame format: This represents IEEE 802.3-2012 MAC frame format.
- UNI Max. service frame size: This represents the UNI max. service frame size in a positive integer, which is 1,522 or greater.
- Service multiplexing: This could be either enabled for supporting multiple EVCs at the UNI or disabled otherwise. Note that when multiple EVCs are supported at a UNI, each of those EVCs could be of any type and do not need to be the all-in-one type.
- Customer edge CE-VLAN ID: This identifies the EVC that resides on the UNI. This attribute is not applicable when all-to-one bundling attribute is enabled.
- CE-VLAN ID/EVC map: This mapping table maps C-VLAN-IDs to existing EVCs at a UNI. Untagged and priority tagged service frames are organized into the same C-VLAN-ID value between 1 and 4,094. Note that multiple C-VLAN-IDs may be mapped to a single EVC.
- Max. number of EVCs: This value defines the maximum number of EVCs that can be supported by the UNI and it should be at least one.
- Bundling: This could be either enabled or disabled. When enabled, the UNI can support multiple CE-VLAN IDs be mapped to a single EVC and the CE-VLAN ID preservation attribute must be enabled at the UNI.
- All-to-one bundling: This could be either enabled or disabled. When enabled, all CE-VLAN IDs are mapped into a single EVC at the UNI and all

corresponding UNIs in the same EVC need to have this attribute enabled as well.

- Token share: This could be either enabled or disabled. When enabled, it indicates that the UNI can share tokens with multiple bandwidth profile flows in an envelope. Otherwise, the UNI has only one bandwidth profile flow per envelope.
- Envelopes: An envelope is a container of n bandwidth flows, each of which has a rank between 1 and n. This attribute tabulates a list of these envelopes in 3 tuple (envelope ID, CF, n) at the UNI. The value of n is an integer that represents the number of bandwidth profile flows in the envelope.
- Ingress bandwidth profile per UNI: This represents the list of parameters and the corresponding values for the ingress bandwidth profile at the UNI. The list includes CIR, CBS, EIR, EBS, CF, and CM.
- Egress bandwidth profile per UNI: This represents the list of parameters and the corresponding values for the egress bandwidth profile at the UNI. The list includes CIR, CBS, EIR, EBS, CF, and CM.
- Link OAM: This could be either enabled or disabled. When enabled, CEN supports active DTE mode as in the IEEE 802.3 at the UNI.
- UNI MEG: This could be either enabled or disabled. When enabled, the CEN can apply the UNI MEG service.
- E-LMI: This could be either enabled or disabled. When enabled, CEN can apply E-LMI and adopt the mandatory requirements for UNI-N in MEF 16.
- UNI L2CP address set: This specifies either CTB, or CTB-2, or CTA as referred to MEF 45.
- UNI L2CP peering: This represents the peering for L2CP. It either tabulates a list of destination address and protocol identifier or destination address, protocol identifier, and link identifier to be peered at the UNI as specified in MEF 45.

6.4.3.2 Common EVC Attributes

There are specific EVC attributes that are common for all Ethernet services and valid for all UNIs that the EVC delivers service. The following is the list of these parameters with some specific values:

- EVC type: This specifies the type of the EVC. The values are point-to-point, multipoint-to-multipoint, or rooted-multipoint, as explained in MEF 10.3.
- EVC ID: This is an arbitrary string that uniquely identifies the EVC within the CEN. The EVC ID is provided and managed by the service provider.
- UNI list: This is a list of UNIs that are associated by the EVC in the UNI ID and UNI role format. The role is reported as either root or leaf. If the EVC type is point-to-point or multipoint-to-multipoint, then the roles for all UNIs in the EVC are reported as root.

- Max. number of UNIs: Two or more depending on the EVC type. For example, the value is two for a point-to-point EVC.
- Unicast service frame delivery: This defines how the unicast service frames are handled by the EVC; the values are discard, deliver unconditionally, or deliver conditionally.
- Multicast service frame delivery: This defines how the multicast service frames are handled by the EVC; the values are discard, deliver unconditionally, or deliver conditionally.
- Broadcast service frame delivery: This defines how the broadcast service frames are handled by the EVC; the values are discard, deliver unconditionally, or deliver conditionally.
- CE-VLAN ID preservation: This could be either enabled or disabled. When enabled, the CE-VLAN ID that is originated at the ingress UNI is preserved at the egress UNI(s).
- CE-VLAN CoS preservation: This could be either enabled or disabled. When enabled, the PCP field of the CE-VLAN tag in the service field at the egress UNI is kept identical to the value that is originated at the ingress UNI.
- EVC performance: This specifies the list of performance metrics for the EVC. The content of this list may vary depending on the service-level specification (SLS) that is negotiated between the service provider and the subscriber of the service. The list may contain the following service frame performance metrics; one-way frame delay, one-way frame delay range, one-way mean frame delay, one-way interframe delay variation, one-way frame loss, one-way availability, and so forth; refer to MEF 10.3, Section 8.8, for a complete list and specifications. This attribute may be null, if there is no SLS defined for the service.
- EVC Max. service frame size: This specifies the maximum service frame size in bytes. The value must be larger than 1,522 bytes. Any frame that has larger size than this value is discarded at the ingress UNI.

6.4.3.3 Common EVC per UNI Attributes

There are specific attributes for an EVC that may differ from UNI to UNI. The list of these attributes is given here:

- UNI EVC ID: This identifies an EVC at the UNI and is used for management and control. The value of the attribute is a string that is the concatenation of the EVC and the UNI identifiers.
- CoS identifier for data service frames: This identifies how CoS is supported for the service. The possible values are EVC, CE-VLAN, and IP. The value is EVC if each CoS is dedicated to an entire EVC, where the performance metrics associated to the CoS are applied to the delivery of the EVC. In the CE-VLAN option, the priority code point (PCP) field in the CE-tag is used along with the EVC to specify a CoS. The last IP option uses DHCP bits at the IP level again along with EVC to specify a CoS.

6.4 EVC Services and Attributes

- CoS identifier for L2CP service frame: The attribute lists each L2CP defined in the EVC to a CoS name.
- CoS identifier SOMA service frame: This is the same as data service frames.
- Color identifier for service frame: The attribute dictates the color identification for the service frame. The colors that apply to the classification are green and yellow. When the bandwidth profile is color-aware, the bandwidth profile algorithm differentiates the handling of service frames based on their colors. The color identification can be implemented in multiple ways. An implicit way is to map all service frames with the same color in to a specified EVC. Other options are to use: (1) CE-VLAN CoS value, (2) CE-VLAN Tag DEI field, and (3) DSCP value at the IP layer. One restriction is that both CoS and color identification should use the similar options; for example, both must use either the PCP field in CE-tag in MAC level or the DHCP field in the IP level.
- Egress equivalence class identifier for data service frames: The attribute dictates the equivalence class at the egress UNI for the EVC data service frames. There are multiple options to make this identification. One is to use CE-VLAN CoS where the CoS identification also determines the equivalence class. Other option is to use IP-level DSCP bits carried by IPv4 or IPv6 packets. It is possible for an EVC not to have an egress profile to apply at a specific UNI. In this case, no attribute would have a meaning.
- Egress equivalence class identifier for Layer-2 control protocol service frames: The attribute lists each L2CP defined in the EVC to an equivalence class. It is possible for an EVC not to have an egress profile to apply at a specific UNI. In this case, no attribute would have a meaning.
- Egress equivalence class identifier for SOAM service frames: These are the same as data service frames.
- Ingress bandwidth profile per EVC: There may be multiple bandwidth profile flows defined one for each EVC at a UNI in a case of multiple EVCs at a UNI. In this case, each bandwidth profile is determined by this attribute for an EVC. The collection of bandwidth profiles for all EVCs is captured under a logical container, called an envelope.
- Egress bandwidth profile per EVC: Similar to an ingress bandwidth profile per EVC attribute, this attribute determines the bandwidth profile for each EVC within the envelope.
- Ingress bandwidth profile per CoS identifier: When there are multiple CoS levels defined in an EVC at a UNI, the corresponding bandwidth profiles for these CoS levels need to be defined by using this attribute. The attribute constructs a list of CoS levels with corresponding bandwidth profile parameters for each CoS level. The overarching logical container that contains all Bandwidth profiles and flows per each EVC is called envelope.
- Egress bandwidth profile per egress equivalence class: When there are multiple equivalence classes defined for an EVC at an egress UNI, the corresponding bandwidth profiles for these CoS levels need to be defined by using this attribute. The attribute constructs a list of equivalence classes with corresponding bandwidth profile parameters for each equivalence class. The

overarching logical container that contains all bandwidth profiles and flows per each EVC is called an envelope.

- Source MAC address limit: Source MAC address could be limited to a number, say N, with a time window parameter. If this attribute is enabled, then a list of source MAC address and arrival time pairs are maintained at the UNI for the EVC. The size of the list is not to exceed N and the pairs should age based on the arrival time against the specified window parameter. For example, if the window parameter is 100 ms, then all pairs that have arrival times that are older than the recent 100 ms will be removed from the list. This way, a limiting factor can be applied to the MAC addresses.
- Test MEG: This could be enabled or disabled. When enabled, the CEN should accommodate the requirements of the test MEG as described in MEF 30.1 [17].
- Subscriber MEG MIP: This could be enabled or disabled. When enabled, the CEN is allowed to set up a subscriber-level MIP as described in MEF 30.1.

6.4.4 E-Line Services Attributes and Parameters

6.4.4.1 EPL Service Attributes and Parameters

There are some service attributes that require specific values for the EPL service. These include UNI service attributes, EVC service attributes, and EVC per UNI service attributes.

The UNI service attributes specific for EPL service are:

- The service multiplexing attribute is disabled, since there is no service multiplexing at the UNI for EPL service that is port based.
- The bundling attribute is disabled.
- All-to-one bundling attribute is enabled to bundle all CE-VLAN IDs into the port-based EVC.
- CE-VLAN ID and EVC Map attribute dictates that all CE-VLAN ISs are mapped to the port-based EVC.
- The maximum number of EVCs attribute is set to one.

EVC service attributes specific for EPL service are:

- EVC type attribute is set to be point-to-point.
- The UNI list has two UNIs with root types.
- The maximum number of UNIs is set to two representing point-to-point connection.
- Unicast, multicast, and broadcast service frame delivery attributes are set to unconditional.
- CE-VLAN ID and CoS preservation attributes are enabled to preserve the CE-VLAN tag values and CoS classification values at both ingress and egress sides of the EVC.

- EVC performance attribute includes all performance metrics for a CoS label as specified in MEF 23.1 [18].

EVC per UNI service attributes specific for EPL service are:

- The egress bandwidth profile per egress equivalence class is set to no.
- The source MAC address limit attribute is disabled.

6.4.4.2 EVPL Service Attributes and Parameters

There are some service attributes that require specific values for EVPL service. These include UNI service attributes, EVC service attributes, and EVC per UNI service attributes.

The UNI service attributes specific for EVPL service are:

- The service multiplexing attribute is enabled, since there are multiple EVC services being multiplied out of a UNI.
- All-to-one bundling is disabled, since there may be only a subset of CE-VLAN IDs being mapped to an EVC.

The EVC service attributes specific for EPL service are:

- The EVC type attribute is set to be point-to-point.
- The UNI list has two UNIs with root types.
- The maximum number of UNIs is set to two representing point-to-point connection.
- Unicast, multicast, and broadcast service frame delivery attributes are set to unconditional.
- The EVC performance attribute includes all performance metrics for a CoS label as specified in MEF 23.1.

The EVC per UNI service attribute specific for EPL service is that the source MAC address limit attribute is disabled.

6.4.5 E-LAN Services Attributes and Parameters

6.4.5.1 EP-LAN Service Attributes and Parameters

There are some service attributes that require specific values for EP-LAN service. These include UNI service attributes, EVC service attributes, and EVC per UNI service attributes.

The UNI service attributes specific for the EP-LAN service are:

- The service multiplexing attribute is disabled, since there is no service multiplexing at the UNI for EP-LAN service that is port-based.
- The bundling attribute is disabled.

- All-to-one bundling attribute is enabled to bundle all CE-VLAN IDs into the port-based EVC.
- CE-VLAN ID and EVC map attribute dictates that all CE-VLAN ISs are mapped to the port-based EVC.
- The maximum number of EVCs attribute is set to one.

The EVC service attributes specific for EPL service are:

- The EVC type attribute is set to be multipoint-to-multipoint.
- The UNI list has multiple UNIs with root types.
- The maximum number of UNIs is set to a value greater than two representing a multipoint-to-multipoint connection.
- The unicast service frame attribute is set to conditional, since the unicast service frames are subject to bridge learning and filtering as explained in IEEE Std. 802.1Q-2011 for VLAN bridges.
- The broadcast service frame delivery attribute is set to unconditional.
- CE-VLAN ID and CoS preservation attributes are enabled to preserve the CE-VLAN tag values and CoS classification values at both the ingress and egress sides of the EVC.
- The EVC performance attribute includes all performance metrics for a CoS label as specified in MEF 23.1.

6.4.5.2 EVP-LAN Service Attributes and Parameters

There are some service attributes that require specific values for EVP-LAN Service. These include UNI Service attributes, EVC Service attributes, and EVC per UNI service attributes.

The UNI service attributes specific for EVPL service are:

- The service multiplexing attribute is enabled, since there are multiple EVC services being multiplied out of a UNI.
- All-to-one bundling is disabled, since there may be only a subset of CE-VLAN IDs being mapped to an EVC.
- CE-VLAN ID and EVC map attribute dictates that at least one CE-VLAN ID is mapped to one EVC.

The EVC service attributes specific for EVP-LAN service are:

- The EVC type attribute is set to be multipoint-to-multipoint.
- The UNI list has UNIs with root types.
- The maximum number of UNIs is set to a value greater than two representing a multipoint-to-multipoint connection.

- The unicast service frame attribute is set to conditional, since the unicast service frames are subject to bridge learning and filtering as explained in IEEE Std. 802.1Q-2011 for VLAN bridges.
- The broadcast service frame delivery attribute is set to unconditional.
- The EVC performance attribute includes all performance metrics for a CoS label as specified in MEF 23.1.

6.4.6 E-Tree Services Attributes and Parameters

6.4.6.1 EP-Tree Service Attributes and Parameters

There are some service attributes that require specific values for the EP-Tree service. These include UNI service attributes, EVC service attributes, and EVC per UNI service attributes.

The UNI service attributes specific for the EP-Tree service are:

- The service multiplexing attribute is disabled, since there is no service multiplexing at the UNI that is port-based.
- The bundling attribute is disabled.
- The all-to-one bundling attribute is enabled to bundle all CE-VLAN IDs into the port-based EVC.
- CE-VLAN ID and EVC map attribute dictates that all CE-VLAN IDs are mapped to the port-based EVC.
- The maximum number of EVCs attribute is set to one.

The EVC service attributes specific for EP-Tree service are:

- The EVC type attribute is set to be the rooted multipoint service type.
- The UNI list has two or more UNIs with at least one UNI with a root type. The other UNI(s) should be the leaf type.
- The maximum number of UNIs is set to a value greater than two, similar to the multipoint connection.
- The unicast service frame attribute is set to conditional, since the unicast service frames are subject to bridge learning and filtering as explained in IEEE Std. 802.1Q-2011 for VLAN bridges.
- The broadcast service frame delivery attribute is set to unconditional.
- CE-VLAN ID and CoS preservation attributes are enabled to preserve the CE-VLAN tag values and CoS classification values at both the ingress and egress sides of the EVC.
- The EVC performance attribute includes all performance metrics for a CoS label as specified in MEF 23.1.

6.4.6.2 EVP-Tree Service Attributes and Parameters

There are some service attributes that require specific values for the EVP-Tree service. These include UNI service attributes and EVC service attributes.

The UNI service attributes specific for EVP-Tree service are:

- The service multiplexing attribute is enabled, since there are multiple EVC services being multiplied out of a UNI.
- The all-to-one bundling is disabled, since there may be only a subset of CE-VLAN IDs being mapped to an EVC.
- CE-VLAN ID and EVC map attribute dictates that at least one CE-VLAN ID is mapped to one EVC.

The EVC service attributes specific for EVP-LAN service are:

- The EVC type attribute is set to be rooted multipoint.
- The UNI list has two or greater number of UNIs with at least one UNI in the root type. The other UNI(s) should be the leaf type.
- The maximum number of UNIs is set to a value greater than two, similar to a multipoint connection.
- The unicast service frame attribute is set to conditional, since the unicast service frames are subject to bridge learning and filtering as explained in IEEE Std. 802.1Q-2011 for VLAN bridges.
- The broadcast service frame delivery attribute is set to unconditional.
- The EVC performance attribute includes all performance metrics for a CoS label as specified in MEF 23.1.

6.5 OVC Services and Attributes

For Carrier Ethernet services, there are three main entities that are involved in delivering these services: (1) subscriber, (2) service provider (SP), and (3) network operator (operator). The subscriber is the entity (customer) that receives the Ethernet service that connects its multiple sites from the service provider. The service provider provides the Ethernet services (EVCs) to its subscribers by employing OVCs from network operators. The service provider is also responsible for making sure that collective performance of all OVCs meets the subscriber SLA. Network operator (operator) is the entity that provides network connectivity service in terms of OVCs to service providers either as transit or access to service providers' subscribers.

OVC is defined as a network connection service in a single CEN provided by the operator that owns the network. The service provider may employ OVCs from one or multiple CEN operators to stitch together an end-to-end EVC service for a subscriber (see Figure 6.29).

There are three general OVC service types in terms of connectivity:

6.5 OVC Services and Attributes

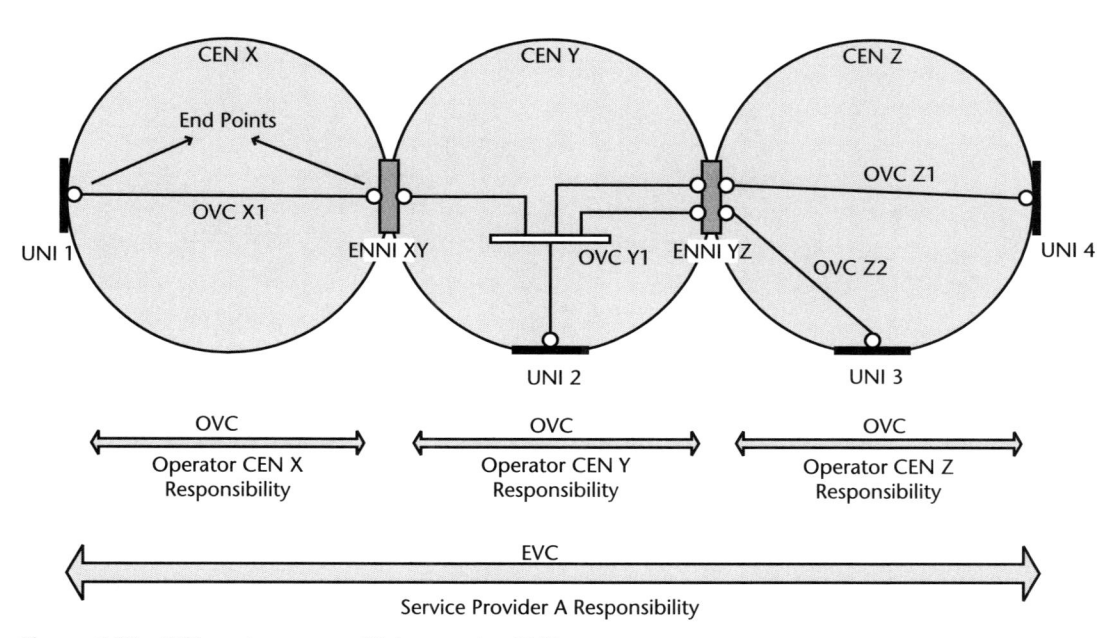

Figure 6.29 EVC service over multiple-operator OVCs.

- O-Line: This uses a point-to-point OVC between two OVC endpoints, one of which is an ENNI. The OVC X1, OVC Z1, and OVC Z2 are all instances of O-Line service type in Figure 6.29. One should note that an O-Line service type could be a part of any EVC service type including E-Line, E-LAN, and E-Tree. To give as an example, all O-Line service instances are part of the E-LAN service being delivered over three operator CENs in Figure 6.29.

- O-LAN: This uses a multipoint-to-multipoint OVC among two or more OVC endpoints, one of which is an ENNI. The OVC Y1 represents an instance of O-LAN service type in Figure 6.29. Unlike the O-Line, O-LAN cannot be a part of an E-Line EVC service; however, it can only be a part of an E-LAN EVC. One should note that there are multiple options to deliver an E-LAN service over a multi-CEN (multi-operator) environment. A combination of O-LAN and O-Line OVC types can be used for delivering the same E-LAN service. For example, the E-LAN service in Figure 6.29 can also be delivered by an alternative combination of OVCs given in Figure 6.30. In this delivery of the same EVC, the service provider buys an O-LAN OVC type, OVC Z3, instead of two O-Line OVCs from the operator who owns the CEN Z (see Figure 6.30).

- O-Tree: This uses a rooted-multipoint OVC among two or more OVC endpoints, one of which is an ENNI. Similar to E-Tree service definition, the UNI endpoints play either root or leaf roles for an O-Tree service. Therefore, for any tree-like service (O-Tree or E-Tree), every endpoint carries a type attribute (conceptually given in the legend portion of Figure 6.31). For ENNI endpoints in the O-Tree service, a third role called the trunk role exists. If an ENNI endpoint provides connectivity to both the leaf and root types of UNIs at the other end of the connection, then the ENNI is called playing a trunk role. For example, the ENNI_YZ endpoints of both OVC Y2 and OVC Z3

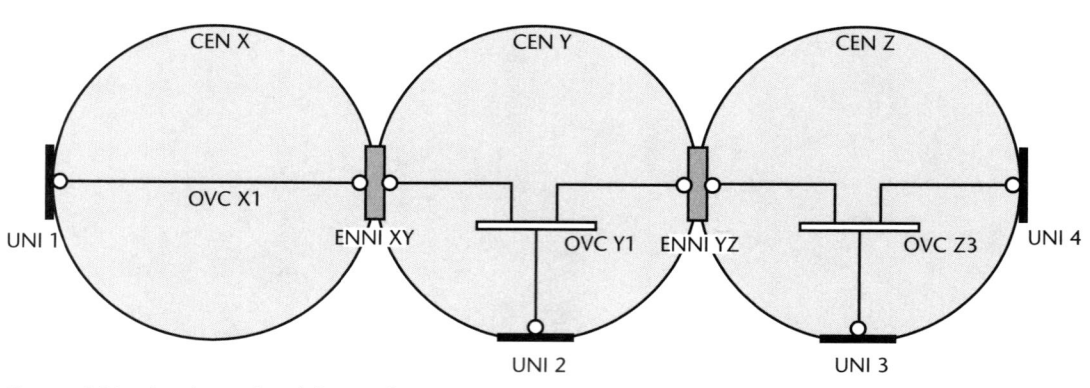

Figure 6.30 An alternative delivery of E-LAN service

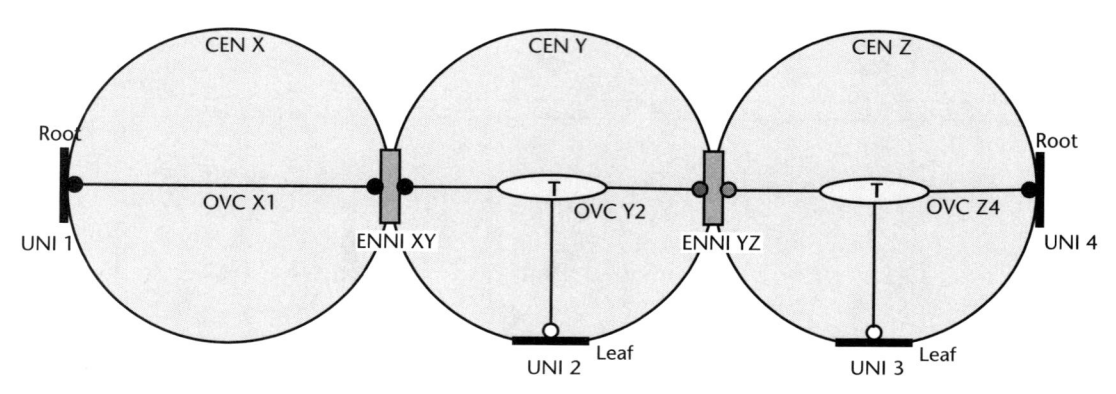

- ● Root OVC End Point
- ○ Leaf OVC End Point
- ◉ Trunk OVC End Point

Figure 6.31 O-Tree services.

O-Tree services in Figure 6.31 play trunk roles, because both root and leaf types of UNIs are being served at the other ends of this E-Tree service. One should note that there are multiple options to deliver an E-Tree service over a multi-CEN (multi-operator) environment. A combination of O-Tree and O-Line OVC types can be used. As an example, one O-Tree in one CEN and multiple O-Line OVC service instances in all other CENs can be used to deliver the E-Tree service as seen in Figure 6.32. As an alternative, one O-Tree instance can be used in each CEN to deliver the same E-Tree service end to end. However, the bridging and where it takes place will change; therefore, the efficiency and performance may differ in these alternatives. To this end, one should compare and contrast all alternatives and choose the best fit to the delivery of the given multi-CEN environment. In such a compare-and-contrast exercise, the number of OVCs to be bought from an operator, the functionality in terms of complexity of control (bridging), and the traffic matrix are all among the issues to be looked at before any decision is made.

OVCs are also classified by looking into their functionalities. Functionality-wise, OVCs play either access or transit role for the EVCs that they serve; therefore,

6.5 OVC Services and Attributes

Figure 6.32 An alternative combination of OVC services to deliver an E-Tree service.

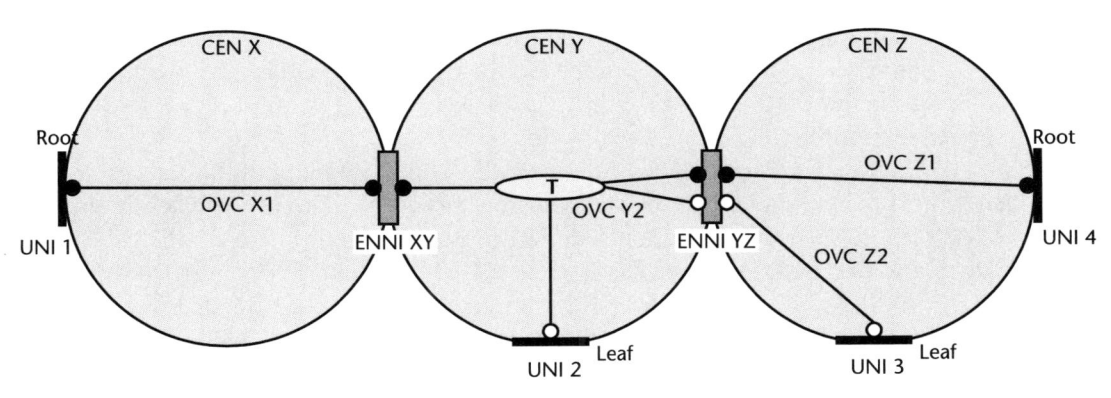

- ● Root OVC End Point
- ○ Leaf OVC End Point
- ● Trunk OVC End Point

an OVC that provides a point-to-point connection between two ENNIs for an EVC that is transiting the operator's CEN is called the transit E-Line OVC. For example, OVC Y4 in Figure 6.33 is a transit E-Line OVC. Following the same taxonomy, an OVC that utilizes the same point-to-point connection between a UNI and an ENNI providing access to one or more subscribers is called the access E-Line OVC; the OVC X1, OVC X2, OVC Z4, and OVC W1 are all access E-Line OVCs. A similar classification applies to multipoint-to-multipoint OVCs. If all endpoints are ENNIs for a multipoint-to-multipoint OVC, then it is called a transit E-LAN OVC. The

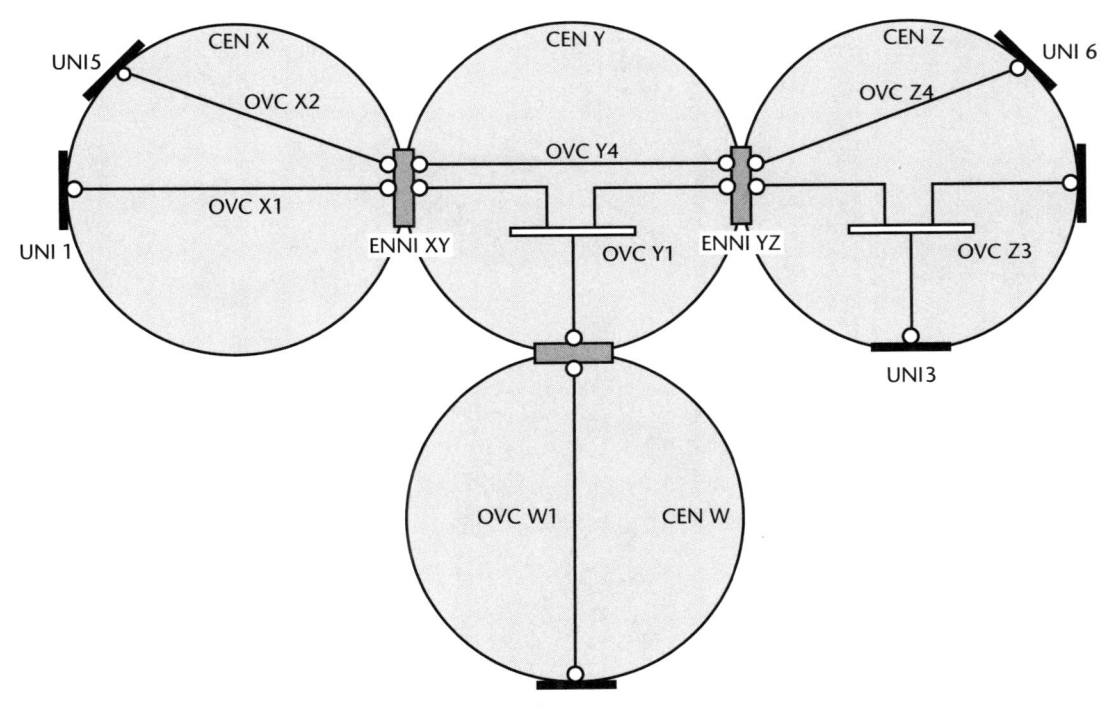

Figure 6.33 Transit/access E-Line/E-LAN OVC services.

OVC Y1 in Figure 6.33 represents an example E-LAN OVC. Having at least one UNI among the multiple OVC endpoints suggests an access to a subscriber and it is called the access E-LAN OVC service. The OVC Z3 is an access E-LAN OVC service in Figure 6.33.

All E-access services perform defined CE-VLAN ID mapping functions at UNI endpoints. These include mapping a single CE-VLAN ID to an OVC endpoint per UNI (see CE-VLAN 112 mapping in Figure 6.34), or multiple (not all) CE-VLAN IDs to an OVC endpoint per UNI (see CE-VLAN 31, 32, 35 mapping to OVC X5), or all CE-VLAN IDs to an OVC endpoint per UNI (see all-to-one mapping at UNI 1 in Figure 6.34). All E-transit and E-access services perform defined S-VLAN ID mapping functions including mapping a single S-VLAN ID to an OVC endpoint per ENNI or multiple S-VLAN IDs per OVC endpoint per ENNI (see S-VLAN mappings in Figure 6.34).

With regard to the management of EVC and OVC services, the existence of OVC services is to support an EVC service; therefore, the collective management of these OVC services that support the delivery of an EVC service is critical. The types of OVCs and EVC should match. For example, the delivery of a multipoint-to-multipoint EVC should require at least one multipoint-to-multipoint OVC in at least one operator's CEN, although it is possible to use more than one in multiple CENs. These options may be compared and contrasted for a better fit in a particular situation. Similarly, it is sufficient to use only point-to-point OVCs for the delivery of a point-to-point EVC. The OVC MTU sizes should also match the EVC values. C-tag preservation attributes for both EVC and corresponding OVCs should be aligned. The service provider needs to carefully negotiate the SLAs for OVCs services such that they collectively support the SLA of the EVC service that is

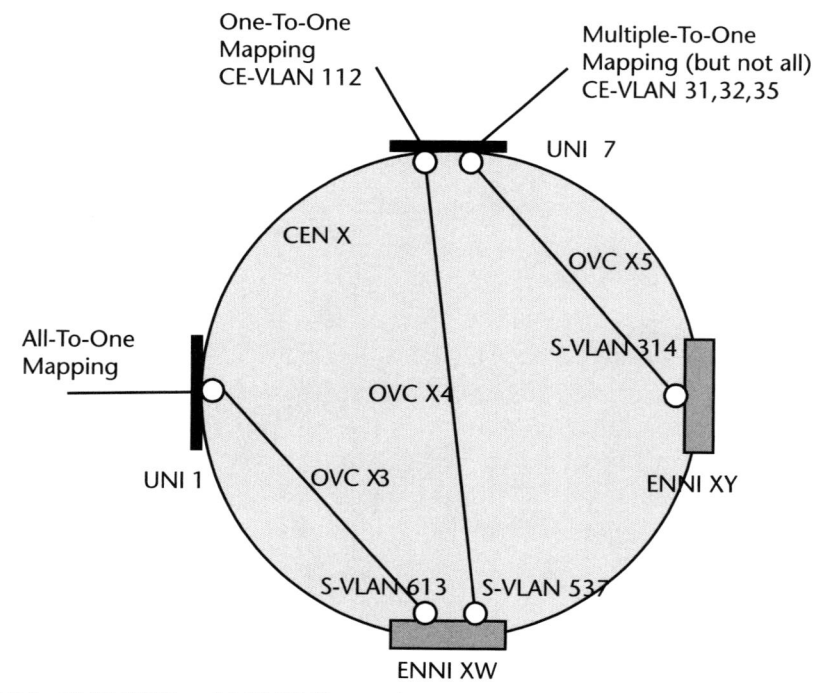

Figure 6.34 CE-VLAN ID and S-VLAN ID mappings.

provided to the subscriber. One should note that CE-VLAN ID mapping attributes for OVCs could be different than for the EVC being served at UNIs. The reason for that is because an OVC endpoint could be shared by multiple EVCs, which leads to a more inclusive mapping for OVC than the EVC at a UNI.

For example, assume that a small private university needs to connect all of its four campus locations in the same large metro area that is covered by multiple operators' CENs as shown in Figure 6.35. For the delivery of this service, the service provider sells the customer (university) a multipoint-to-multipoint EVC service and negotiates the OVC services including the required attributes with operators who own the corresponding CENs in the metro area. During the negotiations, the service provider should also be considering the options to deliver this EVC service. One option is to buy an O-LAN service in the CEN where there are considerably more UNIs and O-Line services for the other CEN(s) as seen in Figure 6.35. Another obvious option is to place multiple the O-LAN services in the metro area and buy O-Line as a transit service to bridge two ends of the metro, as seen Figure 6.36. These options are to be carefully analyzed and compared in both technical feasibility and finances (cost and pricing by the operators). We suggest paying attention to this step while making decisions on employing OVCs to deliver the end-to-end EVC service to the customer.

6.5.1 Operator Services Attributes

Operator-related services will involve the following set of attributes:

- ENNI service attributes;
- UNI service attributes;

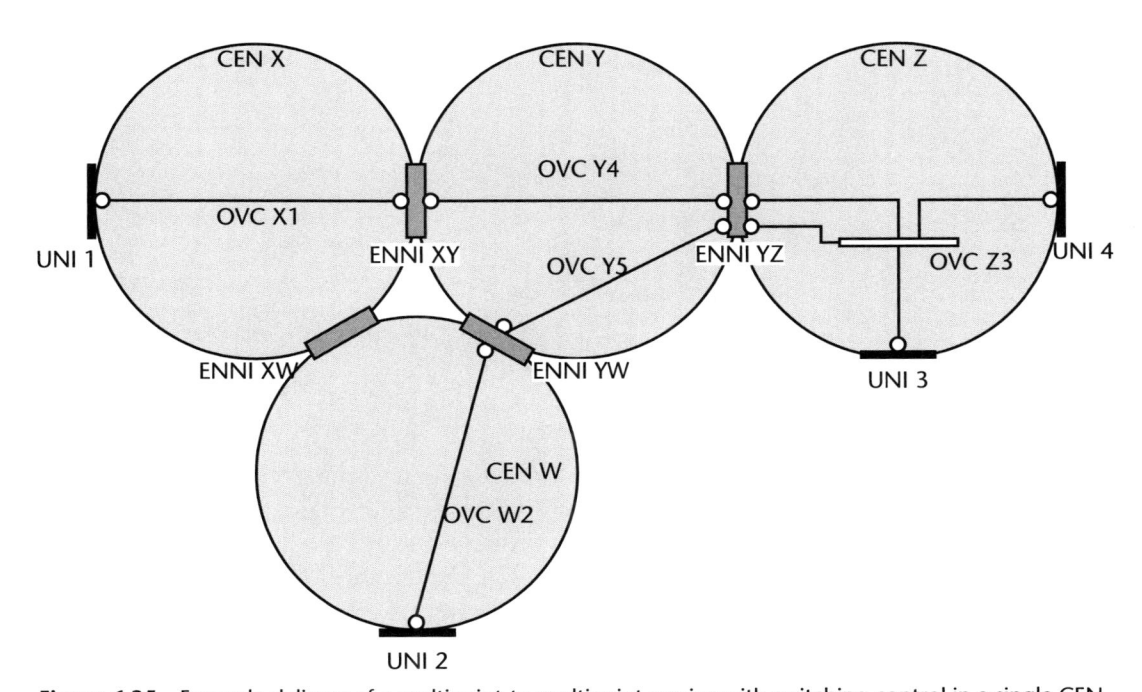

Figure 6.35 Example delivery of a multipoint-to-multipoint service with switching control in a single CEN.

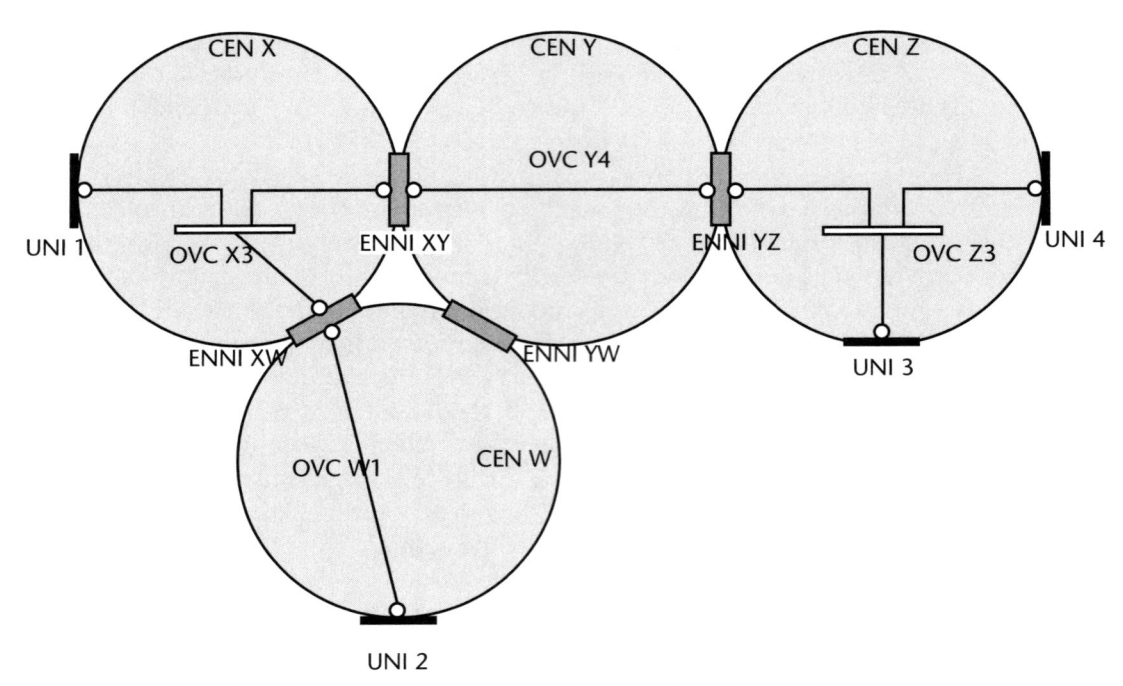

Figure 6.36 Example delivery of a multipoint-to-multipoint service with switching control in multiple CENs.

- OVC service attributes;
- OVC per ENNI service attributes;
- OVC per UNI service attributes.

6.5.1.1 ENNI Service Attributes

ENNI is the demarcation point between two operators that deliver an EVC for a service provider. An ENNI has two sets of attributes, one for each operator that shares the ENNI. These sets may or may not share the same values for each operator. These attributes are as follows:

- ENNI ID: A unique identifier for the ENNI in the operator's network.
- Physical layer: This identifies the physical link that supports the ENNI in the operator's network. Some of the possible values are 1000Base-SX, 1000Base-LX, 1000Base T, 10GBASE-SR, 10GBASE-LX4, 10GBASE-LR, 10GBASE-ER, 10GBASE-SW, 10GBASE-LW, and 10GBASE-EW.
- Frame format: This specifies the frame format at the ENNI. The frames format may have no tag, S-tag, or both C-tag and S-tag. As defined by IEEE 802.1ad-2005, an ENNI frame can have zero or more VLAN tags. When there is a single tag, that tag is an S-tag. When there are two tags, the outer tag is an S-tag (TPID = 0x88A8) and the next tag is a C-tag (TPID= 0x8100).
- Number of links: This specifies the number of links in the ENNI. The value is either 1 or 2.

- Protection mechanism: This specifies the method of protection including LAG.
- ENNI max. transport unit size: This specifies the maximm length in bytes for frames at ENNI.
- Endpoint map: The list that associates S-tag values from each frame with an OVC endpoint at the ENNI. The list has the rows of the form <S-VLAN ID value, endpoint identifier, endpoint type>.
- Max. number of OVC: This specifies the maximum number of OVCs that the operator defines at the ENNI.
- Max. number of OVC endpoints per OVC: This specifies the maximum number of OVC endpoints for an operator OVC at the ENNI.

6.5.1.2 UNI Service Attributes

These attributes are the same as described in the EVC service attributes.

6.5.1.3 OVC Service Attributes

OVC service attributes are as follows:

- OVC ID: This identifies the OVC service uniquely in the operator's network.
- OVC type: This specifies the types of OVCs including point-point, multipoint-to-multipoint, and rooted multipoint.
- OVC endpoint list: This lists the OVC endpoints per OVC service.
- Max. number of UNI OVC endpoints: This specifies the maximum number of OVC endpoints per OVC at different UNIs.
- Max. number of ENNI OVC endpoints: This specifies the maximum number of OVC endpoints per OVC at ENNIs.
- OVC max. transmission unit size: This specifies the maximum length in bytes for frames that are mapped to an OVC endpoint per an OVC.
- CE-VLAN ID preservation: This specifies if C-tags need to be preserved over the OVC service.
- CE-VLAN CoS preservation: This specifies if CoS value of C-tag needs to be preserved over the OVC service.
- S-VLAN ID preservation: This specifies if S-tags need to be preserved over the OVC service.
- S-VLAN CoS preservation: This specifies if CoS value of S-tag needs to be preserved over the OVC service.
- Color forwarding: This specifies if the color of the ingress ENNI frames need to be forwarded to the egress ENNI frames. When color forwarding is enabled for the OVC, the OVC will mark the yellow frame as yellow and cannot promote a frame from yellow to green. The color of the ingress frame will be the same as the color of the ingress frame.

- SLS: This specifies the delivery performance of frames among external interfaces for the OVC service.
- Unicast, multicast, and broadcast service frame delivery: This specifies how ingress frames coming to an OVC endpoint are delivered to other OVC endpoints that belong to the same OVC service.
- Layer-2 control protocol tunneling: This lists the Layer-2 Control Protocols that are tunneled by the defined OVC service.

6.5.1.4 OVC Endpoint Attributes per ENNI

Each OVC endpoint at an ENNI interface has a set of the following attributes:

- OVC endpoint ID: This identifies the OVC endpoint.
- Trunk identifiers: This specifies the S-tag values at an ENNI to differentiate frames between those originating at a root UNI and a leaf UNI.
- CoS identifiers: This specifies the CoS delivery for each S-tagged service frame at ENNI.
- Ingress bandwidth profile per OVC endpoint: This specifies the policing on all ingress ENNI frames per OVC endpoint. The bandwidth profile parameters are <CIR, CBS, EIR, EBS, CM, CF>. The ingress CIR for an OVC at the ENNI should be greater than the corresponding ingress CIR at the UNI due to the presence of the added SVLAN tag (4 bytes) at the ENNI.
- Ingress bandwidth profile per ENNI CoS identifier: This specifies the policing on all ingress ENNI frames per OVC endpoint with a given CoS identifier.
- Egress bandwidth profile per OVC endpoint: This specifies the policing on all egress ENNI frames per OVC endpoint.
- Egress bandwidth profile per ENNI CoS identifier: This specifies the policing on all egress ENNI frames per OVC endpoint with a given CoS identifier.

6.5.1.5 OVC Attributes per UNI

Each OVC at a UNI interface has a set of the following attributes:

- UNI OVC ID: This identifies the OVC at a UNI.
- OVC endpoint map: This lists the C-tags that are associated with the UNI over the OVC service.
- CoS identifiers: This specifies the CoS delivery for each C-tagged service frame at UNI.
- Ingress bandwidth profile per OVC: This specifies the policing on all ingress UNI frames per OVC.
- Ingress bandwidth profile per CoS identifier: This specifies the policing on all ingress UNI frames per OVC with a given CoS identifier.

- Egress bandwidth profile per OVC: This specifies the policing on all egress UNI frames per OVC.
- Egress bandwidth profile per CoS identifier: This specifies the policing on all egress UNI frames per OVC with a given CoS identifier.

6.6 Single CEN and Multi-CEN L2CP (Processing for Single CEN and Multi-CEN)

A Layer-2 Control Protocol (L2CP) frame is uniquely defined by having one of the standardized reserved addresses in Table 6.6 [16] as the destination address. These addresses in Table 6.6 are divided into two blocks: (1) bridge block of protocols, and (2) Multiple Registration Protocol (MRP) block of protocols. The assignment of bridge block addresses is detailed in Table 6.7 [16]. Each address from this block, once it is used as a destination address, will dictate a special forwarding rule at some bridge types. Applying this forwarding rule, called filtering, the bridge makes sure that the frame is not propagated further to any other egress ports of the bridge and does not confuse the other network devices in the network. This forwarding process will isolate the Layer-2 Control Protocol communication to the involved Layer-2 devices only. From the Layer-2 device point of view, each bridge type may filter a different subset of the bridge reserved addresses. For example, end stations, tag-blind MAC bridges, and tag-aware customer and provider edge bridges apply the filtering to the entire block of addresses as seen in Table 6.7. Two port MAC relays and provider bridges that understand S-VLAN tags apply the filtering rules to only a few of those addresses from the reserved block (see Table 6.7). This classification of filtering devices will allow the sender of L2CP frames to choose the intended device types that filter and receive these frames. For example, the frames that have the MAC-specific control protocol address of 01-80-C2-00-00-01 will be filtered by all devices. Frames with the address of 01-80-C2-00-00-00 that indicate the nearest customer bridge will only be filtered by the customer bridge and provider bridges and two port MAC relays will not peer or discard these frames. It is important to note that the address does not specify a L2CP; instead, the protocol identifier uniquely determines the L2CP. Therefore, a specific L2CP can use different addresses from the block to determine the intended region and the recipient of the L2CP frames. For example, one should use the nearest customer bridge address for the RSTP protocol, if the intent is to run the RSTP protocol in a customer network. The address will change to the corresponding one from the reserved block, if the provider network is targeted to run the protocol.

Table 6.6 L2CP Destination MAC Addresses (Standardized) [16]

L2CP Destination Address	Address Block Type
From 01-80-C2-00-00-00 to 01-80-C2-00-00-0F	Bridge reserved addresses
From 01-80-C2-00-00-20 to 01-80-C2-00-00-2F	Multiple Registration Protocol (MRP) reserved addresses

Table 6.7 Bridge Reserved L2CP Addresses [16]

		Filtered By		
Address	Assignment	End station, MAC bridge, customer bridge, provider edge bridge	Provider bridge	Two Port MAC relay
01-80-C2-00-00-00	Nearest customer bridge	√	—	—
01-80-C2-00-00-01	IEEE MAC specific Control Protocols	√	√	√
01-80-C2-00-00-02	IEEE 802 Slow Protocols	√	√	√
01-80-C2-00-00-03	Nearest non-TPMR bridge	√	√	—
01-80-C2-00-00-04	IEEE MAC specific Control Protocols	√	√	√
01-80-C2-00-00-05	Reserved	√	√	—
01-80-C2-00-00-06	Reserved	√	√	—
01-80-C2-00-00-07	MEF Forum ELMI	√	√	—
01-80-C2-00-00-08	Provider Bridge Group	√	√	—
01-80-C2-00-00-09	Reserved	√	√	—
01-80-C2-00-00-0A	Reserved	√	√	—
01-80-C2-00-00-0B	Reserved	√	—	—
01-80-C2-00-00-0C	Reserved	√	—	—
01-80-C2-00-00-0D	Provider Bridge MVRP	√	—	—
01-80-C2-00-00-0E	Nearest bridge, individual LAN scope	√	√	√
01-80-C2-00-00-0F	Reserved	√	—	—

The MRP addresses from the table are used to implement a filtering rule that allows intermediate Layer 2 devices to pass the frame if they do not understand the protocol ID as opposed to discarding them.

The general behavior of the CEN for L2CP frames has two distinct models regarding interfaces. One model is from the subscriber and service provider (S/SP) point of view, which involves only UNIs. The other model, which is potentially for multiple CENs, is from the operator and service provider (O/SP) point of view and involves both UNIs and ENNIs. In O/SP model that runs on multi-CEN networks, there are actions taken on L2CP frames at UNNIs by multiple operators, as well as at UNIs. By instituting this model, the subscriber L2CP frames can be treated differently than the provider L2CP frames.

To implement these behavior models, a conceptual entity called the decision point (DP) has been defined to process the filtering rules at external interfaces (EI) (see Figure 6.37) [16]. These DPs are introduced at each UNI, ENNI, and VUNI in a CEN to make decisions about the incoming L2CP frames and dictate if they need to be discarded, peered with the corresponding protocol instance, or passed to an EVC or OVC in the CEN. These three actions that are illustrated in Figure 6.37 are decided upon the destination address and the protocol identifier information that come in within the frame and the L2CP service attributes that are configured for that particular UNI, VUNI, or ENNI at a location. The set of DP service attributes that will be consulted to make a decision are tabulated in Table 6.8 depending on the type of interface at a location.

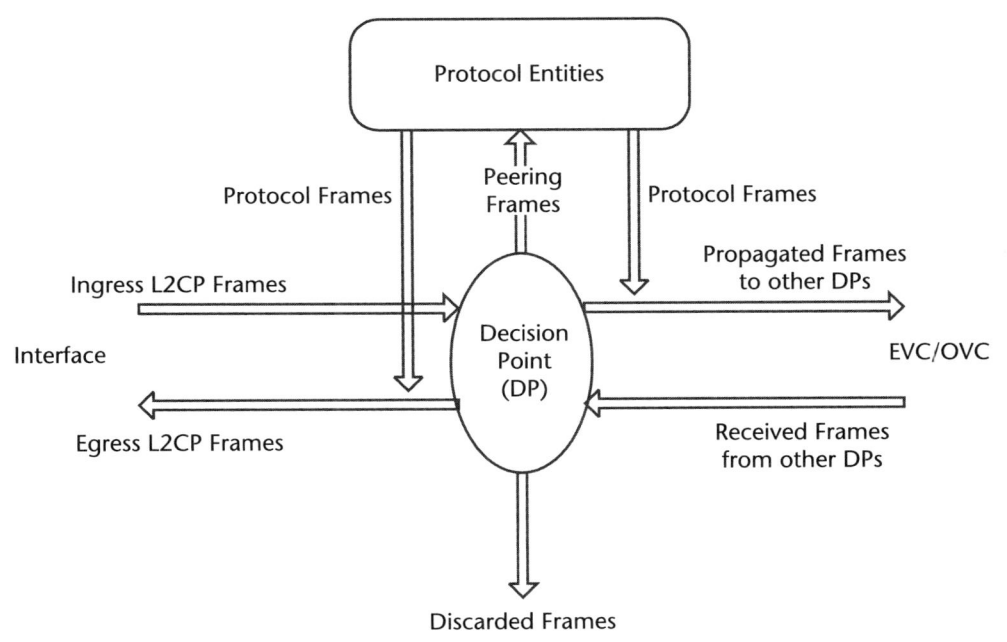

6.6 Single CEN and Multi-CEN L2CP (Processing for Single CEN and Multi-CEN)

Figure 6.37 L2CP decision point [16].

Table 6.8 DP Service Attributes [16]

DP Located at	L2CP Service Attributes used by DP
UNI	UNI L2CP peering service attribute and UNI L2CP address set service attribute
VUNI	VUNI L2CP peering service attribute and VUNI L2CP address set service attribute
ENNI	ENNI L2CP peering service attribute, ENNI tagged L2CP frame processing service attribute, and OVC L2CP address set service attribute (for each OVC with an OVC endpoint at the ENNI)

In any decision, the behavior may change; therefore, should be analyzed separately for ingress and egress bound frames.

If the decision is made for discarding an ingress frame at an EI, then it should not be propagated to any EVC/OVC or should not be peered with any protocol entity either (see Figure 6.37). A separate decision should be made for the egress frame coming from an EVC or OVC into an EI. If that is also decided to be a discard, then the incoming frame is not propagated to any EI or not peered either.

If the decision is made for peering ingress or an egress frame, then the frame is handed off to the corresponding protocol instance that is running at the node. The protocol instance will be determined by much other input including the protocol identifier and the address used in the frame. The specifics of the protocol and how the frame will be treated after the hand-off are dependent on the particular protocol.

If the decision is made for passing the L2CP frame without any peering or discarding, then the frame is treated as a data frame with a multicast destination address. Ingress frames are mapped to the corresponding EVC or OVC by following the same mapping rules applied to regular data frames. Egress frames are simply propagated to the corresponding EI.

In a single CEN environment, the L2CP behavior model is constructed over an EVC that could have two or more UNIs. Each UNI that participates in an EVC has a decision point. The behavior that is implemented between any pair of UNIs is the product of the decisions made by both the ingress and egress decision points (see Figure 6.38). For example, if one of the DPs discards the frame, then the frame will be off the network. If any DP decides to peer the frame, then the frame will not be propagated to any EI and the corresponding protocol activity will take place at the peering DP. This activity may trigger a set of actions depending on the specifics of the protocol.

In a multi-CEN environment where multiple network operators are involved, an EVC may span over two or more CENs. In addition to UNIs, the EVC involves multiple OVCs (each of which belongs to an operator) that connect via ENNIs. Figure 6.39 illustrates an example EVC that spans over two CENs with a single ENNI in between two operators' CENs. DPs reside in each UNI and each side of an ENNI in the corresponding CEN. Therefore, there is one DP in the CEN 1 side of the ENNI and another on the CEN 2 side. The overall filtering/forwarding behavior for an L2CP frame is determined by all these DPs in the multi-CEN environment. For example, when the UNI DP in CEN 1 operator network decides to pass the ingress L2CP frame, the frame is mapped to the serving OVC in CEN 1 network and propagated to the egress ENNI DP in CEN 1 operator network. Once it receives, the egress DP at ENNI in CEN 1 processes the frame and decides to discard, peer, or pass this frame to the ingress ENNI DP of the CEN 2 operator network. The CEN 2 operator network DPs repeat the same procedures and make the necessary decisions according to the L2CP service attributes. It is important to note that an L2CP frame that show up at the ingress ENNI of an operator's CEN might have been originated at the UNI or an egress ENNI of an earlier CEN in transit. Therefore, there are both subscriber and operator L2CP frames in a multi-CEN environment.

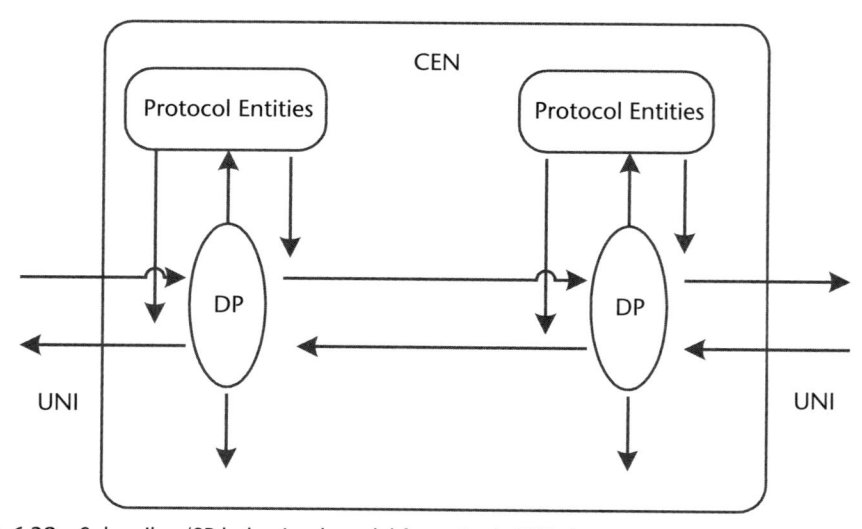

Figure 6.38 Subscriber/SP behavioral model for a single EVC that spans over a pair of UNIs [16].

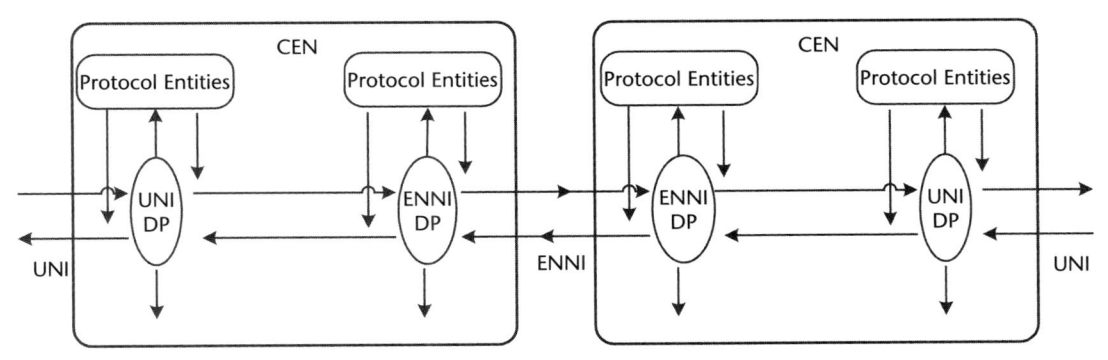

6.6 Single CEN and Multi-CEN L2CP (Processing for Single CEN and Multi-CEN)

Figure 6.39 Operator/SP behavioral model for multi-CEN [16].

The L2CP address set service attribute specifies the filtering rules for a subset of bridge reserved address list. Depending on the attribute, the L2CP frame is either peered or discarded, but not passed, if a specified address is used for the frame. There are three values for the L2CP address set service attribute:

- CTA: The C-VLAN tag-aware value is assigned when CE-VLAN ID is used to map a frame to a service.
- CTB: The C-VLAN tag-blind value is assigned when CE-VLAN is not used to map a frame to a service.
- CTB-2: The C-VLAN tag-blind option 2 value is assigned when point-to-point, port-based services support the EPL option 2 L2CP processing.

Every UNI, VUNI, and OVC entity in a single and multi-CEN environment has a defined value for the L2CP address set service attribute to process the L2CP frames at the DPs. Assigning these values to the attributes has some constraints depending on the type of the agreement. The subscriber-to-service provider agreement dictates the mapping of frames to an EVC at an UNI. Under this agreement, a UNI will have CTB as the value of the attribute, when all-to-one bundling is applied. The value will be CTB-2, if all-to-one bundling is applied and EPL service uses option 2 processing. Otherwise, this attribute will be CTA, when all-to-one bundling is disabled. The operator-to-service provider agreement dictates the mapping of frames to an OVC at an UNI or VUNI when a multi-CEN environment is concerned. Under this agreement, similar constraints exist; if not all CE-VLAN IDs map to a single OVC at a UNI or VUNI, then the attribute value should be CTA.

A L2CP peering service attribute is assigned to each UNI, VUNI, and ENNI and constitutes a list of L2CP that should be peered at the corresponding DP at those interfaces. Those frames that are not listed as "to-be-peered" will be either discarded or passed. In Table 6.9 an L2CP peering service attribute is illustrated for Spanning Tree protocols (RSTP/MSTP), as an example.

Table 6.9 An Example L2CP Peering Service Attribute List

Peering Protocol	Protocol ID	L2CP Frame Destination Address
Spanning Tree (RSTP/MSTP)	0x82	01-80-C2-00-00-00

There is also an ENNI-tagged L2CP frame processing attribute for all ENNIs that indicates whether or not the ENNI can process the S-VLAN tag in the L2CP frame as it is described in the IEEE 802.1 standard. The binary values for this attribute are defined as either 802.1-compliant, if the DP at ENNI can process the L2CP frame according to the 802.1, or 802.1-noncompliant.

The decision points at UNIs and VUNIs interfaces will process both the ingress and egress L2CP frames as seen in the flow chart given in Figure 6.40. The interface that hosts the DP is assumed to have the L2CP address set attribute and peering service attribute assigned. Then the DP basically inspects the destination address and the protocol identifier of the incoming frame and makes a filtering decision by comparing those with the assigned attributes of the interface. The first decision, which can be seen in the flow diagram, is about peering the L2CP frame. The DP checks to see if the destination address and the protocol identifier from the frame have a match in the peering attribute list. If there is a match, then the frame will be

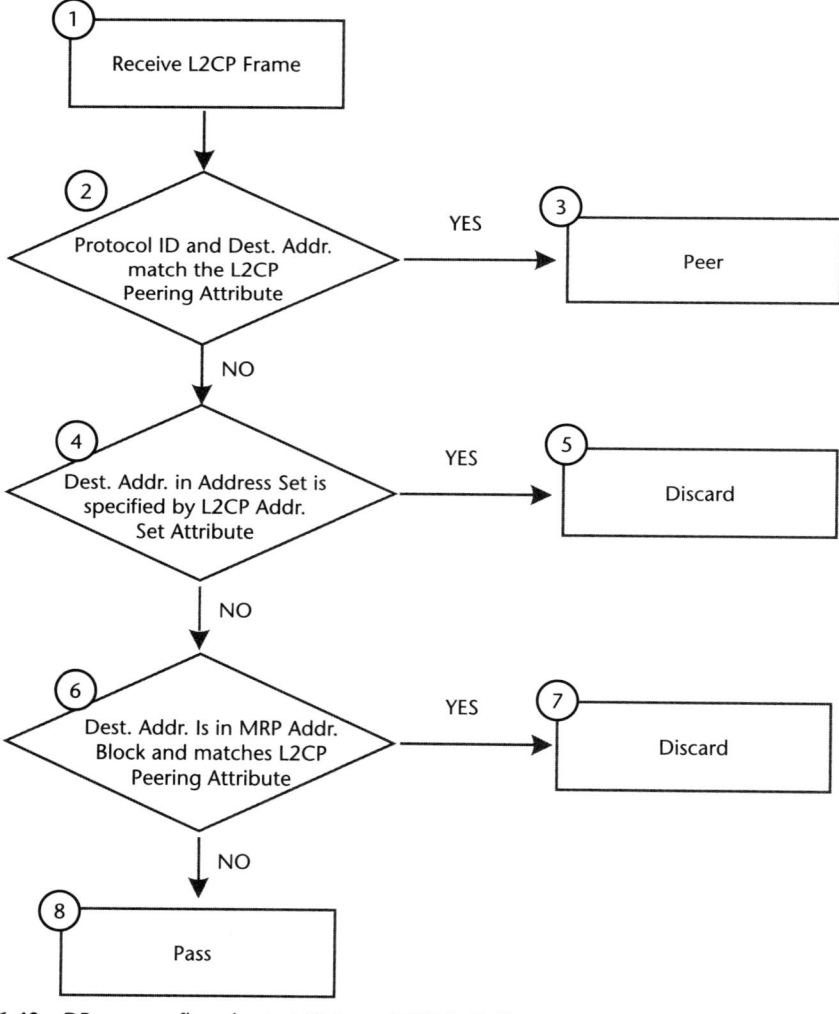

Figure 6.40 DP process flowchart at UNIs and VUNIs [16].

peered at this interface. In the second decision, the DP will check the destination address against the L2CP address set attributes to see if the frame should be discarded. Lastly, the destination address will be checked if it is one of the addresses in the MRP reserved address block and also matches a destination address in peering service attribute list. These matching frames will be discarded and all others will be passed. The reason for this last check is for filtering all MRP address frames that have a peering destination address but have a different protocol identifier.

The decision points at UNNI interfaces follow a different process to filter the L2CP frames; see the extended flowchart in Figure 6.41. In addition to the peering service attribute and L2CP address set attribute, the ENNI has the tagged L2CP frame processing attribute which states the compliancy of S-tags to the 802.1 standard. The DP at the particular ENNI interface will inspect the S-tag in addition to the destination address and the protocol identifier from the L2CP frame and make a decision to peer, discard, or pass the frame by comparing the information extracted from the frame with the assigned attributes of the interface. The DP will start with checking if the frame is s-tagged. If the frame is without the S-tag, then it will try to peer the frame by matching the destination address and the protocol identifier with the peering attributes of the ENNI interface. If the frame cannot be peered, then it should be discarded because the port-based L2CP frames cannot be mapped to an OVC at an ENNI. Those frames with an S-tag that map to an OVC endpoint with an option 2 (CBT-2) will be passed. The next decision point will examine if the ENNI supports the 802.1-compliant processing of frames. If not compliant, the frame will be passed. Those compliant frames that are not passed will be checked for peering by matching the destination address and protocol identifier with the peering attributes of the ENNI interface. If the frame is not peered, then its destination address will be checked against the addresses that are listed as port-based service (CTB) in the L2CP address set service attribute list. If there is a match, then the frame will be discarded. The last decision in the filtering process will be made for the MRP reserved address block, which is the same filtering done by the DPs at UNIs and VUNIs. Again, the frames with the destination addresses that match any address in UNNI peering service attribute list will be discarded. Any frame that comes out negative from this last check by this ENNI decision point will be passed.

As an example to walk through the ENNI DP flow chart, assume that a Spanning Tree (STP/RSTP/MSTP) frame shows up at an OVC end point, where it is evaluated by an egress ENNI L2CP DP first. The DP will first check the S-VID and determine that the frame is mapped to a nonzero S-VID. Later, the DP will examine the OVC address set service attribute and pass the frame if attribute value is CTB-2. For those frames that do not pass, there will be a check for 802.1 compliancy. If the attribute value is noncompliant, the DP will pass the frame. The destination address for the Spanning Tree is given as the nearest customer bridge address; therefore, there should not be any matching line in the ENNI L2CP peering service attribute for this frame. The frame does not carry any MRP address either. As a result, the frame cannot be discarded and will be passed.

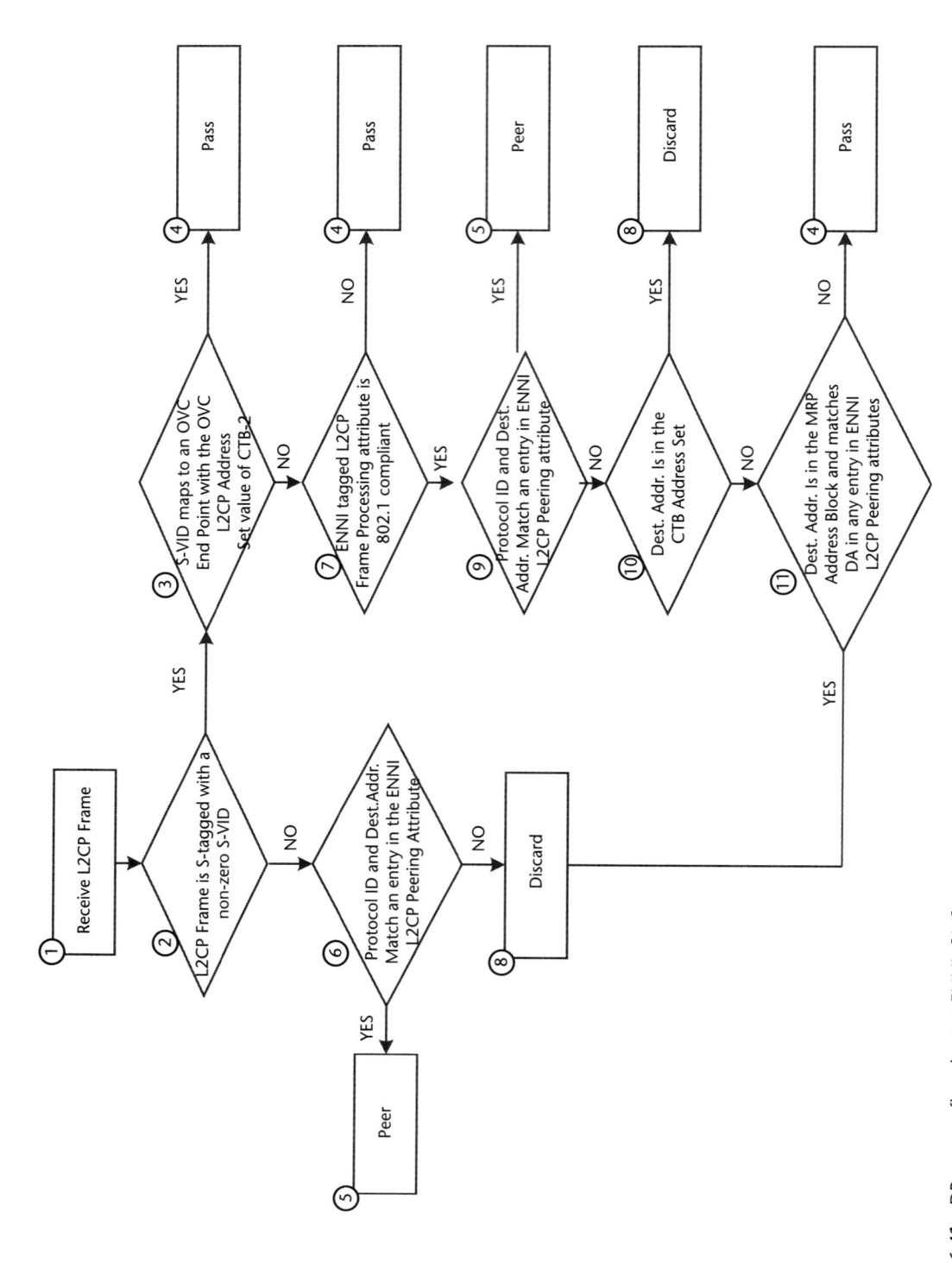

Figure 6.41 DP process flowchart at ENNIs [16].

6.7 Conclusion

For a long time, the Ethernet has been a dominant technology in enterprise networks, allowing bandwidth partitioning, user segregation, and traffic prioritization, introduced by IEEE 802.1d. It is scalable from 10 Mbps to 100 Gbps. Various services have been defined by MEF that can be used to support numerous user applications.

In this chapter, we have defined Carrier Ethernet fundamentals in terms of architectures and services that can be used in service deployments on a single network as well as multiple networks.

References

[1] MEF, "Carrier Ethernet Services Overview," 2011.
[2] "Understanding Carrier Ethernet Service Assurance – Part I/Part II," MEF White Paper, September 2016.
[3] "Carrier Ethernet and NFV," MEF Whitepaper, July 2015.
[4] G8032, "Ethernet Ring Protection Switching," 2012.
[5] IEEE 1588v2, "IEEE Standard for a Precision Clock Synchronization Protocol for Networked Measurement and Control Systems," 2008.
[6] ITU-T G. 8261, "Timing and Synchronization aspects in Packet Networks," 2008
[7] ITU-T G.8010, "Architecture of Ethernet Layer Networks," 2004.
[8] ITU-T G.809, "Functional Architecture of Connectionless Layer Networks," 2003.
[9] IEEE 802.3-2005, "Information Technology – Telecommunications and Information Exchange Between Systems – Local and Metropolitan Area Networks -Specific Requirements – Part 3: Carrier Sense Multiple Access with Collision Detection," 2005.
[10] MEF 11, "User Network Interface (UNI) Requirements and Framework," November 2004.
[11] MEF 13, "User Network Interface (UNI) Type 1 Implementation Agreement," November 2005.
[12] MEF 20, "User Network Interface (UNI) Type 2 Implementation Agreement," 2008.
[13] MEF 10.3, "Ethernet Services Attributes Phase 3," October 2013.
[14] MEF 28, "External Network-Network Interface (ENNI) Support for UNI Tunnel Access and Virtual UNI," October 2010.
[15] MEF 6.2, "EVC Ethernet Services Definitions Phase 3," August 2014.
[16] MEF 45, "Multi-CEN L2CP," August 2014.
[17] MEF 30.1, "Service OAM Fault Management Implementation Agreement Phase 2," April 2013.
[18] MEF 23.2, "Class of Service Phase 3 Implementation Agreement," August 2016.

CHAPTER 7

Carrier Ethernet Traffic Management

7.1 Introduction

The data plane for Carrier Ethernet consists of media access control (MAC), frame header parsing, frame classification for forwarding, quality of service (QoS), data encapsulation for multiple protocols, and traffic management including policing, queuing, and scheduling of frames. Traffic management function is also called packet conditioning.

The classification function identifies packet flows using filters that have been statically configured in the node. It may consists of a series of filter rules specifying values of various fields in the Layer-2, Layer-3 (IPv4 or IPv6), or Layer-4 (TCP, UDP, or ICMP) headers. Some of these fields are MAC source and destination addresses, 802.1p priority bits, VLAN ID, IPv4 protocol field, source, and destination addresses, DSCP field, IPv6 source and destination address, class, flow label fields, TCP/UDP port, flag and mask fields, and ICMP type and code fields.

Packets are conditioned after classification. Once the classification processing has identified a particular flow to which the given packet belongs, the matching filter will specify a chain of actions to execute on the packet flow to yield appropriate conditioning. In Layer-2 and Layer-3 networks, the conditioning actions include:

1. Counting bytes and packets that pass and drop;
2. Policing packet flow according to predetermined burst/flow rate (includes both color-aware and color-unaware);
3. Setting color;
4. Setting 1P/DSCP/TOS priority value;
5. Remarking/remapping 1P/DSCP/TOS based on a predefined re-marking table;
6. Setting next hop for policy-based routing (PBR);
7. Setting forwarding information base (FIB);
8. Shaping packets to conformance to predetermined flow rate;
9. Sending packets to a particular queue.

In the upstream (i.e., customer to network) direction, the conditioning unit maps frames arriving on a UNI to one or more EVCs. Frames in EVCs are assigned

163

to a flow and a flow is assigned to a policer for optional rate limiting. A flow can consist of all the frames on a UNI or some subset of frames, based on VLAN and priority values. The resulting policed flows are then combined into QoS queues that are scheduled for transmission on the network EVCs. QoS queues can consist of one or more policed flows from a UNI. In the downstream direction (i.e., network to customer), frames arrive from the network port and are classified into EVCs on the basis of VLAN tags. Each flow can consist of multiple classes of service that are mapped onto schedulers for transmission on a client UNI.

Ingress frames arriving on a UNI to one or more EVCs are mapped to a CoS flow for optional policing. A CoS flow may consist of:

- All frames belonging to the same UNI.
- A subset of frames belongs to one or more VLANs on the same UNI. Untagged frames can be assigned to a default port VLAN.
- A subset of frames belongs to a VLAN with one or more priority (802.1p or TOS or DSCP) levels. A single VLAN can have multiple policing flows in order to police the rate of different classes of service independently.

Once classified to a CoS flow, ingress frames are then policed according to the CoS bandwidth profile assigned to the flow. Following policing, a CoS flow is then assigned to a QoS queue for scheduling.

Frames in a CoS flow that are not discarded by the policer proceed to the next step, where they are queued on a QoS flow queue associated with the destination EVC. The system determines whether to queue the frame or discard based on the color marking of the frame and the queue fill ratio. When the queue reaches a certain level, let's say 70 or 80%, yellow frames will be discarded.

It is desirable, within the same EVC, to support multiple priorities to support applications such as VoIP to coexist in the same flow with other data services.

Priority classification is enabled at the UNI (port) level by the user. Priority classifications can be defined by 801.2p, TOS, and DSCP markings. Ingress frames may then be mapped to one of the CoS priorities in the system. In addition to assigning a CoS priority level, the table can be used to re-mark the priority level of the incoming frame's 802.1p field.

In the following sections, we describe the key components of Carrier Ethernet traffic engineering including CoS and associated SLAs, bandwidth profiles, burst size and TCP impact, and token sharing.

7.2 Policing

CoS flows are policed according to CIR, CBS, EIR, EBS, CF, and CM. Policers employ the leaky bucket method [1, 2] to rate limit traffic. Leaky bucket method is used to support single-rate, three-color marking (SRTCM) [3] or two-rate, three-color marking (TRTCM) [4].

Bandwidth profile parameters are committed information rate (CIR), excess information rate (EIR), committed burst size (CBS), excess burst size (EBS), coupling flag (CF), and color mode (CM).

CIR defines the average rate (bps) of ingress traffic up to which the device guarantees the delivery of the frames while meeting the performance objectives related to traffic loss, frame delay, and frame delay variations.

EIR defines the excess rate (bps) of ingress traffic up to which the device may deliver the frames without any performance objectives. EIR traffic may be oversubscribed on the WAN interface and may be dropped if the network cannot support it. Traffic offered above the EIR is dropped at the UNI.

CBS defines the maximum number of bytes available for a committed burst of ingress frames sent at the UNI line rate for which the device guarantees delivery.

EBS defines the maximum number of bytes available for a burst of excess ingress frames sent at the UNI line rate for which the device may deliver.

Policers employ the token bucket method to rate limit traffic. The system can operate in color-aware (CA) mode or color-blind mode. The CF is used to share bandwidth between CIR and EIR settings on a flow.

Allocated tokens never exceed CBS + EBS. Token rates determine policing granularity. For example, if 1 token represents 64 bits and tokens are refreshed every millisecond, then we can have a scheduler granularity of 64 kbps. An ingress frame is accepted if there is at least one token available (i.e., token count can become negative).

Most data traffic on typical networks exhibits a burst-like nature, consisting of short bursts of packets at line rate, interspersed with intervals of silence. The resulting average traffic rate is often much less than the line rate.

Traffic that exceeds the CIR setting but is less than or equal to the EIR setting is transported if bandwidth is available. The aggregate of CIR+EIR traffic on a device may exceed the line rate of the network interface.

Through combinations of CIR, EIR, CBS, and EBS settings, the system can be tuned to provide different QoS characteristics to match the bandwidth, traffic loss, latency, and jitter requirements for each traffic class.

The policing algorithm determines whether a customer frame is accepted from the UNI and if it is to be marked as green or yellow. The algorithm is executed for each received frame.

CBS and EBS values depend on various parameters, including CIR, EIR, loss, and the delay to be introduced. In [5], CBS is required to be ≥12,176 bytes.

7.3 Queuing, Scheduling, and Flow Control

Frames that are marked green by the policer are always queued, and frames that are marked yellow are only queued if the fill level of the queue is less than a threshold such as 70%. This ensures that green frames are always forwarded.

Within a QoS flow, unused tokens at one priority level cascade to lower priority levels. Committed tokens flow down to committed tokens, and excess tokens flow down to excess tokens.

Each CoS flow queue is assigned a set of buffers from a buffer pool. The size of the queue can be tuned to satisfy the performance characteristics of the associated flow. Buffers can be allocated on a per-queue basis and are not shared between queues or between services. Each EVC instance can be assigned multiple queues for ingress and each client UNI can be assigned multiple queues for egress to support prioritized traffic scheduling.

Flow control can be enabled on ports. As traffic arrives from client LAN ports, system buffers may begin to fill. At some point, the buffer utilization may cross a pause threshold. If the service definition has enabled 802.3x [6] flow control, the system generates a PAUSE control frame to the sender. The sender, interpreting the PAUSE control frame, halts transmission for a brief period of time to allow the traffic in the bucket to flow onto the network. This active flow control function permits the sending device (customer) to manage traffic priority, discard, and traffic shaping according to its own policies.

7.4 Three-CoS Model

The integration of real-time and nonreal-time applications over data networks require differentiating packets from different applications and provide differentiated performance according to the needs of each application. This differentiation is referred to as class of service (CoS).

There are three classes of services [7]. Applications such as e-mail, Internet access, control traffic, VoIP, video, and pricing considerations due to distance differences such as New York-Washington, D.C., versus New York-San Francisco or New York-London may require more than three classes. Three classes are defined as H, M, and L. Their priorities are indicated by PCP or DSCP or TOS bytes. Each CoS label has its own performance parameters.

- MEF Class H is intended for real-time applications with tight delay/jitter constraints such as VoIP.
- MEF Class M is intended for time-critical data.
- MEF Class L is intended for nontime-critical data such as e-mail.

Tables 7.1 and 7.2 list the parameters of these three classes [8].

Performance parameters do not have to be uniform across CoS labels, PTs, or EVC/OVC types. For example, the T associated with FLR may be different from the T associated with FD.

Delay and jitter values above are valid for PT-to-PT connections. Those constraints for multipoint connections have not been defined by MEF yet. We expect similar figures to hold for multipoint connections as well.

Bandwidth profile for CoS label L allows CIR or EIR = 0. When CIR = 0, there will be no performance objectives while the case of CIR > 0 will require conformance with performance attribute objectives.

At ENNI [9], DEI is used to represent yellow; therefore, there is no need for DSCP/PCP values to represent color.

A CoS instance for a service frame is identified either by the EVC or by the combination of the EVC and the user priority field in tagged service frames (i.e., PCP) or DSCP values.

Table 7.1 Parameters of H, M, and L Classes

Performance Attribute	H	M	L
FD	≥99.9th percentile of frame delay (FD) values monitored over a month	≥99th percentile of FD values monitored over a month	≥95th percentile of FD values monitored over a month
MFD	Mean frame delay values calculated over a month	Mean frame delay values calculated over a month	Mean frame delay values calculated over a month
IFDV	≥99.9th percentile of FD values monitored over a month, with 1 second between frame pairs	≥99th percentile of FD values monitored over a month, with 1 second between frame pairs or not supported	Not supported
FDR	≥99th percentile corresponding to minimum delay	≥99th percentile corresponding to minimum delay or not supported	Not supported
FLR	Frame loss ratio values calculated over a month	Frame loss ratio values calculated over a month	Frame loss ratio values calculated over a month
Availability	To be defined by MEF	To be defined by MEF	To be defined by MEF
High loss interval	To be defined by MEF	To be defined by MEF	To be defined by MEF
Consecutive high loss interval	To be defined by MEF	To be defined by MEF	To be defined by MEF

7.5 Service-Level Agreements

Service-level agreements (SLAs) between a service provider and subscriber for a given EVC are defined in terms of delay, jitter, loss, and availability performances. Service-level specification (SLS) is also used instead of SLA. In the following sections, we will describe each of them.

7.5.1 Frame Delay Performance

Frame delay performance is a measure of the delays experienced by different service frames belonging to the same CoS instance.

The one-way frame delay for an egress service frame at a given UNI in the EVC is defined as the time elapsed from the reception at the ingress UNI of the first bit of the corresponding ingress service frame until the transmission of the last bit of the service frame at the given UNI. This delay definition as illustrated in Figure 7.1 is the one-way delay that includes transmission delays across the ingress and egress UNIs as well as that introduced by the Carrier Ethernet network (also called Metro Ethernet network).

Frame delay performance is defined for a time interval T for a class of service instance on an EVC that carries a subset of ordered pairs of UNIs. Each frame delay performance metric is defined as follows:

$D_T^{\langle i,j \rangle}$, the set of one-way frame delay values for all qualified service frames at UNI j resulting from an ingress service frame at UNI$_I$, can be expressed as $D_T^{\langle i,j \rangle} = \left\{ d_1^{\langle i,j \rangle}, d_2^{\langle i,j \rangle}, \ldots, d_{N_{\langle i,j \rangle}}^{\langle i,j \rangle} \right\}$, where $d_k^{\langle i,j \rangle}$ is the one-way frame delay of the kth service frame. Define $\overline{d}_T^{\langle i,j \rangle}$ for $P > 0$ as

Table 7.2 MEF Three-CoS Bandwidth Constraints and PCP and DSCP Mapping

CoS Label	Ingress EI Band-width Profile Constraints	CoS and Color Identifiers			PHB (DSCP)			S-Tag PCP Color Green/ CoS with DEI	Color Yellow	CoS-Only Identifiers				Example Applications
		C-Tag PCP			Color Green	Color Yellow				C-Tag PCP	PHB (DSCP)		S-Tag PCP	
		Color Green	Color Yellow											
H	CIR>0 EIR≥0 CF=0	5	Undefined		EF (46)	Undefined	5	Undefined		5	EF (46)		5	VoIP and mobile backhaul control
M	CIR>0 EIR≥0	3	2		AF31 (26)	AF32 (28) or AF33 (30)	3	2		2–3	AF31–33 (26, 28, 30)		2–3	Near-real-time or critical data apps
L	CIR≥0 EIR≥0	1	0		AF11 (10)	AF12 (12), AF13 (14) or default (0)	1	0		0–1	AF11–13 (10, 12, 14) or default (0)		0–1	Noncritical data apps

7.5 Service-Level Agreements

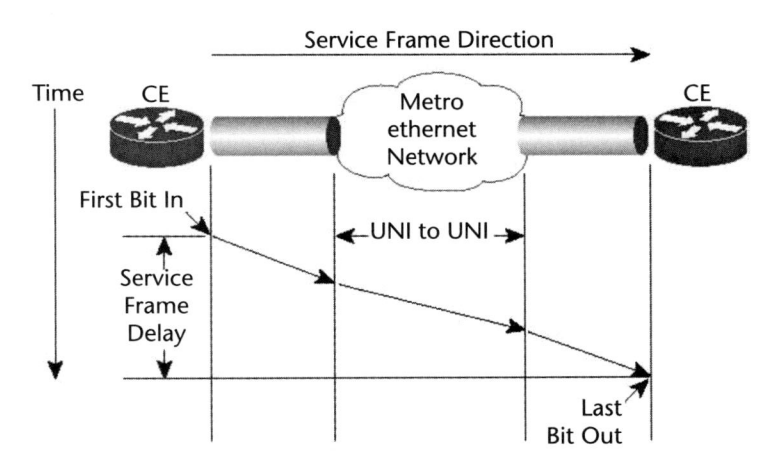

Figure 7.1 Frame delay for service frame.

$$\overline{d}_T^{\langle i,j \rangle} = \begin{cases} \min\left\{ d \mid P \leq \dfrac{100}{N_{\langle i,j \rangle}} \sum_{k=1}^{N_{\langle i,j \rangle}} I(d, d_k^{\langle i,j \rangle}) \right\} & \text{if } N_{\langle i,j \rangle} \geq 1 \\ \text{Undefined} & \text{otherwise} \end{cases}$$

where $I(d, d_k) = \begin{cases} 1 & \text{if } d \geq d_k \\ 0 & \text{otherwise} \end{cases}$

and N(i,j) is the number of frames involved in the measurement.

$\overline{d}_T^{\langle i,j \rangle}$ is the minimal delay during the time interval T that P percent of the frames do not exceed.

One-way frame delay performance metric for an EVC can be expressed as

$$\overline{d}_{T,S} = \begin{cases} \max\left\{ \overline{d}_T^{\langle i,j \rangle} \mid \langle i,j \rangle \in S \text{ and where } N_{\langle i,j \rangle} > 0 \right\} \\ \text{Undefined when all } N_{\langle i,j \rangle} = 0 \mid \langle i,j \rangle \in S \end{cases}$$

where S is a subset of ordered UNI pairs in the EVC.

7.5.2 Frame Delay Range Performance

In addition to FD and IFDV, frame delay range performance and mean delay performance are defined in [7]. Frame delay range is the difference between the frame delay performance values corresponding to two different percentiles. Frame delay range performance is a measure of the extent of delay variability experienced by different service frames belonging to the same CoS instance.

The difference between the delay performance of two selected percentiles, P_x and P_y, can be expressed as

$$\overline{d}_{Tyx}^{\langle i,j \rangle} = \begin{cases} (\overline{d}_{Ty}^{\langle i,j \rangle} - \overline{d}_{Tx}^{\langle i,j \rangle}) & \text{if } N_{\langle i,j \rangle} > 0 \\ \text{Undefined} & \text{if } N_{\langle i,j \rangle} = 0 \end{cases}$$

Then a one-way frame delay range performance metric for an EVC can be expressed as

$$\bar{d}_{TyxS} = \begin{cases} \max\{\bar{d}_{Tyx}^{\langle i,j\rangle} \mid \langle i,j\rangle \in S \text{ and where } N_{\langle i,j\rangle} > 0\} \\ Undefined \text{ when all } N_{\langle i,j\rangle} = 0 \mid \langle i,j\rangle \in S \end{cases}$$

7.5.3 Mean Frame Delay Performance

The minimum one-way delay is an element of $D_T^{\langle i,j\rangle}$, where $d_{\min}^{\langle i,j\rangle} \leq d_k^{\langle i,j\rangle}$ (for all $k = 1$, 2, ..., $N_{\langle i,j\rangle}$), and is a possible selection as one of the percentiles. The minimum delay represents the $N_{\langle i,j\rangle}-1$th percentile and all lower values of P as $P \to 0$.

Another one-way frame delay attribute is the arithmetic mean of $D_T^{\langle i,j\rangle}$, which can be expressed as

$$\bar{\mu}_T^{\langle i,j\rangle} = \begin{cases} \dfrac{1}{N_{\langle i,j\rangle}} \sum_{k=1}^{N_{\langle i,j\rangle}} \left(d_k^{\langle i,j\rangle}\right) & \text{if } N_{\langle i,j\rangle} > 0 \\ Undefined & \text{if } N_{\langle i,j\rangle} = 0 \end{cases}$$

Then a one-way mean frame delay performance metric for an EVC can be expressed as

$$\bar{\mu}_{TS} = \begin{cases} \max\{\bar{\mu}_T^{\langle i,j\rangle} \mid \langle i,j\rangle \in S \text{ and } \bar{\mu}_T^{\langle i,j\rangle} \text{ where } N_{\langle i,j\rangle} > 0\} \\ Undefined \text{ when all } N_{\langle i,j\rangle} = 0 \mid \langle i,j\rangle \in S \end{cases}$$

For a point-to-point EVC, S may include one or both of the ordered pairs of UNIs in the EVC. For a multipoint-to-multipoint EVC, S may be any subset of the ordered pairs of UNIs in the EVC. For a rooted-multipoint EVC, S must be such that all ordered pairs in S contain at least one UNI that is designated as a root.

7.5.4 Frame Delay Variation

Interframe delay variation (IFDV) is the difference between the one-way delays of a pair of selected service frames. For a particular class of service identifier and an ordered pair of UNIs in the EVC, IFDV performance is applicable to qualified service frames.

The IFDV performance is defined as the P-percentile of the absolute values of the difference between the frame delays of all qualified service frame pairs given that the difference in the arrival times of the first bit of each service frame at the ingress UNI was exactly Δt.

The choice of the value for Δt can be related to the application timing information. As an example for voice applications where voice frames are generated at regular intervals, Δt may be chosen to be few multiples of the interframe time.

Let be the time of the arrival of the first bit of the ith service frame at the ingress UNI, then the two frames i and j are selected according to the selection criterion:

7.5 Service-Level Agreements

$$\{a_j - a_i = \Delta t \quad and \quad j > i\}$$

Let r_i be the time frame i is successfully received (last bit of the frame) at the egress UNI, then the difference in the delays encountered by frame i and frame j is given by $d_i - d_j$. Define

$$\Delta d_{ij} = |d_i - d_j| = |(r_i - a_i) - (r_j - a_j)| = |(a_j - a_i) - (r_j - r_i)|$$

If either or both frames are lost or not delivered due to, for example, FCS violation, then the value Δ_{ij}^d is not defined and does not contribute to the evaluation of the interframe delay variation.

Figure 7.2 shows a depiction of the different times that are related to interframe delay variation performance.

7.5.5 Frame Loss Ratio

Frame loss ratio is a measure of the number of lost frames between the ingress UNI and the egress UNI. Frame loss ratio is expressed as a percentage.

One-way frame loss ratio performance is defined for a time interval T of a class of service instance on an EVC that carries a subset of ordered pairs of UNIs. One-way frame loss ratio performance metric is defined as follows:

- Let $I_T^{\langle i,j \rangle}$ denote the number of ingress service frames at UNI i whose first bit arrived at UNI i during the time interval T, whose ingress bandwidth profile compliance was green, and that should have been delivered to UNI j according to the service frame delivery service attributes.
- Let $E_T^{\langle i,j \rangle}$ denote the number of such service frames delivered to UNI j.
- Define $FLR_T^{\langle i,j \rangle} = \begin{cases} \left(\dfrac{I_T^{\langle i,j \rangle} - E_T^{\langle i,j \rangle}}{I_T^{i,j}} \right) \times 100 & \text{if } I_T^{\langle i,j \rangle} \geq 1 \\ Undefined & \text{otherwise} \end{cases}$
- Then the one-way frame loss ratio performance metric is defined as:

$$FLR_{T,S} = \begin{cases} \max\{FLR_T^{\langle i,j \rangle} \mid \langle i,j \rangle \in S \text{ and where } I_T^{\langle i,j \rangle} \geq 1\} \\ Undefined \text{ when all } I_T^{\langle i,j \rangle} = 0 \mid \langle i,j \rangle \in S \end{cases}$$

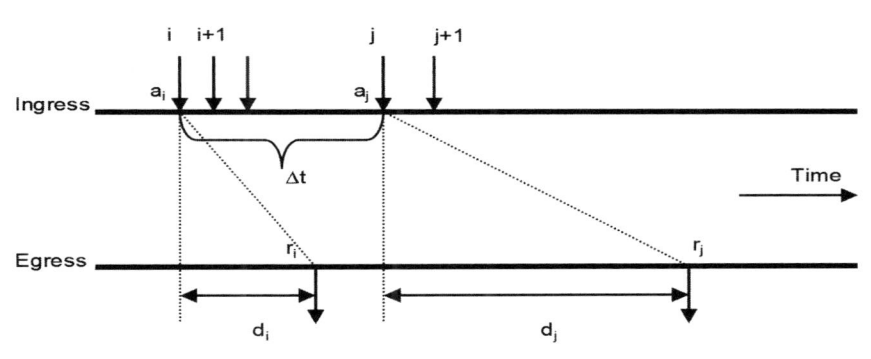

Figure 7.2 Interframe delay variation.

Given T, S, and a one-way frame loss ratio performance objective, the one-way frame loss performance is defined as met over the time interval T for the subset S if and only if $FLR_{T,S} \le \hat{L}$.

Recall that if the one-way frame loss ratio performance is undefined for time interval T and ordered pair $\langle i,j \rangle$, then the performance for that ordered pair is excluded from calculations on the performance of pairs in S.

For a point-to-point EVC, S may include one or both of the ordered pairs of UNIs in the EVC.

For a multipoint-to-multipoint EVC, S may be any subset of the ordered pairs of UNIs in the EVC.

For a rooted-multipoint EVC, all ordered pairs in S contain at least one UNI that is designated as a root.

7.6 SLSs

MEF 23.1 [8] defined four performance tiers (PTs) that associates SLSs with distance. MEF 23.2 [5] added a new PT ("PT0.3" or "City") with more stringent CoS performance objectives (CPOs) than PT 1, to support additional applications. PT0.3 mainly covers a subset of the mobile backhaul traffic with smaller FD/MFD value. MEF 23.2 [5] also adds CPOs for multipoint services in all performance tiers:

- PT0.3 (City PT): Derived from distances less than metro in extent (<75 km, 0.6 ms).
- PT1 (Metro PT): Derived from typical metro distances (<250 km, 2 ms).
- PT2 (Regional PT): Derived from typical regional distances (<1,200 km, 8 ms).
- PT3 (Continental PT): Derived from typical national/continental distances (<7,000 km, 44 ms).
- PT4 (Global PT): Derived from typical global/intercontinental distances (<27,500 km, 172 ms).

The tables of CPOs define objectives for multipoint services that differ from point-to-point services. These performance values apply to multipoint services with 100 or fewer external interfaces. This document does not specify objectives and parameters for multipoint services larger than 100 external interfaces.

These multipoint objectives also do not apply in time periods where the focused overload condition is present. The focused overload condition is described next.

Tables 7.3 through 7.7 list SLS figures for these PTs. In order to meet the CPO of an EVC formed of multiple OVCs, alignment of CBS between operators and/or shaping at the ENNI is necessary. Otherwise, the EVC CPOs in Tables 7.3 through 7.7 may not be met even if CoS Label mapping is aligned.

In these tables, a user or carrier may choose to support FD or MFD for delay, and IFDV or FDR for jitter.

7.6 SLSs

Table 7.3 PT0.3 CPOs [5]

SLS Parameters Measured over a Month	CoS Label H Point-Point	CoS Label H Multipoint	CoS Label M Point-Point	CoS Label M Multipoint	CoS Label L Point-Point	CoS Label L Multipoint
FD (ms)	≤3 for 99.9th percentile	≤3 for 99.9th percentile	≤6 for 99.9th percentile	≤6 for 99.9th percentile	≤11 for 95th percentile	≤11 for 95th percentile
MFD (ms)	≤2	≤2	≤4	≤5	≤9	≤10
IFDV (ms)	≤1 for 99.9th percentile with 1-second pair interval	≤1 for 99.9th percentile with 1-second pair interval	≤2.5 for 99.9th percentile with 1 s pair interval or not specified	≤2.5 for 99.9th percentile with 1-second pair interval or not specified	Not specified	Not specified
FDR (ms)	≤1.25	≤1.25	≤3 or not specified	≤3 or not specified	Not specified	Not specified
FLR (percent)	≤0.001% i.e., 10^{-5}	≤0.001% i.e., 10^{-5}	≤0.001% i.e., 10^{-5}	≤0.001% i.e., 10^{-5}	≤0.1% i.e., 10^{-3}	≤0.1% i.e., 10^{-3}
Availability, high loss interval (HLI), consecutive HLI (CHLI), one-way group availability	Not specified	Not specified	Not specified	Not specified	Not specified	Not specified

*Ingress bandwidth profile parameters may be chosen such that no frames are subject to SLS.

Table 7.4 PT1 CPOs [5]

SLS Parameters Measured over a Month	CoS Label H Point-Point	CoS Label H Multipoint	CoS Label M Point-Point	CoS Label M Multipoint	CoS Label L* Point-Point	CoS Label L* Multipoint
FD (ms)	≤10 for 99.9th percentile	≤10 for 99.9th percentile	≤20 for 99.9th percentile	≤20 for 99.9th percentile	≤37 for 95th percentile	≤37 for 95th percentile
MFD (ms)	≤7	≤9	≤13	≤15	≤28	≤30
One-way IFDV (ms)	≤3 for 99.9th percentile with 1-second pair interval	≤3 for 99.9th percentile with 1-second pair interval	≤8 for 99.9th percentile with 1-second pair interval or not specified	≤8 for 99.9th percentile with 1-second pair interval or not specified	Not specified	Not specified
FDR (ms)	≤5	≤5	≤10 or not specified	≤10 or not specified	Not specified	Not specified
FLR (percent)	≤0.01% i.e., 10^{-4}	≤0.01% i.e., 10^{-4}	≤0.01% i.e., 10^{-4}	≤0.01% i.e., 10^{-4}	≤0.1% i.e., 10^{-3}	≤0.1% i.e., 10^{-3}
Availability, HLI, CHLI, one-way group availability	Not specified	Not specified	Not specified	Not specified	Not specified	Not specified

*Ingress bandwidth profile parameters may be chosen such that no frames are subject to SLS.

7.6.1 Multipoint CoS Performance Objectives

The CPOs for multipoint services are defined as less stringent in comparison to point-to-point services due to the additional processing required to achieve one-to-many connectivity, mainly frame replication and address table lookup.

Table 7.5 PT2 CPOs [5]

SLS Parameters Measured over a Month	CoS Label H Point-Point	CoS Label H Multipoint	CoS Label M Point-Point	CoS Label M Multipoint	CoS Label L* Point-Point	CoS Label L* Multipoint
FD (ms)	≤25 for 99.9th percentile	≤25 for 99.9th percentile	≤75 for 99.9th percentile	≤75 for 99.9th percentile	≤125 for 95th percentile	≤125 for 95th percentile
MFD (ms)	≤18	≤20	≤30	≤32	≤50	≤52
One-way IFDV (ms)	≤8	≤8	≤40 or not specified	≤40 or not specified	Not specified	Not specified
FDR (ms)	≤10	≤10	≤50 or not specified	≤50 or not specified	Not specified	Not specified
FLR (percent)	≤0.01% i.e., 10^{-4}	≤0.01% i.e., 10^{-4}	≤0.01% i.e., 10^{-4}	≤0.01% i.e., 10^{-4}	≤0.1% i.e., 10^{-3}	≤0.1% i.e., 10^{-3}
Availability, HLI, CHLI, one-way group availability	Not specified	Not specified	Not specified	Not specified	Not specified	Not specified

*Ingress bandwidth profile parameters may be chosen such that no frames are subject to SLS.

Table 7.6 PT3 CPOs [5]

SLS Parameters Measured over a Month	CoS Label H Point-Point	CoS Label H Multipoint	CoS Label M Point-Point	CoS Label M Multipoint	CoS Label L* Point-Point	CoS Label L* Multipoint
FD (ms)	≤77 for 99.9th percentile	≤77 for 99.9th percentile	≤115 for 99.9th percentile	≤115 for 99.9th percentile	≤230 for 95th percentile	≤230 for 95th percentile
MFD (ms)	≤70	≤72	≤80	≤82	≤125	≤127
One-way IFDV (ms)	≤10 for 99.9th percentile with 1-second pair interval	≤10 for 99.9th percentile with 1-second pair interval	≤40 for 99.9th percentile with 1-second pair interval or not specified	≤40 for 99.9th percentile with 1-second pair interval or not specified	Not specified	Not specified
FDR (ms)	≤12	≤12	≤50 or not specified	≤50 or not specified	Not specified	Not specified
FLR (percent)	≤0.025% i.e., 2.5 × 10^{-4}	≤0.025% i.e., 2.5 × 10^{-4}	≤0.025% i.e., 2.5 × 10^{-4}	≤0.025% i.e., 2.5 × 10^{-4}	≤.1% i.e., 10^{-3}	≤.1% i.e., 10^{-3}
Availability, HLI, CHLI, one-way group availability	Not specified	Not specified	Not specified	Not specified	Not specified	Not specified

*Ingress bandwidth profile parameters may be chosen such that no frames are subject to SLS.

For example, one of the common implementations of multipoint services is VPLS which replicates flooded frames at ingress, placing the entire processing burden on a single node. Hierarchical VPLS (H-VPLS) as discussed in Chapter 14 or point-to-multipoint LSPs can be considered to reduce the effect of replication processing.

7.6 SLSs

Table 7.7 PT4 CPOs [5]

SLS Parameters Measured over a Month	CoS Label H		CoS Label M		CoS Label L*	
	Point-Point	*Multipoint*	*Point-Point*	*Multipoint*	*Point-Point*	*Multipoint*
FD (ms)	≤230 for 99.9th percentile	≤230 for 99.9th percentile	≤250 for 99.9th percentile	≤250 for 99.9th percentile	≤390 for 95th percentile	≤390 for 95th percentile
MFD (ms)	≤200	≤202	≤220	≤222	≤240	≤242
One-way IFDV (ms)	≤32 for 99.9th percentile with 1-second pair interval	≤32	≤40 for 99.9th percentile with 1-second pair interval or not specified	≤40 for 99.9th percentile with 1-second pair interval or not specified	Not specified	Not specified
FDR (ms)	≤40	≤40	≤50 or not specified	≤50 or not specified	Not specified	Not specified
FLR (percent)	≤0.05% i.e., 5×10^{-4}	≤0.05% i.e., 5×10^{-4}	≤0.05% i.e., 5×10^{-4}	≤0.05% i.e., $5\ 10^{-4}$	≤0.1% i.e., 10^{-3}	≤0.1% i.e., 10^{-3}
Availability, HLI, CHLI, one-way group availability	Not specified	Not specified	Not specified	Not specified	Not specified	Not specified

*Ingress bandwidth profile parameters may be chosen such that no frames are subject to SLS.

Replication is needed for not only frames requiring flooding, but also unknown unicast frames for which the MAC forwarding table does not have a matching destination address entry.

A significant increase in delay could be observed under the following conditions [9]:

- EVCs comprising 100 UNIs;
- Flooding traffic reaching 80% of link speed (1-Gbps links tested).

The relaxed objectives defined in Tables 7.8 through 7.12 are recommended for EVCs comprising 100 or fewer UNIs. As with all CoS IA performance objectives, operators can always define more stringent objectives.

7.6.2 Focused Overload

The focused overload condition occurs when the sum of network traffic from ingress external interfaces that are members of one or more multipoint EVCs exceeds the available capacity of an egress external interface or a CEN internal link. Point-to-point services can introduce similar conditions, but only in a hub-and-spoke architecture.

When frames are discarded due to focused overload of egress traffic at a UNI for a multipoint-to-multipoint or a rooted-multipoint EVC, MEF 10.3 [10] provides the option to exclude those discarded frames from the availability and frame loss ratio performance.

Table 7.8 An Example for IP Service Classes Configured with DiffServ and Mapping of Applications to These Classes [11]

Service Class	DSCP Name	DSCP Binary (Decimal) Value	Application Examples
Network Control	CS6	110000 (48)	Network routing
Telephony	EF	101110 (46)	IP telephony bearer
Signaling	CS5	101000 (40)	Videoconferencing
Real-time Interactive	CS4	100000 (32)	Interactive control (e.g., CAM), real-time e-learning, games, e-arts
Broadcast Video	CS3	011000 (24)	Broadcast TV and live events
Multimedia Streaming	AF31, AF32, AF33	011010 (26), 011100 (28), 011110 (30)	Streaming video and audio on demand
Low Latency Data	AF21, AF22, AF23	010010 (18), 010100 (20), 010110 (22)	Transactional applications, database access, interactive data applications
OAM	CS2	010000 (16)	OAM (e.g., SNMP, Ethernet CFM, proprietary NMS traffic)
High throughput data	AF11, AF12, AF13	001010 (10), 001100 (12), 001110 (14)	
Standard	DF (CS0)+Other	000000 (0)	Undifferentiated applications

Table 7.9 IP Service Classes and Example Applications [14]

Table 2/Y.1541 – Guidance for IP QoS classes

QoS Class	Applications (examples)	Node Mechanisms	Network Techniques
0 (Highest Priority)	Real-time, jitter sensitive, high interaction (VoIP, VTC)	Separate queue with preferential servicing, traffic grooming	Constrained routing and distance
1	Real-time, jitter sensitive, interactive (VoIP, video teleconference)		Less constrained routing and distances
2	Transaction data, highly interactive (signaling)	Separate queue, drop priority	Constrained routing and distances
3	Transaction data, interactive		Less constrained routing and distances
4	Low loss only (short transactions, bulk data, video streaming)	Long queue, drop priority Any route/path	Any route/path
5 (Lowest Priority)	Traditional applications of default IP networks	Separate queue (lowest priority)	Any route/path

Figure 7.3 shows an example multipoint EVC, where the combined traffic of UNI-1, UNI-2, and UNI-3 focused onto UNI-4 will overload its 10 Mbps access with as high as 12 Mbps of CIR traffic.

While the focused overload condition cannot be entirely avoided in multipoint service deployments, an operator might choose to place multipoint services in different queues than point-to-point services

Table 7.10 IP Service Classes and Example Applications

Network Performance Parameter	Network Performance Objective	Class 0	Class 1	Class 2	Class 3	Class 4	Class 5	Class 6	Class 7
Applications examples		Real-time, jitter sensitive, high interaction (VoIP, VTC)	Real-time, jitter sensitive, interactive (VoIP, VTC)	Transaction data, highly interactive (signaling)	Transaction data, interactive	Low loss, only (short transactions, bulk data, video streaming	Traditional applications of default IP networks	TV transport, high capacity TCP transfers, and TDM circuit emulation	TV transport, high capacity TCP transfers, and TDM circuit emulation
IPTD	Upper bound on the mean IPTD	100 ms	400 ms	100 ms	400 ms	1 second	Undefined	100 ms	400 ms
IPDV	Upper bound on the 1–10^{-3} (1–10^{-5} for Class 5 and 6) quantile of IPTD minus the minimum IPTD	50 ms	50 ms	Undefined	Undefined	Undefined	Undefined	50 ms	50 ms
IPLR	Upper bound on the packet loss probability	10^{-3}	10^{-3}	10^{-3}	10^{-3}	10^{-3}	Undefined	10^{-5}	10^{-5}
IPER	Upper bound	10^{-4}	10^{-4}	10^{-4}	10^{-4}	10^{-4}	Undefined	10^{-6}	10^{-6}
IPRR	Upper bound	Undefined	Undefined	Undefined	Undefined	Undefined	Undefined	10^{-6}	10^{-6}

Table 7.11 Service Classes for 3G Traffic [13]

Traffic Class	Conversational Class Conversational RT	Streaming Class Streaming RT	Interactive Class Interactive Best Effort	Background, Background Best Effort
Fundamental characteristics	Preserve time relation (variation) between information entities of the stream, conversational pattern (stringent and low delay)	Preserve time relation (variation) between information entities of the stream	Request response pattern, preserve payload content	Destination is not expecting the data within a certain time, preserve payload content
Example of the application	Voice	Streaming video	Web browsing	Background download of e-mails

7.7 Application-CoS-Priority Mapping

Defining number of priorities to be supported has been a challenge for equipment vendors and service providers. More priorities may increase the cost of equipment and operations systems.

MEF picked three priorities, H, M, and L. However, the number of priorities recommended by ITU, IETF and 3GPP/3GPP2 [13] are much higher. In this section, we summarize application prioritization and related performance parameters for these standards bodies.

An example by IETF for recommended Class of Services-DSCP Code points-Applications mapping given by [4] is described in Table 7.8. This table suggests 10 CoS.

Similarly, ITU-T defines IP service classes, maps them onto applications, and identifies performance parameters for each in [14]. In Table 7.9, six classes are defined.

Furthermore, ITU-T [14] adds two more classes and defines performance parameters for each as depicted in Table 7.10. Classes 6 and 7 are intended to support performance requirements of applications with stringent loss/error constraints.

Similar effort has taken place in wireless standards. Table 7.11 describes CoS defined in 3GPP for 3G traffic [13].

ITU-T performance constraints in [14] are also considered for mobile applications. Table 7.12 summarizes that. The IPRR parameter of ITU-T is ignored.

7.7.1 CoS Identification

CoS identification is obtained by CoS label and color identifier. CoS label is a name for the CoS identified by a CoS identifier at UNI and ENNI.

CoS identifier at UNI can be EVC, EVC + PCP (i.e., CE-VLAN service frame priority), EVC + DSCP, or EVC+L2CP. Clearly EVC ID is not carried by service frame, therefore, when CoS ID is an EVC ID, operator may use string within EVC ID indicating the CoS or tables in OSS (operations support system) grouping EVCs according to their CoS.

Untagged service frames without a CoS identifier will have the same CoS identifier as an ingress data service frame with PCP = 0. L2CP frames whose CoS is not

7.7 Application-CoS-Priority Mapping

Table 7.12 Service Classes Defined by ETSI for Mobile Applications Aligned with 3GPP/3GPP2 [12–14]

Network Performance Parameter	Nature of Network Performance Objective	Class 0	Class 1	Class 2	Class 3	Class 4	Class 5	Class 6	Class 7
		Real-Time, Jitter Sensitive, High Interaction (VoIP, VTC)	Real-Time, Jitter Sensitive, Interactive (VoIP, VTC)	Transaction Data, Highly Interactive (Signalling)	Transaction Data, Interactive	Low Loss Only (Short Transactions, Bulk Data, Video Streaming)	Traditional Application of Default IP Networks	Television transport, high-capacity TCP transfers, and TDM circuit emulation	Television transport, high-capacity TCP transfers, and TDM circuit emulation
IPTD	Upper bound on the mean IPTD	100 ms	400 ms	100 ms	400 ms	1 s	U	100 ms	400 ms
IPDV	Upper bound on the $1\text{-}10^{-3}$ quantile of IPTD minus the minimum IPTD	50 ms	50 ms	U	U	U	U	50 ms	50 ms
IPLR	Upper bound on the packet loss probability	$1*10^{-3}$	$1*10^{-3}$	$1*10^{-3}$	$1*10^{-3}$	$1*10^{-3}$	U	$1*10^{-5}$	$1*10^{-5}$
IPER	Upper Bound	$1*10^{4}$					U	U	U

Table of Ingress BWP parameters

UNI	UNI Speed	EVC CIR	EVC EIR
1	10 Mbps	4 Mbps	6 Mbps
2	10 Mbps	4 Mbps	6 Mbps
3	10 Mbps	4 Mbps	6 Mbps
4	10 Mbps	4 Mbps	6 Mbps

● = focused overload

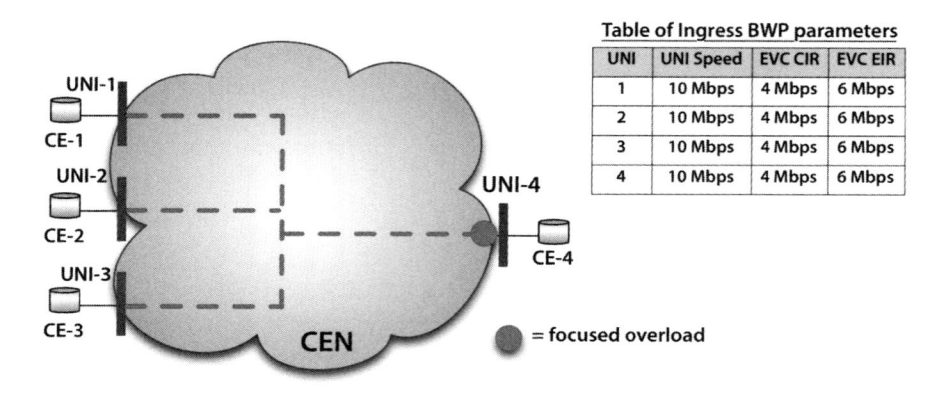

Figure 7.3 Example of focused overload.

determined using the EVC may have a CoS label determined by their MAC DA and/or EtherType (protocol type).

Color for a service frame at UNI and ENNI is indicated by a color identifier when the EVC is in color-aware mode, as part of the bandwidth profile. Color identification is accomplished via the PCP or DSCP at the UNI and via the PCP or DEI at the E-NNI. Thus, the PCP or DSCP may convey both CoS and color. When frames are untagged at the UNI, DSCP can be used to indicate color.

In service VLAN tag (S-TAG), the drop eligible parameter can be encoded in and decoded from the drop eligible indicator (DEI) [15].

At the color-aware E-NNI, color is indicated using either the PCP field of the S-TAG, or indicated separately using the S-TAG DEI field. The PCP field of the S-TAG must be used when the CoS identifier type is EVC+PCP. Use of DEI may free up additional values of the PCP.

PCP code point usage to indicate color and CoS is described in [5] as depicted in Table 7.13. For example, in the first row, there are 8 priorities; therefore, there is no PCP bit representing discard eligibility (DE) (i.e., yellow-colored frames). In the second row, 5 PCPs represent priorities; thus, 3 PCPs are available to represent discard eligibility.

At UNI, the service provider can choose to recognize service frame colors (i.e., color-aware mode) or not to recognize service frame colors (i.e., color-blind mode).

When the color mode (CM) parameter of the ingress bandwidth profile of a CoS instance is set to color-aware at the UNI, the color mode parameter of all egress bandwidth profiles at the UNIs and ingress/egress bandwidth profiles at the

Table 7.13 PCP Values for QoS and Color

PCP Allocation		PCP Values and Traffic Classes						
# PCP Priorities	# PCP Drop Eligible	PCP = 7	6	5	4	3	2	1	0
8	0	IEEE Traffic Class = 7	6	5	4	3	2	1	0
5	3	IEEE Traffic Class = 7	6	4	4 DE	2	2 DE	0	0 DE

E-NNIs associated with the CoS instance must also be set to color-aware, in order to support color awareness end-to-end.

When the CM parameter of the ingress bandwidth profile of a CoS instance is set to color-blind at the UNI, the color mode parameter for all egress bandwidth profiles at the UNIs and ingress/egress bandwidth profiles at the ENNIs associated with the CoS instance should be set to color-aware.

7.7.2 PCP and DSCP Mapping

For a single-CoS EVC, all ingress data service frames mapped to the EVC, regardless of PCP or DSCP value, will have the same CoS identifier. For a multi-CoS EVC where EVC+PCP is used for CoS identification, all possible PCP values (i.e., 0 to 7) need to be mapped to the CoS supported on a given EVC at the UNI. For a multi-CoS EVC where EVC+DSCP is used for CoS identification, all possible DSCP values (i.e., from 0 to 63) must be fully mapped to the CoS supported on a given EVC at the UNI.

Table 7.14 is an example of PCP mapping to MEF H, M, and L service categories. The mapping in this table ignores color of user traffic.

7.8 Bandwidth Profile

The bandwidth profile defines long-term average bandwidth limits on the amount of data burst by specifying rate and buffering for the given data stream and how the Carrier Ethernet network should treat a service frame, depending on its level of compliance. It consists of CIR, CBS, EIR, excess burst size (EBS), coupling flag (CF), which is either 0 or 1, and color mode.

- Committed information rate: The average rate in bits per second of service frames, which will be declared green by the bandwidth profile.
- Committed burst size: Dictates the maximum number of bytes available for a burst of service frames, which will be declared green by the bandwidth profile.
- Excess information rate: Average rate in bits per second of service frames, which will be declared yellow by the bandwidth profile.

Table 7.14 Example for PCP Mapping for MEF H, M, and L Service Categories in Color-Blind Mode

MEF CoS Combination Supported on EVC	PCP Mapping per Class of Service		
	H	M	L
H, M, L	5	2–4, 6, 7	0, 1
H, M	5	0–4, 6, 7	NA
H, L	5	NA	0–4, 6, 7
M, L	NA	2–7	0, 1

- Excess burst size: Dictates the maximum number of bytes available for a burst of service frames, which will be declared yellow by the bandwidth profile.
- Coupling flag: Dictates if the overflow tokens that are not used for green service frames should be used for yellow service frames or not.
- Color mode: Dictates if the bandwidth profile is color-aware or color-blind.

The compliance of the service frames is generically determined through two leaky token buckets method (Figure 7.4).

Each bucket holds up to CBS and EBS tokens, for C bucket and E bucket, respectively, with 1 token representing 1 byte. Tokens are added to C and E buckets at the rate of CIR/8 and EIR/8, respectively. When the bucket becomes full, additional tokens overflow. Depending on the specific bandwidth profile model, these overflow tokens are either used by rerouting them to another bucket or dropped. In generic terms, when an ingress service frame is classified, its length is compared to the tokens in the C bucket. If the number of tokens is at least equal to the length of the service frame, the frame is declared green and tokens equal to the frame length are removed from the bucket. If there are not sufficient tokens in the C bucket, the service-frame length is compared to the tokens in the E bucket and the process is repeated. If there are sufficient tokens in the E bucket, the frame is declared to be yellow; otherwise, it is declared to be red.

Bandwidth profiles are defined and implemented at both ingress and egress ends of the service. The ingress bandwidth profile regulates the amount of incoming service traffic that is entering to the network at a UNI. The egress bandwidth profile similarly regulates the amount of service traffic that is being received at the egress end of the service. Two concepts, namely bandwidth profile flow and envelope, have been introduced to implement the overall bandwidth profile [10]. The bandwidth profile flow defined the bandwidth profile for a specific flow of service frames that are arriving at a UNI regardless of EVCs and/or CoS specification. It also could be defined to map service frames to EVC. The last option is to map service frames to EVC by its CoS specification. The envelope is defined to contain "n" number of bandwidth profile flows of the same kind (either based on EVC or EVC/CoS). These n flows in an envelope are ranked from 1 to n, where 1 represents the lowest and n the highest ranks.

Based on the flow and envelope concepts there are three bandwidth profile modes: (1) bandwidth profile per UNI; (2) bandwidth profile per EVC; and (3) bandwidth profile per CoS per EVC.

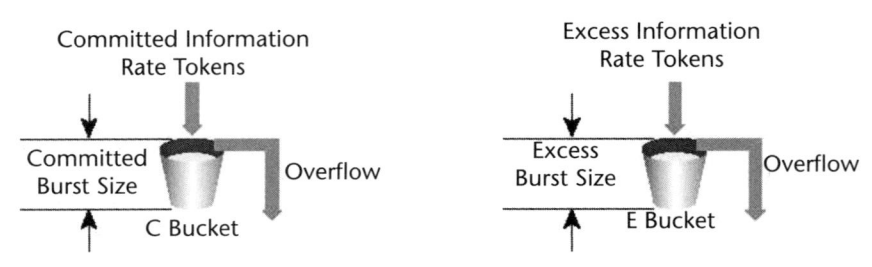

Figure 7.4 Two leaky buckets (i.e., C bucket is for CIR and E bucket is for EIR).

7.8 Bandwidth Profile

- Bandwidth profile per UNI: For bandwidth profile over a single UNI, there is only a single-bandwidth profile flow defined in a single envelope. An example of 4 EVCs given in Figure 7.5 treats all traffic coming from EVCs nondiscriminately within a single-bandwidth profile flow. This means that there is no regulation between EVC traffic; some EVCs may end up using more bandwidth than others.
- Bandwidth profile per EVC: Traffic for each EVC is defined as a bandwidth profile flow with its own dedicated parameters (CIR, CBS, EIR, EBS, and so forth). These multiple-bandwidth profile flows from multiple EVCs share a single envelope; see Figure 7.6 as an example for 4 EVCs sharing an envelope over an UNI.
- Bandwidth profile per CoS identifier: For bandwidth per CoS identifier, the traffic based on its CoS identifier is mapped to a bandwidth profile flow. There may be multiple CoS identifiers assigned to the same CoS level (CoS name) and this CoS level has its own bandwidth profile flow. These bandwidth profile flows based on CoS identifier share an envelope, see Figure 7.7 as an example for 2 CoS levels over an EVC sharing an envelope.

7.8.1 Bandwidth Profile Algorithms

The bandwidth profile dictates a set of traffic parameters that are applied to the service frames. The way that these parameters are applied among flows within an envelope is determined by the bandwidth profile algorithm.

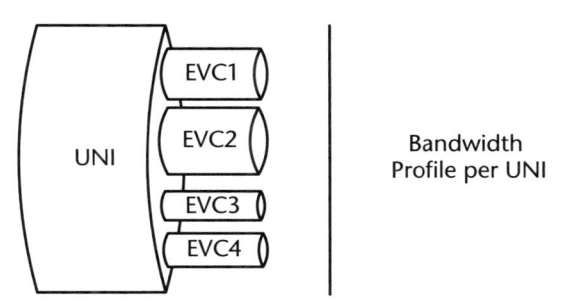

Figure 7.5 Bandwidth profile per UNI.

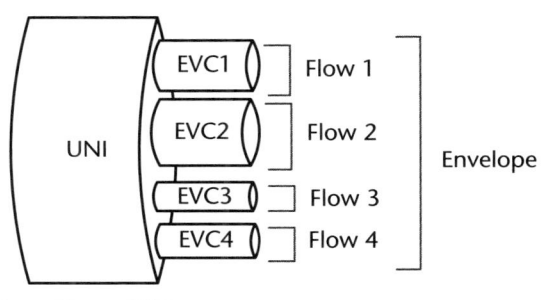

Figure 7.6 Bandwidth profile per EVC.

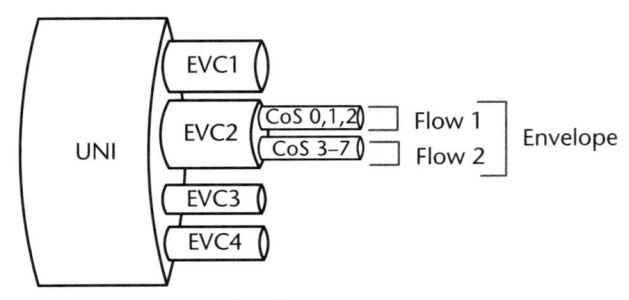

Figure 7.7 Bandwidth profile per CoS identifier.

As seen in Figure 7.7, multiple envelopes each of which contains multiple flows can be defined at a UNI. Each service frame can belong to only one flow within only one envelope.

The bandwidth profile algorithm has flags and switches that can be configured such that the unused bandwidth (capacity or tokens) that was allocated to a flow can be reused by other flows within the same envelope. The priority of this reuse among flows within an envelope is defined by ranking. In envelope ranking, each flow has a unique rank (priority number) that suggests the preference for reallocating the unused bandwidth. In implementation, higher ranking number suggests higher preference.

Based on the parameters of all ranks (CIR, CIRmax, EIR, CBS, EBS, CF, CM, ER), the algorithm determines how service frame will be treated in terms of complying with the bandwidth profile. As a result, each service frame is classified as green, yellow, or red. The bandwidth profile algorithm has two modes defined for each flow: (1) color-aware mode, where the algorithm acknowledges the color identifier of the service frame and decides the compliance level accordingly; and (2) color-blind mode, where the color-identifier of the service frame is ignored.

The color mode for each Flow_i is configured (determined) by the parameter CM^i. Based on the leaky token principle, the algorithm maintains a pair of committed (C bucket) and Excess (E bucket) token buckets for each flow. The sizes of these buckets for Flow_i are given by CBS^i and EBS^i, respectively. The flow of service frames over these buckets is generated by tokens. The tokens are sourced at a given rate (IR) and consumed by the Service frames as they become compliant to the bandwidth profile. The tokens are sourced at a rate of CIR^i and EIR^i for the C bucket and E bucket of Flow_i, respectively. The maximum rate for these buckets is capped at $CIR^i max$ and $EIR^i max$, which constitute the maximum rate at which these buckets should be sourced. Each service frame that belongs to a certain Flow_i is declared green, yellow, or red, depending on the number of tokens that exist in the buckets. One token is needed for a byte of information in the service frame. Once a service frame is declared green, the number of tokens that is equal to the length of service frame in bytes will be removed from the C bucket. If the service frame is declared yellow, similar reduction of tokens will take place from E bucket.

To regulate processing of service frames among the flows for the envelope, the following variables are defined for each Flow_i, where $i = 1, 2, .., n$.

7.8 Bandwidth Profile

$B^i{}_C(t)$ represents the number of C (committed) tokens in C bucket. $T^i{}_C(t_1, t_2)$ is the number of tokens that could be added to the C bucket within the time frame between t_1 and t_2.

$$T^i{}_C(t_1, t_2) = (CIR^i/8) * (t_2 - t_1) + S^{i+1}{}_C(t_1, t_2)$$

The total number of tokens to be potentially added to the C bucket is of two types: (1) the tokens that are collected during the time period between t_1 and t_2 at the rate of CIR^i, and (2) the overflow unused C bucket tokens that are contributed by the flow at a higher rank $(i + 1)$, if the sharing is allowed. The overflow unused tokens, represented by $O^{i+1}{}_C(t_1, t_2)$ with the coupling flag (CF^{i+1}) that controls the allowance of the sharing from the rank above is given here:

$$S^{i+1}{}_C(t_1, t_2) = (1 - CF^{i+1}) * O^{i+1}{}_C(t_1 + t_2)$$

Note that $S^{i+1}{}_C(t_1, t_2) = 0$ because there is no rank above rank n that would contribute tokens.

The C bucket of Flow_i has a limit and a restriction for the number of tokens that it can receive at a given time. The limit is its size represented by CBS^i; it cannot receive more tokens than its size. The restriction is the maximum rate that it can take in tokens, represented by $CIR^i{}_{max}$. The number of tokens to be received cannot exceed this rate within the same duration defined between t_1 and t_2. The actual number of tokens that C bucket can take is

$$A^i{}_C(t_1, t_2) = \min\{T^i{}_C(t_1, t_2), CBS^i - B^i{}_C(t_1), (CIR^i{}_{max}/8) * (t_2 - t_1)\}$$

The number of unused tokens that are not added to the C bucket can be calculated as the overflow and given here:

$$O^i{}_C(t_1, t_2) = T^i{}_C(t_1, t_2) - A^i{}_C(t_1, t_2)$$

Similar calculations apply to the excess tokens and E bucket for a Flow_i where $i = 1, 2, \ldots, n$.

$B^i{}_E(t)$ represents the number of E (excess) tokens in E bucket. The number of tokens that could be added to the E bucket within the time frame between t_1 and t_2, denoted as $T^i{}_E(t_1, t_2)$, include: (1) the tokens that are sourced during the time period between t_1 and t_2 at the rate of EIR^i, (2) the overflow unused E bucket tokens that are contributed by the flow at a higher rank $(i + 1)$, and (3) the overflow unused C bucket tokens that are contributed by the same flow, if sharing is not allowed among different flows.

The overflow unused E bucket tokens from the higher rank flow is given by:

$S^{i+1}{}_E(t_1, t_2) = O^{i+1}{}_E(t_1 + t_2)$, when $i < n$, and $S^{i+1}{}_E(t_1, t_2) = CF^0 * O^1{}_C(t_1 + t_2)$, when $i = n$.

Similar to C bucket, E bucket has the same limiting and restricting factors, namely EBS^i and $EIR^i{}_{max}$. Applying these factors, the actual number of tokens that an E bucket can take is:

$$A^i{}_E(t_1, t_2) = \min\{T^i{}_E(t_1, t_2), EBS^i - B^i{}_E(t_1), (EIR^i{}_{max}/8) * (t_2 - t_1)\}$$

With the help of these definitions, the algorithm that declares service frames in different compliance level is given in Figure 7.8.

In case, where there is only one flow defined for the envelope and when CIR^i_{max} and EIR^i_{max} are represented by large values, then the general algorithm shown in Figure 7.8 is reduced to a simpler form, given in Figure 7.9.

7.8.2 Models and Use Cases for Bandwidth Profiles

The definitions of flows and envelopes have been very instrumental to bring flexibility and efficiency for the implementation of the bandwidth profile for capacity use. Introducing multiple flows within an envelope can allow sharing of unused C bucket and/or E bucket tokens from a higher-rank flow by a lower-rank flow [16]. If token sharing is not necessary for an application, then a single flow may be defined for an envelope. In this way, there is no ranking between different service traffic; all frames will be treated equally. Another option is to allocate fixed bandwidth profiles for each service traffic, where peaks in bandwidth usually are covered by over provisioning and inefficiencies occur in bandwidth utilization.

When token sharing is enabled in general, there are multiple ways of sharing tokens for each flow by setting or resetting its CF^i parameter. One can also enable

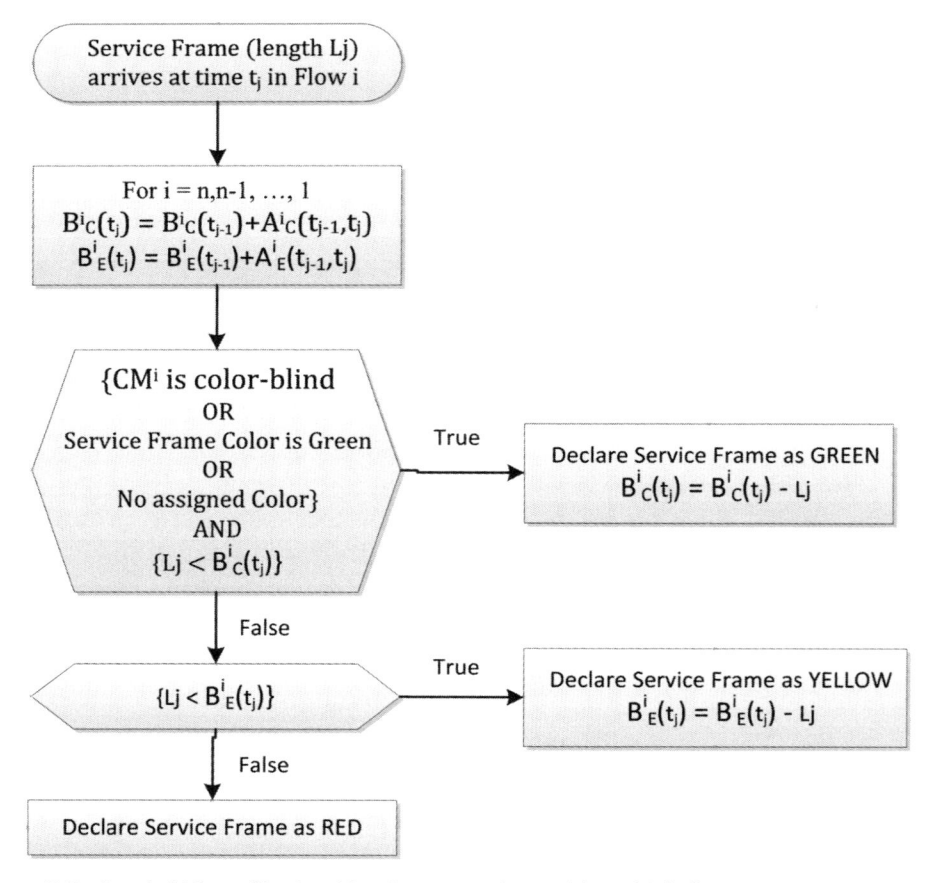

Figure 7.8 Bandwidth profile algorithm for an envelope with multiple flows.

7.8 Bandwidth Profile

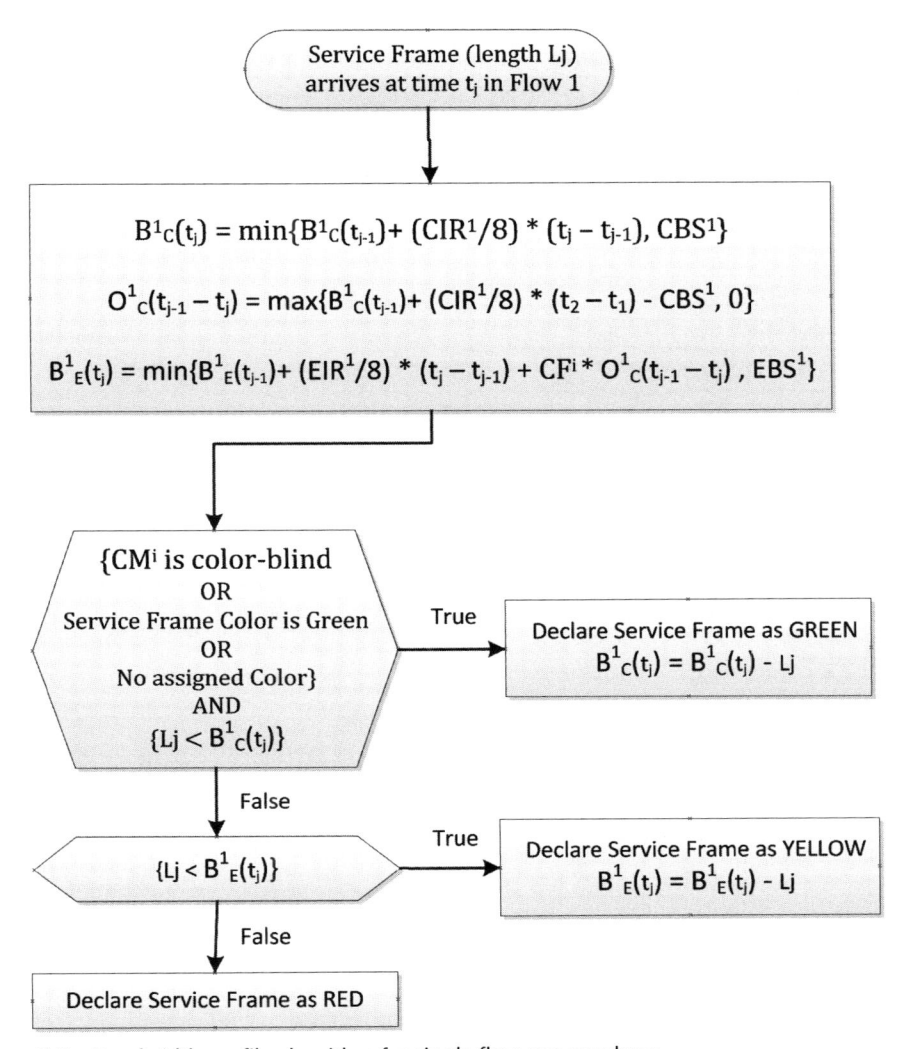

Service Frame (length Lj) arrives at time t_j in Flow 1

$$B^1{}_C(t_j) = \min\{B^1{}_C(t_{j-1}) + (CIR^1/8) * (t_j - t_{j-1}), CBS^1\}$$

$$O^1{}_C(t_{j-1} - t_j) = \max\{B^1{}_C(t_{j-1}) + (CIR^1/8) * (t_2 - t_1) - CBS^1, 0\}$$

$$B^1{}_E(t_j) = \min\{B^1{}_E(t_{j-1}) + (EIR^1/8) * (t_j - t_{j-1}) + CF^i * O^1{}_C(t_{j-1} - t_j), EBS^1\}$$

{CM^i is color-blind OR Service Frame Color is Green OR No assigned Color} AND {$L_j < B^1{}_C(t_j)$}

— True → Declare Service Frame as GREEN: $B^1{}_C(t_j) = B^1{}_C(t_j) - L_j$

— False → {$L_j < B^1{}_E(t_j)$}

— True → Declare Service Frame as YELLOW: $B^1{}_E(t_j) = B^1{}_E(t_j) - L_j$

— False → Declare Service Frame as RED

Figure 7.9 Bandwidth profile algorithm for single flow per envelope.

recirculation of unused tokens from the bottom rank flow back to the top rank flow for reuse by setting $CF^0 = 1$. Another control point is either to choose sourcing only C bucket tokens ($CIR^i > 0$; $EIR^i = 0$) or both C and E bucket tokens ($CIR^i > 0$; $EIR^i > 0$). By configuring all these parameters, multiple models of bandwidth profiles can be implemented and used.

One example would be sourcing only C bucket tokens ($CIR^i > 0$; $EIR^i = 0$) and enabling token sharing only by recirculating the unused C bucket tokens from the bottom rank flow back to the top rank flow ($CF^i = 0$; $CF^0 = 1$). By applying this model, C bucket tokens are shared among all C buckets first observing priority by ranks, and then only overall unused C bucket tokens are fed into the top-most rank E bucket and will be used for yellow service frames only if they are available. This implementation can be seen in Figure 7.10.

Another example could be a configuration where there is no circulation through the E buckets but sourcing excess bucket tokens is configured for E buckets. This

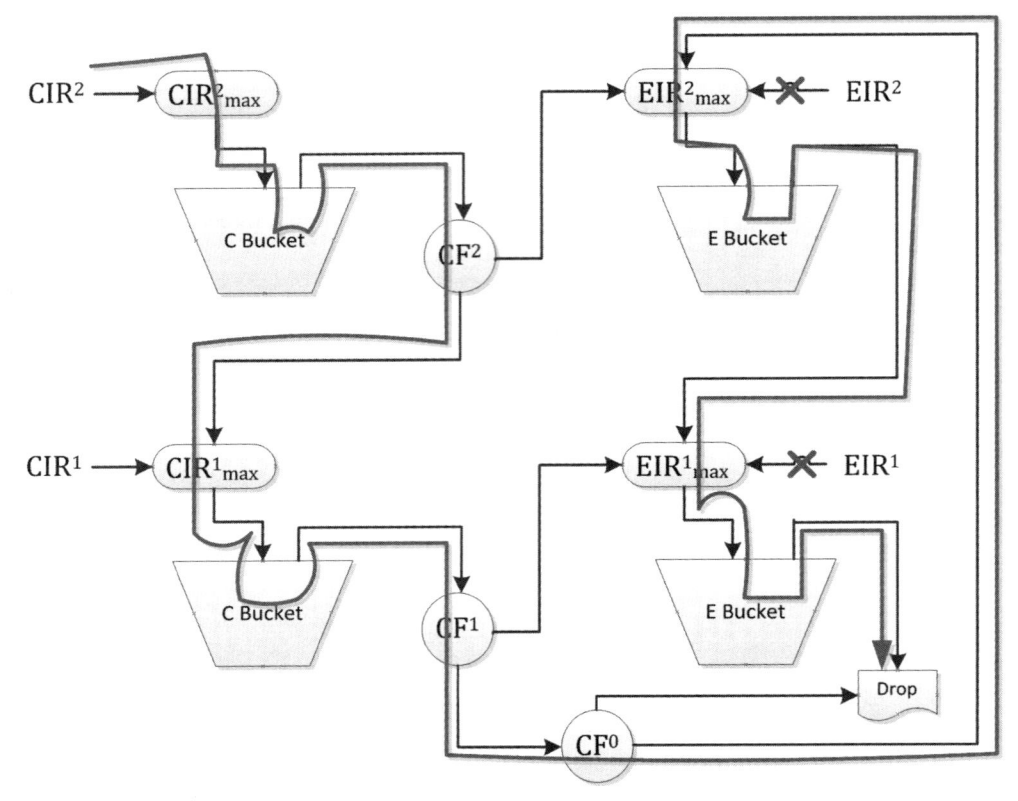

Figure 7.10 Recirculating tokens use case.

would require a successful provisioning of CIR and EIR for a flow (see Figure 7.11).

7.9 CBS Values, TCP, and Shaping

The minimum size of CBS is required to be greater than the maximum service frame (MSF) size per MEF 10.3 [10] and MEF 19 [17]. The minimum value of MSF is 1,522 bytes. Per MEF 13 [18] and MEF 34 [19], the minimum size of CBS is required to be greater than 8*MFS = 12,176 bytes. However, MEF 37 [20] requires CBS to be greater than 2*ENNI MSF. As we can see, the required CBS values are not consistent in MEF specifications. Furthermore, the author observed substantially higher CBS values than 8*MFS required in order to support CIR values for various traffic patterns.

Reference [5] suggested a policer CBS value in the vicinity of 80 KB (50*MFS) is a reasonable upper bound for most situations. The author's experience is 64 KB adequate for most of the nonreal-time traffic.

There is an obvious fact that if one keeps the same traffic pattern, the increasing rate reduces CBS. This is formulated by [1].

Given a CIR and UNI bit rate r, what is the minimum value of CBS that allows n, back to back service frames of length L bytes to be declared green assuming that the token bucket was full at time $t = 0$.

7.9 CBS Values, TCP, and Shaping

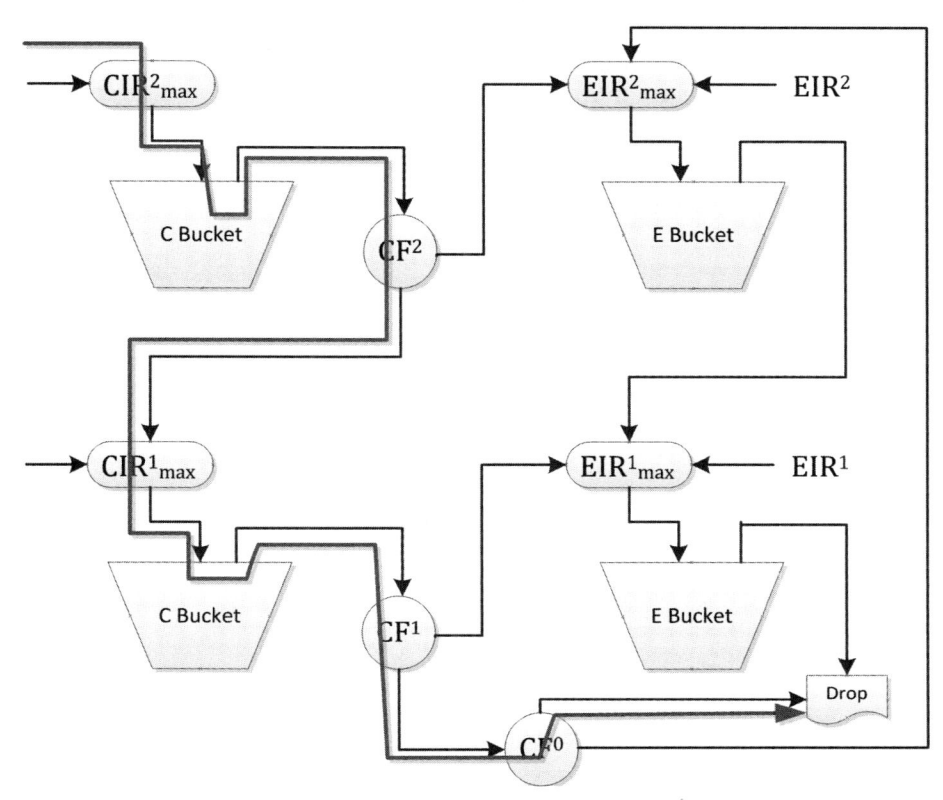

Figure 7.11 No token recirculation use case.

The time to transmit a service frame plus overhead is Δt.
The number of tokens in the token bucket is (t)
The time to transmit a service frame plus overhead is

$$\Delta t = \frac{L+20}{r/8} = \frac{8(L+20)}{r}$$

Let $B(t)$ be the number of tokens in the token bucket at time t. Then

$$B(0) = CBS$$
$$B(\Delta t) = CBS - L$$
$$B(2\Delta t) = B(\Delta t) - L + \frac{\Delta tCIR}{8} = CBS - 2L + \frac{\Delta tCIR}{8}$$
$$B(n\Delta t) = B((n-1)\Delta t) - L + \frac{\Delta tCIR}{8} = CBS - nL + \frac{(n-1)\Delta tCIR}{8}$$

In order to determine CBS, let's set $B(n\Delta t)$:

$$CBS_{min} = nL - \frac{(n-1)\Delta tCIR}{8} = nL - \frac{(n-1)CIR}{8}\left[\frac{8(L+20)}{r}\right] = nL - (L+20)\frac{CIR}{r}$$

Figure 7.12 shows CBS_{min} versus CIR for a 100-Mbps UNI and $n = 10$. Looking at the 2,000-byte service frame size line reveals that a 10 service frame burst of 20,000 bytes can be accommodated by CBS ≈ 11,000 bytes when CIR = 50 Mbps.

There is another obvious fact that is shaping traffic before the policer reduces the burst and as a result reduces CBS. MEF 13 [18] recommends that the UNI-C shapes its traffic to the contracted BWP in order to receive the contracted QoS commitments.

Most applications run over Transport Control Protocol (TCP). It is important to analyze TCP performance without and with a shaper, in addition to policing [21].

A TCP source dynamically adjusts its data transmission rate to match the bandwidth available at the point between the source and destination where the bandwidth is most constrained. Therefore, if a TCP flow goes through an EVC where the bandwidth is restricted by a bandwidth profile (i.e., ingress or egress) with a given CIR, TCP will adjust its transmission rate such that the average throughput equals the CIR. This may not happen unless the bandwidth profile is also configured with an unexpectedly large value for the CBS.

References [6, 7] examined the interaction of TCP traffic with functions that constrain that traffic to meet the bandwidth profile specifications of an EVC or OVC. Without shaping, it is possible to configure the ingress bandwidth profile such that TCP throughput matches the CIR. In many cases doing so requires very

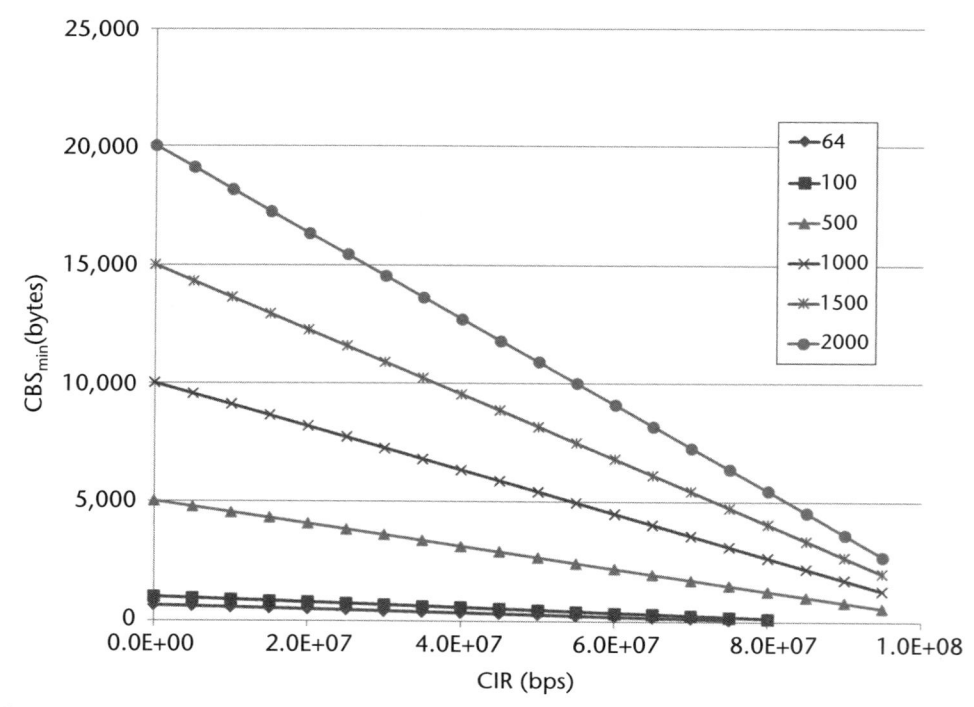

Figure 7.12 CIR versus CBS.

7.9 CBS Values, TCP, and Shaping

large values for CBS. However, shaping the traffic to CIR allows significantly smaller values for CBS and provides better predictability of TCP throughput when large values of CBS cannot be accommodated.

When bandwidth of a TCP flow is enforced by a policer without shaping, the TCP throughput is a function of the CIR and CBS parameters of the bandwidth profile. Figure 7.13 shows simulation results of TCP source transmitting to a TCP receiver through an EVC with an ingress bandwidth policer and no traffic shapers. [In the simulation the TCP source has an unlimited amount of data to send, has a minimum retransmission timeout (RTO) value of 250 ms, and uses selective acknowledgment (SACK) information for retransmitting lost segments. The receiver window size is 60 KB and the round-trip time (RTT) of the TCP connection is 5 ms. These parameters are the same for all of the simulation results in this section.] As can be seen, the TCP throughput is typically much smaller than the CIR until large values of CBS are reached.

The TCP throughput is the amount of data transmitted in RTT divided by RTT of the flow. TCP will attempt to maximize utilization of the available bandwidth by ramping up the amount of data transmitted on the flow until the flow capacity is reached [5]. The flow capacity can be calculated as the bottleneck bandwidth times the average RTT of the flow. The available bandwidth is maximized when the throughput equals the bottleneck bandwidth. TCP probes for changes in the flow capacity by gradually increasing the amount of data transmitted until it detects that the capacity has been exceeded by detecting a packet loss. When TCP detects loss of one or more packets, it reduces the transmission rate by a factor of 2, retransmitting the lost packet(s), and then begins to ramp up the rate for transmission again. However, when multiple packets are lost within a short time interval, TCP waits for a retransmission timeout (RTO) period before restarting transmission [5].

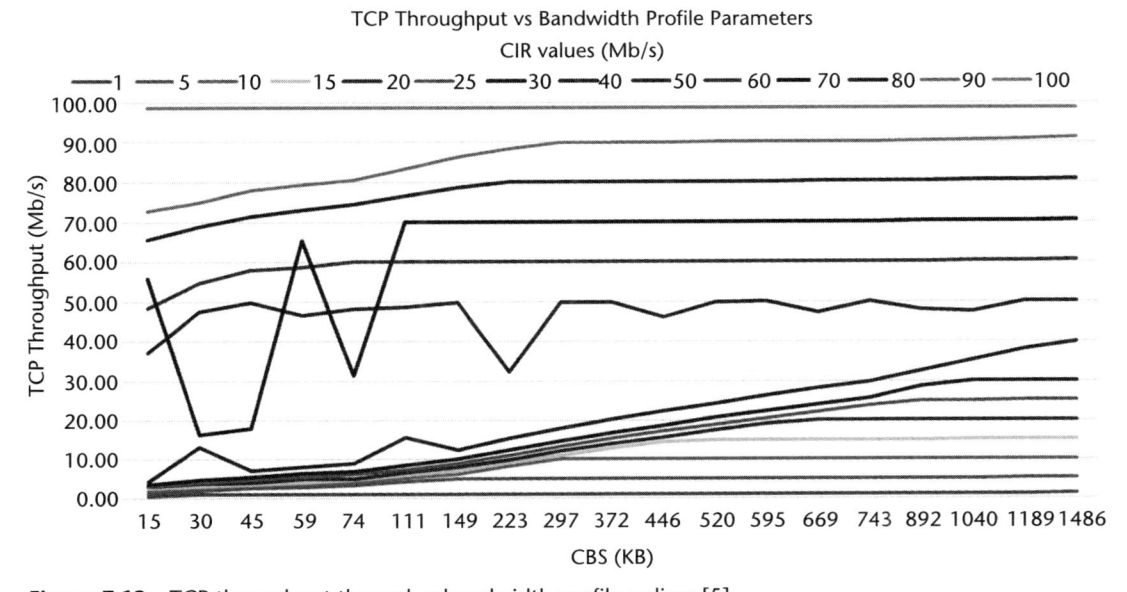

Figure 7.13 TCP throughput through a bandwidth profile policer [5].

This limits the TCP throughput to approximately CBS worth of data transmitted every timeout interval and typically results in TCP throughput well below the CIR.

To get TCP throughput close to the service rate, CBS must be increased to be greater than or equal to CIR multiplied by the timeout interval. With a minimum timeout value typically at 250 ms, this results in exceedingly large CBS values.

For a flow with a bandwidth profile of CIR, EIR, CBS, and EBS, it will allow traffic to pass at EIR+CIR while there are tokens available in the bucket, but when the bucket is depleted, the amount of traffic allowed to pass is limited to CIR. While there are tokens in the bucket, TCP will ramp up transmissions to utilize the full line rate. When the bucket is depleted and the available rate suddenly drops to CIR, there are typically multiple TCP packets lost and TCP reacts with a timeout. During the timeout interval, the token bucket gets a chance to refill, and the process repeats when TCP resumes transmission. The result is that a first approximation of the overall throughput of a single TCP flow is roughly one CBS worth of data every RTO interval, or CIR, whichever is less.

$$\text{Individual TPC Flow Throughput} \sim \min\left\{CIR, \frac{CBS}{RTO}\right\} \quad (7.1)$$

TCP dynamically adjusts the RTO based on a weighted moving average of the measured flow RTT; however, there is a minimum value for the RTO. Although RFC 6298 recommends a minimum RTO value of 1 second, most common implementations reduce to this value to between 200 and 250 milliseconds. Using 250 ms for the minimum RTO, the minimum value of CBS necessary for the TCP throughput to equal CIR can be calculated as

$$CBS \geq CIR \times RTO = CIR \times 250\text{ms} \quad (7.2)$$

Figure 7.14 shows simulation results of a TCP flow (with unlimited data to send) when restricted by an ingress bandwidth profile policer on a 100-Mbps UNI. The figure plots the aggregate amount of data acknowledged by the receiver over time, with a policer configured with a CIR of 10 Mbps and four different values of CBS. The data received at the receiver and CIR are measured in units of TCP segment, which is 1,460 bytes for TCP data, 1,522 bytes in a policer due to the addition of TCP/IP and MAC headers, VLAN tag, and FCS, or 1,542 bytes due to the addition of preamble and accounting for a minimum interpacket gap.

The average TCP throughput for each CBS value is the slope of a line fit to the data set corresponding to that value. A gray line with a slope of CIR is shown as a reference for the target TCP throughput. The periodic transmit-then-timeout pattern is clearly visible. The TCP throughput is well below CIR for CBS equal to 10 segments. At a CBS of 100 segments, the throughput is closer to CIR because the increased CBS allows more data to be transmitted during each cycle. At CBS equal to 200 segments, the throughput matches CIR and at any given time T the cumulative data received is between $CIR * T$ and $CBS + CIR * T$. Increasing CBS still further does not increase the TCP throughput because the average throughput cannot exceed CIR, so the plot for CBS equal to 700 segments has a slope equal

7.9 CBS Values, TCP, and Shaping

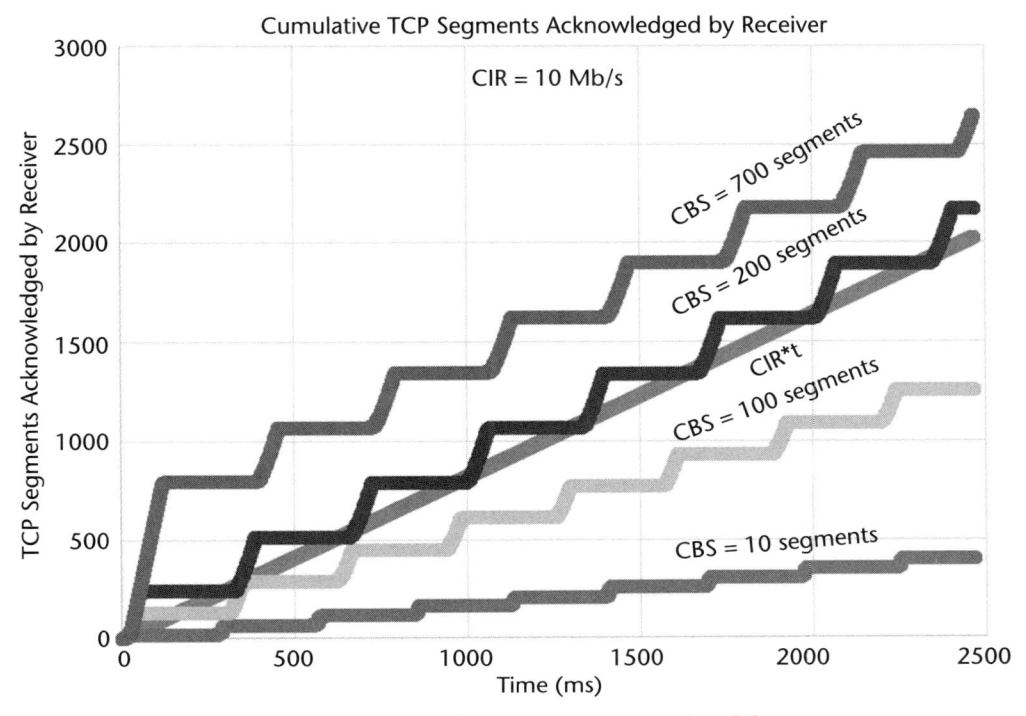

Figure 7.14 TCP segments received over time through a BWP policer [9].

to CIR. The plot is elevated above the CIR line because the initial condition of a full token bucket causes a large initial burst. This does not affect a long-term average because the token bucket never completely fills again so the initial burst size is never repeated.

Using equation (7.2) to calculate the minimum value of CBS that results in TCP throughput equal to a CIR of 10 Mbps with RTO of 250-ms results in a CBS equal to 205 segments. This correlates well with the plot for CBS equal to 200 segments in Figure 7.14. Using equation (7.1) to calculate the expected TCP throughput for CBS equal to 10 segments results in a throughput of 0.49 Mbps. This makes sense since this equation predicts a linear relationship between CBS and throughput for a given CIR. However the slope of the line in the simulation results for CBS equal to 10 segments is 2.1 Mbps, which is significantly higher than predicted. The source of the discrepancy is that the derivation of this equation implicitly assumes the TCP transmitter sends CBS of data instantly at the start of each cycle, and no more for a timeout interval. Actually it takes time for the transmitter to send the data, during which time the bandwidth profile is also adding tokens to the token bucket at a rate of CIR. Furthermore, the transmitter does not stop transmitting when the first packet is discarded at the bandwidth profile, but continues sending up to a TCP window size worth of additional data, some of which is received. Both of these factors are significant when CBS is small, but diminish as CBS approaches the value where TCP throughput equals CIR. Nonetheless the predictions of equation (7.1) are sufficiently close to the simulation results to justify a conclusion that the actual TCP throughput will be well below CIR for small values of CBS, and the predictions of equation (7.2) correlate very well with simulation results in Figure 7.15 for the value of CBS when the TCP throughput reaches CIR, at least for values of CIR up to 40 Mbps.

In Figure 7.13, even though the value of CBS where TCP throughput reaches CIR is much lower than predicted by equation (7.1), it is still much larger than might be expected assuming expectations are on the order of the TCP window size or the bandwidth delay product (CIR * RTT). In this simulation the TCP window size is 42 segments (62 KB) and the bandwidth delay product for a CIR of 90 Mbps is 38 segments, but the CBS required to have the TCP throughput reach CIR is 200 segments (297 KB). The reason for this is that even though the transmit pattern does not involve a timeout, it still involves a cycle of ramping up the allowable number of packets in flight until the token bucket empties and packets are lost, then cutting the allowable number of packets in flight in half and ramping it up again. When the allowable number of packets in flight is small the transmission rate is less than CIR and the number of tokens in the bucket increases. As the allowable number of packets in flight increases the transmission rate will increase above CIR. Whether the average throughput is equal to CIR or less than CIR depends upon the CBS value. To maintain an average rate equal to CIR the CBS has to be large enough to hold all the tokens that accumulate during the period when the transmission rate is below CIR so that this much extra data can be sent when the transmission rate is greater than CIR. With a small CBS the token bucket will overflow, with each token that overflows representing a transmit opportunity that cannot be made up, resulting in a throughput less than CIR.

Shaping traffic prior to the ingress bandwidth profile policer substantially changes the TCP behavior [5]. Figure 7.16 shows the TCP segments acknowledged

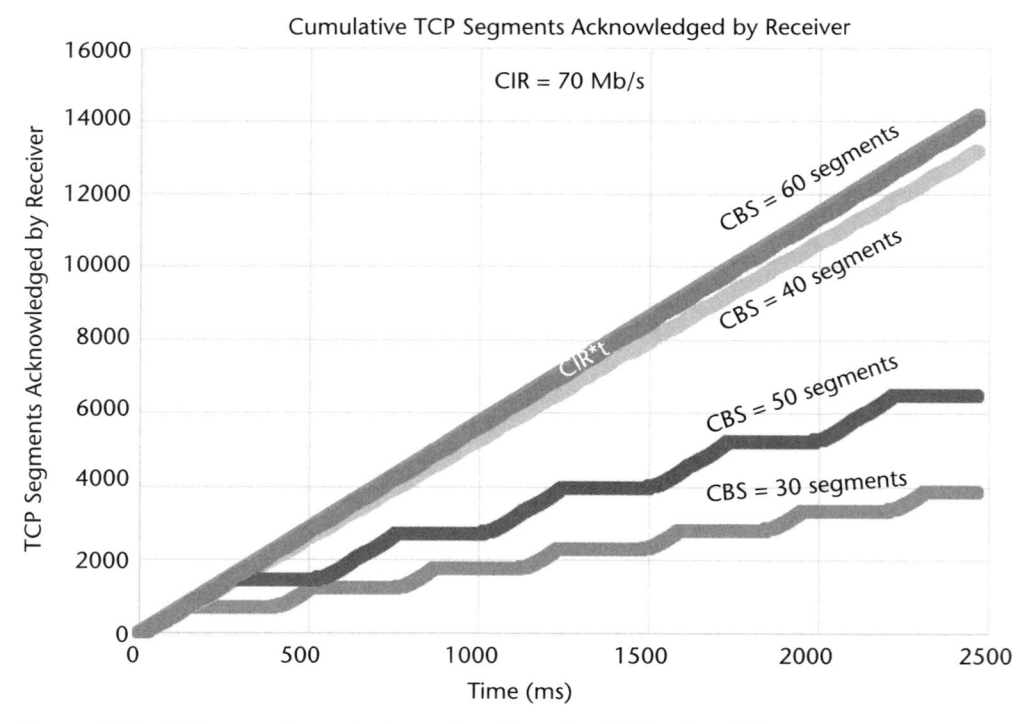

Figure 7.15 TCP segments received over time through a BWP policer [6].

7.9 CBS Values, TCP, and Shaping

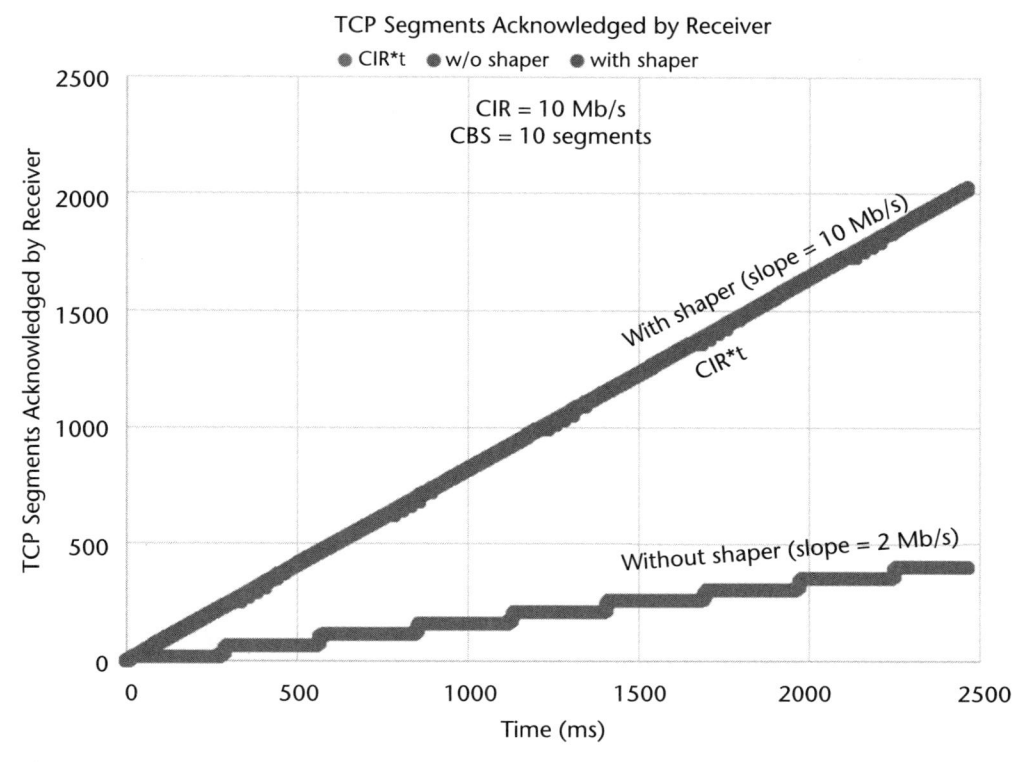

Figure 7.16 TCP behavior with and without a shaper [5].

by the receiver over time when a shaper is added prior to the policer and both are configured with CIR equal to 10 Mbps and CBS equal to 10 segments. With a shaper, the transmit-then-timeout pattern disappears and the TCP throughput is equal to CIR. The same result occurs with any CIR and CBS values as long as the CIR of the shaper (CIR$_{shaper}$) is less than or equal to the CIR of the bandwidth profile policer (CIR$_{BWP}$), and the CBS of the shaper (CBS$_{shaper}$) is less than or equal to the CBS of the bandwidth profile policer (CBS$_{BWP}$). In this case, the TCP bottleneck moves from the policer to the shaper, and when TCP ramps up its transmission rate above the bandwidth profile the shaper buffers the nonconformant packets rather than discarding them. The shape of the traffic reaching the policer is now conformant with the ingress bandwidth profile and so the policer does not discard any packets. However, introducing a buffer at the shaper means the effect of the buffer on TCP behavior needs to be considered.

When the TCP bottleneck is at a shaper, TCP behavior is dominated by the capacity of the shaper buffer. In most cases the effect of the buffer is seen in the delay of the packets, not throughput, although for small buffers the TCP throughput may be affected.

When the bottleneck bandwidth is at a shaper and the capacity of the shaper buffer is less than the sum of the receiver window size for all TCP flows, each flow will ramp up the amount of data in flight until the shaper buffer fills and packets are dropped. TCP responds to the lost packets by reducing the amount of data allowed to be in flight by a factor of two, which forces the transmitter to wait until the number of packets currently in flight is reduced by a factor of 2 before transmitting new data. This gives the shaper buffer an opportunity to drain. When TCP resumes transmission, it will ramp up the amount of data in flight again until the shaper buffer fills again and packets are dropped again. The cycle repeats, creating the sawtooth pattern in the buffer depth over time that is characteristic of TCP [5].

For a single TCP flow, throughput is maximized when the shaper buffer depth equals the capacity of the TCP flow. The capacity is the bandwidth delay product given by the bottleneck bandwidth times the RTT of the TCP flow. A larger shaper buffer depth results in there being residual data in the shaper buffer when TCP transmissions resume. This establishes a minimum buffer depth that adds delay to each packet but does not increase throughput. A smaller buffer depth results in the shaper buffer emptying before the TCP transmitter resumes. This means the TCP flow is not kept filled to capacity, which reduces throughput. Therefore, given an estimate of the flow RTT, the ideal shaper buffer depth can be calculated as

$$CBS^* = CIR \times RTT \qquad (7.3)$$

For multiple TCP flows, all flows will experience packet loss when the shaper buffer is full. This tends to synchronize the TCP behavior resulting in a sawtooth pattern in the buffer depth over time with the same amplitude but higher frequency as that for a single TCP flow. Equation (7.3) can be applied to cases with multiple TCP flows where the RTT is the average RTT of all the flows, RTT_{avg}.

One example is that research suggests equation (7.3) holds for small numbers of flows ($N < 100$), but tends to overestimate the ideal shaper buffer depth for large numbers of flows ($N > 500$). The rationale is that as the number of flows gets large it is not possible for all flows to have a packet arriving at the shaper at the point when the sawtooth pattern reaches the maximum shaper depth. The result is that flows cannot synchronize so the sawtooth pattern breaks down and the difference between the maximum and minimum shaper buffer depth decreases. The conclusion is that ideal shaper depth can be reduced by the square root of the number of TCP flows when there is reason to assume a large number of long-lived flows [5]:

$$CBS^* = \frac{CIR \times RTT_{avg}}{\sqrt{N}} \text{ for large } N \qquad (7.4)$$

In summary, the use of shaping prior to policing allows TCP to get a high utilization of the bandwidth profile CIR while limiting the size of the CBS value.

It should be noted that even though the TCP behavior is the same whether shaping occurs at an ingress or egress UNI, there will be a significant difference in the performance metrics of the EVC. An ingress shaper or ingress bandwidth profile policer will delay or discard frames before they are admitted to the EVC and before they are declared to be qualified frames for performance monitoring. Therefore the performance metrics of the EVC will not reflect this delay or frame loss. With the bottleneck bandwidth at an egress shaper, however, it will be qualified frames that are delayed or discarded as TCP adapts its transmission rate, and this will affect the EVC performance metrics.

When the shaping is done after (or in conjunction with) the ingress bandwidth profile policer in the CEN, the CBS value of the ingress bandwidth profile policer needs to be greater than or equal to the TCP bandwidth delay product (otherwise, the policer becomes the primary constraint on TCP rather than the shaper).Given above, CBS values depend on several factors such as the application which dictates the traffic pattern, whether traffic shaping is performed by the customer edge equipment before the traffic enters into CEN, the class of service and performance objectives, and so forth.

Large CBS values result in accepting traffic to CEN at rates higher than their CIRs. This can cause larger queues in internal network links, increased delay, delay variation, and frame loss that are not desirable for flows carrying real-time traffic with tight objectives for frame delay (FD or MFD) and/or frame delay variation (FDR or IFDV).

It is not possible to admit a maximum-sized frame if CBS is less than MFS. Furthermore, CBS should never be equal to MFS since this does not allow for the common case of back-to-back MFS frames. It is suggested that CBS be $\geq 3*$MFS, which is 5 KB for 1,522-byte MFS and 6 KB for 2,000-byte MFS [5]. The author believes that 3*MFS should be good for real-time applications such as circuit emulation. References [18, 19] also suggested a maximum of 8*MFS for heavily interactive applications such as telnet, ssh, and database transactions.

CBS should, in most cases, be inversely proportional to CIR / <line rate>. When the ratio is small (low CIR and/or high line rate), the token bucket can be emptied quickly but is filled slowly. As the ratio increases, the bucket is filled more quickly. In the limit, when the ratio =1, the bucket cannot be emptied faster than it is being filled.

Given above, due to the interaction between the policer and TCP's congestion control algorithm, reasonable performance can only be achieved by using a shaper before admitting traffic into the network at ingress and before delivering traffic to the customer at egress in the path of the session.

7.10 Conclusion

As users and carriers become familiar with Carrier Ethernet technology, we expect applications with strict performance requirements and best-effort applications to be mixed in the same port for economic reasons. As a result, we would need to prioritize and segregate one flow from another. Each flow will have its own bandwidth parameters and associated SLSs (i.e., delay, jitter, loss, and availability). In this chapter, we have summarized the guidelines to support applications requiring multi-CoS.

Furthermore, we have provided guidelines for defining CBS values for TCP traffic, SLSs for multipoint EVCs, and token sharing techniques to improve link utilization.

References

[1] Berger, A. W., "Performance Analysis of a Rate Control Throttle Where Tokens and Jobs Queue," *IEEE INFOCOM'90*, June 1990, pp. 30–38.

[2] Chao, H. J., "Design of Leaky Bucket Access Control Schemes in ATM Networks," *ICC'91*, 1991, pp. 180–187.

[3] RFC 2697, "A Single Rate Three Color Marker," September 1999.

[4] RFC 2698, "A Two Rate Three Color Marker," September 1999.

[5] MEF 23.2, "Carrier Ethernet Class of Service – Phase 3 Implementation Agreement," August 2016.

[6] IEEE 802.3x-1997, "IEEE Standards for Local and Metropolitan Area Networks: Supplements to Carrier Sense Multiple Access with Collision Detection (CSMA/CD) Access Method and Physical Layer Specifications - Specification for 802.3 Full Duplex Operation and Physical Layer Specification for 100 Mb/s Operation on Two Pairs of Category 3 or Better Balanced Twisted Pair Cable (100BASE-T2)," 1997.

[7] MEF 23, "Carrier Ethernet Class of Service – Phase 1, Implementation Agreement," June 2009.

[8] MEF 23.1, "Carrier Ethernet Class of Service – Phase 2, Implementation Agreement-Draft," January 2012.

[9] MEF 26.2, "External Network-Network Interfaces (ENNI) and Operator Service Attributes," August 2016.

[10] MEF 10.3, "Ethernet Services Attributes Phase 3," October 2013.

[11] RFC 4594, "Configuration Guidelines for DiffServ Service Classes," August 2006.

[12] MEF 33, "Ethernet Access Services Definition," January 2012.

[13] 3GPP TS 23.107 v3.0 (1999-10), "QoS Concept and Architecture," October 1999.

[14] ITU-T Y.1541, "Network Performance Objectives for IP-Based Services," 2011.

[15] IEEE 802.1ad-2005, "Local and Metropolitan Area Networks Virtual Bridged Local Area Networks," 2005.

[16] Bjorkman, B., "Models for Bandwidth Profiles with Token Sharing," Amendment to MEF 23.2, August 2016

[17] MEF 19, "Abstract Test Suite for UNI Type 1," April 2007.

[18] MEF 13, "User Network Interface (UNI) Type 1 Implementation Agreement," November 2005.

[19] MEF 34, "Abstract Test Suite for Ethernet Access Services," February 2012.

[20] MEF 37, "Abstract Test Suite for ENNI," January 2012.

[21] Haddock, S., "Shaper Considerations and Comparisons," https://wiki.mef.net/dosearchsite.action.queryStrong=shaper_considerations, February 2015.

Bibliography

MEF 10.2, "Ethernet Services Attributes Phase 2," October 2009.

MEF 10.3, "Ethernet Services Attributes Phase 3," October 2013.

Klessig, B., "Impact of CIR on Burst Size," MEF, October 2013.

Carrier Ethernet Operation, Administration, Management, and Performance

8.1 Introduction

The advent of Ethernet as service rather than transport in metropolitan and wide-area networks has accelerated the need for a new set of in-band operation, administration, and management (OAM) protocols. End-to-end Ethernet service, namely Ethernet virtual circuit (EVC), often involves one or more operators in addition to a service provider.

An EVC is initiated with a service order from user. The order consists of various EVC information including UNI locations, bandwidth, CoS, and SLAs. After provisioning the EVC, service provider and operator conducts turn-up and testing to ensure that EVC is operational and supports the user contract. Wthe EVC is in use, all the parties involved in EVC (i.e., subscriber, operators, and service provider) want to monitor the same EVC to ensure performance to service-level agreements (SLAs) such as delay, jitter, loss, throughput, and availability. Service provider and operators need to identify service anomalies in advance through measurements and threshold crossing alerts (TCAs) and take appropriate actions. When there is a problem identified through various alarms, the service provider and operators need to isolate the problem and identify faulty components to repair them within availability constraints.

IEEE, ITU, and MEF standards provide the tools to support the goals above. Ethernet OAM consists of fault management, performance management, testing, and service monitoring capabilities to support Ethernet services. Measurements, events/alarms, and provisioning attributes for interfaces (i.e., UNI, VUNI, ENNI), EVC, and OVC are defined.

Ethernet OAM can be categorized as:

- Link layer OAM defined by IEEE 802.3ah OAM [1],
- Service layer OAM defined by IEEE 802.1ag [2], ITU Y.1731 [3] and MEF implementation agreements [4–6];

- OAM via Ethernet local management interface (E-LMI) defined by MEF 16 [7].

Figure 8.1 depicts how these three protocols complement each other in fault monitoring of Ethernet networks. Figure 8.2 depicts operating boundaries of these protocols for an EVC.

IEEE 802.1ag-2007 and ITU-T Recommendation Y.1731 [2, 3] allow service providers to manage each customer service instance, EVC, individually and enables them to know if an EVC has failed, and if so, provides the tools to rapidly isolate the failure.

IEEE Std. 802.3ah [1] enables service providers to monitor and troubleshoot a single Ethernet link. Remote end MAC address discovery, alarms for link failures, and loopback are supported.

E-LMI protocol, defined in [7], operates between the customer edge (CE) device and the network edge (NE) of a service provider. It enables the service provider to automatically configure the CE device to match the subscribed service. This automatic provisioning of the CE device not only reduces the effort to set up the service, but also reduces the amount of coordination required between the service provider and enterprise customer. Furthermore, the enterprise customer does not have to learn how to configure the CE device, reducing barriers to adoption and greatly decreasing the risk of human error. In addition to automatic provisioning of the CE device, E-LMI can provide EVC status information to the CE device. If an EVC fault is detected, the service provider edge device can notify the CE device of the failure so that traffic can be rerouted to a different path quickly. However, ELMI is not widely deployed in industry today to take advantage of its capabilities. Figure 8.3 summarizes OAM functions provided by these standards.

In addition to in-band procedures defined by these protocols, network management through element management system (EMS), network management system (NMS), and operations support system (OSS) via in-band and out-of-band communication channels are also needed timely management of Carrier Ethernet networks.

In the following sections, we will describe link and EVC level OAM functions and messages, performance measurement and monitoring, availability, and testing in detail.

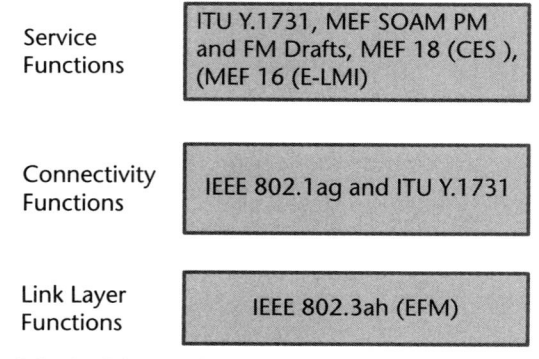

Service Functions	ITU Y.1731, MEF SOAM PM and FM Drafts, MEF 18 (CES), (MEF 16 (E-LMI)
Connectivity Functions	IEEE 802.1ag and ITU Y.1731
Link Layer Functions	IEEE 802.3ah (EFM)

Figure 8.1 Elements of Carrier Ethernet OAM.

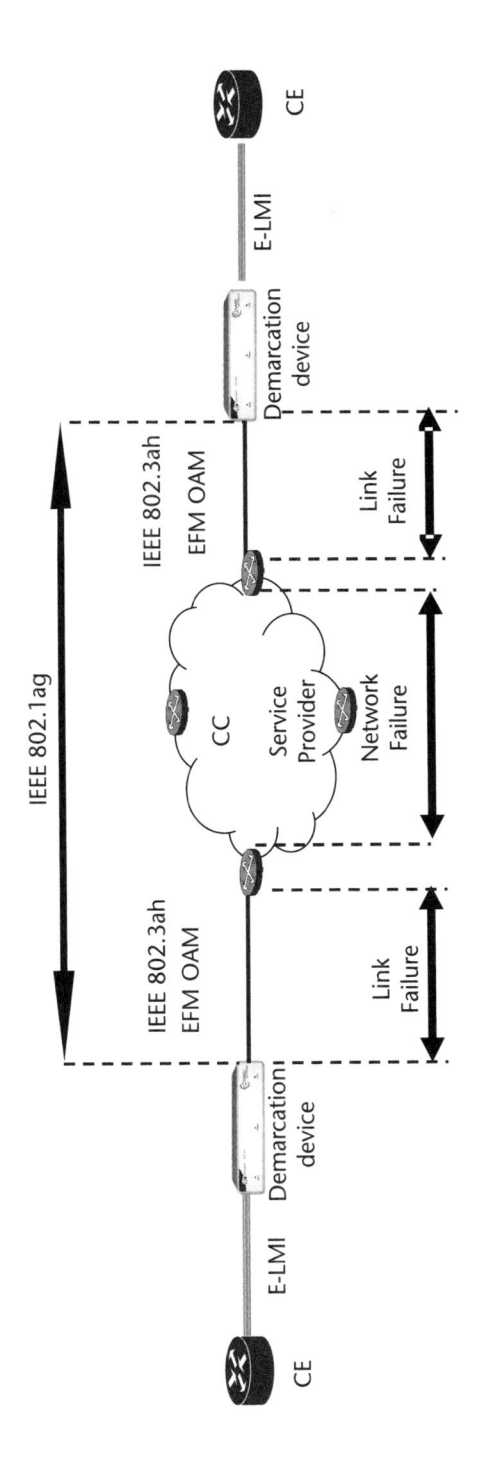

Figure 8.2 OAM process operating boundaries.

Services and Performance (ITU Y.1731/MEF)	Basic Connectivity (IEEE 802.1ag, ITU)	Transport/Link (802.3ah EFM)
Discovery	Discovery	Discovery
Continuity check (keep alive)	Continuity check	Remote failure indication: Dying gasp, link fault & critical event
Loopback (non-intrusive and intrusive)	Loopback	Remote, local loopback
AIS/RDI/Test		Fault isolation
Link Trace	Link Trace	Performance monitoring with threshold alarms
Performance management		Status monitoring

Figure 8.3 Summary of Ethernet OAM standards features.

8.2 Link OAM

Link OAM, as defined in IEEE 802.3ah, provides mechanisms to monitor link operation and health and improves fault isolation. It can be implemented on any full-duplex point-to-point or emulated point-to-point Ethernet link. OAM protocol data units (PDUs) cannot propagate beyond a single hop within an Ethernet network and not forwarded by bridges. OAM data is conveyed in untagged 802.3 slow protocol frames; therefore, OAM frame transmission rate is limited to a maximum of 10 frames per second. The major features covered by this protocol are:

- Discovery of MAC address of the next hop;
- Remote failure indication for link failures;
- Power failure reporting via dying gasp;
- Remote loopback.

OAM PDUs are identified by MAC address and Ethernet Length/Type/subtype field. This protocol uses a protocol sublayer between physical and data link layers. EFM OAM is enabled/disabled on a per-port basis.

Discovery is the first phase of link layer OAM. It identifies the devices at each end of the link along with their OAM capabilities. Discovery allows provider edge switch to determine the OAM capability of the remote demarcation device (Figure 8.4). If both ends support OAM, then the two ends exchange state and configuration information.

- Mode: active versus passive;
- OAMPDU size;
- Identity;

8.2 Link OAM

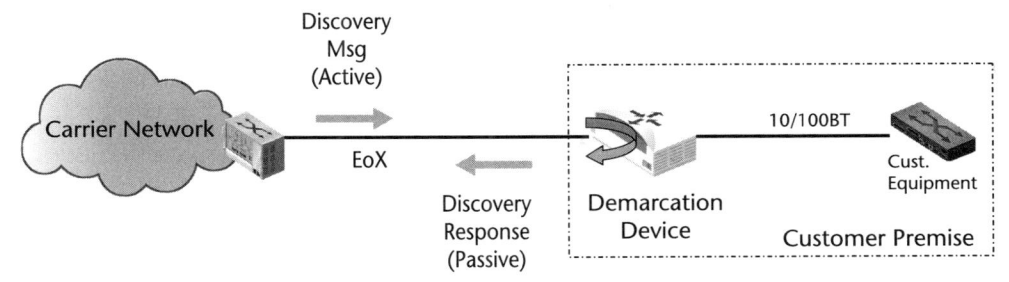

Figure 8.4 OAM discovery.

- Loopback support.

Link monitoring OAM serves for detecting and indicating link faults under a variety of conditions. It provides statistics on the number of frame errors (or percent of frames that have errors) as well as the number of coding symbol errors.

Failure conditions of an OAM entity are conveyed to its peer via OAMPDUs. For example, the failure conditions can be communicated via a loss of signal in one direction on the link (Figure 8.5).

Ethernet AIS/RDI is used to suppress downstream alarms and eliminate alarm storms from a single failure (Figure 8.6). This is similar to AIS/RDI in legacy services like SONET, TDM, and frame relay.

Demarcation devices can be configured to propagate link failures downstream and vice versa. In other words, WAN port failures can be propagated to LAN ports and vice versa (Figure 8.7). The fault propagation may be used for link level protec-

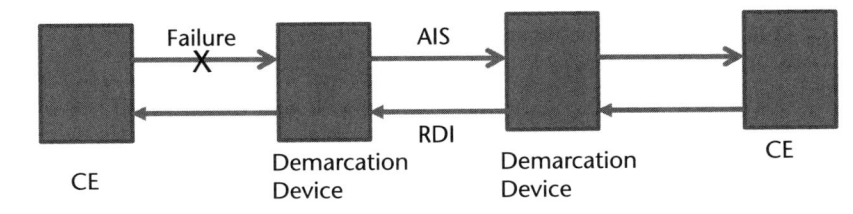

Figure 8.5 AIS/RDI.

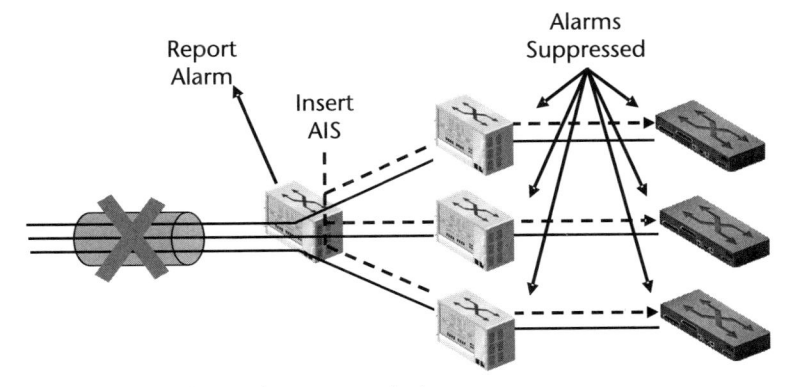

Figure 8.6 Alarm propagation to downstream devices.

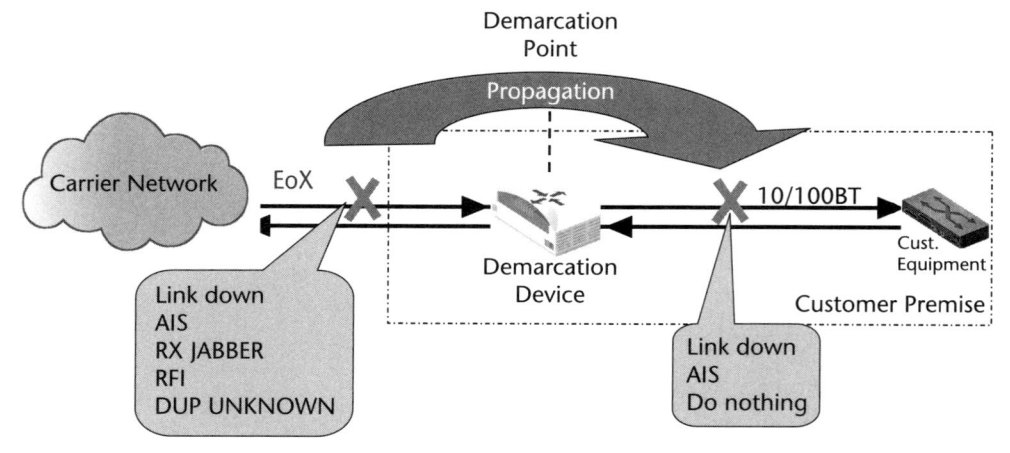

Figure 8.7 Alarm propagation from network to customer access.

tion schemes such as Hot Standby Routing Protocols (HSRP) allowing transparent fail-over [23].

Dying gasp message is generated for ac or dc power failures, sent immediately and continuously as depicted in Figure 8.8.

An OAM entity can put its remote peer, which is at the other side of a link, into a loopback mode using the loopback control OAMPDU. In the loopback mode, every frame received is transmitted back unchanged on the same port (except for OAMPDUs, which are needed to maintain the OAM session). Port-level testing is used in service turn-up and when troubleshooting.

This feature can be configured such that the service provider device can put the customer device into loopback mode, but not conversely.

In the loopback mode, a central test head or testing software embedded in the device can perform RFC 2544 tests to measure throughput, delay, jitter, and dropped packets (Figure 8.9).

Loopback can be facility or terminal where terminal loopback signal travels through the device. Applying both loopbacks will identify whether problem is with facility or device (Figure 8.10).

In addition to IEEE 802.3ah [1], some vendors support time domain reflectometer (TDR) for cable integrity testing to identify opens, shorts, and imped-

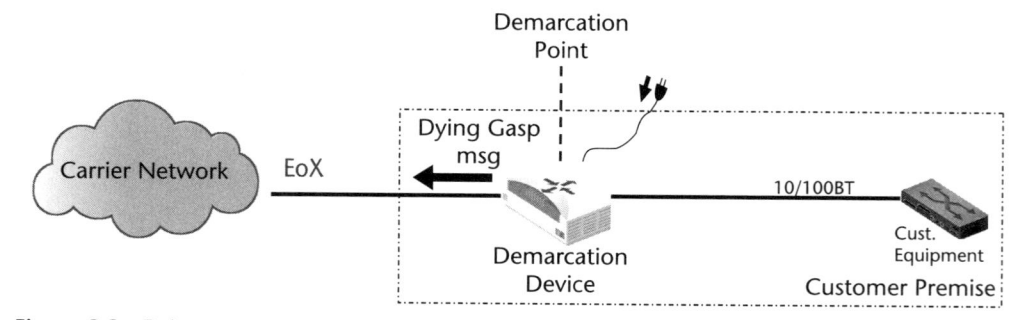

Figure 8.8 Dying gasp message propagation.

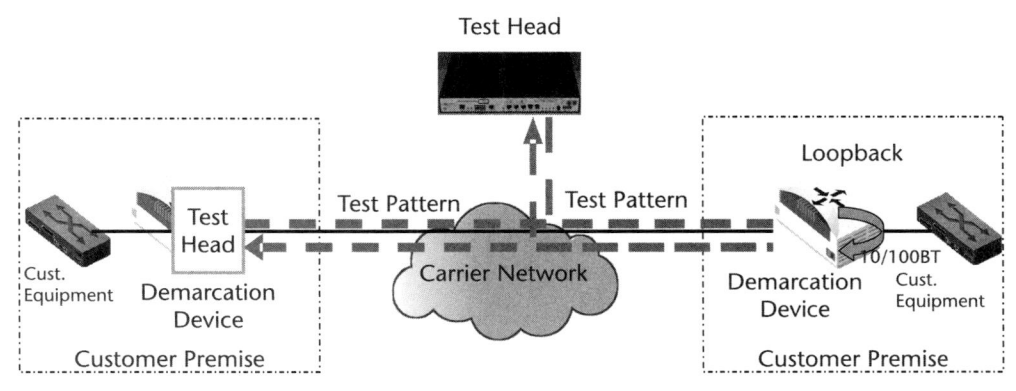

Figure 8.9 Loopback test using a test head.

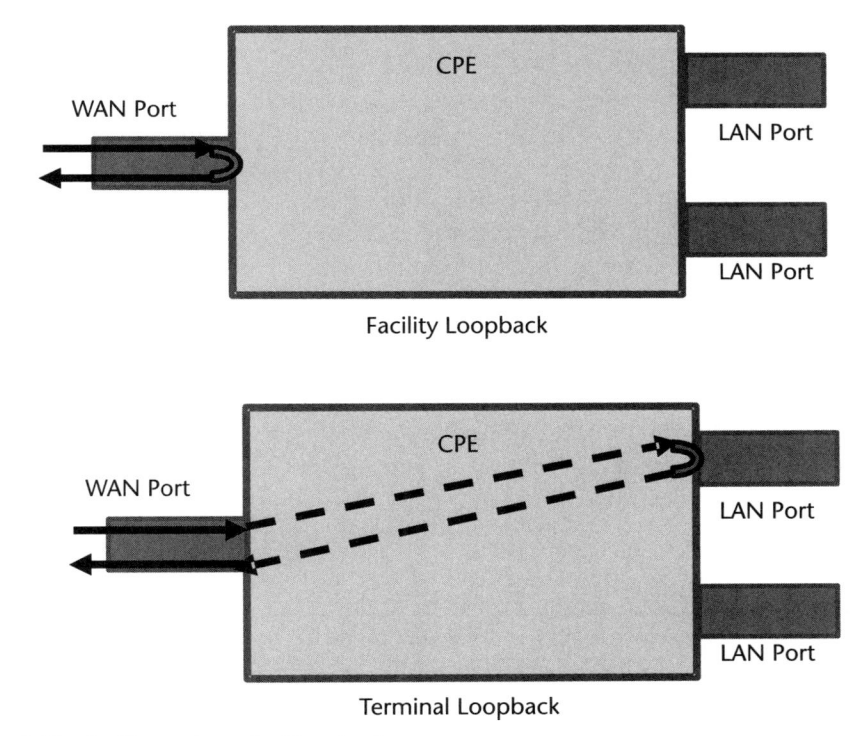

Figure 8.10 Facility and terminal loopbacks.

ance problems with CAT-5 cable on customer premise (Figure 8.11). This testing method helps to identify patch-panel and large percent of customer-induced issues.

8.3 Service OAM

The service OAM (SOAM) is addressed by IEEE 802.1ag (Connectivity Fault Management) [2], ITU Y.1731 [3], RFC 2544 [8], Y.1564 [9], and MEF specifications [5, 10, 11]. Continuity check, loopback, link trace, delay/jitter measurements, loss

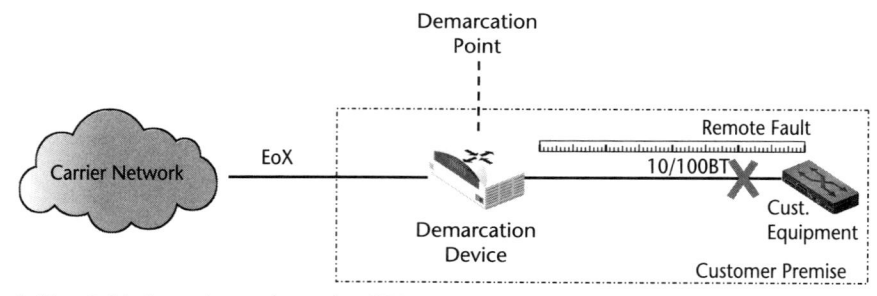

Figure 8.11 Cable integrity testing using TDR.

measurements, and in-service and out-of-service testing protocols are defined in these standards to monitor SLAs [i.e., service-level specifications (SLSs)], identify service level issues, and debug them.

Connectivity fault management (CFM) defined in [2] creates a maintenance hierarchy by defining maintenance domain and maintenance domain levels where maintenance endpoints (MEPs) determine domain boundaries. Maintenance points (MPs) that are between these two boundary points, MEPs, are called maintenance intermediate points (MIPs). Maintenance entities defined for CFM is depicted in Figure 8.12.

An EVC service between two subscribers from UNI to UNI may involve in multiple operators, in addition to a service provider. Customers purchase Ethernet service from service providers. Service provider domain covers the span of the service offered by the service provider. Service providers may use their own networks or the networks of other operators to provide connectivity for the requested service. Operator domain covers the service monitored by the network operator. Customers themselves may be service providers, for example, a customer may be an Internet service provider that sells Internet connectivity or a cloud service broker orchestrating Ethernet cloud carrier access. Subscriber OAM domain covers the portion of the service from subscriber equipment to the other subscriber equipment. The test OAM domain is used to test the connectivity to UNI-C by the service providers.

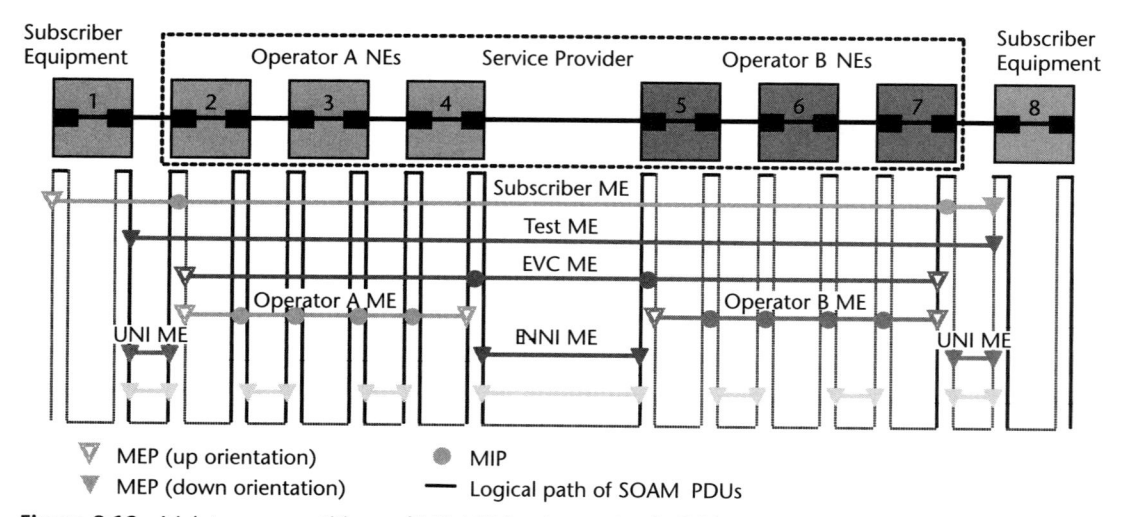

Figure 8.12 Maintenance entities and MA-MD level mapping in Table 8.1.

The ENNI domain covers the portion of the service between operators at the ENNI interfaces. Similarly, UNI domain covers the portion of the service between the networks side of the UNI (UNI-N) and the customer side of the UNI (UNI-C).

In all scenarios, the service provider and operators want to monitor and maintain the part of service that they are responsible from and subscriber wants to monitor the service he/she pays for. In order to have all parties to operate without interfering each other, maintenance domains and maintenance domain levels are defined. A maintenance domain is an administrative domain for the purpose of managing and administering a network. There are eight domain levels. Each domain is assigned a unique maintenance level by the administrator. Maintenance domains may nest or touch, but cannot intersect. If two domains nest, the outer domain must have a higher maintenance level than the one inside.

In this chapter we will describe OAM, service monitoring, and testing capabilities.

8.4 Maintenance Entities

An entity that requires management is called management entity (ME) in ITU-T and MEF. A set of MEs that satisfy the following conditions form a maintenance entity group (MEG) [3, 12]:

- MEs in an MEG exist in the same administrative domain and have the same MEG level.
- MEs in an MEG belong to the same service provider VLAN (S-VLAN). In ITU-T terminology, this is a point-to-point or multipoint Ethernet connection.

For a point-to-point Ethernet (ETH) connection, a MEG contains a single ME. For a multipoint Ethernet connection, a MEG contains n*(n-1)/2 MEs, where n is the number of Ethernet connection endpoints.

An MEP is a maintenance functional entity that is implemented at the ends of an ME. It generates and receives OAM frames. An ME represents a relationship between two MEPs.

A MEP can be up MEP or down MEP. Up MEP is an MEP residing in a bridge that transmits CFM PDUs towards, and receives them from, the direction of the bridge relay entity. Down MEP is an MEP residing in a bridge that receives CFM PDUs from, and transmits them towards, the direction of the LAN. The MEP that notifies the ETH layer MEPs upon failure detection is called a server MEP.

An MIP is located at intermediate points along the end-to-end path where Ethernet frames are bridged to a set of transport links. It reacts and responds to OAM frames.

In case MEGs are nested, the OAM flow of each MEG has to be clearly identifiable and separable from the OAM flows of the other MEGs. This is accomplished by MEG level/maintenance domain and maintenance domain levels.

Eight MEG levels can be shared among customer, provider, and operator to distinguish between OAM frames belonging to nested MEGs of customers, providers, and operators. The default MEG level assignment among customer, provider, and operator roles is:

- Customer role is assigned three MEG Levels: 7, 6, and 5.
- Provider role is assigned two MEG Levels: 4 and 3.
- Operator role is assigned three MEG Levels: 2, 1, and 0.
- MEF FM IA [11] recommends the default values in Table 8.1.

The service network is partitioned into customer, provider, and operator maintenance levels. Providers have the end-to-end service responsibility. Operators provide service transport across a subnetwork.

The OAM architecture is designed such that a MEP at a particular MEG level transparently passes SOAM traffic at a higher MEG level, terminates traffic at its own MEG level, and discards SOAM traffic at a lower MEG level. This results in a nesting requirement where a MEG with a lower MEG Level cannot exceed the boundary of a MEG with a higher MEG level [5].

The domain hierarchy provides a mechanism for protecting an MP (MEP or MIP) from other MPs with which the MP is designed to communicate. However, it is possible for an MP to flood one or more of its peer MPs with SOAM PDUs. This can result in a denial of service by forcing the receiving MPs to use computing resources for processing the SOAM PDUs from the flooding MP. Therefore, an NE supporting MPs needs to limit the number of SOAM PDUs per second that are processed at the local node.

8.5 Maintenance Points

Any port of a bridge is referred to as a maintenance point (MP). MPs are unique for a given MD level. An MP may be classified as a maintenance endpoint or an MIP. According to [2], MEPs and MIPs are the short form of maintenance association (MA) endpoints and intermediate points. They are functionally equivalent to ITU-T MEPs and MIPs.

MEG endpoint (MEP) marks the end point of an ETH MEG, which is capable of initiating and terminating OAM frames for fault management and performance

Table 8.1 MEF Recommendation for MA-MD Level Mapping and MEP Types

MA (Maintenance Entity in Y.1731)	Default MD Level (i.e., Maintenance Entity Group Level in Y.1731)	MEP Type
Subscriber-MA	6	Up
Test-MA	5	Down
EVC-MA	4	Up
Operator-MA	2	Up
UNI-MA	1	Down
E-NNI-MA	1	Down

8.5 Maintenance Points

monitoring. The OAM frames are distinct from the transit ETH flows and subject to the same forwarding treatment as the transit ETH flows being monitored.

A MEP does not terminate the transit ETH flows, although it can observe these flows, for example, count the frames. It can generate and receive CFM PDUs and track any responses. It is identified by maintenance association endpoint identifier (MEPID), which is a small integer, unique over a given MA.

A server MEP represents both the server layer termination function and server/ETH adaptation function which is used to notify the ETH layer MEPs upon failure detection by the server layer termination function or server/ETH adaptation function.

On fault detection, AIS is generated in the upstream direction at the maintenance level of the client layer. The transport layer can be am MPLS pseudowire path, a SONET path, or an Ethernet link. At this layer it is necessary to implement an Ethernet OAM AIS generation function for the Ethernet client layer at the appropriate MA level.

A server MEP needs to support ETH-AIS function, where the server/ETH adaptation function is required to issue frames with ETH-AIS information upon detection of defect at the Server layer by the server layer termination and/or adaptation function.

MEG intermediate point (MIP) is an intermediate point in a MEG that is capable of reacting to some OAM frames. A MIP does not initiate OAM frames. An MIP takes no action on the transit ETH flows. MP functions are listed in Table 8.2.

An MIP will forward CFM packets unless it is a loopback or link trace destined for that intermediate point, while MEPs do not forward CFM frames because they must keep them within the domain.

OAM frames belonging to an administrative domain originate and terminate in MEPs present at the boundary of that administrative domain. The MEP allows OAM frames from outside administrative domains belonging to higher-level MEs to pass transparently while it blocks OAM frames from outside administrative domains belonging to same or lower level MEs. The MEP at an edge of a domain filter by maintenance level, therefore, OAM messages at their own level and at higher levels of the MEPs, are filtered. With this, OAM frames can be prevented from leaking.

An MEP must be provisioned with information about its peer MEPs. This information can be potentially discovered. MEPs can proactively discover other MEPs by CCM messages. A multicast loopback can be used to discover other MEPs on an on-demand basis as well. MIPs can be discovered by using link trace.

Table 8.2 Maintenance Point Functions

Functions	MEP	MIP	Transparent Point
Initiate CFM messages related to continuity check, loopback, and link trace	Yes	No	No
Respond to loopback and link trace messages	Yes	Yes	No
Catalog continuity-check information received	Yes	Yes	No
Forward CFM messages	No	Yes	Yes

The principal operational issue for Ethernet OAM is scalability. CCM can be sent as fast as every 3.3 ms. There can be 4,094 VLANs per port and up to eight maintenance levels. This yields a worst-case CCM transmission rate of 9.8 million CCMs per second.

The main operational issue for linktrace is Ethernet MAC address learning and aging. When there is a network fault, the MAC address of a target node can age out in several minutes (e.g., typically 5 minutes). Solutions are to launch linktrace within the age-out time or to maintain a separate target MEP database at intermediate MIPs. However, this requires a MIP CCM database.

Fault notification and alarm suppression is accomplished by using SNMP notifications and AIS/RDI. AIS can provide both alarm suppression and upstream notification. RDI provides downstream notification. The main issues with AIS are multipoint service instances, and the potential interaction with Ethernet Spanning Tree Protocol (STP) loop prevention and recovery. An STP-based network reconfiguration may result in AIS interruption or redirection. The issues with RDI are multipoint service instances and bidirectional faults, which would block RDI downstream transmission.

The customer and service provider can independently use the all eight MEG levels, as well as mutually agree how to share the MEG levels.

The customer must send OAM frames as VLAN-tagged or priority-tagged frames to utilize all eight MEG levels independently. However, if the customer uses untagged OAM frames, the MEG levels may not be independent anymore and the customer and provider MEG levels need to be mutually agreed between the customer and the service provider.

Figure 8.12 illustrates the MDs and maintenance entities (MEs) by the MEF where pairs of MEPs (thus MEs) are communicating across various MDs that are hierarchically related to each other.

The scope of a MD is restricted to its associated VLAN, which has implications when VLAN identifiers are stacked. For example, if an ingress subscriber frame with a C-tag is stacked with an S-tag by the service provider, then the subscriber and test MDs are in a different VLAN than the EVC and operator MDs. Therefore, MD levels may not need to be coordinated between the subscriber and service provider.

An MA is identified by a maintenance association identifier (MAID), which is unique over the domain. The MAID has two parts: the maintenance domain name and the short MA name. The MD name has the *null* format and the short MA name has the *text* format. Using the text format for the short MA name allows for a maximum length of 45 octets for the short MA name (see Table 8.3).

The MAID is carried in CFM PDUs to identify inadvertent connections among MEPs. A small integer, the MEPID uniquely identifies each MEP among those configured on a single MA.

A set of MEPs, each configured with the same MAID and MD Level, established to verify the integrity of a single service instance, is called MA. An MA can also be thought of as a full mesh of maintenance entities among a set of MEPs so configured.

8.5 Maintenance Points

Table 8.3 Terminology Mapping

Y.1731 Term	802.1ag Term	Comments
MEG	MA	
MEG ID	MAID (Domain Name+Short MA Name)	Unlike 802.1ag, the MEG ID does not imply a split between domain name and a short MEG name in Y.1731
MEG level	MA level	

A MEP is associated with exactly one MA and identified for management purposes by MAID identifying the MA and MEPID where the MEPID is an integer in the range 1 to 8,191. It provides each MEP with the unique identity required for the MEPs to verify the correct connectivity of the MA. The MEP is identified for data forwarding purposes by a set of VIDs, including a primary VID, inherited from the MA. From the MA, the MEP inherits a maintenance domain, and from this it inherits a maintenance domain level (MD Level), inherited from its maintenance domain, and a primary VID, inherited from its MA (see Tables 8.4 and 8.5).

Table 8.4 Maintenance Domain Name Format

Maintenance Domain Name Format Field	Value
Reserved for IEEE 802.1	0
No maintenance domain name present	1
Domain name-based string denoting the identity of a particular maintenance domain)	2
MAC address + 2-octet integer	3
Character string (an IETF RFC 2579 DisplayString, with the exception that character codes 0 to 31 are not used)	4
Reserved for IEEE 802.1	5–31
Defined by ITU-T Y.1731	32–63
Reserved for IEEE 802.1	64–255

Table 8.5 Short MA Name Format

Short MA Name Format Field	Value
Reserved for IEEE 802.1	0
Primary VID	1
Character string (an IETF RFC 2579 DisplayString, with the exception that character codes 0 to 31 are not used)	2
2-octet integer	3
RFC 2685 VPN ID	4
Reserved for IEEE 802.1	5–31
Defined by ITU-T Y.1731	32–63
Reserved for IEEE 802.1	64–255

8.6 OAM Addressing and Frame Format

OAM Frames are identified by a unique EtherType value, which is 0x8902. OAM frames processing and filtering at a MEP is based on the OAM EtherType and MEG level fields for both unicast and multicast DA (destination address).

IEEE 802.1ag supports two addressing modes, the bridge port model and the master port model. In the bridge port model, MEPs and MIPs assume the same MAC address as the bridge port. For the master port model, MEPs are implemented in a logical bridge master port such as a CPU. All MEPs use the same master port MAC address. These master port MEPs have an ambiguity in the identification of a MEP to which a loopback reply is destined. This model is for legacy devices that cannot support port-based SOAM.

The source MAC address for all OAM frames is always a unicast MAC address. The destination MAC address may be either a unicast or a multicast address. Two types of multicast addresses are required depending on the type of OAM function:

- Multicast DA Class 1: OAM frames that are addressed to all MEPs in a MEG (e.g., CCM, Multicast LBM, AIS);
- Multicast DA Class 2: OAM frames that are addressed to all MIPs and MEPs associated with a MEG (e.g., LTM).

A multicast DA could also implicitly carry the MEG level, which will result in 8 distinct addresses for each of the multicast DA classes 1 and 2 for the 8 different MEG levels.

Multicast addresses for class 1 and multicast addresses for class 2 are 01-80-C2-00-00-3x and 01-80-C2-00-00-3y respectively where x represents MEG level with x being a value in the range 0 to 7 and y represents the MEG level with y being a value in the range 8 to F.

The following lists the usage of unicast DA and multicast DA by each OAM messages per Y.1731.

- CCM frames can be generated with a specific multicast class 1 DA or unicast DA. When a multicast DA is used, CCM frames allow discovery of MAC addresses associated with MEPs and detection of misconnections among domains.
- LBM frames can be generated with unicast or multicast class 1 DAs.
- LBR frames are always generated with unicast DAs.
- LTM frame is generated with a multicast class 2 DA. A multicast DA is used instead of unicast DA for LTM frames since in current bridges, ports do not look at the EtherType before looking at the DA; therefore, the MIPs would not be able to intercept a frame with a unicast DA which was not their own address and would forward it.
- LTR frames are always generated with unicast DAs.
- AIS frame can be generated with a multicast class 1 DA in a multipoint EVC and with unicast DA for point-to-point EVCs if unicast DA of the downstream MEP is configured on the MEP transmitting AIS.

8.6 OAM Addressing and Frame Format 213

- LCK frame can be generated with a multicast class 1 DA in a multipoint EVC and with unicast DA for point-to-point EVCs if unicast DA of the downstream MEP is configured on the MEP transmitting LCK.
- TST frames are generated with unicast DAs. TST frames may be generated with multicast class 1 DA if multipoint diagnostics are desired.
- APS (Automatic Protection Switching, defined in G.8031) frames can be generated with a specific multicast class 1 DA or unicast DA.
- Maintenance communication channel (MCC) frames are generated with unicast DAs.
- LMM frames are generated with unicast DAs. LMM frames may be generated with multicast class 1 DA if multipoint measurements are desired.
- LMR frames are always generated with unicast DAs.
- 1DM frames are generated with unicast DAs. 1DM frames may be generated with multicast class 1 DA if multipoint measurements are desired.
- DMM frames are generated with unicast DAs. DMM frames may be generated with multicast class 1 DA if multipoint measurements are desired.
- DMR frames are always generated with unicast DAs.

Figure 8.13 illustrates the common Ethernet service OAM frame format. Each specific OAM message type will add additional fields to the common PDU format.

As with all Ethernet frames, the destination and source MAC address (DA/SA), is preceded by a 7-octet preamble and a 1-octet start of frame delimiter. The frame may or may not include a customer VLAN tag. However, it can include an S-tag.

IEEE 802.1ag also supports the LLC/SNAP encoded frame format, which includes an LLC header in addition to the OAM EtherType.

The following information elements are common across the OAM PDUs:

- MEG level (i.e., MD level) is a 3-bit field. It contains an integer value that identifies the MEG level of the OAM PDU. Its value ranges from 0 to 7. The MD level corresponds to the administrative domains shown in Figure 8.12. MD levels 5 through 7 are reserved for customer domains, MD levels 3 and

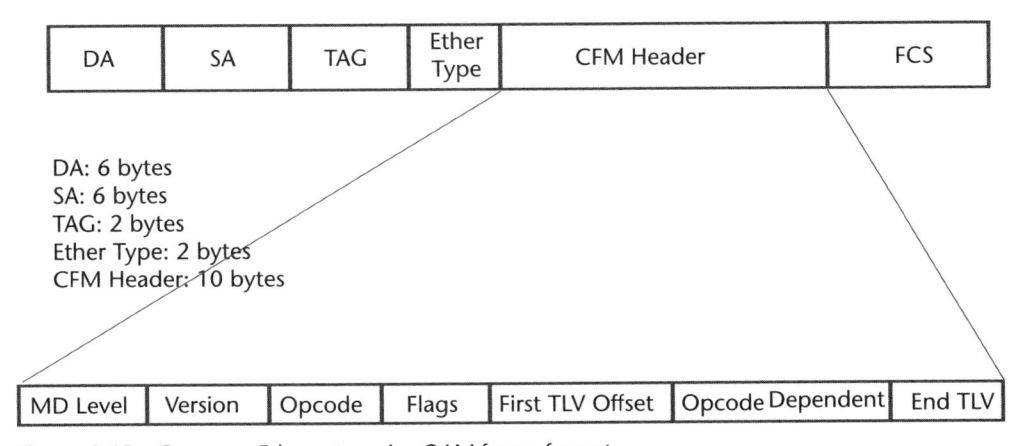

| DA | SA | TAG | Ether Type | CFM Header | FCS |

DA: 6 bytes
SA: 6 bytes
TAG: 2 bytes
Ether Type: 2 bytes
CFM Header: 10 bytes

| MD Level | Version | Opcode | Flags | First TLV Offset | Opcode Dependent | End TLV |

Figure 8.13 Common Ethernet service OAM frame format.

4 are reserved for service provider domains, and MA levels 0 through 2 are reserved for operator domains. Relationships among EVCs, maintenance domains, maintenance associations, and MEPs are depicted in Figure 8.14.

- Version is a 5-bit field containing an integer value that identifies the OAM protocol version. In Y.1731 the version is always 0.
- OpCode is a 1-octet field containing an OpCode that identifies an OAM PDU type (i.e., CCM, LTM, LBM). OpCode is used to identify the remaining content of an OAM PDU. Continuity check, loopback, and link trace use OpCodes 0 to 31 while Y.1731 performance management functions use OpCodes 32 to 63 (see Table 8.6)
- Flags are an 8-bit field and their usage is dependent on the OAM PDU type.
- TLV (Type, Length, and Value) offset is a 1-octet field. It contains the offset to the first TLV in an OAM PDU relative to the TLV offset field. The value of this field is associated with an OAM PDU type. When the TLV offset is 0, it points to the first octet following the TLV offset field.

The following is a set of TLVs defined by IEEE 802.1ag for various CFM PDU types. Each TLV can be identified by the unique value assigned to its type field. Some type field values are reserved.

- TLVs applicable for continuity check message (CCM): End TLV, sender ID TLV, port status TLV, interface status TLV, organization-specific TLV;
- TLVs applicable for loopback message (LBM): End TLV, sender ID TLV, data TLV, organization-specific TLV;
- TLVs applicable for loopback reply (LBR): End TLV, sender ID TLV, data TLV, organization-specific TLV;
- TLVs applicable for linktrace message (LTM): End TLV, LTM egress identifier TLV, sender ID TLV, organization-specific TLV;
- TLVs applicable for linktrace reply (LTR): End TLV, LTR egress identifier TLV, reply ingress TLV, reply egress TLV, sender ID TLV, organization-specific TLV.

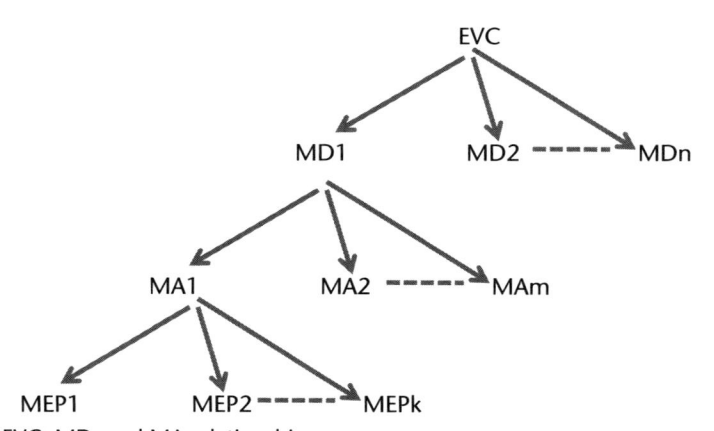

Figure 8.14 EVC, MD, and MA relationship.

8.7 Continuity Check Message (CCM)

Table 8.6 Op Codes for OAM PDUs

OpCode Value	OAM PDU Type	OpCode Relevance for MEPs/MIPs
1	CCM	MEPs
3	LBM	MEPs and MIPs (connectivity verification)
2	LBR	MEPs and MIPs (connectivity verification)
5	LTM	MEPs and MIPs
4	LTR	MEPs and MIPs
33	AIS	MEPs
35	LCK	MEPs
37	TST	MEPs
39	Linear APS	MEPs
40	Ring APS	MEPs
41	MCC	MEPs
43	LMM	MEPs
42	LMR	MEPs
45	1DM	MEPs
47	DMM	MEPs
46	DMR	MEPs
49	EXM	
48	EXR	
51	VSM	
50	VSR	

The general format of TLVs is shown in Figure 8.15. Type values are specified in Table 8.7: type (1 byte), length (2–3 bytes), and value, which is optional (4 bytes).

In an end TLV, both length and value fields are not used.

Priority and drop eligibility of a specific OAM frame are not present in OAM PDUs, but conveyed in frames carrying OAM PDUs.

8.7 Continuity Check Message (CCM)

The OAM PDU used for Ethernet continuity check (ETH-CC) information is continuity check message (CCM). Frames carrying the CCM PDU are called CCM frames.

Continuity checks are "heartbeat" messages issued periodically by MEPs (Figure 8.16). They are used for proactive OAM to detect loss of continuity (LOC) among MEPs and discovery of each other in the same domain. MIPs will discover

| Type (1 byte) |
| Length (2-3 bytes) |
| Value which is optional (4 bytes) |

Figure 8.15 Generic TLV format.

Table 8.7 Type Value

Type Value	TLV Name
0	End TLV
3	Data TLV
5	Reply ingress TLV
6	Reply egress TLV
7	LTM egress identifier TLV
8	LTR egress identifier TLV
32	Test TLV
33–63	Reserved

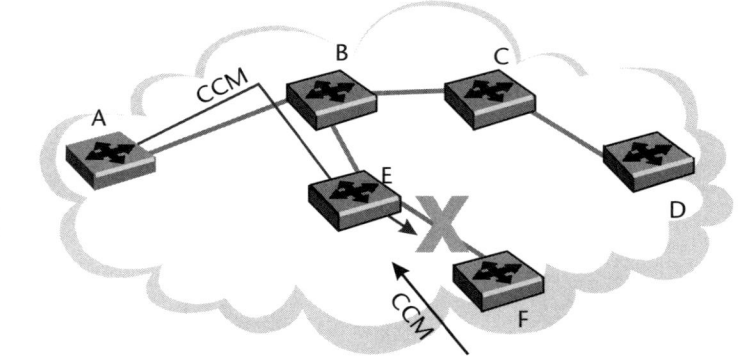

Alarm generated for CCM Timeout due to link failure between E and F

Figure 8.16 CCM.

MEPs that are in the same domain using CCMs as well. In addition, CCM can be used for loss measurements and triggering protection switching.

CCM also allows detection of unintended connectivity between two maintenance domains (or MEGs), namely mismerge; unintended connectivity within the MEG with an unexpected MEP; and other defect conditions such as unexpected MEG level and unexpected period. CCM is applicable for fault management, performance monitoring, or protection switching applications.

When CCM transmission is enabled, the MEP periodically transmits frames to all other MEPs in the same domain. The CCM transmission period must be the same for all MEPs in the same domain. When a MEP is enabled to generate frames with CCM information, it also expects to receive frames with ETH-CC information from its peer MEPs in the MEG.

A CCM has the following attributes (see Figure 8.17 and Table 8.8) to be configured:

- MEG ID (i.e., MAID) is 48 octets and identifies the MEG to which the MEP belongs. The MAID used by CCM must be unique across all MAs at the same MD level. The default MA values of CCM are untagged UNI-MA, untagged ENNI-MA, and EVC-MA.
- MEP ID is a 2-octet field whose 13 least significant bits identify the MEP within the MEG.

8.7 Continuity Check Message (CCM)

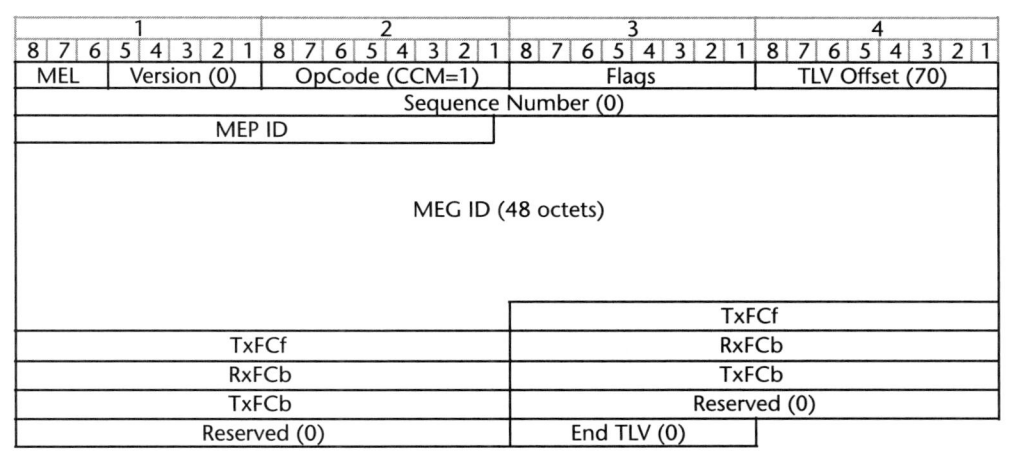

1		2		3		4	
8 7 6 5 4 3 2 1		8 7 6 5 4 3 2 1		8 7 6 5 4 3 2 1		8 7 6 5 4 3 2 1	
MEL	Version (0)	OpCode (CCM=1)		Flags		TLV Offset (70)	
Sequence Number (0)							
MEP ID								
MEG ID (48 octets)								
				TxFCf			
TxFCf				RxFCb			
RxFCb				TxFCb			
TxFCb				Reserved (0)			
Reserved (0)				End TLV (0)			

Figure 8.17 CCM PDU.

Table 8.8 Continuity Check Message Format Octet

Field	Length/Comment
Common CFM header	1–4
Sequence number	5–8
Maintenance association endpoint identifier	9–10
Maintenance association identifier (MAID)	11–58
Defined by ITU-T Y.1731	59–74
Reserved for definition in future versions of the protocol	Zero length
Optional CCM TLVs	First TLV Offset + 5 (Octet 75 for transmitted CCMs.)
End TLV (0)	First TLV offset + 5 if no optional CCM TLVs are present

- List of peer MEP IDs in the MEG: For a point-to-point MEG with a single ME, the list would consist of a single MEP ID for the peer.
- MEG level at which the MEP exists.
- CCM transmission period is encoded in the 3 least significant bits of the flags field. The default transmission period for fault management, performance monitoring, and protection switching are 1 second, 100 ms, and 3.33 ms, respectively. Ten-second, 1-minute, and 10-minute transmission periods are also used. If the port associated with an MEP experiences a fault condition, the MEP will encode RDI in the flags field.
- Priority of frame with ETH-CC information: By default, the frame with ETH-CC information is transmitted with the highest priority available to the data traffic.

- Drop eligibility of the frame with ETH-CC information that is always marked as drop ineligible.

However, a MIP is transparent to the CCM information and does not require any configuration information to support CCM.

When CCM is used to support dual-ended loss measurement, the PDU includes the following (Figures 8.17 through 8.19):

- TxFCf is a 4-octet field that carries the value of the counter of in-profile data frames transmitted by the MEP towards its peer MEP, at the time of CCM frame transmission.
- RxFCb is a 4-octet field that carries the value of the counter of data frames received by the MEP from its peer MEP, at the time of receiving the last CCM frame from that peer MEP.
- TxFCb is a 4-octet field that carries the value of the TxFCf field in the last CCM frame received by the MEP from its peer MEP.

A sender ID TLV, if included, indicates the management address of the source of the CCM frame. This can be used to identify the management address of a remote device rather than just its MAC address, making the identification of the device possible in a large network where MAC addresses are not well-known.

The port status and interface status TLVs indicate the status of various parameters on the sender of the CCM. The interface status TLV can indicate to the far end that a problem has occurred on the near end. When applied to the EVC MA, the interface status TLV can be used to indicate to the far end that the local UNI interface is down, even though the CCMs on the EVC are successful.

MEPs are configured (MEG ID and MEP ID) with a list of all the other MEPs in their maintenance level. Every active MEP maintains a CCM database. As an MEP receives CCMs, it catalogs them in the database indexed by MEP ID. Exchange of CCMs between MEPs in a MEG allow for the detection of the following defects:

- LOC condition: If no CCM frames from a peer MEP are received within the interval equal to 3.5 times the receiving MEP's CCM transmission period, loss of continuity with peer MEP is detected, as a result of hard failures

MSB							LSB
8	7	6	5	4	3	2	1
RDI	Reserved (0)				Period		

Figure 8.18 Format of CCM PDU flags.

MSB														LSB
octet 9								octet 10						
8	7	6	5	4	3	2	1	8	7	6	5	4	3	2	1
0	0	0						MEP ID							

Figure 8.19 MEP ID format in CCM PDU.

8.7 Continuity Check Message (CCM)

(e.g., link failure, device failure) or soft failures (e.g., memory corruption, misconfigurations). LOC state is exited during an interval equal to 3.5 times the CCM transmission period configured at the MEP, if the MEP receives at least 3 CCM frames from that peer MEP.

There is a direct correlation between the CCM frame transmission periods supported and the level of resiliency a network element can offer a specific EVC. Three CCM messages must be lost before a failure is detected across a specific MA. A failure must be detected before any protection switching mechanisms can enable a new path through the network. For example, to enact a protection switching mechanism that claims a maximum correction time of 50 ms and uses CCMs to detect the failure, the CCM frame transmission period must be 10 ms or less. Otherwise, just detecting the failure would take more than 50 ms.

- Unexpected MEG condition: If a CCM frame with a MEG level lower than the receiving MEP's MEG level is received, unexpected MEG level is detected.

- Mismerge condition: If a CCM frame with same MEG level but with an MEG ID different than the receiving MEP's own MEG ID is received, mismerge is detected. Such a defect condition is most likely caused by misconfiguration, but could also be caused by a hardware/software failure in the network. Mismerge state is exited when the MEP does not receive CCM frames with incorrect MEG ID during an interval equal to 3.5 times the CCM transmission period configured at the MEP.

- Unexpected MEP (UnexpectedMEP) condition: An MEP detects UnexpectedMEP when it receives a CCM frame from a remote MEP, which is not in the remote MEP list of this local MEP mostly due to misconfiguration, but has a correct MEG level with an unexpected MEP ID. Determination of unexpected MEP ID is possible when the MEP maintains a list of its peer MEP IDs. Unexpected MEP state is exited when the MEP does not receive CCM frames with an unexpected MEP ID during a specified interval such as equal to 3.5 times the CCM transmission period configured at the MEP.

- Unexpected period (UnexpectedPeriod) condition: If a CCM frame is received with a correct MEG level, a correct MEG ID, a correct MEP ID, but with a period field value different than the receiving MEP's own CCM transmission period, unexpected period is detected. Such a defect is most likely caused by misconfiguration. UnexpectedPeriod state is exited when the MEP does not receive CCM frames with incorrect period field value during an interval equal to 3.5 times CCM transmission period configured at the MEP.

- Signal fail (SignalFail) condition: An MEP declares signal fail condition upon detection of defect conditions including loss of continuity, mismerge, unexpected MEP, and unexpected MEG level. In addition, signal fail condition may also be declared by the server layer termination function to notify the server/ETH adaptation function, about defect condition in the server layer.

A receiving MEP notifies the equipment fault management process when it detects the above defect conditions.

8.8 Loopback and Reply Messages (LBM and LBR)

The Ethernet loopback function (ETH-LB) is an on-demand OAM function that is used to verify connectivity of a MEP with a peer MP(s). Loopback is transmitted by an MEP on the request of the administrator to verify connectivity to a particular maintenance point. Loopback indicates whether the destination is reachable or not; it does not allow hop-by-hop discovery of the path. It is similar in concept to ICMP Echo (Ping).

An ETH-LB session is defined as a sequence that begins with management initiating the transmission of (*n*) periodic LBM frames from an MEP to a peer MP and ends when the last LBR frame is received or incurs a timeout.

The number of LBM transmissions in an ETH-LB session is configurable in the range of 0 to 3,600. Two to three LBM messages are common. The period for LBM transmission is configurable in the range 0 to 60 seconds, but 1 second is more common.

LBM and LBR message sizes can be from 64 bytes to maximum MTU sizes of the EVC.

There are two ETH-LB types, unicast ETH-LB and multicast ETH-LB. In order to dynamically discover the MAC address of the remote MEP(s) on an MA, in addition to supporting a configurable destination to any unicast MAC address, an MEP may also support multicast MAC addresses corresponding to the reserved CFM multicast addresses for CCM.

Unicast ETH-LB can be used to verify bidirectional connectivity of an MEP with an MIP or a peer MEP and to perform a bidirectional in-service or out-of-service diagnostics test between a pair of peer MEPs. This includes verifying bandwidth, detecting bit errors, and so forth.

When unicast ETH-LB used to verify bidirectional connectivity, a MEP sends a unicast frame with ETH-LB request information and expects to receive a unicast frame with ETH-LB reply information from a peer MP within a specified period of time. The peer MP is identified by its MAC address. This MAC address is encoded in the DA of the unicast request frame. If the MEP does not receive the unicast frame with ETH-LB reply information within the specified period of time, loss of connectivity with the peer MP can be declared.

For performing bidirectional diagnostics tests, an MEP sends unicast frames with ETH-LB request information to a peer MEP. This ETH-LB request information includes test patterns. For out-of-service diagnostic tests, MEPs are configured to send frames with ETH-LCK information.

For bidirectional connectivity verification, an MEP transmits a unicast LBM frame addressed to the remote peer MP with a specific transaction ID inserted in the transaction ID/sequence number field. After unicast LBM frame transmission, an MEP expects to receive a unicast LBR frame within 5 seconds. The transmitted transaction ID is therefore retained by the MEP for at least 5 seconds after the unicast LBM frame is transmitted. A different transaction ID must be used for every unicast LBM frame, and no transaction ID from the same MEP may be repeated within 1 minute.

The maximum rate for frames with unicast ETH-LB should be at a value that does not adversely impact the user traffic.

MEP attributes to be configured to support unicast ETH-LB are the following:

8.8 Loopback and Reply Messages (LBM and LBR)

- MEG level at which the MEP exists.
- Unicast MAC address of remote peer MP.
- A test pattern and an optional checksum, which is optional. The test pattern can be pseudorandom bit sequence (PRBS) $(2^{31} - 1)$, all 0 pattern, various sizes of frames, and so forth. For bidirectional diagnostic test application, a test signal generator and a test signal detector associated with the MEP need to be configured.
- Priority of frames with unicast ETH-LB information.
- Drop eligibility of frames with unicast ETH-LB information to be discarded when congestion conditions are encountered.

A remote MP, upon receiving the unicast frame with ETH-LB request, responds with a unicast frame with ETH-LB reply information.

Unicast frames carrying the LBM PDU are called unicast LBM frames. Unicast frames carrying the LBR PDU are called unicast LBR frames. Unicast LBM frames are transmitted by an MEP on an on-demand basis.

Whenever a valid unicast LBM frame is received by an MIP or MEP, an LBR frame is generated and transmitted to the requesting MEP. A unicast LBM frame with a valid MEG level and a destination MAC address equal to the MAC address of receiving MIP or MEP is considered to be a valid unicast LBM frame. Every field in the unicast LBM frame is copied to the LBR frame with the following exceptions:

- The source and destination MAC addresses are swapped.
- The OpCode field is changed from LBM to LBR.

An LBR is valid if the MEP receives the LBR frame within 5 seconds after transmitting unicast LBM frame and the LBR has the same MEG level. Otherwise, the LBR frame addressed to it is invalid and is discarded.

When an MEP configured for a diagnostics test receives an LBR frame addressed to it with the same MEG level as its own MEG level, the LBR frame is valid. The test signal receiver associated with MEP may also validate the received sequence number against expected sequence numbers. If a MIP receives an LBR frame addressed to it, such an LBR frame is invalid and the MIP should discard it.

For diagnostic tests, an MEP transmits a unicast LBM frame addressed to the remote peer MEP with a test TLV. The test TLV is used to carry the test pattern generated by a test signal generator associated with the MEP. When the MEP is configured for an out-of-service diagnostic test, the MEP also generates LCK frames, at the client MEG level in a direction opposite to the direction where LBM frames are issued.

Multicast ETH-LB is an on-demand OAM function that is used to verify bidirectional connectivity of a MEP with its peer MEPs. Multicast frames carrying the LBM PDU are called multicast LBM frames. When a multicast ETH-LB function is invoked on a MEP, the MEP returns to the initiator of multicast ETH-LB a list of its peer MEPs with whom the bidirectional connectivity is detected.

When multicast ETH-LB is invoked on an MEP, a multicast frame with ETH-LB request information is sent from an MEP to other peer MEPs in the same MEG.

The MEP expects to receive unicast frame with ETH-LB reply information from its peer MEPs within a specified period of time. Upon reception of a multicast frame with ETH-LB request information, the receiving MEPs validate the multicast frame with ETH-LB request information and transmit a unicast frame with ETH-LB reply information after a randomized delay in the range of 0 to 1 second.

In order for an MEP to support multicast ETH-LB, the MEG level at which the MEP exists and priority of multicast frames with ETH-LB request information need to be configured.

Multicast frames with ETH-LB request information are always marked as drop ineligible; therefore, it is not configured.

However, an MIP is transparent to the multicast frames with ETH-LB request information and therefore does not require any information to support multicast ETH-LB.

After transmitting the multicast LBM frame with a specific transaction ID, the MEP expects to receive LBR frames within 5 seconds. The transmitted transaction ID is therefore retained for at least 5 seconds after the multicast LBM frame is transmitted. A different transaction ID must be used for every multicast LBM frame, and no transaction ID from the same MEP may be repeated within 1 minute.

Whenever a valid multicast LBM frame is received by an MEP, an LBR frame is generated and transmitted to the requesting MEP after a randomized delay in the range of 0 to 1 second (see Figure 8.20). The validity of the multicast LBM frame is determined based on the correct MEG level. Every field in the multicast LBM frame is copied to the LBR frame with the following exceptions:

- The source MAC address in the LBR frame is the unicast MAC address of the replying MEP. The destination MAC address in the LBR frame is copied from the source MAC address of the multicast LBM frame, which should be a unicast address.
- The OpCode field is changed from LBM to LBR.

When an LBR frame is received by an MEP with an expected transaction ID and within 5 seconds of transmitting the multicast LBM frame, the LBR frame is valid. If an MEP receives an LBR frame with a transaction ID that is not in the list

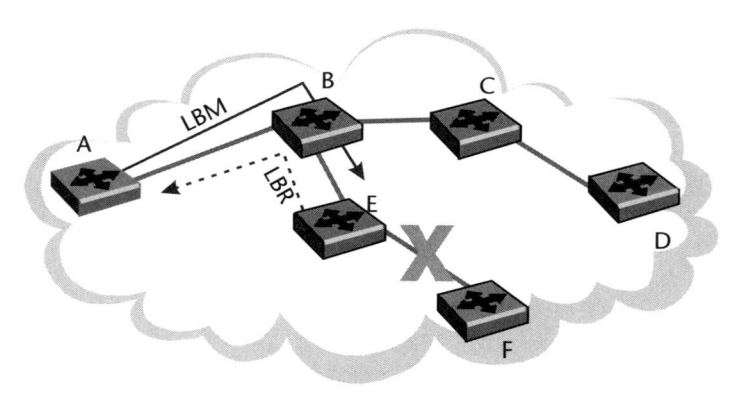

Figure 8.20 LBM/LBR.

8.8 Loopback and Reply Messages (LBM and LBR)

of transmitted transaction IDs maintained by the MEP, the LBR frame is invalid and is discarded.

If an MIP receives an LBR frame addressed to it, such an LBR frame is invalid and the MIP should discard it.

For the connectivity test, LBMs can be transmitted with either a unicast or a multicast DA. This address can be learned from the CCM database. The unicast DA can address either an MEP or an MIP while the multicast DA is only used to address MEPs. The LBM includes a transaction ID/sequence number, which is retained by the transmitting MEP for at least 5 seconds. After unicast LBM frame transmission, a MEP expects to receive a unicast LBR frame, with the same transaction ID/sequence number within 5 seconds.

When an LBM is received by a remote MEP/MIP that matches its address, an LBR will be generated. Every field in the LBM is copied to the LBR with the exception that: (1) the source and destination MAC addresses are swapped, and (2) the OpCode field is changed from LBM to LBR. The transaction ID/sequence number and data TLV fields are returned to the originating MEP unchanged. These fields are verified by the originating MEP. For multipoint loopback, each MEP returns an LBR after a randomized delay. A loopback diagnostic test is only for point-to-point applications between MEPs and uses unicast destination MAC addresses. The LBM includes a test pattern and the LBR returns the same test pattern.

As with CCMs, the sender ID TLV can be used to identify the management address of the MEP to its peers, thus providing IP correlation to the MAC address across a large network. However, for security reasons, the management domain and address can be empty in the sender ID TLV.

For an LB session, the number of LBMs transmitted, the number of LBRs received, and the percentage of responses lost (timed out) are reported by the initiating MEP.

For an LB session, the round-trip time (RTT) min/max/average statistics must be supported by the initiating MEP.

LBM and LBR PDUs, as depicted in Figures 8.21 through 8.24, contain transaction ID/sequence number is mandatory and data/test pattern TLV is optional. Transaction ID/sequence number is a 4-octet field that contains the transaction ID/sequence number for the LBM. The length and contents of the data/test pattern TLV are determined by the transmitting MEP.

In LBM PDU, Version is 0, OpCode is 3, and flags are all 0s.

Let us consider an example where a service provider is using the networks of two operators to provide service (Figure 8.25). The customer level allows the customer to test connectivity (using connectivity checks) and isolate issues (using loopback and link trace).

The customer could use CFM loopback or link trace to isolate a fault between the MEP on the CPE and the MIP on CE. By definition, the link between the CPE and CE is a single hop and, therefore, the customer would know which link has the fault. However, if the fault is between the two MIPs, the customer will need to rely on the service provider to determine between which MEPs or MIPs the fault has occurred. Even then, the service provider may simply isolate the fault to a specific operator's network, and will in turn rely on the operator to isolate the fault to a specific link in its network.

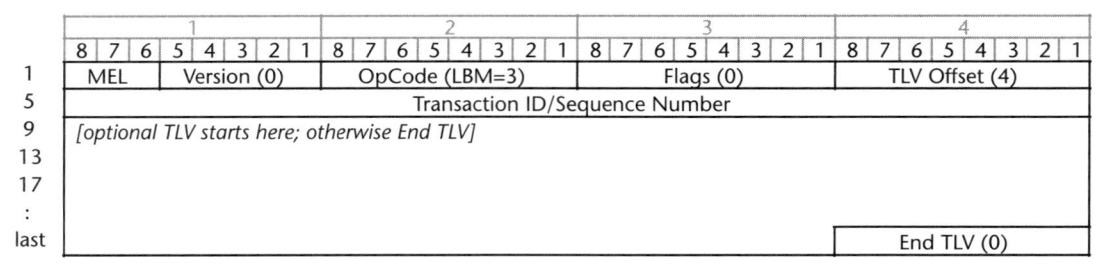

Figure 8.21 LBM PDU format.

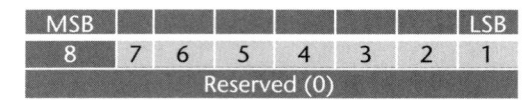

Figure 8.22 Flags format in LBM PDU.

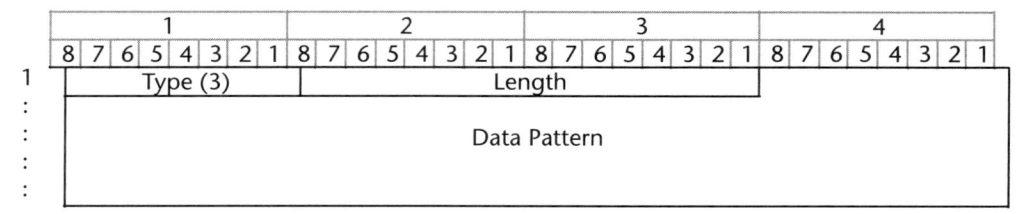

Figure 8.23 Data TLV format.

Figure 8.24 LBR PDU format.

Therefore, all the parties have the ability to isolate the fault within their domain boundaries, without the service provider having to share its network information to the customer, or the operator having to share its network information to the service provider.

8.9 Link Trace and Reply Messages (LTM and LTR)

The Ethernet link trace function (ETH-LT) is an on-demand OAM function initiated in an MEP on the request of the administrator to track the path to a destination MEP. They allow the transmitting node to discover connectivity data about the

8.9 Link Trace and Reply Messages (LTM and LTR)

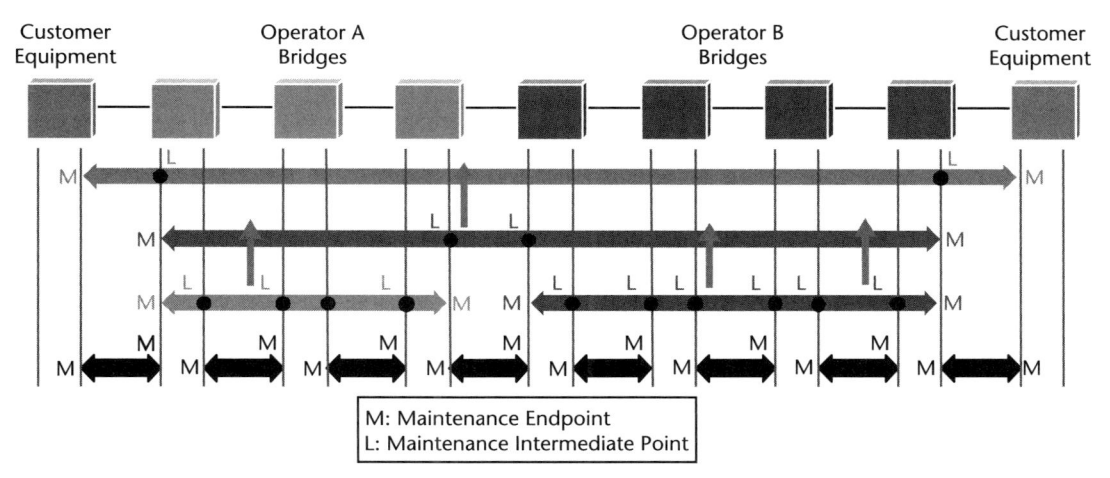

Figure 8.25 Maintenance points and maintenance domains.

path. The PDU used for ETH-LT request information is called LTM and the PDU used for ETH-LT reply information is called LTR. Frames carrying the LTM PDU are called LTM frames. Frames carrying the LTR PDU are called LTR frames.

As each network element containing the MP needs to be aware of the Target-MAC address in the received LTM frame and associates it to a single egress port, in order for the MP to reply, a unicast ETH-LB to the TargetMAC address could be performed by an MEP before transmitting the LTM frame. This ensures that the network elements along the path to the TargetMAC address have information about the route to the TargetMAC address if the TargetMAC address is reachable in the same MEG.

During a failure condition the information about the route to the TargetMAC address may age out after a certain time. The ETH-LT function has to be performed before the age out occurs in order to provide information about the route.

If an LTM frame is received by an MP, it forwards the LTM frame to network element's ETH-LT responder, which performs the following validation:

- If LTM frames have the same MEG level of the receiving MP;
- If the time to live (TTL) field value is 0, the LTM frame (i.e., invalid frame) is discarded;
- If the LTM egress identifier TLV is present. The LTM frame without LTM egress identifier TLV is discarded.

If the LTM frame is valid, the ETH-LT responder performs the following:

- It determines the destination address for the LTR frame from the Origin-MAC address in the received LTM frame.
- If the LTM frame terminates at the MP, an LTR frame is sent backwards to the originating MEP after a random time interval in the range of 0 to 1 second.

- If the above condition applies and LTM frame does not terminate at the MP and the TTL field in LTM frame is greater than 1, the LTM frame is forwarded towards the single egress port. All the fields of the relayed LTM frame are the same as the original LTM frame except for TTL, which is decremented by 1.

Link trace is similar in concept to UDP Traceroute. It can be used for retrieving adjacency relationship between an MEP and a remote MP. The result of running ETH-LT function is a sequence of MIPs from the source MEP until the target MIP or MEP where each MP is identified by its MAC address. The link trace also can localize faults when a link and/or a device failure or a forwarding plane loop occurs. The sequence of MPs will likely be different from the expected one. A difference in the sequences provides information about the fault location.

On an on-demand basis, an MEP will multicast LTM on its associated Ethernet connection. The multicast destination MAC address is a class 2 multicast address. The transaction ID, TTL, origin MAC address, and target MAC address are encoded in the LTM PDU. The target MAC address is the address of the MEP at the end of the EVC/OVC, which is being traced. It can be learned from CCM. The origin MAC address is the address of the MEP that initiates the linktrace. The transaction ID/sequence number and target MAC address are retained for at least 5 seconds after the LTM is transmitted. This is for comparison with the linktrace reply.

As with CCMs, the sender ID TLV in the LTM/LTR can be used to identify the management address of the MEP to its peers.

After transmitting the LTM frame with a specific transaction number, the MEP expects to receive LTR frames within a specified period of time, usually 5 seconds. A different transaction number must be used for every LTM frame, and no transaction number from the same MEP may be repeated within 1 minute. If an MEP receives an LTR frame with a transaction number that is not in the list of transmitted transaction numbers maintained by the MEP, the LTR frame is invalid. If an MIP receives an LTR frame addressed to it, the MIP should discard it, but an MIP may relay the frame with ETH-LT request information.

The configuration information required by a MEP to support ETH-LT is the following:

- MEG level at which the MEP exists.
- Priority of the frames with ETH-LT request information.
- Drop eligibility of frames. ETH-LT is always marked as drop ineligible.
- Target MAC address for which ETH-LT is intended.
- TTL to allow the receiver to determine if frames with ETH-LT request information can be terminated. TTL is decremented every time frames with ETH-LT request information are relayed. Frames with ETH-LT request information with TTL ≤ 1 are not relayed.

The LTR frame contains the LTR egress identifier TLV with last egress identifier field, which identifies the network element that originated or forwarded the LTM frame for which this LTR frame is the response. This field takes the same value as the LTM egress identifier TLV of that LTM frame. The LTR egress identifier TLV also contains the next egress identifier field, which identifies the network element that transmitted this LTR frame, and can relay a modified LTM frame to the next hop. This field takes the same value as the LTM egress identifier TLV if the relayed modified LTM frame, if any.

LTM PDU and LTR PDU are depicted in Figures 8.26 through 8.29.

8.10 Ethernet Alarm Indication Signal (ETH-AIS)

The PDU used for ETH-AIS information is AIS PDU (Figure 8.30). Frames carrying the AIS PDU are called AIS frames.

An MEP, upon detecting a defect condition, can transmit AIS frames periodically in a direction opposite to its peer MEP(s) immediately following the detection of a defect condition. An MIP is transparent to AIS frames and therefore does not require any information to support ETH-AIS functionality.

The periodicity of AIS frame transmission is based on the AIS transmission period, which is encoded as a CCM period. The transmission period can be in the range of 3.3 ms to 10 minutes, but 1 second is recommended. An MEP continues to transmit periodic frames with ETH-AIS information until the defect condition is removed.

ETH-AIS is used to suppress alarms following detection of defect conditions at the server (sub) layer. Due to independent restoration capabilities provided within the Spanning Tree Protocol (STP) environments, ETH-AIS is not expected to be ap-

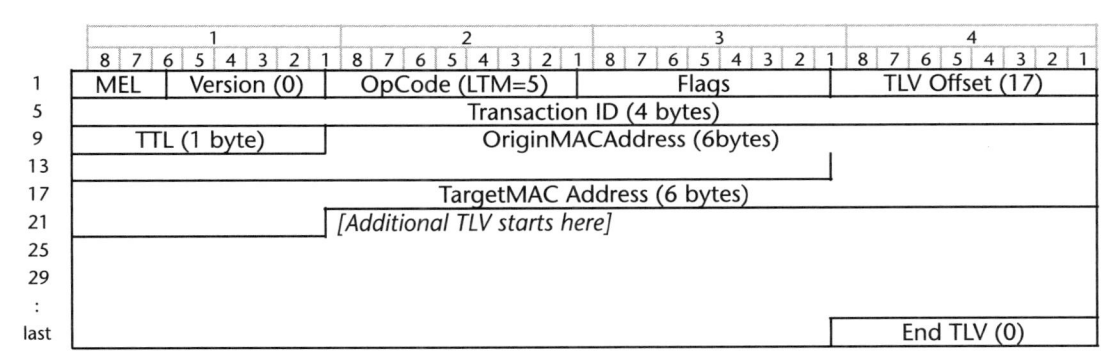

Figure 8.26 LTM PDU format.

Figure 8.27 LTR PDU format.

MSB							LSB
8	7	6	5	4	3	2	1
HWonly	Reserved (0)						

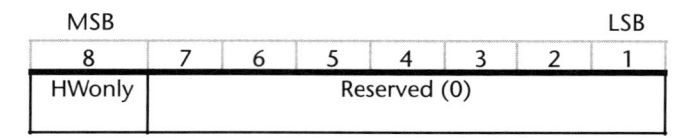

Figure 8.28 Flag form in LTM PDU.

	1	2	3	4
	8 7 6 5 4 3 2 1	8 7 6 5 4 3 2 1	8 7 6 5 4 3 2 1	8 7 6 5 4 3 2 1
1	Type (7)	Length		
2	Egress Identifier			
3				

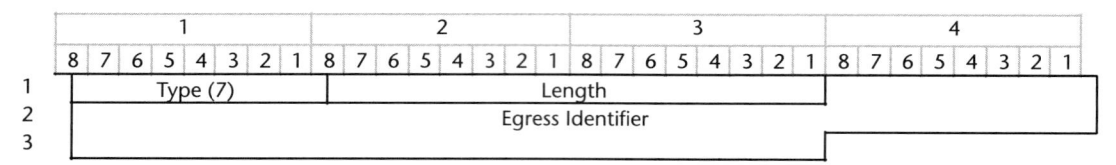

Figure 8.29 LTM egress identifier TLV format.

	1	2	3	4
	8 7 6 5 4 3 2 1	8 7 6 5 4 3 2 1	8 7 6 5 4 3 2 1	8 7 6 5 4 3 2 1
1	MEL Version (0)	OpCode (AIS=33)	Flags	TLV Offset (0)
5	End TLV (0)			

Figure 8.30 AIS PDU format.

plied in the STP environments. Transmission of frames with ETH-AIS information can be enabled or disabled on an MEP (or on a server MEP).

The client layer may consist of multiple MEGs that should be notified to suppress alarms resulting from defect conditions detected by the server layer MEP. The server layer MEP, upon detecting the signal fail condition, needs to send AIS frames to each of these client layer MEGs. The first AIS frame for all client layer MEGs must be transmitted within 1 second of defect condition. If the system is stressed when issuing AIS frames every second potentially across all 4,094 VLANs, AIS transmission period may be configured to in minutes. An AIS frame communicates the used AIS transmission period via the period field.

AIS is generally transmitted with a class 1 multicast DA. A unicast DA is allowed for point-to-point applications.

Upon receiving an AIS frame, the MEP examines it to ensure that its MEG level corresponds to its own MEG level and detects an AIS defect condition. The period field indicates the period at which the AIS frames can be expected. Following the detection of the AIS defect condition, if no AIS frames are received within an interval of 3.5 times the AIS transmission period indicated in the AIS frames received before, the MEP clears the AIS defect condition.

CoS for the AIS frame must be the CoS, which yields the lowest frame loss performance for this EVC.

An MEP detects AIS when it receives an AIS frame that is caused by detection of signal fail condition at a server layer or reception of AIS at a server layer MEP where the MEP does not use the ETH-CC function.

The AIS state is entered when an MEP receives an AIS and exited the MEP but does not receive AIS frames or when the ETH-CC is used, upon clearing of LOC defect at MEP, during an interval equal to 3.5 times the AIS transmission period indicated in the AIS frames received earlier.

Frames with ETH-AIS information can be issued at the client MEG level by an MEP, including a server MEP, upon detecting defect conditions. The defect conditions may include:

- Signal fail conditions in the case that ETH-CC is enabled;
- AIS condition or LCK condition in the case that ETH-CC is disabled.

Since a server MEP does not run ETH-CC, a server MEP can transmit frames with ETH-AIS information upon detection of any signal fail condition. For multipoint ETH connectivity, an MEP cannot determine the specific server layer entity that has encountered defect conditions upon receiving a frame with ETH-AIS information. Therefore, upon reception of a frame with ETH-AIS information, the MEP will suppress alarms (i.e., loss of continuity) for all peer MEPs whether there is still connectivity or not.

For a point-to-point ETH connection, an MEP has only a single peer MEP. Therefore, there is no ambiguity regarding the peer MEP for which it should suppress alarms when it receives the ETH-AIS information.

Configuration parameters for an MEP to support ETH-AIS transmission are:

- Client MEG level at which the most immediate client layer MIPs and MEPs exist;
- ETH-AIS transmission period;
- Priority of frames with ETH-AIS information;
- ETH-AIS is always marked as drop ineligible.

However, for multipoint EVCs/OVCs, a client layer MEP, upon receiving an AIS, cannot determine which of its remote peers have lost connectivity. It is recommended that for multipoint the client-layer MEP should suppress alarms for all peer MEPs. Use of AIS is not recommended for environments that utilize the spanning tree protocol, which provides an independent restoration capability. Due to the spanning tree and multipoint limitation associated with AIS, IEEE 802.1 committee has chosen not to support AIS in 802.1ag.

When we have CCM capability, the need for AIS is questioned. The fact is that AIS is generated immediately when the failure is detected. If there is AIS capability, LOC is generated after not receiving three CCMs. Therefore, AIS would provide faster failure reporting than CCM.

8.11 Ethernet Remote Defect Indication (ETH-RDI)

ETH-RDI is used only when ETH-CC transmission is enabled. When a downstream MEP detects a defect condition, such as receive signal failure or AIS, it will send an RDI in the opposite upstream direction to its peer MEP or MEPs. This informs the upstream MEPs that there has been a downstream failure.

RDI is subject to the same multipoint issue as AIS. A MEP that receives an RDI cannot determine what subset of peer MEPs have experienced a defect. For Y.1711 [13], RDI is encoded as a bit in the flags field in CC messages. IEEE 802.1ag does not support RDI.

A MEP, upon detecting a defect condition with its peer MEP, sets the RDI field in the CCM frames for the duration of the defect condition. CCM frames are transmitted periodically based on the CCM transmission period, when the MEP is enabled for CCM frames transmission. When the defect condition clears, the MEP clears the RDI field in the CCM frames in subsequent transmissions.

The configuration information required by a MEP to support ETH-RDI function is the following:

- MEG level at which the MEP exists;
- ETH-RDI transmission period configured to be the same as ETH-CC transmission period, depending on application;
- Priority of ETH-RDI frames, which are the same as ETH-CC priority;
- Drop eligibility of ETH-RDI frames that are always marked as drop ineligible.

An MIP is transparent to frames with ETH-RDI information.

The PDU used to carry ETH-RDI information is CCM. Eight bit in CCM PDU is set to 1 to indicate RDI; otherwise, it is set to 0 (Figure 8.31).

8.12 Ethernet Locked Signal (ETH-LCK)

LCK condition is caused by an administrative/diagnostic action at a server (sub) layer MEP that results in disruption of the client data traffic.

An MEP continues to transmit periodic frames with ETH-LCK information at the client MEG level until the administrative/diagnostic condition is removed. An MEP extracts frames with ETH-LCK information (Figure 8.32) at its own MEG level and detects a LCK condition, which contributes to the signal fail condition of the MEP. The signal fail condition may result in the transmission of AIS frames to its client MEPs.

MSB							LSB
8	7	6	5	4	3	2	1
RDI	Reserved (0)				Period		

Figure 8.31 Flags format in CCM PDU.

1	2	3	4
8 7 6 5 4 3 2 1	8 7 6 5 4 3 2 1	8 7 6 5 4 3 2 1	8 7 6 5 4 3 2 1
MEL / Version (0)	OpCode (LCK=35)	Flags	TLV Offset (0)
End TLV (0)			

Figure 8.32 LCK PDU format.

LCK state is entered when a MEP receives a LCK frame and exited during an interval equal to 3.5 times the LCK transmission period indicated in the LCK frames received earlier, the MEP does not receive LCK frames.

The LCK transmission period encoded the same as a CCM period. An example of an application that would require administrative locking of an MEP is out-of-service ETH-test.

The configuration information required by a MEP to support ETH-LCK transmission is the following:

- Client MEG level at which the most immediate client layer MIPs and MEPs exist;
- ETH-LCK transmission period that determines transmission periodicity of frames with ETH-LCK information;
- Priority that identifies the priority of frames with ETH-LCK information;
- Drop eligibility of ETH-LCK frames that are always marked as drop ineligible.

An MIP is transparent to the frames with ETH-LCK information.

An MEP, when administratively locked, transmits LCK frames in a direction opposite to its peer MEP(s). The periodicity of LCK frames transmission is based on the LCK transmission period. The LCK transmission period is the same as the AIS transmission period.

Upon receiving an LCK frame, a MEP examines it to ensure that its MEG level detects an LCK condition. If no LCK frames are received within an interval of 3.5 times the LCK transmission period, the MEP clears the LCK condition.

It is used for client-layer alarm suppression and enables client MEPs to differentiate between the defect conditions and intentional administrative/diagnostic actions at the server layer MEP. This capability is only supported in Y.1731 [3].

8.13 Performance Measurements

Collecting measurements proactively (i.e., continuously and periodically) or an on-demand basis for each management entity such as UNI, ENNI, CoS, EVC, and OVC is important in managing networks. These measurements are used in determining if contractual agreements (SLAs) are satisfied, as well as the need for capacity and location of network failures. The continuous measurements are coupled with service-level monitoring in accomplishing these tasks.

At the physical port, the following media attachment unit (MAU) termination performance data set for each MAU transport termination is counted [8, 14, 15]:

- Number of time the MAU leaves the available state;
- Number of times the MAU enters the jabbering state;
- Number of false carrier events during idle.

Traffic measurements on a per UNI [6] and ENNI basis are counted by NE. They are Octets Transmitted OK, Unicast Frames Transmitted OK, Multicast Frames Transmitted OK, Broadcast Frames Transmitted OK, Octets Received OK, Unicast Frames Received OK, Multicast Frames Received OK, and Broadcast Frames Received OK. In addition, the following anomalies at UNI and ENNI should be counted: undersized frames, oversized frames, fragments, frames with FCS or alignment errors, frames with invalid CE-VLAN ID, and frames with S-VLAN ID (at ENNI only).

Measurements on a per UNI, ENNI, COS per UNI, CoS per ENNI, OVC or CoS per OVC, EVC, or COS per EVC basis for each entity that enforces traffic management at ingress direction (CE to MEN at UNI and ENNI gateway to ENNI gateway at ENNI) are counted:

- Green frames sent by the ingress UNI to the MEN;
- Yellow frames sent by the ingress UNI to the MEN;
- Red (discarded) frames at the ingress UNI;
- Green octets sent by the ingress UNI to the MEN;
- Yellow octets sent by the ingress UNI to the MEN;
- Red (discarded) octets at the ingress UNI.

At the egress, for the same entities, the following is counted:

- Green frames received by the egress UNI from the MEN;
- Yellow frames received by the egress UNI from the MEN;
- Green octets received by the egress UNI from the MEN;
- Yellow octets received by the egress UNI from the MEN.

At both ingress and egress, for the same entities, the following is counted:

- Green frames discarded due to congestion;
- Yellow frames discarded due to congestion;
- Green octets discarded due to congestion;
- Yellow octets discarded due to congestion.

Counts for the current 15-minute interval and the past 32 15-minute intervals are stored in NE counters that can store a value of at least $2^{64} - 1$. If the counter reaches its maximum value, it will either keep its maximum value duration of the interval (interval counter) or wrap around (continuous counter). A timestamp for measurement interval with 1-second accuracy is expected.

Specific threshold values can be assigned to counters by NMS. When a threshold counter exceeds its threshold during a current measurement interval, the NE sends a threshold crossing alert to the managing system within 60 seconds of the occurrence of the threshold crossing.

Measurements can be stored in measurement intervals or bins. A measurement interval is an interval of time and stores the results of performance measurements conducted during that interval. A measurement bin is a counter and stores the results of performance measurements falling within a specified range. Figure 8.33 is an example that illustrates the relationship between measurement intervals and measurement bins.

The number of measurement bins per measurement interval depends on application. The minimum number is 2. For example, the bins for a measurement of a 12-ms end-to-end delay can be as depicted here:

- bin 1 = 0 ms (range is 0 ms ≤ measurement < 4 ms);
- bin 2 = 4 ms (range is 4 ms ≤ measurement < 8 ms);
- bin 3 = 8 ms (range is 8 ms ≤ measurement < 12 ms);
- bin 4 = 12 ms (range is 12 ms ≤ measurement < timeout).

The suspect flag should be set when these measurements are not reliable due to service provisioning or service failure or time-of-day clock adjustment.

8.14 Performance Monitoring

There are two types of measurements: periodic and on-demand measurements. Periodic measurements are collected by polling counters of corresponding devices at a polling rate of not slower than 1 per minute. They are available in 15-minute bins.

In-service measurements are needed to monitor health of the network as well as user SLAs. Out-of-service measurements are needed before turning up the service, isolating troubles, and identifying failed components during failures.

On-demand measurements are mostly needed during failures and service degradation while proactive measurements help to identify possible problems in advance.

As we described in CoS section, each CoS/EVC has its own performance parameters, namely SLA/SLS. The SLS is usually part of the contract between service provider and subscriber. The service provider monitors the EVC to ensure that contractual obligations are satisfied while subscribers might want to monitor the

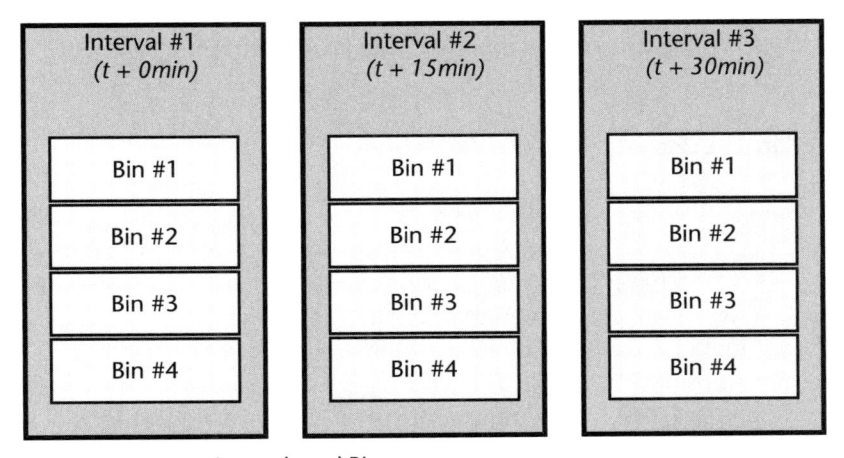

Figure 8.33 Measurement Intervals and Bins

EVC for the same reasons. SLS parameters to be monitored are delay, jitter, loss, and availability for the given EVC/CoS.

For a point-to-point EVC, the performance counts encompass service frames in both directions of the UNI pair, while for a multipoint EVC, they encompass service frames in one direction over all or a subset of ordered UNI pairs. Since there are multiple ingress and egress points as well as the potential for frame replication, counters in a multipoint EVC may not be directly correlated. Periodic synthetic frames can be used for delay measurements. By counting and measuring the one-way FLR of uniform synthetic traffic, statistical methods can be used to estimate the one-way FLR of service traffic as well.

In general, the performance measurements can be performed by using service frames as well as synthetic OAM test frames, although the accuracy of the measurements with synthetic frames depends on how closely synthetic frames emulate the service frames.

Delay measurement can be performed using the 1DM (one-way delay measurement) or DMM/DMR (delay measurement message/delay measurement reply) PDUs. Loss measurement can be performed by counting service frames using the LMM/LMR (loss measurement message/loss measurement reply) PDUs as well as by counting synthetic frames via the SLM/SLR (synthetic loss message/synthetic loss reply) PDUs.

For delay measurements, DMM frames are sent from a controller MEP to a responder MEP, which in turn responds with DMR frames. With time-of-day clock synchronization one-way measurements are taken. Two-way measurements (i.e., round-trip measurements) do not require time-of-day clock synchronization.

For loss measurements with synthetic frames, SLM are sent from an originating MEP to a responder MEP which in turn responds with SLR frames. One-way and two-way measurements of FLR and availability are always taken with this mechanism.

The test with synthetic frames must generate enough test frames to be statistically valid. The test frames must also be similar to the service frames carried by the EVC; in particular, the OAM test frames must have representative frame length, be generated by a representative arrival process, and be treated by the network elements implementing the network between the MEPs in the same way that service frames are treated.

PM functions are described further in the following sections.

8.15 Frame Loss Measurements

Frame loss is measured in terms of frame loss ratio (FLR) per UNI, ENNI, EVC, or OVC. For UNI, the FLR can be defined as ((Number of frames delivered at the egress of receiving UNI − number of frames transmitted at the ingress of transmitting UNI)/(number of frames transmitted at the ingress of transmitting UNI)) in a given time interval according to a user-service provider contract. This time interval could be a month. An MIP is transparent to frames with ETH-LM information and therefore does not require any information to support ETH-LM (Ethernet loss measurement) functionality.

Measuring the one-way FLR of service frames between two measurement points requires transmission and reception counters, where the ratio between them corresponds to the one-way FLR. Collection of the counters needs to be coordinated for accuracy. If counters are allocated for green and yellow frames, it is possible to find FLR for each color. If there are counters for the total count, it would be possible to identify FLR for the total frames.

Two types of FLR measurement are possible, dual-ended (i.e., one-way) and single-ended (i.e., round trip or two-way). Dual-ended loss measurement can be accomplished by exchanging CCM OAM frames as well as ETH-LM.

When CCM OAM frames are used, they include counts of frames transmitted and received, namely, TxFCf, RxFCb, and TxFCb where

- TxFCf: Value of the local counter TxFCl at the time of transmission of the CCM frame;
- RxFCb: Value of the local counter RxFCl at the time of reception of the last CCM frame from the peer MEP;
- TxFCb: Value of TxFCf in the last received CCM frame from the peer MEP.

These counts do not include OAM frames at the MEP's MEG level.

When configured for proactive loss measurement, a MEP periodically transmits CCM frames with a period value equal to the CCM transmission period configured for performance monitoring application at the transmitting MEP. The receiving MEP detects an unexpected period defect condition if the CCM transmission period is not the same as the configured value. In this case, frame loss measurement is not carried out. When a MEP detects a loss-of-continuity defect condition, it ignores loss measurements during the defect condition and assumes 100% losses.

Near-end and far-end FLR are calculated across pairs of consecutive frames as:

Frame Loss (far-end) = |TxFCb[tc] − TxFCb[tp]| − |RxFCb[tc] − RxFCb[tp]|
Frame Loss (near-end) = |TxFCf[tc] − TxFCf[tp]| − |RxFCl[tc] − RxFCl[tp]|

where
    tc = Counter values for the current CCM frame
    tp = Counter values for the previous CCM frame

For a MEP, near-end frame loss refers to frame loss associated with ingress data frames while far-end frame loss refers to frame loss associated with egress data frames.

An MEP maintains the following two local counters for each peer MEP and for each priority class being monitored in a point-to-point ME for which loss measurements are to be performed:

- TxFCl: Counter for in-profile data frames transmitted towards the peer MEP;
- RxFCl: counter for in-profile data frames received from the peer MEP.

The TxFCl and RxFCl counters do not count the OAM frames (except proactive OAM frames for protection) at the MEP's MEG level. However, the counters do count OAM frames from the higher MEG levels that pass through the MEPs in a manner similar to the data frames.

Loss measurement can be performed with ETH-LM (Figure 8.34) as well. The level of accuracy in the loss measurements is dependent on how frames with ETH-LM information are added to the data stream after the counter values are copied in the ETH-LM information. For example, if additional data frames get transmitted and/or received between the time of reading the counter values and adding the frame with ETH-LM information to the data stream, the counter values copied in ETH-LM information become inaccurate. In order to increase accuracy, a hardware-based implementation that is able to add frames with ETH-LM information to the data stream immediately after reading the counter values is necessary.

MP configuration information to support ETH-LM is the following:

- MEG level at which the MEP exists.
- ETH-LM transmission period is the frequency of LMM which is configurable to values 100 ms, 1 second, and 10 seconds; 1 second is more common.
- Priority that identifies the priority of the frames with ETH-LM information.
- Drop eligibility indicating if ETH-LM can be dropped during congestion. ETH-LM is always marked as drop ineligible.

Whenever a valid LMM frame is received by an MEP, an LMR frame is generated and transmitted to the requesting MEP. An LMM frame with a valid MEG level and a destination MAC address equal to the receiving MEP's MAC address is considered to be a valid LMM frame. An LMR frame contains the following fields and their values:

- TxFCf, which is a 4-octet field: Its value is copied from the LMM frame.

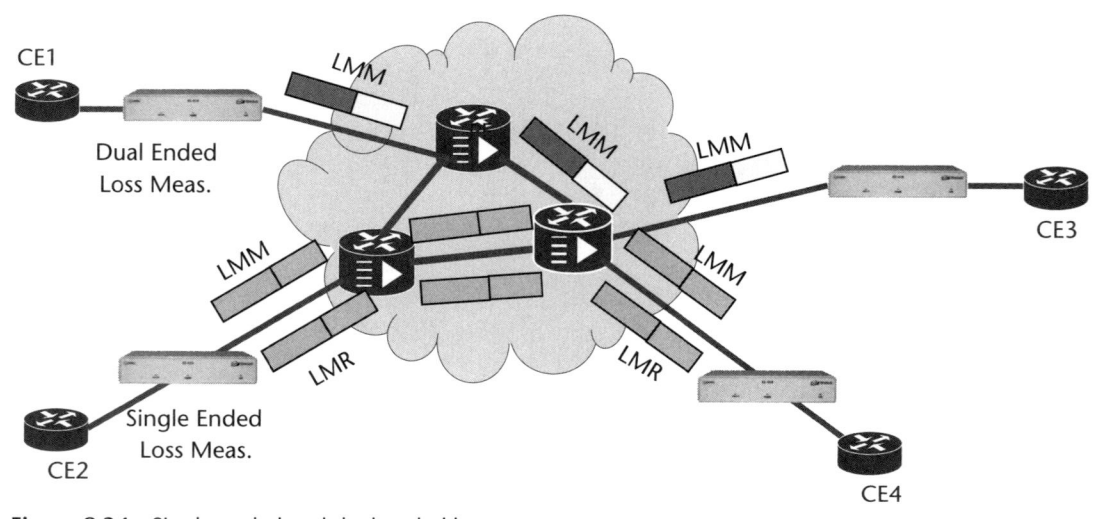

Figure 8.34 Single-ended and dual-ended loss measurements.

8.15 Frame Loss Measurements

- RxFCf, which is a 4-octet field: Its value is the value of local counter RxFCl at the time of LMM frame reception.
- TxFCb, which is a 4-octet field: Its value is the value of local counter TxFCl at the time of LMR frame transmission.

Upon receiving an LMR frame, near-end and far-end loss measurements at the MEP can be calculated as:

Frame Loss(far-end) = |TxFCf[tc] − TxFCf[tp]| - |RxFCf[tc] − RxFCf[tp]|
Frame Loss(near-end) = |TxFCb[tc] − TxFCb[tp]| - |RxFCl[tc] − RxFCl[tp]|

where

- TxFCf is a 4-octet field of LMR, which carries the value of the TxFCf field in the last LMM PDU received by the MEP from its peer MEP.
- TxFCb is a 4-octet field of LMR, which carries the value of the counter of in-profile data frames transmitted by the MEP towards its peer MEP at the time of LMR frame transmission.
- RxFCf is a 4-octet field of LMR, which carries the value of the counter of data frames received by the MEP from its peer MEP, at the time of receiving last LMM frame from that peer MEP.

Single-ended loss measurement is accomplished by the on-demand exchange of LMM and LMR OAM frames. These frames include appropriate counts of frames transmitted and received. Near-end and far-end FLR are calculated at the end that initiated the loss measurement request.

FLR measurement is performed on a point-to-point basis using a unicast DA. ITU Y.1731 [3] also allows the use of class 1 multicast DA to support multipoint testing.

LMM and LMR PDU formats used by a MEP to transmit LMM information are shown in Figures 8.35 and 8.36.

The location of measurement points is also important. The synthetic frames should be inserted after the traffic conditioning point for the ingress traffic.

Octet	1	2	3	4
1	MEL / Version (0)	OpCode (LMM=43)	Flags (0)	TLV Offset (12)
5	TxFCf			
9	Reserved for RxFCf in LMR			
13	Reserved for TxFCb in LMR			
17	End TLV (0)			

Figure 8.35 LMM PDU format.

	1	2	3	4
	8 7 6 5 4 3 2 1	8 7 6 5 4 3 2 1	8 7 6 5 4 3 2 1	8 7 6 5 4 3 2 1
1	MEL / Version	OpCode (LMR=42)	Flags	TLV Offset
5	TxFCf			
9	RxFCf			
13	TxFCb			
17	End TLV (0)			

Figure 8.36 LMR PDU format.

8.16 Availability

Service availability is one of the customer SLA parameters. It is usually calculated per month or per year. The 99.999% (i.e., five nines) availability per year, which is equivalent to approximately 5 minutes down time in a year, for carrier-class systems and services is common in the industry.

Service availability is the availability of the EVC for a given CoS between UNI-C at location A and UNI-C at location Z. If both locations A and Z are on one network, the EVC consists of:

- EVC segment over the access network of location A;
- EVC segment over backbone network;
- EVC segment over the access network of location Z.

If location A is in an operator's footprint while the location Z is in another operator's footprint, the EVC consists of:

- OVC_1, OVC_2, ..., OVC_n of EVC segments in MEN_1, MEN_2, ..., MEN_n, respectively;
- EVC segment over link between $ENNI_1$ and $ENNI_2$, EVC segment over link between $ENNI_2$ and $ENNI_3$..., EVC segment over link between $ENNI_{n-2}$ and $ENNI_{n-1}$.

The service availability is the product of the availabilities of all the components making the EVC. In order to determine the availability of the EVC, availability of each of these components needs to be monitored and correlated.

A number of consecutive FLR measurements are needed to evaluate the availability/unavailability status of an entity. The default number is 10 [16].

A configurable availability threshold C_a and a configurable unavailable threshold C_u are used in evaluating the availability/unavailability status of an entity. The thresholds can range from 0 to 1 where the availability threshold is less than or equal to the unavailability threshold. However, $C_a = C_u = C$ is a recommended equality.

Whenever a transition between available and unavailable occurs in the status of an adjacent pair of availability indicators, it should be reported to management systems (i.e., EMS and NMS).

Availability is calculated via sliding windows based on a number of consecutive FLR measurements [16–18]. The availability for a CoS/EVC/OVC of a

8.16 Availability

point-to-point connection during Δt_k is based on the frame loss ratio during the short interval and each of the following short intervals and the availability of the previous short time interval (i.e., sliding window of width $n\Delta t$).

Figure 8.37 presents an example of the determination of the availability for the small time intervals Δt_k within T with a sliding window of 10 small time intervals, as we determine the availability of an EVC from UNI i to UNI j for a time interval T, excluding the small time intervals that occur during a maintenance interval.

- Δt is a time interval much smaller than T.
- C is a loss ratio threshold which if exceeded suggests unavailability.
- is the number of consecutive small time intervals, Δt, over which to assess availability.

Each Δt_k in T is defined to be either available or unavailable and this is represented by $A_{\langle i,j \rangle}(\Delta t_k)$ where $A_{\langle i,j \rangle}(\Delta t_k) = 1$ means that Δt_k is available and $A_{\langle i,j \rangle}(\Delta t_k) = 0$ means that Δt_k is unavailable. The definition of $A_{\langle i,j \rangle}(\Delta t_k)$ is based on the following frame loss ratio function.

Let $I_{\Delta t}^{\langle i,j \rangle}$ be the number of ingress service frames at UNIi to be delivered to UNIj for a given CoS where the first bit of each service frame arrives at UNI i within the time interval , and subject to an ingress bandwidth profile with either no color identifier or a color identifier that corresponds to green.

Let $E_{\Delta t}^{\langle i,j \rangle}$ be the number of unique (not duplicate) unerrored egress service frames where each service frame is the first egress service frame at UNI j that results from a service frame counted in $I_{\Delta t}^{\langle i,j \rangle}$.

Then

$$flr_{\langle i,j \rangle}(\Delta t) = \begin{cases} \left(\dfrac{I_{\Delta t}^{\langle i,j \rangle} - E_{\Delta t}^{\langle i,j \rangle}}{I_{\Delta t}^{\langle i,j \rangle}} \right) & \text{if } I_{\Delta t}^{\langle i,j \rangle} \geq 1 \\ 0 & \text{otherwise} \end{cases}$$

Δt_0 is the first short time interval agreed upon by the service provider and subscriber at or after turn-up of the EVC. $A_{\langle i,j \rangle}(\Delta t_k)$ for $k = 0$ is

Available $A_{\langle i,j \rangle}(\Delta t_k) = 1$ Unavailable $A_{\langle i,j \rangle}(\Delta t_k) = 0$ Available $A_{\langle i,j \rangle}(\Delta t_k) = 1$ Time

$n = 10$

■ $flr_{\langle i,j \rangle}(\Delta t_m) > C$

□ $flr_{\langle i,j \rangle}(\Delta t_m) \leq C$

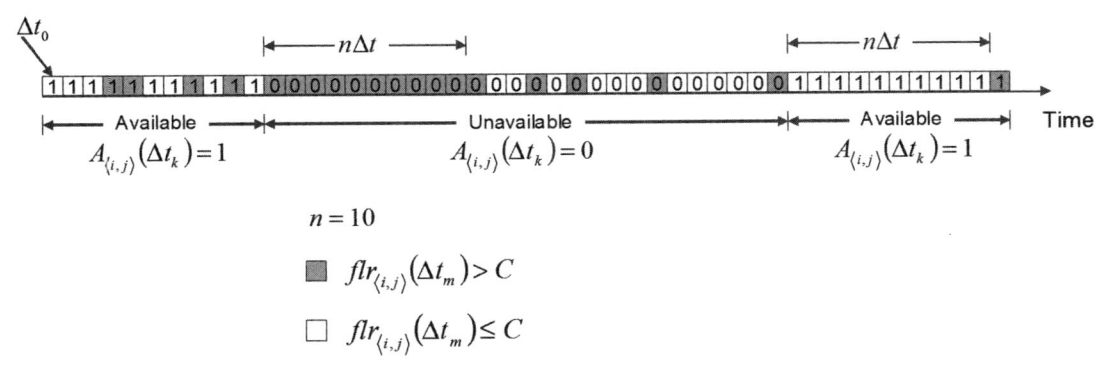

Figure 8.37 Determining availability via sliding window.

$$A_{\langle i,j\rangle}(\Delta t_0) = \begin{cases} 0 & \text{if } flr_{\langle i,j\rangle}(\Delta t_m) > C, \forall m = 0,1,\ldots n-1 \\ 1 & \text{otherwise} \end{cases}$$

and for $k = 1, 2, \ldots$ is

$$A_{\langle i,j\rangle}(\Delta t_0) = \begin{cases} 0 & \text{if } flr_{\langle i,j\rangle}(\Delta t_m) > C, \forall m = 0,1,\ldots n-1 \\ 1 & \text{otherwise} \end{cases}$$

Multipoint-to-multipoint EVC availability may be calculated by a scheme agreed upon by service provider and customer. Theoretically, one may have to monitor simultaneously all possible point-to-point connections for the given multipoint connection and multiply availability of all the point-to-point connections. However, this is impractical to monitor and calculate in the real world. Instead, it may be practical to monitor and calculate FLR using synthetic frames for a subset of point-to-point connections simultaneously.

FLR-based availability calculation is still not widely practiced in the industry due to difficulty with loss calculations. Instead failure intervals are measured and the availability is calculated by dividing unavailability time by the total service time.

For wireless applications, it is suggested that availability should include delay and jitter in addition to loss. How often an EVC meets or exceeds the frame delay, interframe delay variation (i.e., jitter), and frame loss over a time interval T in a given direction should be measured. This is called one-way composite performance metric (CPM) for an EVC [18].

CPM is expressed as the percentage of time intervals contained in time period T that are deemed to have acceptable performance. Acceptable performance for a short time interval,, is based on combinations of three service frame delivery characteristics (frame delay, jitter, and frame loss) and thresholds in the SLS. The combination of three service frame delivery characteristics for the short time interval,, is denoted as a composite performance indicator, $D(\Delta t_k)$

For an ordered pair of UNIs, <i, j>, in the EVC, each Δt_k in T is defined to be either acceptable or unacceptable. In other words, $cA_{\langle i,j\rangle}(\Delta t_k) = 1$ if Δt_k is acceptable and $cA_{\langle i,j\rangle}(\Delta t_k) = 0$ if is unacceptable. $cA_{\langle i,j\rangle}(\Delta t_k)$ is based on the composite performance indicator, $D_{\langle i,j\rangle}(\Delta t_k)$, over n consecutive short time intervals.

For each ingress qualified service frame, there are three frame delivery characteristics:

- Frame loss characteristic

$$fl_{\langle i,j\rangle}(m) = \begin{cases} 1 & \text{if the } m\text{th ServiceFrame gets lost} \\ 0 & \text{otherwise} \end{cases}$$

- Frame delay characteristic

8.16 Availability

$$fd_{\langle i,i\rangle}(m) = \begin{cases} 1 & \text{if } fl_{\langle i,i\rangle}(m) \text{ and } fd_m(m) > DL \\ 0 & \text{otherwise} \end{cases}$$

$fd_m(m)$ is the one-way frame delay of the mth service frame. DL is the one-way frame delay threshold parameter in the SLS.

- Interframe delay variation characteristic

$$fdv_{\langle i,i\rangle}(m) = \begin{cases} 1 & fl_{\langle i,i\rangle}(m) \neq 1 \text{ and } fl_{\langle i,i\rangle}(m-1) \neq 1 \text{ and} \\ & fdv_m(m) > Jt, 2 \leq m \leq M_k \\ 0 & \text{otherwise} \end{cases}$$

$fdv_m(m) = |fd_m(m) - fd_m(m-1)|$, $m > 1$, is the interframe delay variation between two consecutive service frames: mth frame and $(m-1)$th frame. Jt is the one-way interframe delay variation threshold parameter in the SLS.

$D_{\langle i,i\rangle}(\Delta t_k)$ is defined as the ratio of the number of those service frame delivery characteristics with the value of 1 (i.e., failed characteristics) to the number of all frame delivery characteristics counted during the short time interval, Δt_k.

$$D_{\langle i,i\rangle}(\Delta t_k) = \begin{cases} 0 & \text{if } M_k = 0 \\ 1 & \text{if } \left(W_{fl} = 0 \text{ and } M_{ke} = 0\right) \text{ or if } \left(W_{fl} = W_{fd} = 0\right) \& \left(\sum_{m=2}^{M_k} v_{\langle i,i\rangle}(m) = 0\right) \\ \dfrac{\sum_{m=1}^{M_k}\left(fl_{\langle i,i\rangle}(m)*W_{fl} + fd_{\langle i,i\rangle}(m)*W_{fd} + fdv_{\langle i,i\rangle}(m)*W_{fdv}\right)}{M_k*W_{fl} + \left(M_k - \sum_{m=1}^{M_k}fl_{\langle i,i\rangle}(m)\right)*W_{fd} + \left(\sum_{m=2}^{M_k}v_{\langle i,i\rangle}(m)\right)*W_{fdv}}, & \text{otherwise} \end{cases}$$

where

$$v_{\langle i,i\rangle}(m) = \begin{cases} 1 & \text{if } fl_{\langle i,i\rangle}(m) \neq 1 \text{ and } fl_{\langle i,i\rangle}(m-1) \neq 1 \\ 0 & \text{othewise} \end{cases}$$

is the indicator and $v_{\langle i,i\rangle}(m) = 1$ means that the interframe delay variation characteristic between mth and $(m-1)$th frames is counted in the number of all frame delivery characteristics during the short time interval, Δt_k; W_{fl}, the indicator for frame loss characteristic which equals 0 or 1; W_{fd}, the indicator for frame delay characteristic which equals 0 or 1; and W_{fdv}, the indicator for interframe delay variation characteristic which equals 0 or 1.

One of the three indicators W_{fl}, W_{fd}, and W_{fdv} must be 1.
From above, availability can be expressed as

$$cA_{\langle i,j \rangle}(\Delta t_0) = \begin{cases} 0 & \text{if } D_{\langle i,j \rangle}(\Delta t_l) > U, \forall l = 0, 1, \ldots, n-1 \\ 1 & \text{otherwise} \end{cases}$$

and for $k = 1, 2, \ldots$, is

$$cA_{\langle i,j \rangle}(\Delta t_0) = \begin{cases} 0 \text{ if } cA_{\langle i,j \rangle}(\Delta t_{k-1}) = 1 \text{ and } D_{\langle i,j \rangle}(\Delta t_l) > U, \forall l = k, k+1, \ldots, k+n-1 \\ 1 \text{ if } cA_{\langle i,j \rangle}(\Delta t_{k-1}) = 0 \text{ and } D_{\langle i,j \rangle}(\Delta t_l) \leq U, \forall l = k, k+1, \ldots, k+n-1 \\ cA_{\langle i,j \rangle}(\Delta t_{k-1}) \text{ othewise} \end{cases}$$

8.17 Frame Delay Measurements

There are two types of frame delay (FD) measurements, one-way and two-way [or round-trip delay (RTD)]. RTD is defined as the time elapsed since the start of transmission of the first bit of the frame by a source node until the reception of the last bit of the loop backed frame by the same source node, when the loopback is performed at the frame's destination node.

One-way FD is measured by MEPs periodically sending 1DM frames (Figure 8.38), which include appropriate transmit time stamps. The FD is calculated at the receiving MEP by taking the difference between the transmit time stamp and a receive time stamp, which is created when the 1DM frame is received.

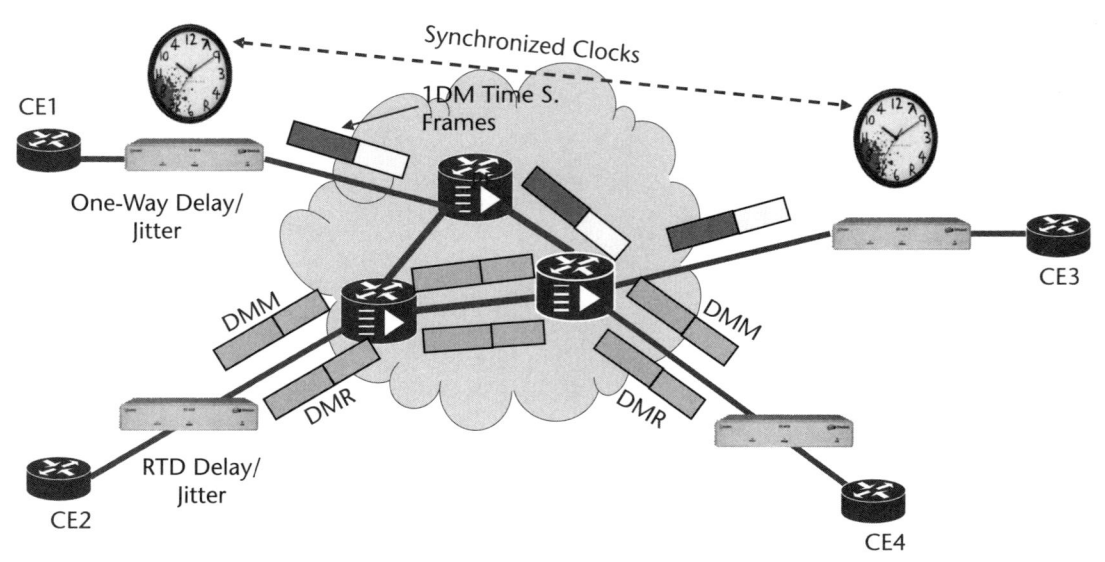

Figure 8.38 One-way delay and RTD measurements.

The 1DM frame can be sent periodically with a period of 100 ms, 1 second, and 10 seconds where 1 second is more common. The 1DM frames are discard ineligible if tagging is used. Before initiating one-way delay measurements between NEs, a Boolean clock synchronization flag must indicate that the clock synchronization is in effect.

Two-way DM measures round trip delay and does not require synchronized clocks. It is accomplished by MEPs exchanging DMM and DMR frames. Each of these DM OAM frames includes transmit time stamps. Y.1731 [3] allows an option for inclusion of additional time stamps such as a receive time stamp and a return transmit time stamp. These additional time stamps compensate for DMR processing time.

The one-way delay is calculated by *Frame Delay = RxTimef – TxTimeStampf* where

RxTimef = Time that the 1DM PDU was received
TxTimeStampf = Time stamp at the time the 1DM PDU was sent

Two-way delay measurement avoids the clock synchronization issue, but could incur inaccuracy due to the DMM to DMR processing in the target MEP. Consequently, [13] allowed for two options in the measurement of two-way delay. If the target MEP turnaround delay is not considered significant, then the round-trip delay can be calculated by

$$Frame\ Delay = RxTimeb - TxTimeStampf$$

where

RxTimeb = time that the DMR PDU is received by the initiating MEP.

A more accurate two-way delay measurement can be achieved if the target MEP turnaround delay is subtracted out. In this cast, the round-trip delay can be calculated by

$$Frame\ Delay = (RxTimeb - TxTimeStampf) - (TxTimeStampb - RxTimeStampf)$$
TxTimeStampb = time that the DMR PDU is sent by the target MEP
RxTimeStampf = time that the DMM PDU is received by the target MEP

The unicast destination MAC address for point-to-point services is configurable. The priority of DMM frame transmission is configurable so that its priority corresponds to the EVC CoS to be monitored. The period values for DMM are 100 ms, 1 second, and 10 seconds where 1 second is more common. Also, DMR frame pairs for FDV calculation can be configured. An offset range of 1 to 10 can be set.

Frame sizes for DMM and 1DM frame transmissions are configurable. The range of frame sizes from 64 to MTU size in 4 octet increments is recommended.

8.18 Interframe Delay Variation Measurements

Similar to delay measurements, there are two interframe delay variation (IFDV) measurements, round-trip, and one-way IFDV. Round-trip IFDV is calculated exactly as the difference between two consecutive two-way FD measurements while one-way IFDV is the difference between two consecutive one-way delay measurements. It is measured in microseconds.

For one-way Ethernet frame delay measurement, only the receiver MEP collects statistics. For two-way Ethernet frame delay measurement, only the initiator MEP collects statistics.

DMM and DMR PDUs are given in Figures 8.39 and 8.40.

8.19 Testing

Testing is one of the important areas in Carrier Ethernet. RFC 2544 [8] defined a mechanism to perform out-of-service throughput and delay measurements while Y.1731 [3] defined an incomplete mechanism to perform in-service testing. In this section, we cover testing in broader range.

Testing is concerned with the testing of equipment and their resources and transport facilities. Testing may be carried out for the purpose of testing connecting facilities in preparation for installation of new equipment, accepting newly installed interfaces or service assignments, validating trouble reports, supporting fault localization, and verifying repair.

The equipment provides diagnostics that can examine the state of each significant element of hardware, and that can identify faults and isolate failures to with-

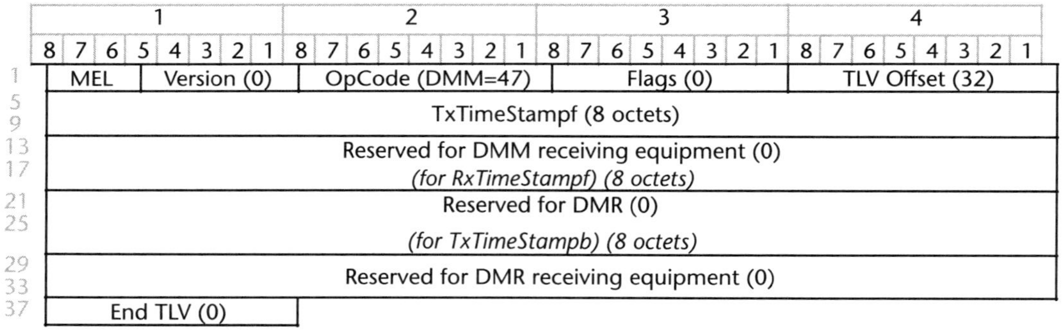

Figure 8.39 DMM PDU format.

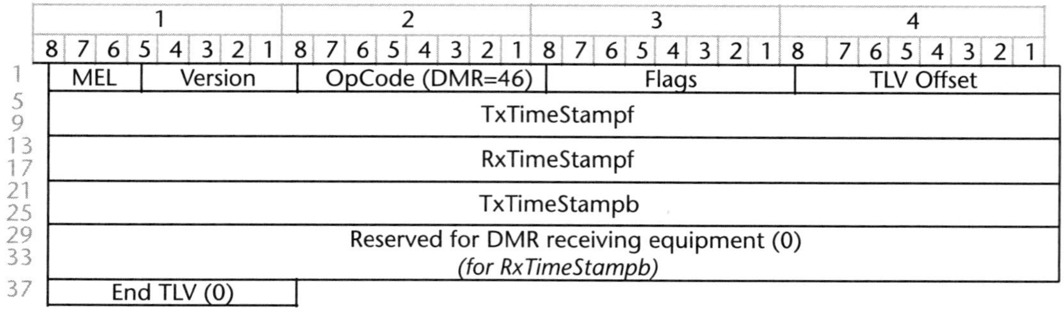

Figure 8.40 DMR PDU format.

in the smallest replaceable unit of hardware. A higher-level management system initiates these diagnostics.

The equipment run diagnostics on hardware, checks on software, and reports the result to the managing system.

8.19.1 Y.1731 Testing

Test signal (TST) is an OAM message exchanged between MEPs in one way only that includes test pattern to test throughput, measure bit errors, or detect frames delivered out of sequence. In general, tests can be performed in service or out of service.

TST OAM messages are generally sent with a unicast DA. The use of a class 1 multicast DA is also allowed for multipoint testing. TST is a one-way diagnostic function. If the diagnostic is performed in service, the repetition rate must not be disruptive of client layer traffic. If the diagnostic is performed out of service the affected MEPs will initiate LCK messages. The TST OAM PDU includes a transaction ID/sequence number and also typically includes pseudorandom test data, which is checked for bit errors by the receiving MEP.

When an in-service ETH-Test function is performed, data traffic is not disrupted and the frames with ETH-Test information are transmitted in such a manner that limited part of the service bandwidth is utilized. MEP attributes to be configured to support ETH-Test are:

- MEG level.
- Unicast MAC address of the peer MEP for which ETH-Test is intended.
- Data pattern with an optional checksum. The length and contents of the data pattern are configurable at the MEP. The contents can be a test pattern and an optional checksum. Test patterns can be pseudorandom bit sequence (PRBS) (231–1) and all 0 pattern, and so forth. At the transmitting MEP, a test signal generator is configured. At a receiving MEP, a test signal detector is configured.
- Priority, which is configurable per operation, identifies the priority of frames with ETH-Test information.
- Drop eligibility, which identifies the eligibility of frames with ETH-Test information to be dropped when congestion conditions are encountered.

Configuration attributes such as the transmission rate of ETH-Test information and the total interval of ETH-Test are undefined.

An MIP is transparent to the frames with ETH-Test information and therefore does not require any configuration information to support ETH-Test functionality.

An MEP inserts frames with ETH-Test information towards a targeted peer MEP at its configured MD level. The receiving MEP detects these frames with ETH-Test information and makes the intended measurements if the MD level matches. The PDU used for ETH-Test information is called TST that support bidirectional testing.

Each TST frame is transmitted with a specific sequence number. A different sequence number must be used for every TST frame, and no sequence number from the same MEP may be repeated within 1 minute. When a MEP is configured for an out-of-service test, the MEP also generates LCK frames at the immediate client MEG level in the same direction where TST frames are transmitted.

Information elements carried in TST PDU (Figures 8.41 and 8.42) are:

- Sequence number, which is a 4-octet field.
- Test, which is an optional field whose length and contents are determined at the transmitting MEP. The contents of test field indicate a test pattern and also carry optional checksum, as mentioned above.

8.19.2 RFC 2544

RFC 2544 [8] and RFC 1242 [19] defined a procedure to test out-of-service throughput, latency (delay) and frame loss rate for a given port or an EVC.

Test setup consists of a tester with both transmitting and receiving ports. Connections are made from the sending ports of the tester to the receiving ports of the device under test (DUT) and from the sending ports of the DUT back to the tester (Figure 8.43).

Frame sizes to be used in testing should include the maximum and minimum sizes and enough sizes in between to be able to get a full characterization of the DUT performance. At least five frame sizes are recommended to be used in each test condition.

For an MTU of 1,522 bytes, frame sizes of 64, 128, 256, 512, 1,024, 1,280, and 1,522 bytes should be considered. For larger frame sizes such as 4,000 bytes and 9,200 bytes, additional frame sizes should be used in the testing.

The stream of test frames should be augmented with 1% frames addressed to the hardware broadcast address to determine if there is any effect on the forwarding rate of the other data in the stream. The broadcast frames should be evenly distributed throughout the data stream, for example, every 100th frame.

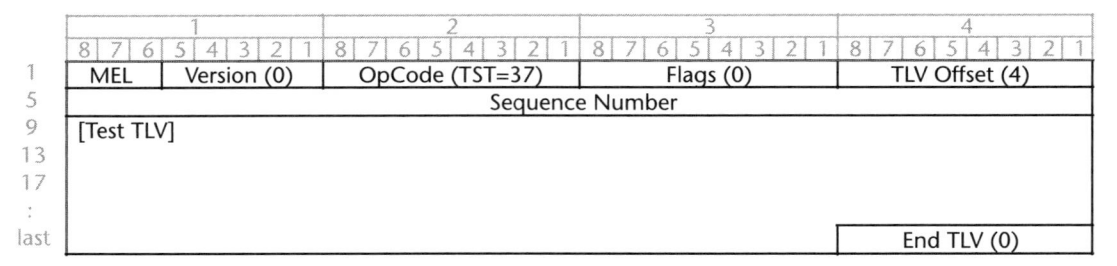

Figure 8.41 TST PDU format.

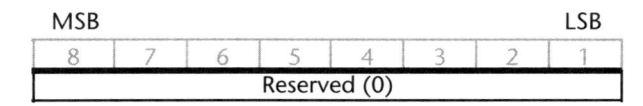

Figure 8.42 Flags format in TST PDU.

8.19 Testing 247

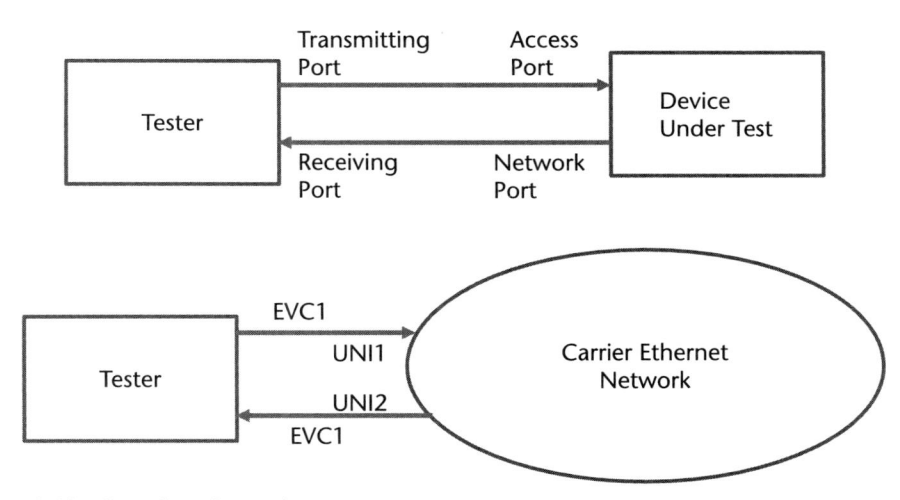

Figure 8.43 Out-of-service testing.

In order to test the bidirectional performance of a DUT, the test series should be run with the same data rate being offered from each direction.

The test equipment should include sequence numbers in the transmitted frames and check for these numbers on the received frames to identify the frames that were received out of order, duplicate frames received, the number of gaps in the received frame numbering sequence, and the number of dropped frames.

In order to simulate real traffic scenarios, some of the tests described next should be performed with both steady-state traffic and with traffic consisting of repeated bursts of frames. The duration of the test should be at least 60 seconds.

In order to test the throughput, which is the fastest rate at which the count of test frames transmitted by the DUT, a specific number of frames at a specific rate through the DUT is sent and the frames transmitted by the DUT are counted. If the count of offered frames is equal to the count of received frames, then fewer frames are received than were transmitted and the rate of the offered stream is reduced by a percentage of maximum rate and the test is rerun. The throughput test results can be plotted frame size versus frame rate as depicted in Figure 8.44.

The procedure for latency is, first the throughput for DUT at each of the frame sizes mentioned above should be determined by sending a stream of frames at a particular frame size through the DUT at the determined throughput rate to a specific destination. The stream should be at least 120 seconds in duration.

An identifying tag should be included in one frame after 60 seconds. The time for frame transmission and receiving are recorded. The latency is the difference between the frame arrival timer and transmission time.

The test must be repeated at least 20 times with the reported value being the average of the recorded values. This test is performed for each of the test frames addressed to a new destination network.

For the frame loss rate, the percentage of offered frames that are dropped are reported. The first trial is run for the frame rate that corresponds to 100% of the maximum rate for the input frame size. The procedure is repeated for the rate that corresponds to 90% of the maximum rate used and then for 80% of this rate. This sequence is continued at reducing 10% intervals until there are two successive

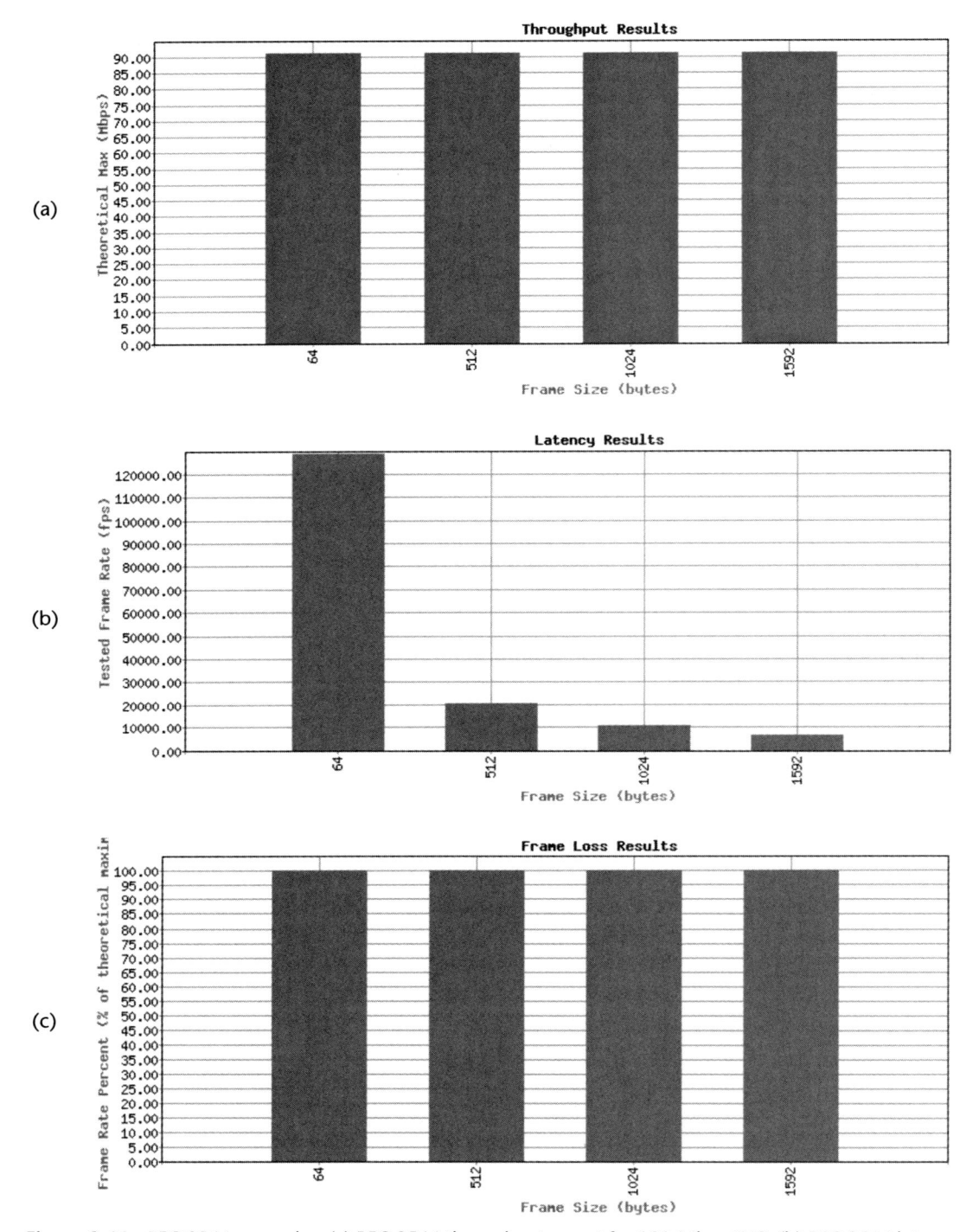

Figure 8.44 RFC 2544 examples: (a) RFC 2544 throughput report for 100-Mbps EVC, (b) RFC 2544 latency report for 100-Mbps point-to-point EVC, and (c) RFC 2544 frame loss report for 100-Mbps point-to-point EVC.

trials in which no frames are lost. Depending on the amount of time to be spent on testing, the maximum granularity of the trials can be chosen to be smaller than the 10%.

The frame loss rate at each point is calculated by

$$((input\_count - output\_count) * 100) / input\_count$$

The results of the frame loss rate test can be plotted as a graph where the x-axis is the input frame rate as a percent of the theoretical rate for the given frame size and the y-axis is the percent loss at the particular input rate.

In order to test DUT for back-to-back frames [19], a burst of frames with minimum interframe gaps to the DUT is sent and the number of frames forwarded by the DUT is counted. If the count of transmitted frames is equal to the number of frames forwarded, the length of the burst is increased and the test is rerun. If the number of forwarded frames is less than the number transmitted, the length of the burst is reduced and the test is rerun.

The back-to-back value is the number of frames in the longest burst that the DUT will handle without the loss of any frames. The trial length must be at least 2 seconds and should be repeated at least 50 times, with the average of the recorded values being reported.

There are various applications today that can generate back-to-back frames. Remote disk servers, data transfers between databases during disaster recovery are among these applications. Therefore, it is necessary to test the service for back-to-back frames.

8.19.3 Complete Service Testing

Complete Carrier Ethernet EVC service testing in a multi-CEN environment includes service activation testing (SAT), service monitoring, and service troubleshooting. The service activation testing generally verifies the configuration of service parameters and validates the required service performance before the service is delivered to the end customer [20]. Service monitoring takes place when the production network is in-service and customers are using the provisioned services. Service monitoring includes fault management and performance monitoring. It constantly monitors the performance of the running service against its SLA and raises a flag when there is an issue. Since it creates an in-band load while actively monitoring the production network, the load should be minimal and must not create service outages. Service monitoring leverages MEGs, MIPs, and MEPs. Service troubleshooting goes in effect when alerted by the service monitoring. It is responsible for diagnosing and correcting problems with the service. It may trigger service activation testing to re-verify the configuration and the performance of the service. Depending on the level of severity of the problem, service troubleshooting may decide to take the service down.

8.19.4 Service Activation Testing

After a Carrier Ethernet device has been installed and the initial configuration parameters have been loaded, setting up a Carrier Ethernet service in multicarrier

Ethernet network environment would require a testing of the service against the service acceptance criteria (SAC), which should be agreeable to all parties including the service provider, the access provider, and the subscriber. This testing checks if the physical connection is set up and all required resources have been configured correctly to deliver the service at the performance level expected by all parties. The result of this activity is the creation of the service activity testing record, which is also considered as the "birth certificate" of the service. This record sets a baseline for the health-check of the service and the troubleshooting and monitoring activities refer to it while the service is in use. As previously stated, service activation testing is made up of two separate activities: configuration verification and performance validation.

In configuration verification testing, the initial connectivity for the service is tested and the initial performance parameters are checked against the values that are agreed in the SLA; these include committed information rates, excess information rates, and so forth. The activity is usually performed by the installation crew in a short testing time span, since it follows the installation and initial configuration before the service is delivered to the subscriber (also called the end customer). After configuration verification is performed, performance validation starts. This activity runs on a longer time span from minutes to hours (instead of seconds) and validates the main performance indicators for the service, including frame delay (FD), FLR, IFDV, and information rate (IR). The numerical results from these testing activities are compared against the service activation criteria list, which sets up a threshold for the acceptability of the service. The acceptability of the service is defined as pass or fail depending on where the results stand compared to the threshold values. The service should target a set of performance objectives, called class of service performance objectives (CPO), which is defined in the service portfolio of the service provider. The service activation criteria for the SAT testing are created based on the CPO for a particular service.

SAT uses different appliances and integrated software, called Ethernet Test Equipment (ETE), to perform testing functions. An Ethernet Test Equipment-Instrument (ETE-I) is an appliance that is used for generating and collecting test frames during testing; these functions, which are performed by software, are called generator test function (GTF) and collector test function (CTF), respectively. The appliance is also able to perform latching loopback function (LLF), which is used to reflect testing frames without doing any generation and/or collection of the frames at certain locations in the network. The ETE-I is an external portable appliance that can be remotely controlled and/or locally hooked up to the testing points in the network. A local connection could be to a customer port at a UNI-U and it can be remotely connected to a UNI-N on the network side. An Ethernet Test Equipment-Application (ETE-A) is implemented in a network element to perform generator test function and collector test function along with latching loopback functionalities. A network element with this application may reside at ENNI-N and UNI-N, and could be instrumental in SAT testing without a technician plugging in and out any testing appliance on site. An Ethernet Test Equipment–Test Head (ETE-TH) is a dedicated testing device that is installed in the network and capable of performing GTF, CTF, and LLF functionalities. These ETE devices and applications are usually controlled by the Ethernet test support system (ETSS) of a service provider, which is a part of the operational support system (OSS).

8.19 Testing

In general, GTF generates and transmits Ethernet frames that are specifically for testing. CTF receives these specific Ethernet test frames and calculates the testing measurements and logs them. The service activation measurement point (SAMP) is a reference point in the network where an Ethernet Test Equipment (ETE) is performing SAT. Depending on the type of ETE, the location of SAMP could be a physical point such as a subscriber port or a logical point such as the inside of an Ethernet device where the ETE-A function has been implemented.

The connection topology for SAT has different options once the service type to be tested has been determined. For example, if an E-Access type of service is to be tested in a multi-CEN network, where there is a service provider and an access provider, the SAT can be performed over different set of SAMPs creating slightly different topologies (see Figure 8.45).

From this figure, there are four possible SAMPs: two at the ENNI (SAMP A on the SP side on ENNI and SAMP B on the AP side of ENNI) and two at the UNI of the AP's network (SAMP C on UNI-N and SAMP D on UNI-U). To perform SAT for the access service, we need one test point on the ENNI and one test point on the UNI of the AP's network. There are multiple options to set up a testing topology. One option is to place an ETE-TH on SAMP A at the SP's ENNI. Another option is to place the ETE-TH on SAMP B at the AP's ENNI-N. As an alternative, an ETE-A can be implemented at an AP's edge device at SAMP B on AP's ENNI. At the customer side of the AP's network, an ETE-I appliance can be attached to the customer port at the UNI-C at SAMP D or an ETE-A can be implemented in the AP's device at SAMP C on the UNI-N. Another solution is to implement an LLF function on the edge device at SAMP C on AP's UNI-N and the ETE-TH at the SP's side will implement GTF/CTF, and generate and collect test frames that are looped back from the customer side of the access network.

Service activation testing mainly verifies the service configuration including the values for UNI, OVC, and EVC attributes, and validates the performance of the service as defined in the SLS. The testing is performed by both service and access providers and starts with setting up the connectivity between ETEs or an ETE and LLF function for loopback. At the end, SAT produces a complete report of the entire testing and its measurement results. These results are compared against

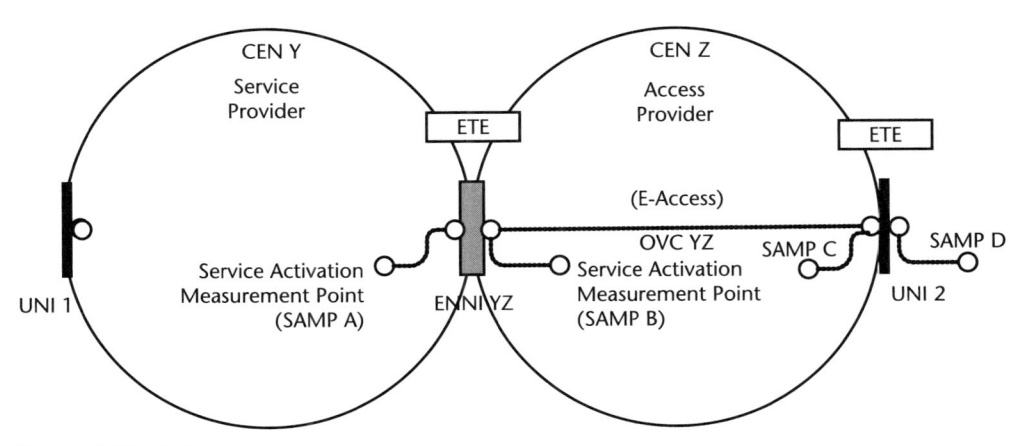

Figure 8.45 E-Access.

the service acceptance criteria for pass or fail of the test. Basically, the criteria set acceptable limits for performance and attribute values of the service. The decision for the criteria has to be made carefully such that a service that passes the criteria should satisfy the class of service performance objectives (CPOs) expected by the customer.

The service configuration test is made up of five subtests and performed by both service provider and access provider in a coordinated manner. These subtests are OVC MTU size test, CE-VLAN ID Preservation test, CE VLAN CoS ID preservation test, broadcast/unicast/multicast frame delivery test, and bandwidth profile (BWP) test (see Figure 8.46). The SAT methodology stems from ITU-T Recommendation Y.1564 documentation.

Among these subtests, frame delivery test is optional for service providers.

A typical access EPL/EVPL service employs an OVC as seen in the Figure 8.45. The OVC has two endpoints; one end point is at UNI, and the other is at the ENNI between the AP and SP networks. There is an ETE mapped to each one of these SAMP endpoints.

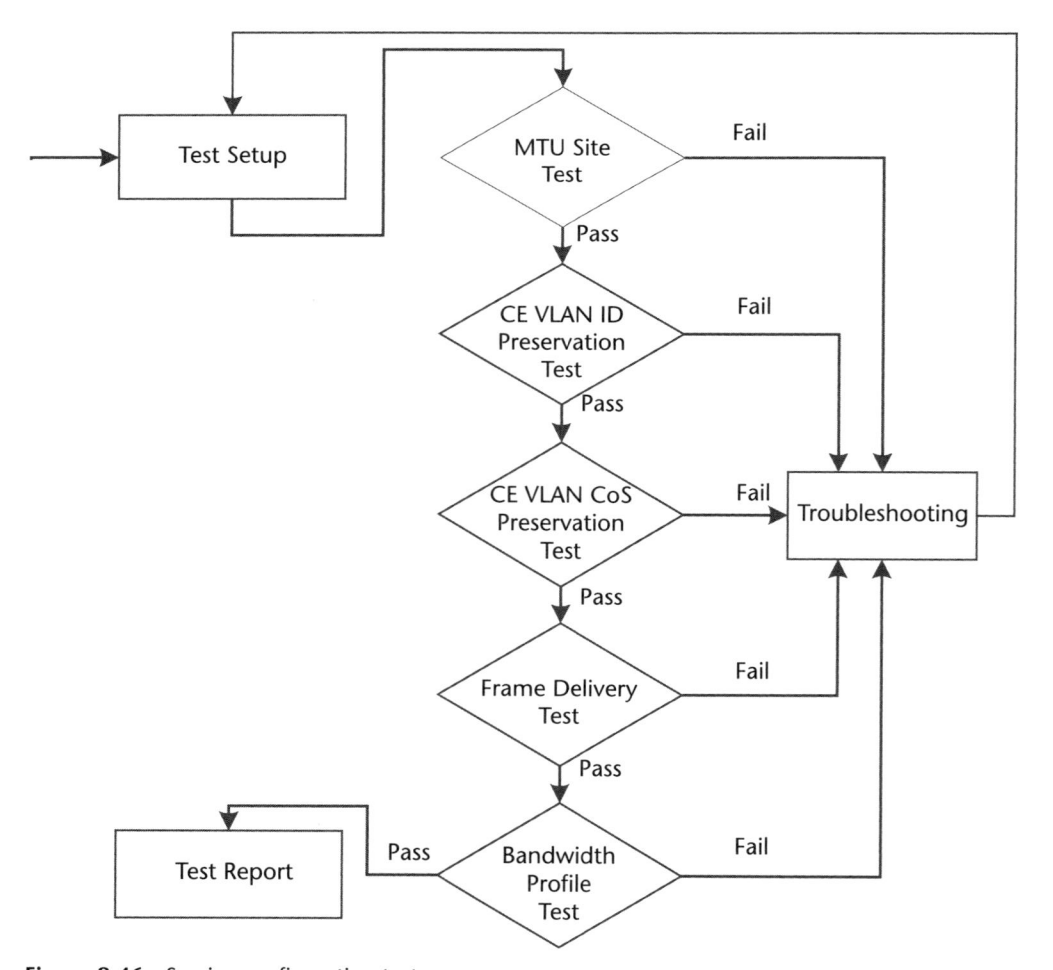

Figure 8.46 Service configuration test process.

To test OVC MTU size, both ETEs at each end simultaneously send each other service frames with the configured OVC MTU frame size as defined by the SP and AP. The receiving ETE checks this service frame and verifies the MTU size. If the verification is positive, then the result of the test is recorded as pass for this test.

For the CE-VLAN preservation test, ETEs simultaneously exchange service frames with CE-VLAN IDs defined by the test procedure and verify them. Usually, the procedure tests four CE-VLAN IDs that are determined by AP and SP in collaboration. The best practice is to pick one at the beginning of the VLAN ID spectrum, one at the end and the other two in the middle of the spectrum. If all four CE-VLAN IDs are verified as defined by the test, then the result is recorded as pass.

To test the CE-VLAN CoS ID preservation, both ETEs simultaneously start sending each other service frames with CE-VLAN ID equal to 65 and CoS ID equal to 0. After this fist CoS ID has been verified, they incrementally check all IDs in the CoS spectrum (1 to 7). Once all CoS IDs are checked and verified, then the overall test result is recorded as pass.

Broadcast, unicast, and multicast frame delivery tests are done in a similar fashion. Both ETEs send each other service frames with CE-VLAN ID equal to 65 and a defined destination MAC address (unicast, multicast, and broadcast). If all tests are verified at the receiving ETEs, then the result is recorded as pass.

The ingress bandwidth profile per OVC endpoint is defined by the operator to control all service frames that are mapped to an OVC at an ingress endpoint [21]. It includes committed information rate (CIR) and excess information rate (EIR). Testing ingress bandwidth profile helps validate both CIR and EIR (this behavior). In this test, both ETEs simultaneously send service frames at a rate that is equal to CIR. ETEs at the receiving end measures the information rate for the duration of the test. Along with information rates, frame loss ratio, frame delay, interframe delay variation, and delay range are also measured during the test. If these measured values are within the limits specified by the service activation criteria list, the test results are recorded as pass. A similar test is done for the EIR. ETEs send service frames at a rate that is equal to CIR + EIR. At the receiving end, the incoming information rate is measured and compared against the criteria list. If the received rate is within the criteria, the test is recorded as pass. Additionally, traffic policing as part of the ingress bandwidth profile can also be tested by the service provider and access provider. In this test, ETEs send service frames at a rate that is equal to CIR + 125% of the EIR. At the receiving end, ETEs measure the incoming information rate, which should be equal to CIR+EIR, if the policing is done successfully. Otherwise, the test is recorded as fail. Testing committed burst size (CBS) and excess burst size (EBS) are optional by SP and AP.

Checking performance related service attributes first by the service configuration SAT tests verify that the services are configured correctly and should deliver the traffic as intended. Service performance tests validate the performance of the configured services and make sure that the quality of service objectives are met over a longer period of time. Instead of seconds like in a typical service configuration test, performance tests take minutes to hours. The service performance tests duration is generally negotiated and decided between the service provider and the access provider. Typical durations are 15 minutes, 2 hours, and 24 hours. The 15-minute test duration is used, if the services are configured for a metro area network. The 2-hour and 24-hour durations are used for multi-CEN long-haul networks and international networks, respectively. In case that multiple virtual services are configured for a UNI and ENNI pair, collective performance for all services are measured simultaneously and evaluated against the service activation criteria. Similar to service CIR/EIR configuration tests, both ETEs send service frames at a rate that is equal to CIR for a specified longer test duration. The receiving end ETE measures the incoming information rate, frame loss ratio, frame delay, frame delay range, and interframe delay variation and compares those measured and calculated values to the list in SAC. If all meet the SAC values, then the test is recorded as pass; otherwise, the test fails and corresponding troubleshoot activities should take place before returning to service configuration tests again for a recheck. Usually, the collected test results are stored in a back-office database application. An SNMP or NETCONF interface could be used to interact with the database to query a result. In addition to storing into a database, a specific ETE device, like ETE-I, can also communicate some important test status on a device display. Some test states that indicate status are as follows:

- Completed: The test has been completed.
- In progress: The test is still in progress.
- Not supported: The test is not supported by the device in use.
- Unable to run: The test cannot be run, due to a failure.
- Aborted: The test has encountered a failure and is aborted; already generated test results are not accepted as valid.
- Failed: The test generated results that are not meeting the criteria.

Both service provider and access provider need to share resources and coordinate timing for accomplishing SAT testing. Their ETE devices and software should be interoperable, since they sink and source each other's traffic. If there is no existing service at any of the testing interfaces, then setting up the testing connection and performing the testing process are relatively easy; however, if this is not the case, then attention needs to be paid such that the traffic load that will be introduced by the SAT testing should not negatively affect the live subscriber traffic ad neither should it cut off the in-service connectivity of any EVC.

8.19.5 SAT Control Protocol and Messages

To make ETE vendors interoperable with each other, the SAT control protocol, messages, and the SAT Test PDUs are standardized. The SAT control protocol defines a set of methods to initiate the test and collect the test results. There are two SAT Test PDUs: frame loss PDU (FL_PDU) is used to measure frame loss and throughput numbers; and frame delay PDU (FD_PDU) is used to calculate frame delay measurements. The details of these test PDUs are provided in MEF 49 SAT Control Protocol and PDU Formats document [22]. The Test PDUs are sourced from the GTF towards CTF in a unidirectional way to perform a one-way performance measurement. The ETSS coordinates the test and provides the parameters that configure a test session. The SAT Control Protocol is a method of procedure that initiates, runs, and terminates a SAT Test session between ETEs. After a successful termination, it also facilitates the collection of the test results. More than one session can be run between the same ETEs; however, each has a unique session ID. A unidirectional test session has two ends: the controller end, and the responder end. The controller end initiates the test session and is responsible to feed the test configuration parameters to the responder end. Since each session is unidirectional, a session is called forward test, if the Test PDUs are sourced from the controller end and received at the responder end. The session that has the reverse direction from the responder end to the controller end is called the backward test. The controller end is responsible to create the GTF and/or CTF depending on either the forward test or the backward test, to send the parameters for the test to the responder end, to start or stop the test session, and to collect the test results from both GTF and CTF. In order to perform these functions, the SAT Control Protocol uses messages in Table 8.9 [22]. These messages are explained as follows:

- Initiate_Session_Request_Message(Session_ID, Type, Parameter_List): Initiates a test session and used by the controller end. It contains all parameters for the test and the type, for either the forward test or the backward test.
- Initiate_Session_Response_Message(Session_ID, status): Used by the responder end and confirms the interoperability to the controller end. Returns status.
- Start_Session_Request_Message(Session_ID): Used by the controller end to start the test session that is confirmed by the responder end.
- Start_Session_Response_Message(Session_ID): Confirms the receipt of the Start_Session_Request_Message().
- Stop_Session_Request_Message(Session_ID): Used by the controller end to stop a testing session.
- Stop_Session_Response_Message(Session_ID): Confirms the receipt of the Stop_Session_Request_Message().
- Abort_Session_Request_Message(Session_ID): Used by the controller end to abort a testing session.

Table 8.9 SAT Control Protocol Messages [22]

#	Message	Parameters
1	Initiate_Session_Request_Message	Session_ID, Type, Parameter_List
2	Initiate_Session_Response_Message	Session_ID, status
3	Start_Session_Request_Message	Session_ID
4	Start_Session_Response_Message	Session_ID
5	Stop_Session_Request_Message	Session_ID
6	Stop_Session_Response_Message	Session_ID
7	Abort_Session_Request_Message	Session_ID
8	Abort_Session_Response_Message	Session_ID
9	Fetch_Session_Results_Request_Message	Session_ID
10	Fetch_Session_Results_Response_Message	Session_ID, Results
11	Delete_Session_Request_Message	Session_ID
12	Delete_Session_Response_Message	Session_ID

- Abort_Session_Response_Message(Session_ID): Either confirms the receipt of the Abort_Session_Request_Message() or indicates that a failure has occurred at the responder end without having to receive a request message.
- Fetch_Session_Results_Request_Message(Session_ID): Used by the controller end to get the results of a testing session after the testing session has stopped.
- Fetch_Session_Results_Response_Message(Session_ID, Results): Used by the controller end to return the results of a test session that has been stopped. If the test session has not been stopped, then a status is sent indicating that the results are not compiled yet.
- Delete_Session_Request_Message(Session_ID): Used by the controller end to inform the end of a running test session to the responder end so that the results and any other resources that have been assigned to the testing session can be released.
- Delete_Session_Response_Message(Session_ID): Confirms the receipt of the Delete_Session_Request_Message().

There are two main SAT Test configurations: (1) forward SAT Test during which GTF is created at the controller end and CTF is created at the responder end; and (2) backward SAT Test during which GTF is created at the responder end and CTF is created at the controller end. Next we review the message exchanges for both forward and backward SAT Test sessions.

- Forward Test Session: Life cycle of a forward test session goes through an initiation phase, a running phase, and a results collection phase in a successful completion. In case that a failure occurs during a test run, an abort has to be initiated by the ETSS to bring the test to a conclusion. Message interaction for forward test session initiation is given in Figure 8.47. The sequence of events and messages are explained here:
  1. ETSS sends an external message to the controller end to initiate a forward test session and provides the required parameters.
  2. The controller end generates a unique session ID and sends an internal message to create GTF for the testing.
  3. The controller end sends the protocol message Initiate_Session_Request_Message(Session_ID, "Forward", parameter_list) to the responder end.
  4. The responder end sends an internal message to create CTF.
  5. The responder end sends Initiate_Session_Response_Message(Session_ID, status) back to the controller end.
  6. The controller end sends an external message to EETS to acknowledge the initiation of the new test session.

After the initiation phase, ETSS has to trigger the start of the SAT Testing run at the controller end. The sequence of events and messages for a successful SAT Testing completion is given as (Figure 8.48):

8.19 Testing

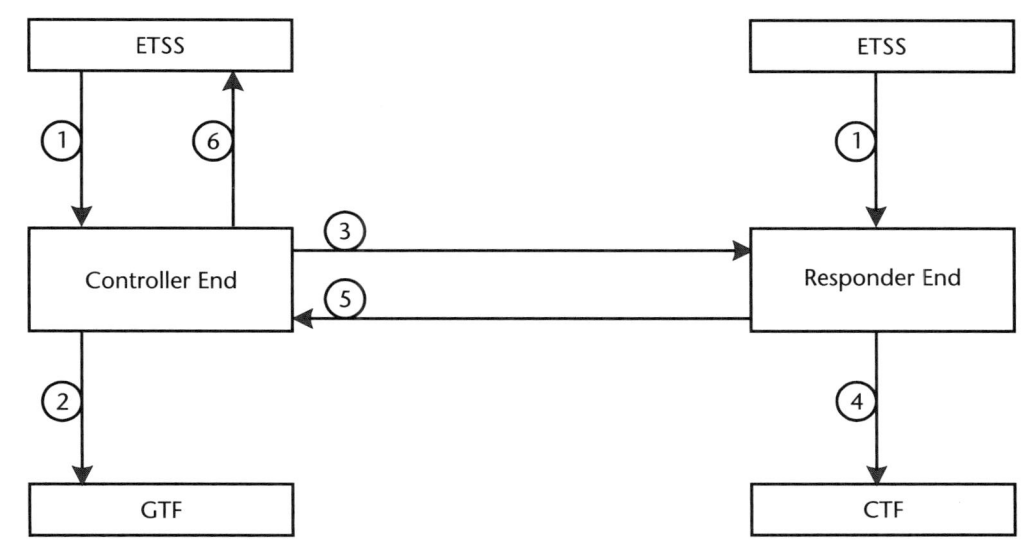

Figure 8.47 Initiation of a test session (both forward and backward).

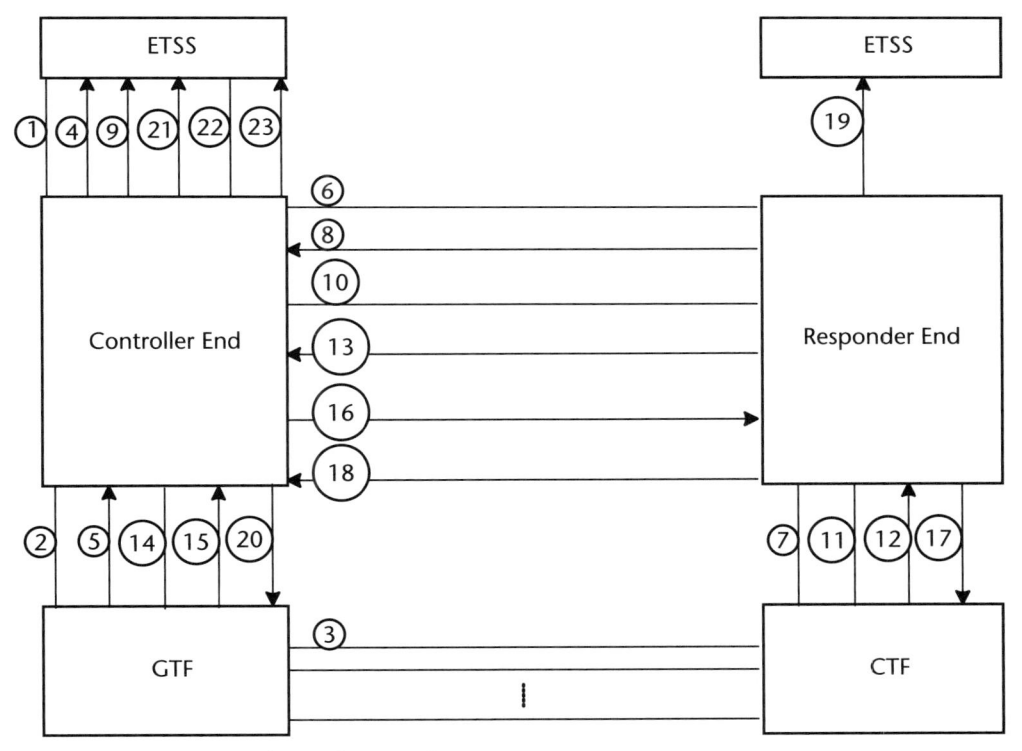

Figure 8.48 Running a forward test session.

1. ETSS sends an external message to the control end to start the initiated test session.

2. The controller end sends an internal message to the created GTF to start sending test frames based on the parameters set at the initiation stage.
3. GTF starts sending test frames to the CTF at the responder end. CTF starts receiving and processing these frames.
4. The controller end sends an external message to ETSS acknowledging the start of a test session run.
5. GTF sends an internal message stating the end of the test run.
6. The controller end sends a protocol message Stop_Session_Request_Message(Session_ID) to the responder end.
7. The responder end sends an internal message to CTF to stop listening and processing the test frames coming from the controller end.
8. Once the responder end is done with stopping the CTF, it sends a protocol message Stop_Session_Response_Message(Session_ID) back to the controller end.
9. The controller end sends an external message to ETSS acknowledging that the test run has been stopped.
10. The controller end asks the responder end to send back the collected test session results by sending the protocol message Fetch_Session_Results_Request_Message(Session_ID).
11. The responder end sends an internal message to CTF to send back the collected results.
12. CTF sends the results back to the responder end.
13. The responder end passes the results back to the controller end by the protocol message Fetch_Session_Results_Response_Message(Session_ID, results).
14. The controller end asks results internally from its own GTF.
15. GTF sends the results for the test session.
16. The controller end sends a protocol message Delete_Session_Request_Message(Session_ID) to delete the session and release its resources.
17. The responder end sends an internal message to CTF to end the session and claim its resources.
18. The responder end acknowledges the controller end about the deletion process at its own end by sending the protocol message Delete_Session_Response_Message(Session_ID).
19. The responder end informs ETSS on its side about the deletion of the test session by sending an external message. This transaction is optional for the responder end.
20. The controller end internally sends a message to delete the testing session on its own GTF.
21. The controller end sends an external message to ETSS acknowledging the entire end of the testing session.
22. ETSS sends an external message to the controller end asking final test session results.
23. The controller end sends all results collected to the ETSS.

8.19 Testing

If the test does not complete successfully, then the test has to be aborted at both sides. The following steps explain the flow of messages to abort a test session (Figure 8.49).

1. ETSS sends an external message to the controller end to start the initiated test session.
2. The control end sends an internal message to the created GTF to start sending test frames based on the parameters set at the initiation stage.
3. GTF starts sending test frames to the CTF at the responder end. CTF starts receiving and processing these frames.
4. The controller end sends an external message to ETSS acknowledging the start of a test session run.
5. ETSS decides to end testing before the session is completed by sending an external abort message to the controller end.
6. The controller end sends an internal message to GTF to abort the test session immediately.
7. The controller end sends the protocol message Abort_Session_Request_Messege(Session_ID) to the responder end.
8. The responder end sends an internal message to CTF to stop listening to the testing frames coming from the GTF.
9. The responder end sends an internal message to CTF to delete the testing session.

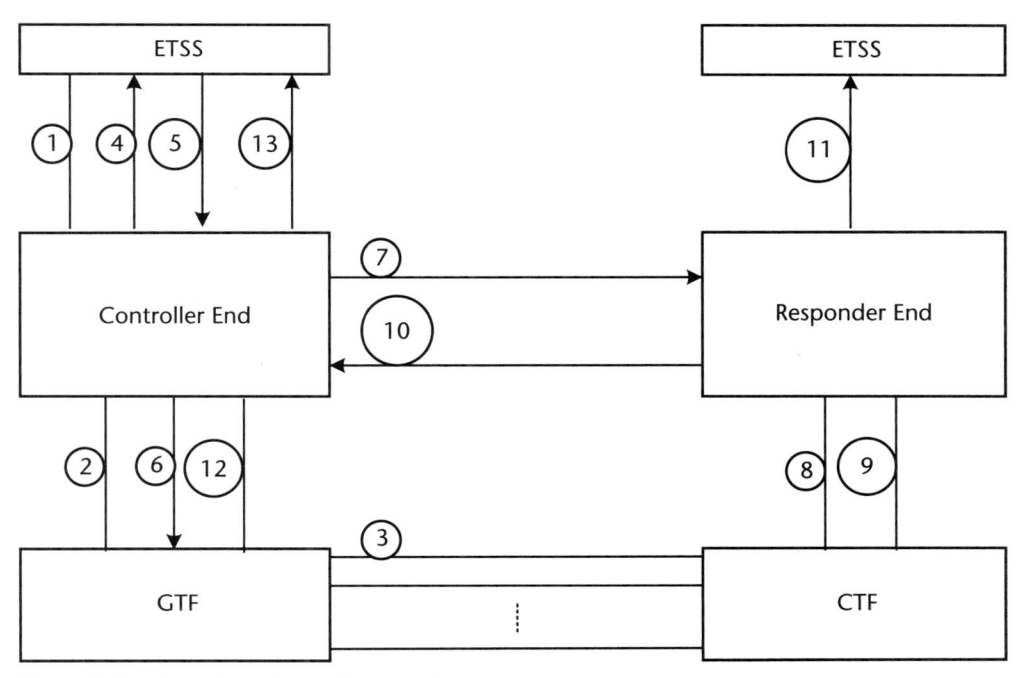

Figure 8.49 Aborting a forward test session.

10. The responder end sends the protocol message Abort_Session_Response_Message(Session_ID) back to the controller end acknowledging the abort request.
11. The responder end sends an external message to its ETSS notifying the test session abort. This action is optional for the responder end.
12. The controller end sends an internal message to GTF to delete the testing session.
13. The controller end sends an external message to its ETSS notifying the test session abort.

- Backward test session: The backward test session also goes through the same phases to initiate, run, and abort a test session; however, the sequence of events and message show a little difference to accomplish the test run. The initiation steps are very similar to the forward test session except the parameter, the type, and functionalities. The steps refer to Figure 8.47 as to the forward test session and are given here:

  1. ETSS sends an external message to the controller end to initiate a backward test session and provides the required parameters.
  2. The controller end generates a unique session ID and sends an internal message to create CTF for the testing.
  3. The controller end sends the protocol message Initiate_Session_Request_Message(Session_ID, "Backward", parameter_list) to the responder end.
  4. The responder end sends an internal message to create GTF.
  5. The responder end sends Initiate_Session_Response_Message(Session_ID, status) back to the controller end acknowledging the session initiation.
  6. The controller end sends an external message to EETS to acknowledge the initiation of the new test session.

Steps to be taken to run and successfully complete a backward test session are shown here (Figure 8.50):

1. ETSS sends an external message to the controller end to start the initiated test session.
2. The control end sends an internal message to the created CTF to start listening to test frames based on the parameters set at the initiation stage.
3. The controller end sends the protocol message Start_Session_Request_Message(Session_ID) to the responder end.
4. The responder end sends an internal message to the created GTF to start sending test frames based on the parameters set at the initiation stage.
5. GTF starts sending test frames to the CTF. CTF starts receiving and processing these frames.

8.19 Testing

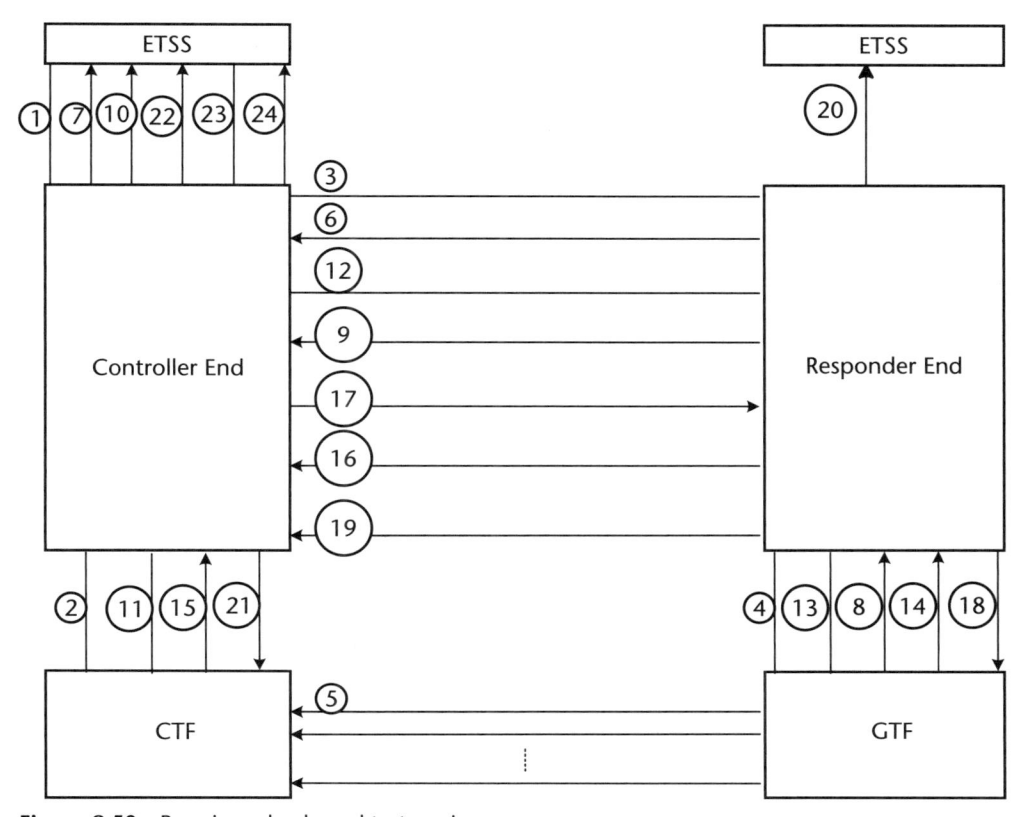

Figure 8.50 Running a backward test session.

6. The responder end sends the protocol message Start_Session_Response_Message(Session_ID) to the controller end acknowledging the start of the session.
7. The controller end sends an external message to ETSS acknowledging the start of the test session.
8. GTF internally informs the responder end stating that the test session has ended.
9. The responder end sends the protocol message Stop_Session_Response_Message(Session_ID) to the controller end.
10. The controller end sends an external message to ETSS acknowledging the end of the test session.
11. The controller end asks results internally from its own CTF.
12. The controller end asks responder end to send back the collected test session results by sending the protocol message Fetch_Session_Results_Request_Message(Session_ID).
13. The responder end sends an internal message to GTF to send back the collected results.
14. GTF sends the results back to the responder end.
15. CTF sends the results back to the controller end.

16. The responder end passes the results back to the controller end by sending the protocol message Fetch_Session_Results_Response_Message(Session_ID, results).
17. The controller end sends a protocol message Delete_Session_Request_Message(Session_ID) to the responder end to delete the session and release its resources.
18. The responder end sends an internal message to GTF to end the session and claim its resources.
19. The responder end acknowledges the controller end by sending the protocol message Delete_Session_Response_Message(Session_ID).
20. The responder end informs ETSS on its side about the deletion of the test session by sending an external message. This transaction is option for the responder end.
21. The controller end internally sends a message to delete the testing session on its own CTF.
22. The controller end sends an external message to ETSS acknowledging the entire end of the testing session.
23. ETSS sends an external message to the controller end asking final test session results.
24. The controller end sends all results collected to the ETSS.

Similar to the forward test session, if the backward test does not complete successfully, then the test has to be aborted at both sides ordered by the ETSS on the controller side. The following steps explain the flow of messages to abort a backward test session (Figure 8.51).

1. ETSS sends an external message to the controller end to start the initiated test session.
2. The control end sends an internal message to the created CTF to start listening to test frames based on the parameters set at the initiation stage.
3. The controller end sends the protocol message Start_Session_Request_Message(Session_ID) to the responder end.
4. The responder end sends an internal message to the created GTF to start sending test frames based on the parameters set at the initiation stage.
5. GTF starts sending test frames to the CTF. CTF starts receiving and processing these frames.
6. The responder end sends the protocol message Start_Session_Response_Message(Session_ID) to the controller end acknowledging the start of the session.
7. The controller end sends an external message to ETSS acknowledging the start of the test session.
8. ETSS decides to end testing before the session is completed be sending an external abort message to the controller end.
9. The controller end sends the protocol message Abort_Session_Request_Messege(Session_ID) to the responder end.
10. The responder end sends an internal message to GTF to stop sending the testing frames to the CTF.

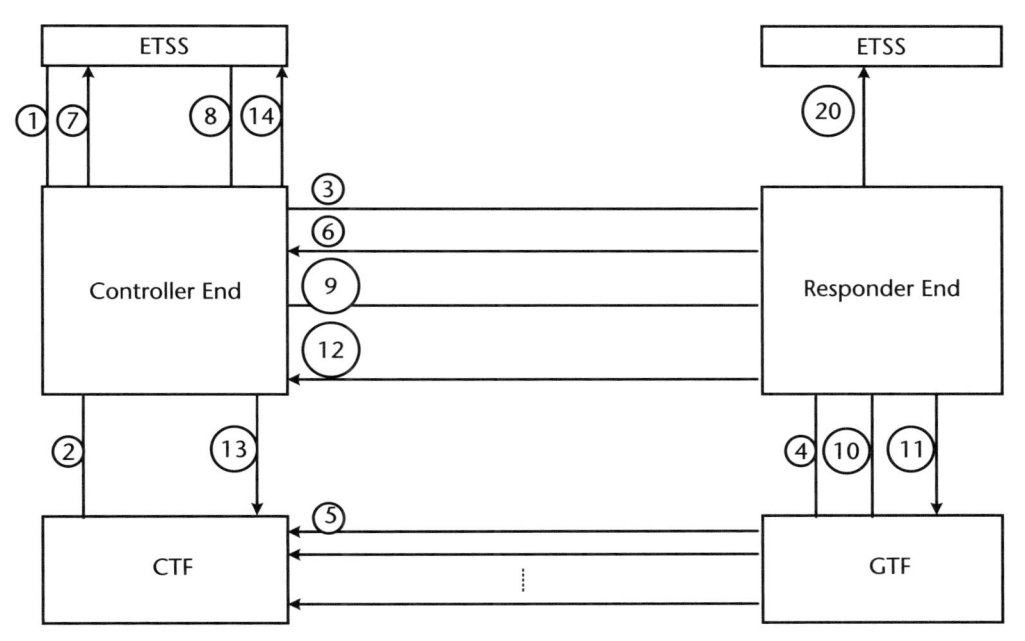

Figure 8.51 Aborting a backward test session.

11. The responder end sends an internal message to GTF to delete the testing session.
12. The responder end sends the protocol message Abort_Session_Response_Message(Session_ID) back to the controller end acknowledging the abort request.
13. The controller end sends an internal message to CTF to delete the testing session.
14. The controller end sends an external message to its ETSS notifying the test session abort.

There may be multiple test session running simultaneously between the same end points (same ETEs). As an example, for a bidirectional SAT Test, there are one forward test session and one backward test session created and run successfully. If one fails, then the entire SAT testing is deemed to be failed and needs to be repeated.

8.20 Performance Monitoring Solutions

A performance-monitoring (PM) solution is a layered functional approach to PM with a specific measurement set up having different set of requirements. There are different components in each layer supporting a PM solution (see Figure 8.52) [5]. At the very bottom of the network element layer, network elements conduct performance measurements by sending and receiving service and/or synthetic frames between MEPs in an MEG. PM solutions use PM functions to monitor the performance of a service in a specific setup (single-ended or double-ended monitoring). The features and capabilities of PM functions are detailed in ITU-T G.8013/Y.1731

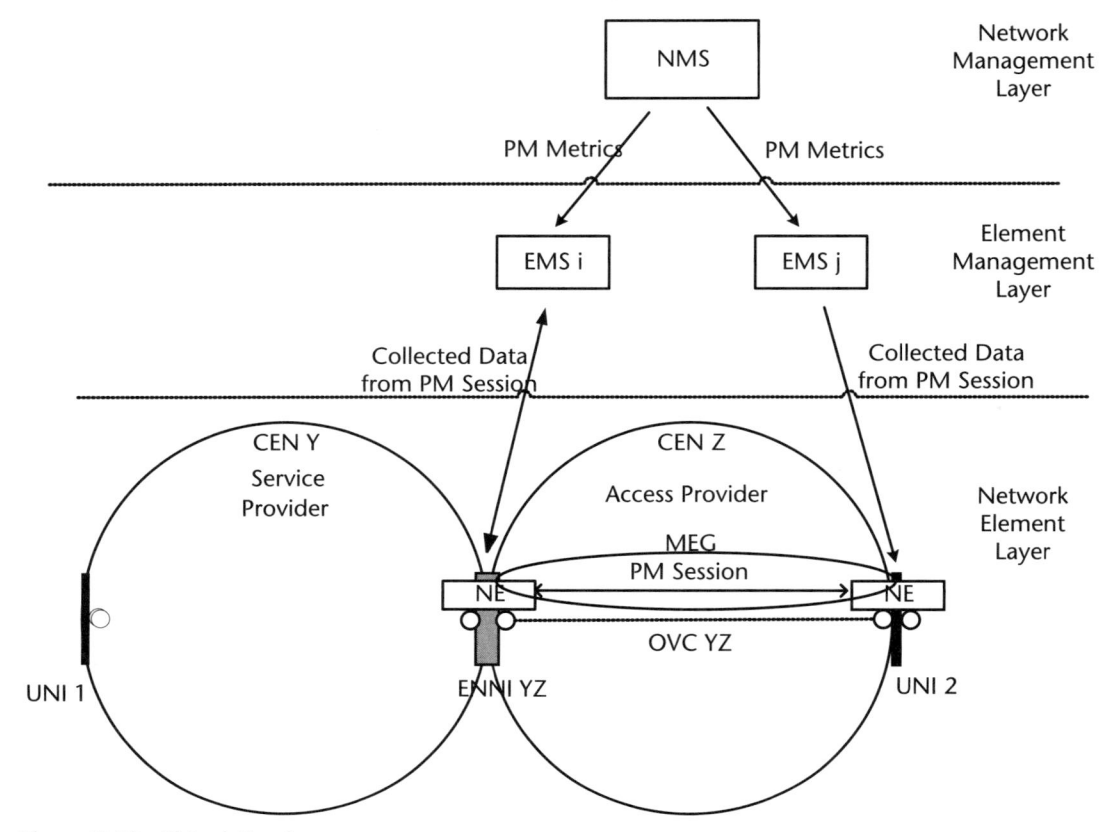

Figure 8.52 PM solution layers

documents. To run a PM function, a PM session needs to be initiated between a pair of MEPs over a dedicated SOAM PM CoS ID. In the element management layer, element management systems configure and collect these measurements data from network elements and compute performance metrics. The network management system in the network management layer manages all network elements and the MEG for the service performance monitoring. It also receives the metrics from multiple EMSs and creates the performance metrics for the service being delivered by the network.

A PM solution may employ one or more PM functions for that particular solution and use point-to-point or multipoint MEG. There are four PM Solutions defined [5].

PM functions perform delay and loss performance monitoring and are of two types: single-ended, and dual-ended. They use both service and synthetic frames.

The single-ended PM function for loss and delay is initiated at one end of the MEG and the other end just responds to the messages received for the PM session. This is used to measure one-way and two-way FD, IFDV, MFD, and FDR metrics using DMM/DMR messages, one-way FLR by using service frames and LMM/LMR messages and using synthetic frames and SLM/SLR messages. It is also used to compute one-way availability metric by using SLM/SLR messages. Dual-ended

8.20 Performance Monitoring Solutions

PM function is initiated at both ends of the MEG and used for only one-way delay, loss, and availability metrics by using 1DM and 1SL messages.

PM-1: Single-Ended Point-to-Point or Multipoint Delay and Synthetic Loss Measurements
The PM-1 solution uses single-ended PM functions and synthetic SOAM PM PDUs to measure FD, mean FD (MFD), frame delay range (FDR), and IFDV, FLR, availability, group availability, and count of high-loss intervals (HLI, CHLI). The PM-1 solution can be used for multipoint MEGs as well as point-to-point MEGs and can run multiple simultaneous PM sessions that could measure and monitor multiple CoS levels for the service. In one-way measurements, PM-1 applies both forward (from the controller to the responder) and backward (from the responder to the controller) measurements. For one-way FD and MDF measurements by using DMM/DMR PDUs, the time-of-day (ToD) clock synchronization is a requirement between the controller and the responder. Note that this requirement does not apply for two-way delay measurements due to the fact that only the clock at the source end is used. In case that there is a lack of ToD clock synchronization, then the one-way delay is estimated from the two-way delay by dividing the value by half.

For frame loss and availability measurements, controller and responder exchanges SLM/SLR messages with a corresponding unicast address of each other's MEP. Since PM-1 exercises frame loss by using synthetic frames, a statistically acceptable number of frames should be exchanged; MEF 35.1 appendices provide some guidance for this statistical consideration. Readers who are to implement and require more information should refer to this reference. One last requirement is to treat synthetic frames just like service frames during the execution of PM-1 regardless of the current load of the production network; this would reveal the performance of the service as close as possible to the one experienced by the end user.

PM-2: Dual-Ended Point-to-Point or Multipoint Delay
PM-2 performs one-way delay measurements in both directions; meaning one PM session is established from the controller MEP to the responder MEP and the other is from the responder to the controller. It uses 1DM PDUs and is an optional PM solution. Since it uses one-way measurements, ToD synchronization requirement is valid for the PM-2 solution. In addition to point-to-point MEGs, the PM-2 can be used in multipoint MEGs as well. In the multipoint case, multicast 1DM PDUs can be used to reduce the number of distinct sessions and the total traffic load. The PM-2 solution is used to measure one-way frame delay performance, one-way mean frame delay, one-way frame delay range, and one-way interframe delay variation performance. Since PM-2 uses synthetic frames for delay measurement, there has to be enough frames generated to be statistically valid and all these synthetic frames need to have a representative frame length and be treated like a service frame, so that the measurement becomes realistic and the representative of the real traffic. Like PM-1, the frame generation for PM-2 should be transparent to the current load; otherwise, measurements would not reflect the existing network performance in a realistic way.

PM-3: Single-Ended Service Loss Measurements

The PM-3 uses only service frames to measure a single-ended service loss performance indicator FLR. Like PM-2, PM-3 is also optional and is used in only point-to-point MEGs; multipoint MEGs are not applicable to the PM-3 solution. During the measurement, the LMM PDU is sent from the controller MEP to the responder MEP with a unicast address and this starts the count of qualified service frames that are sent and received in both ends. The LMR PDU is sent from the responder MEP back to the controller MEP in response and facilitates the calculation of FLR.

PM-4: Dual-Ended Synthetic Loss Measurements

The PM-4 solution is optional and used to measure one-way frame loss ratio performance, one-way availability performance, one-way group availability performance, and one-way resiliency performance including HLI, CHLI by using synthetic frames and 1SL PDUs. It performs the one-way measurements in both directions to match dual-ended mechanism, meaning that one PM session is established from the controller MEP to the responder MEP and the other is from the responder to the controller. In addition to point-to-point MEGs, the PM-4 can also be used in multipoint MEGs. In the multipoint case, multicast 1SL PDUs can be used to reduce the number of distinct sessions and the total traffic load. Since PM-4 uses synthetic frames for loss and availability measurements, there has to be enough frames generated to be statistically valid and all these synthetic frames need to have a representative frame length and be treated like a service frame, so that the measurement becomes realistic and the representative of the real traffic. Frame generation for PM-4 should be transparent to the current load; otherwise, measurements would not reflect the existing network performance in a realistic way.

8.21 Security

Security of physical access for Carrier Ethernet services is the same as that for any other wireline services. Users must identify themselves with their unique user IDs and passwords before performing any actions on the network element and NMS. The network element maintains the identity of all users currently logged on and logs off (or locks) a user ID after a specified time interval of inactivity such as 15 minutes.

If the network element can accept remote operations-related commands, the user can be authenticated via public/private key technology.

At the EVC level, SOAM traffic is not permitted to transit the MEPs of a MA at the same MD level or higher, while SOAM traffic at a higher MD level is transparent to the MA. This domain hierarchy provides a mechanism for protecting maintenance points from other maintenance points with which the maintenance point has not been designed to communicate. However, it is possible for one MP to flood another with OAM PDUs, potentially resulting in a denial of service attack on an MP.

In order to make sure that network elements are not susceptible to a denial of service attack via OAM PDUs, rate limiting of OAM PDU traffic at local node is necessary.

8.22 OAM Bandwidth

An adequate OAM bandwidth for each EVC and each NE should be allocated to ensure timely response to problems in the network. The MEF SOAM PM document requires a network element to be able to process at least 10 times its EVC capacity in OAMPDUs per second.

In other words, for example, if a network element claims to support 1,000 EVCs, it must be able to process at least 10,000 OAMPDUs per second, with any legal or illegal OAMPDU content. Our rule of thumb is to allocate 1% bandwidth for OAM.

8.23 Conclusion

SOAM tools and procedures defined for Carrier Ethernet services are the key to the broader deployment of these services, by saving substantial amount of operational expenses for service providers and allowing users to monitor its service in real time and test. In this chapter, we covered most of these tools and procedures.

References

[1] IEEE Std. 802.3ah-2004, "IEEE Standard for Local and Metropolitan Area Networks - Specific Requirements - Part 3: Carrier Sense Multiple Access with Collision Detection (CSMA/CD) Access Method and Physical Layer Specifications Amendment: Media Access Control Parameters, Physical Layers, and Management Parameters for Subscriber Access Networks," 2004.

[2] IEEE 802.1ag-2007, "Virtual Bridged Local Area Networks – Amendment 5: Connectivity Fault Management," December 2007.

[3] ITU-T Recommendation Y.1731, "OAM Functions and Mechanisms for Ethernet Based Networks," May 2006.

[4] MEF 31, "Service OAM Fault Management Definition of Managed Objects," January 2011.

[5] MEF 35.1, "Service OAM Performance Monitoring Implementation Agreement," May 2015.

[6] MEF 15, "Requirements for Management of Metro Ethernet Phase 1 Network Elements," November 2005.

[7] MEF 16, "Ethernet Local Management Interface (E-LMI)," January 2006.

[8] RFC 2544, "Benchmarking Methodology for Network Interconnect Devices," 1999.

[9] ITU-T Y.1564, "Ethernet Service Activation Test Methodology," March 25, 2011.

[10] MEF 17, "Service OAM Requirements and Framework – Phase 1," April 2007.

[11] MEF 30.1/30.1.1, "Service OAM Fault Management Implementation Agreement, Phase 2 and Amendment to SOAM FM IA," April 2013.

[12] ITU-T G.8010, "Architecture of Ethernet Layer Networks," 2004.

[13] ITU-T Y.1711, "Operation & Maintenance Mechanism for MPLS Networks," 2004.

[14] RFC 3636, "Definitions of Managed Objects for IEEE 802.3 Medium," September 2003.

[15] ITU-T Q.840.1, "Requirements and Analysis for NMS-EMS Management Interface of Ethernet over Transport and Metro Ethernet Network (EoT/MEN)," 2007.

[16] MEF 10.3, "Ethernet Services Attributes- Phase 3," October 2013.

[17] ITU-T Y.1563, "Ethernet Frame Transfer and Availability Performance," 2009.

[18] MEF 10.3.1, "Composite Performance Metric (CPM) Amendment to MEF 10.3," February 2015.
[19] RFC 1242, "Benchmarking Terminology for Network Interconnection Devices," 1991.
[20] MEF 48, "Carrier Ethernet Service Activation Testing (SAT)," October 2014.
[21] MEF 26.2, "External Network-Network Interface (ENNI) and Operator Service Attributes," August 2016.
[22] MEF 49, "Service Activation Testing Control Protocol and PDU Formats," October 2014.
[23] RFC 2218, "Hot Standby Reouter Protocol," Marh 1998.

CHAPTER 9
Circuit Emulation Services

9.1 Introduction

A pseudowire is an emulation of a native service over a packet switched network (PSN). The native service may be low-rate TDM, SDH/SONET while the PSN may be Ethernet, MPLS, or IP. Especially, in mobile backhaul locations, equipment with TDM and SONET/SDH interfaces are common. The CESoETH is the solution to provide low-cost Ethernet transport to these locations.

The CESoETH is offered at UNI. As operators begin offering CESoETH at locations outside of their footprints, the service will be offered at ENNI as well.

In CES, data streams are converted into frames for transmission over Ethernet. At the destination site, the original bit stream is reconstructed when the headers are removed, payload is concatenated, and clock regenerated, while ensuring very low latency. The MEN behaves as a virtual wire which is an application layer function.

TDM data is delivered at a constant rate over a dedicated channel, the native service may have bit errors, but data is never lost in transit. All PSNs suffer to some degree from packet loss, and this must be compensated when delivering TDM over a PSN.

The standard TDM pseudowire operating modes are circuit emulation over PSN CESoPSN, structure agnostic TDM over packet (SAToP), TDMoIP, and HDL-CoPSN (HDLC emulation over PSN).

In SAToP [1], TDM is typically unframed. When it is framed or even channelized, the framing and channelization structure are completely disregarded by the transport mechanisms. In such cases, all structural overhead must be transparently transported along with the payload data, and the encapsulation method employed provides no mechanisms for its location or utilization. However, structure-aware TDM transport may explicitly safeguard TDM structure.

TDM Pseudowire over Ethernet (Ethernet CES) provides emulation of TDM services, such as $N \times 64$ Kbps, T1, T3, and OC-n, across a MEN, but transfers the data across MEN, instead of a traditional circuit-switched TDM network. From the customer perspective, this TDM service is the same as any other TDM service.

From the provider's perspective, two CES interworking functions are provided to interface the TDM service to the Ethernet network. The CES interworking functions (Figure 9.1) are connected via the MEN using point-to-point Ethernet virtual connections (EVCs) [2].

The basic CES is a point to point, constant bit rate service, similar to the traditional leased line type of TDM service. However, service multiplexing can be performed (Figure 9.2) ahead of the CES interworking functions (IWFs), such as aggregation of multiple emulated T1 lines into a single T3 or OC-3 link, creating a multipoint-to-point or even a multipoint-to-multipoint configuration.

In the unstructured emulation mode (SAToP), a service is provided between two service-end-points that use the same interface type. Traffic entering the MEN on one end-point leaves the network at the other end-point and vice versa transparently. The MEN must maintain the bit integrity, timing, and other client-payload specific characteristics of the transported traffic, without causing any degradation. All the management, monitoring, and other functions related to that specific flow must be performed without changing or altering the service payload information or capacity.

The unstructured emulation mode is suitable for leased line services or any other transfer-delay sensitive (real-time) applications. For example, the DS1 unstructured constant bit rate (CBR) service is intended to emulate a point-to-point DS1 circuit, transparently carrying any arbitrary 1.544-Mbps data stream. The end-user timing source for these interface signals is not necessarily traceable to a primary reference source (PRS).

In structured emulation mode, a service provided between two service-end-points use the same interface type. Traffic entering the MEN on one end-point is handled as overhead and payload. The MEN maintains the bit integrity, timing information and other client-payload specific characteristics of the transported traffic without causing any degradation that would exceed the requirements for the given service. All the management, monitoring, and other functions related to that specific flow must be performed without changing or altering the service payload information or capacity.

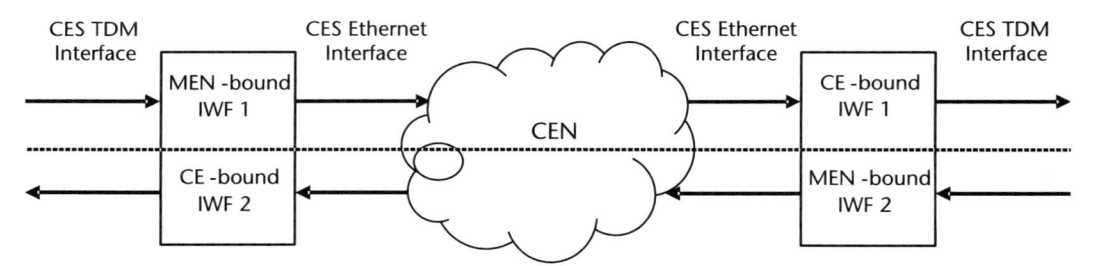

Figure 9.1 CES architecture.

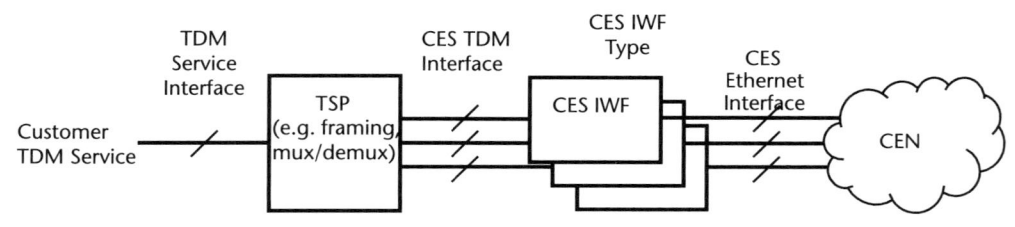

Figure 9.2 CES architecture with multiplexing.

A fractional DS1 service, where the framing bit and unused channels are stripped, and the used channels transported across the MEN as an $N \times 64$ Kbps service, is an example to structured CES.

MEF 8 [2] defines the CES interworking function (CES-IWF), which allows communication between non-Ethernet circuits (such as T1, E1, E3, and T3) and Ethernet UNI interfaces. CES is typically implemented on NID, but it can be implemented on edge routers as well.

In the simplest case, a pair of IWFs connected by an Ethernet private line (EPL) service. However, in many cases several IWFs may share an Ethernet UNI, and are connected via Ethernet virtual private line (EVPL) service (Figure 9.3).

CES is widely used in Carrier Ethernet backhaul where TDM/SONET is still a popular interface for base stations (BSs) and radio network controllers (RNCs), as depicted in Figure 9.4.

9.2 Circuit Emulation Functions

CES functional blocks are depicted in Figure 9.5. The TDM service processor (TSP) may terminate framing overhead, or multiplex several customer TDM services into a single service to be emulated.

One or multiple CES IWFs may use one Ethernet interface. When there are multiple CES IWFs, the flows of the IWFs are multiplexed and demultiplexed using the ECDX (emulated circuit de/multiplexing function). The output of ECDX which is in the packet domain interfaces with the EFTF (Ethernet flow termination function) that handles the MAC layer functions.

The CES IWF is the adaptation function that interfaces the CES application to the Ethernet layer, by handling emulation of the service presented at the CES TDM interface. The CES IWF is responsible for all the functions required for the emulated service:

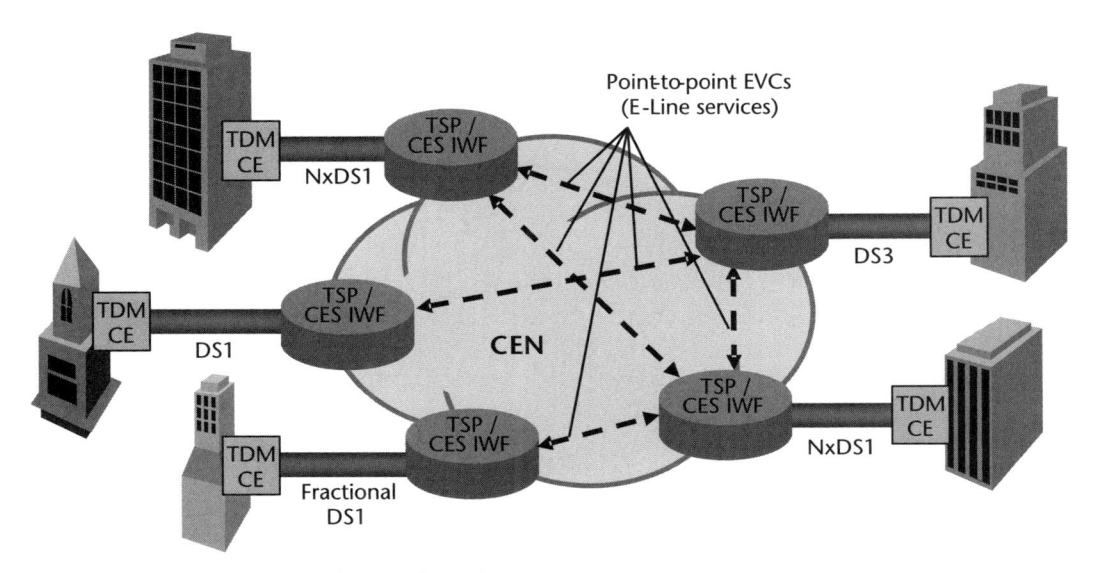

Figure 9.3 TDM virtual private line configurations.

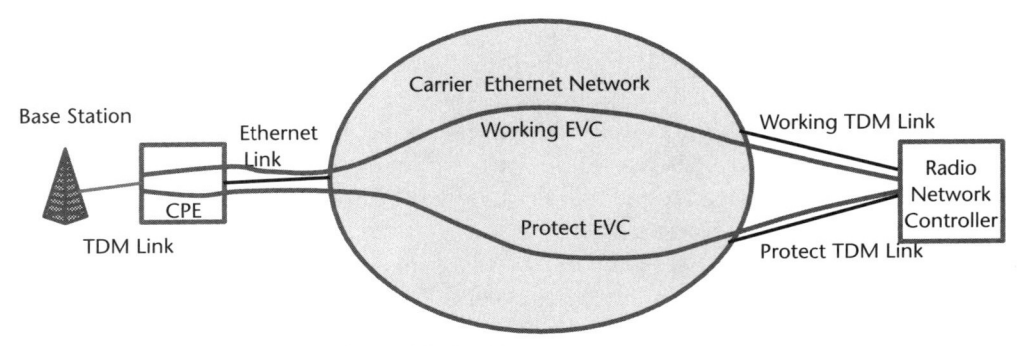

Figure 9.4 TDM over Ethernet for mobile backhaul.

- Encapsulation and decapsulation;
- Payload formation and extraction;
- Synchronization;
- Carriage of TDM signaling and alarms;
- Error response and defect behavior;
- TDM performance monitoring.

The interfaces to the CES IWF can be DS1, E1, DS3, E3, or $N \times 64$ Kbps TDM interfaces or packetized TDM payload, CES control word, and RTP header.

The ECDX prepends an emulated circuit identifier (ECID) and assigns the EtherType field (0x88d8) to every Ethernet frame sent to the MEN. In the TDM-bound direction, the ECDX determines the destination CES IWF of each Ethernet frame from the ECID value and strips the EtherType and ECID fields, before handing off the CES payload to the CES IWF.

The interfaces to the ECDX consist of the following:

- CES payload (i.e., packetized TDM payload, CES control word, optional RTP header) [3];
- Adapted payload (i.e., the CES payload, plus the ECID and EtherType fields).

Ethernet flow termination function (EFTF) takes an adapted payload from the ECDX and then creates an Ethernet service frame by adding, MAC destination and source addresses, optional VLAN tag (if required) and associated tag ID and user priority information, any padding required to meet the minimum Ethernet frame size, and the frame check sequence (FCS).

In the TDM-bound direction, the EFTF takes in an Ethernet frame from the MEN, and checks the FCS, discarding the frame if it is incorrect. It determines whether it contains CES payload from the EtherType field, and forwards it to its associated ECDX function, for passing to the appropriate CES IWF.

9.2 Circuit Emulation Functions 273

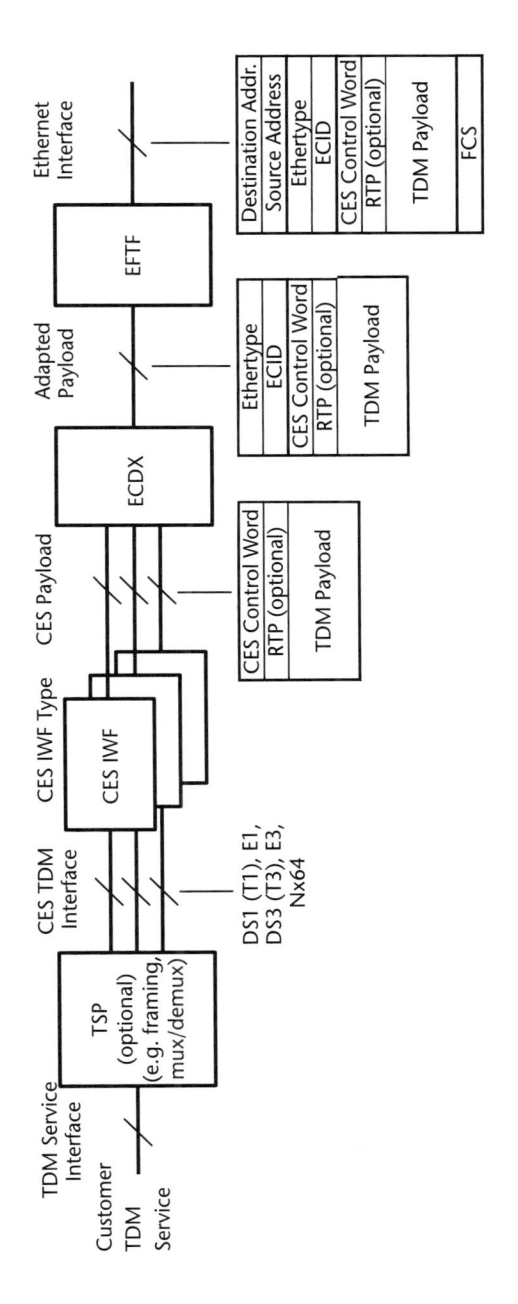

Figure 9.5 Functional components and interface types.

9.3 Adaptation Function Headers

CES over Ethernet adaptation function operates directly on top of the Ethernet layer. There are three components to the adaptation function header [2]:

- ECID that identifies the emulated circuit being carried is added by ECDX. This separates the identification of the emulated circuit from the Ethernet layer, allowing the MEN operator to multiplex several emulated circuits across a single EVC where required.

The ECID consists of a single, 20-bit unsigned binary field containing an identifier for the circuit being emulated. This is added by the ECDX, and shown in Figure 9.6.

ECIDs have local significance only, and are associated with a specific MAC address. Therefore, an emulated circuit may be given different ECIDs for each direction of the circuit. ECIDs are selected during the creation of an emulated circuit.

The length of ECID is the same as an MPLS label size to ease interworking with an MPLS-based circuit emulation service.

- CESoETH (circuit emulation services over Ethernet) control word providing sequencing and signaling of defects such as AIS of the TDM circuit, or packet loss detected in the MEN, is added by the CES IWF. The CESoETH control word added by the TDM-bound IWF (Figure 9.7) consists of the following:
- L bit indicating local TDM failure impacting TDM data when set. For structure-agnostic emulation, an MEN-bound IWF is set the L bit to one when loss of signal (LOS) is detected on the TDM service interface. For structure-aware emulation, an MEN-bound IWF is set the L bit to one where the TDM circuit indicates LOS, alarm indication signal (AIS), or loss of frame alignment (LOF). An MEN-bound IWF clears the L bit as soon as the defect condition is rectified.

For structure-agnostic emulation, on reception of CESoETH frames marked with the L bit set to 1, the TDM-bound IWF discards the payload, and plays out AIS code for the scheduled duration of the CESoETH frame.

For structure-aware emulation, on reception of CESoETH frames marked with the L bit set to 1, the TDM-bound IWF discards the payload, and plays out AIS for the scheduled duration of the CESoETH frame.

0	19	20	31
Emulated Circuit Identifier (ECID) (20 bits)		Reserved (set to 0x102)	

Figure 9.6 ECID.

0 1 2 3 4	5	6 7	8 9	10 15	16 31
Reserved set to zero (12 bits)	L	R	M (2 bits)	FRG (2 bits)	LEN (Length) (6 bits)	SN (Sequence Number) (16 bits)

Figure 9.7 Structure of the CESoETH control word.

- R bit representing remote loss of frames indication when set by an MEN-bound IWF. In other words, it indicates that its local TDM-bound IWF is not receiving frames from the MEN, and consequently has entered a loss of frames state (LOFS), perhaps due to congestion or other network-related faults.

A TDM-bound IWF enters an LOFS following detection of a locally preconfigured number of consecutive lost (including late frames that are discarded) CESoETH frames, and exits LOFS following reception of a locally preconfigured number of consecutive CESoETH frames. An MEN-bound IWF sets the R bit to 1 on all frames transmitted into the MEN while its local TDM-bound IWF is in the LOFS.

On detection of a change in state of the R bit in incoming CESoETH frames, a TDM-bound IWF reports it to the local management entity.

- M bits that are called modifier bits and set by the MEN-bound IWF to supplement the meaning of the L bit, as shown in Table 9.1.

When an MEN-bound IWF is not capable of detecting the conditions described in Table 9.1, it clears the M field to zero on frames to be transmitted into the MEN. However, a TDM-bound IWF discards a CESoETH frame where the M field is set to a value it does not support.

- FRG (fragmentation) bits that are used for fragmenting multiframe structures into multiple CESoETH frames (Table 9.2).
- LEN (length) is an unsigned binary number indicating the length of the payload, which is the sum of the size of the CESoETH control word (4 octets), size of the optional RTP header (12 octets, if present) and size of the TDM payload. The LEN does not include the size of ECID.

Table 9.1 Meaning of the Local TDM Failure Modification Bits

L	M		Indicates
bit 4	bit 6	bit 7	
0	0	0	No local TDM defect detected.
0	1	0	Receipt of RDI at the TDM input to the MEN-bound IWF. When this indication is received, a TDM-bound IWF may generate RDI in the local TDM circuit. This is only applicable to structure-aware emulation.
0	1	1	CESoETH frames containing non-TDM data such as signaling frames.
1	0	0	A TDM defect that triggers AIS generation at the TDM-bound IWF, such as LOS or LOF in structure-aware operation.

Table 9.2 Meaning of the Fragmentation Bits

FRG		Indicates
bit 8	bit 9	
0	0	Entire multiframe structure is carried in a single CESoETH frame, or no multiframe structure is being indicated
0	1	Packet carrying the first fragment of the multiframe structure
1	0	Packet carrying the last fragment of the multiframe structure
1	1	Packet carrying an intermediate fragment of the multifram structure.

A nonzero LEN indicates the presence of padding. LEN is set to zero where the length equals or exceeds 42 octets.

- SN (sequence number) is a 16-bit unsigned binary number which increments by one for each frame transmitted into the MEN with the same ECID, including those frames that are fragments of multiframe structures. The receiving IWF uses it primarily to detect frame loss and to restore frame sequence. The initial value of the SN is random.

Optional RTP (Real-time Transport Protocol) header providing timing and sequencing where approriate. CESoETH uses the fields of the RTP header as described in Figure 9.8.

- The RTP header in CESoETH can be used in conjunction with at least the following modes of timestamp generation:
    - *Absolute mode:* The MEN-bound IWF sets timestamps using the clock recovered from the incoming TDM circuit. As a consequence, the timestamps are closely correlated with the sequence numbers. All CESoETH implementations supporting RTP support this mode.
    - *Differential mode:* The two IWFs at either end of the emulated circuit have access to the same high-quality synchronization source, and this synchronization source is used for timestamp generation.

9.4 Synchronization

In any circuit emulation scheme, the clock that is used to play out the data at the TDM-bound IWF must be the same frequency as the clock that is used to input the data at the MEN-bound IWF. Unsynchronized clocking will result in frame slips.

For clocking, the TDM-bound IWF can use:

- TDM line timing, which is the clock from the incoming TDM line;
- External timing, which is the clocking from an external reference clock source
- Free run timing, which is the clocking from a free-running oscillator;

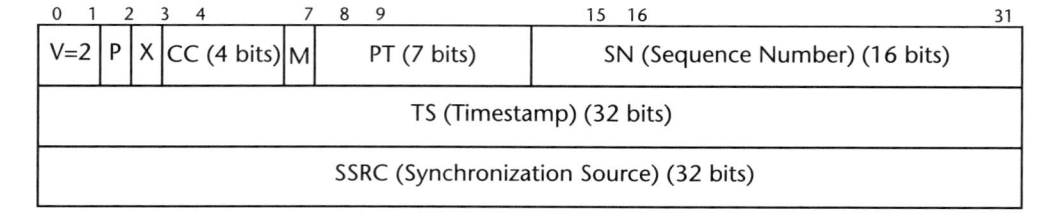

0 1 2 3 4 7 8 9	15 16	31
V=2 | P | X | CC (4 bits) | M | PT (7 bits)	SN (Sequence Number) (16 bits)	
TS (Timestamp) (32 bits)		
SSRC (Synchronization Source) (32 bits)		

Version number (V) is always be set to 2
Padding field (P) is always be set to 0
Header extension (X) is always be set to 0
CSRC (contributingsource) count (CC) field is always be set to 0
Marker field (M) is always be set to 0.
PT=PayloadType

Figure 9.8 RTP header structure.

- Ethernet line timing, which is recovering the clock from the Ethernet interface.

Ethernet line timing covers all methods where information is extracted from the Ethernet, including:

- Adaptive timing, where the clock is recovered from data in the CESoETH frames, and the arrival time of the frames;
- Differential timing, where the clock is recovered from a combination of data contained in the CESoETH frames, and knowledge of a reference clock common to both the MEN-bound and TDM-bound IWFs. Such a reference may be distributed by a variety of means (Figure 9.9).

9.5 TDM Application Signaling

Circuit emulation applications interconnected over a CESoETH service may exchange signaling in addition to TDM data. The typical example is telephony applications that exchange their off-hook/on-hook state in addition to TDM data carrying PCM-encoded voice.

With structure-agnostic emulation, it is not required to intercept or process CE signaling. Signaling is embedded in the TDM data stream, and hence it is carried end-to-end across the emulated circuit.

With structure-aware emulation, transport of common channel signaling (CCS) may be achieved by carrying the signaling channel with the emulated service such as channel 23 for DS1. However, channel associated signaling (CAS) such as DS1 robbed bit signaling requires knowledge of the relationship of the timeslot to the interface multiframe structure, which is indicated by the framing bits.

A generic method for extending the $N \times 64$ Kbps basic service by carrying CE signaling (CAS or CCS) in separate signaling packets is independent of the TDM circuit type.

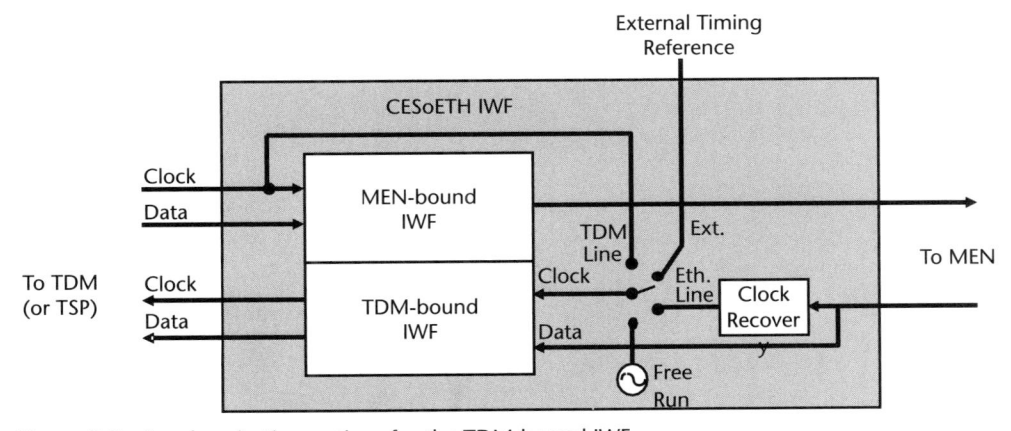

Figure 9.9 Synchronization options for the TDM-bound IWF.

CESoETH data frames and their associated signaling frames have the same destination MAC address, EtherType, and usage of the RTP header. CESoETH data frames and their associated signaling frames use different ECID values. However, CESoETH data frames and their associated signaling frames use a separate sequence number space.

When the RTP header is used:

1. Data frames and associated signaling frames use a different payload type value (both allocated from the range of dynamically allocated RTP payload types).
2. Data frames and associated signaling frames use a different SSRC value.
3. The timestamp values for the data and associated signaling frames are identical at any given time.

In the case where the CE application interconnected by a basic $N \times 64$ kbps CESoETH service is a telephony application using CAS, the payload of each signaling frame consists of N 32-bit words, where N is the number of timeslots in the service. The ith word of the payload contains the current "ABCD" value of the CAS signal corresponding to the ith timeslot (Figure 9.10).

Signaling frames are sent three times at an interval of 5 ms on any of the following events:

1. Setup of the emulated circuit
2. A change in the signaling state of the emulated circuit.
3. Loss of frames defect has been cleared
4. Remote loss of frames indication has been cleared.

A signaling frame is sent every 5 seconds in the absence of any of the events, except when there is a failure of the local TDM circuit leading to the L flag being set in the associated data frames.

9.6 CESoETH Defects and Alarms

CESoETH defects can be caused by the TDM interface interworking or adaptation section and EVC section provided by MEN. Defects caused by the TDM interface are LOS, AIS, and LOF. Defects caused by interworking function are malformed frames, jitter buffer overrun and underrun. Defects caused by MEN are misconnection of CESoETH frames, reordering and loss of frames, and late arriving CESoETH frames.

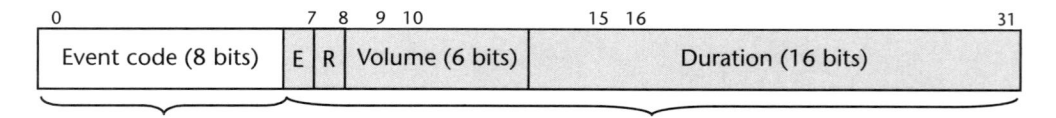

```
0                         7 8  9 10          15 16                        31
┌──────────────────────────┬───┬──────────────┬───────────────────────────┐
│   Event code (8 bits)    │E│R│Volume (6 bits)│     Duration (16 bits)    │
└──────────────────────────┴───┴──────────────┴───────────────────────────┘
```

ABCD signaling value   Not required–set to zero
(codes 144-159)

Figure 9.10 Encoding format for CAS "ABCD" values

Misconnection is caused by wrongly directed frames. They are discarded. If the percentage of stray frames persists above a defined level for a configurable period of time, 2.5 seconds by default, the misconnection alarm is reported.

Detection of out-of-sequence or lost CESoETH frames is accomplished by use of the sequence number, either from the RTP header or the CESoETH control word. Out-of-sequence frames are reordered where they arrive in time to be played out. Out-of-sequence CESoETH frames that cannot be reordered are discarded.

If loss of one or more CESoETH frames is detected by the TDM-bound IWF, it generates exactly one replacement octet for every lost octet of TDM data. If the frame loss ratio (FLR) is above a defined threshold for a configurable period of time, 2.5 seconds by default, a LOF alarm is generated. The LOF alarm is cleared if FLR remains below a defined threshold for a configurable period of time, 10 seconds by default.

When the frames arrive after their scheduled playout time (i.e., late arriving frames), even the jitter buffer is not full, they are considered as lost frames.

If the percentage of frames arriving too late to be played out exceeds a defined level for a configurable period of time, 2.5 seconds by default, a late frame alarm is generated.

CESoETH frames can be malformed if the PT value in its RTP header does not correspond to one of the PT values allocated for this direction of the emulated circuit. If a malformed frame is received by the TDM-bound IWF in time to be played out to the TDM interface, it discards the malformed frame and generates exactly one replacement octet for every lost octet of TDM data. If the percentage of malformed CESoETH frames is above a defined level for a configurable period of time, 2.5 seconds by default, a malformed frames alarm is generated. The malformed frames alarm is cleared if no malformed packets have been detected for a configurable period of time, 10 seconds by default.

The TDM-bound IWF contains a jitter buffer that accumulates data from incoming CESoETH frames to smooth out variation in arrival time of the CESoETH frames. Data is played out of the jitter buffer onto the TDM service at a constant rate. The delay through this buffer needs to be as small as possible, in order to reduce latency of the TDM service, but large enough to absorb the interframe delay variation (IFDV). The jitter buffer overrun condition occurs when the jitter buffer at the TDM-bound IWF cannot accommodate the newly arrived valid CESoETH frame due to insufficient buffer space.

The jitter buffer underrun condition occurs when there is no correctly received CESoETH payload ready to be played out on the TDM interface due to frames lost in MEN or discarded due to error conditions, and replacement data must be played out instead.

If a CESoETH frame arrives that cannot be stored in the jitter buffer due to a jitter buffer overrun condition, the TDM-bound IWF discards the frame. If the percentage of frames causing jitter buffer overruns is above a defined level for a configurable period of time, 2.5 seconds by default, a jitter buffer overrun alarm is generated. The jitter buffer overrun alarm is cleared if no cases of jitter buffer overrun have been detected for a configurable period of time, 10 seconds by default.

9.7 Performance Monitoring of CESoETH

CESoETH service supporting DS1 circuit using ESF framing may monitor messages carried in the facility data link (FDL).

All events in the TDM-bound IWF which lead to replacement data being played out (except when a result of receiving a CESoETH packet with an AIS or idle code indication) result in errored seconds or potentially severely errored seconds.

The collective sum of all these errors can be aggregated into frame error ratio (FER), as the number of errored (i.e., lost or discarded) CESoETH frames, over the total number of CESoETH frames sent. This includes situations where the CESoETH frame fails to arrive at the TDM-bound IWF (i.e., lost in MEN), arrives too late to be played out, arrives too early to be accommodated in the jitter buffer, and arrives with bit errors causing the frame to be discarded.

An MEN-bound CES IWF maintains number of CESoETH frames transmitted and number of payload octets transmitted. A TDM-bound CES IWF maintains the number of CESoETH frames received, number of payload octets received, number of lost frames detected, number of frames received that are out-of-sequence, but successfully reordered, number of transitions from the normal to the LOFS, number of malformed frames received, number of jitter buffer overruns, and number of jitter buffer underruns,

9.8 CESoETH Service Configuration

CESoETH is offered at UNI and ENNI interfaces via E-line services. For the CESoETH, UNI, EVC, ENNI, and OVC attributes are configured. In addition, the following parameters need to be assigned when an emulated circuit is set up.

- Service type such as structure-agnostic DS1 and unstructured DS1;
- TDM bit rate, which is applicable to structure-aware N × 64 Kbps services. The bit rate is defined by N. It is the same for each direction of the emulated circuit.
- Payload size defined as the number of octets of TDM payload carried in each CESoETH frame. It is the same for each direction of the emulated circuit.
- ECID;
- TDM clocking option for the TDM-bound IWF;
- Whether the RTP header is to be used or not.;
- Whether the generic TDM application signaling method is to be used or not;
- Whether the control word is to be extended from 32 bits to 64; where an extended control word is selected, the extension applies to both directions of the emulated circuit.

If RTP is used, the following additional parameters need to be assigned:

- Payload type;
- Timestamp mode, absolute or differential.

- Timestamp resolution parameter that encodes the bit rate of the clock used for setting the timestamps in RTP headers as a multiple of the basic 8-kHz rate;
- SSRC value.

If generic TDM signaling is used, ECID for signaling frames and payload type are to be set.

In addition to parameters above, control word and defect and alarm parameters need to be configured.

9.9 Conclusion

CESoETH is one of the key Carrier Ethernet services, which is widely deployed at mobile backhaul networks. Synchronization of both ends is necessary for the service to work. Delay, jitter, and loss for MBH [4] are also tight.

In this chapter, we described the service, how it works, and the service parameters.

References

[1] RFC 5086, Structure-Aware Time Division Multiplexed (TDM) Circuit. Emulation Service over Packet Switched Network (CESoPSN), December 2007.

[2] MEF 8, Implementation Agreement for the Emulation of PDH Circuits over Metro Ethernet Networks, October 2004.

[3] RFC 3550, A Transport Protocol for Real-Time Applications, July 2003.

[4] MEF 22.2, Mobile Backhaul Implementation Agreement Phase 3, January 2016.

Carrier Ethernet Local Management Interface

10.1 Introduction

The Ethernet Local Management Interface (E-LMI) Protocol, defined in [1], operates between the customer edge (CE) device and the network edge (NE) of the service provider. It defines the protocol and procedures allowing auto-configuration of the LCE device by the service provider's NE device (Figure 10.1). The E-LMI Protocol also provides the means for notification of the status of an EVC and enables the CE to request and receive status and service attributes information [2] from the CEN so that it can configure itself to access Carrier Ethernet services.

E-LMI is a protocol that is terminated by the UNI-C on the CE side of the UNI and by the UNI-N on the MEN side of the UNI (Figure 10.2).

The E-LMI protocol includes the following procedures:

1. Notification to the CE device of the addition of an EVC: The respective CPEs are informed of the availability of a new EVC once the service provider turns on the service. Furthermore, the service endpoints are notified of the corresponding CE-VLAN ID to be used by a given service.
2. Notification to the CE device of the deletion of an EVC.
3. Notification to the CE device of the availability (active/partially active) or unavailability (inactive) state of a configured EVC so that the CE device can take some corrective action, such as rerouting traffic to a different EVC or other WAN service, when informed that an EVC has become inactive.
4. Notification to the CE device of the availability of the remote UNI so that the CE device can take some corrective action, such as rerouting traffic to a different EVC or other WAN service, when informed that the remote UNI is down.
5. Communication of UNI and EVC attributes to the CE device:
    - The network informs the CE device as to which VLAN ID is used to identify each EVC to avoid the possibility of a VLAN mismatch between the service provider and CE.

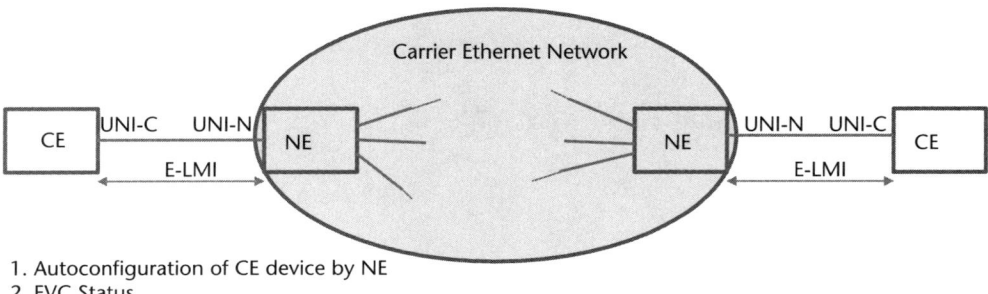

Figure 10.1 E-LMI.

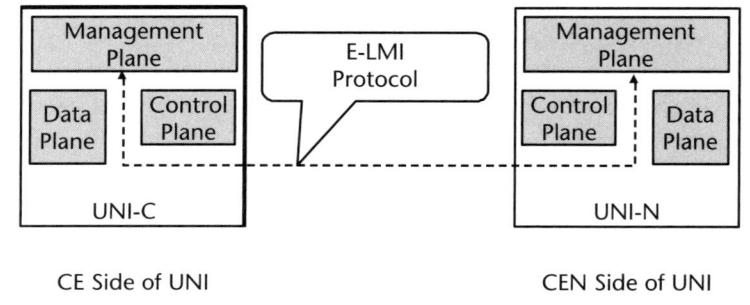

Figure 10.2 E-LMI termination in functional components.

- Remote UNI identification to ensure that the right endpoints have been connected by an EVC.
- Autoconfiguration of bandwidth profiles by CE device.

In order to transfer E-LMI messages between the UNI-C and the UNI-N, a framing or encapsulation mechanism is needed. The E-LMI frame structure is based on the IEEE 802.3 untagged MAC-frame format where the E-LMI messages are encapsulated inside Ethernet frames. The E-LMI framing structure is presented in Figure 10.3.

The destination address and E-LMI EtherType must be 01-80-C2-00-00-07 and 88-EE, respectively. The (01-80-C2-00-00-07) DA can be used if there is no 802.1Q-compliant component between UNI-C and UNI-N.

EVC status can be new, active, not active, and partially active:

- New-EVC has just been added to the CE-VLAN ID/EVC map.
- Active-EVC is in the CE-VLAN ID/EVC map and is fully operational between the UNIs in the EVC.
- Not Active-EVC is in the CE-VLAN ID/EVC map but not capable of transferring traffic among any of the UNIs in the EVC.

10.1 Introduction

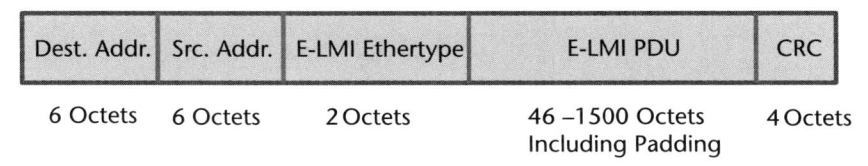

Dest. Addr.	Src. Addr.	E-LMI Ethertype	E-LMI PDU	CRC
6 Octets	6 Octets	2 Octets	46–1500 Octets Including Padding	4 Octets

Figure 10.3 E-LMI framing structure.

- Partially Active is applicable for multipoint-to-multipoint EVCs. When a multipoint-to-multipoint EVC is partially active, it is capable of transferring traffic among some but not all of the UNIs in the EVC.

Tables 10.1 and 10.2 detail the possible combinations of point-to-point EVC status and multipoint-to-multipoint EVC status.

The E-LMI protocol employs two messages, STATUS and STATUS ENQUIRY. Every message of the E-LMI protocol consists of protocol version, message type, report type, and other information elements and subinformation elements. The E-LMI message parts 0, 0, and 0 are common to all the E-LMI messages where each message may have additional information and subinformation elements. The E-LMI message organization example and E-LMI protocol information elements (IEs) are shown in Figure 10.4 and Table 10.3, respectively.

The values of the subinformation elements used for the E-LMI protocol are shown in Table 10.4.

The coding of the information elements other than protocol version and message type is detailed in MEF 16 [1].

Table 10.1 Possible Status Combinations for a Point-to-Point EVC

New	Active	Not Active
X	X	—
X	—	X
—	X	—
—	—	X

Table 10.2 Possible Status Combinations for a Multipoint-to-Multipoint EVC

New	Active	Not Active	Partially Active
X	X	—	—
X	—	X	—
X	—	—	X
—	X	—	—
—	—	X	—
—	—	—	X

8	7	6	5	4	3	2	1	Octet
Protocol Version information element								1
Message Type information element								2
Report Type								3, 4, 5
Other information and sub-information elements as required								6, ...

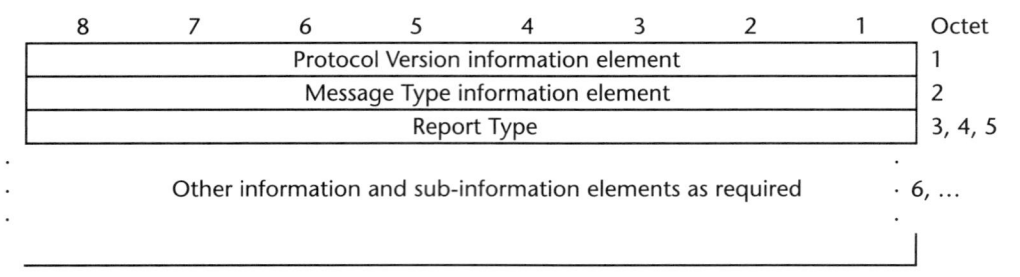

Figure 10.4 General E-LMI message organization example.

Table 10.3 Information Element Identifiers

Information element	Identifier bits 8 7 6 5 4 3 2 1
Protocol version	Not applicable
Message type	Not applicable
Report type	0 0 0 0 0 0 0 1
Sequence numbers	0 0 0 0 0 0 1 0
Data instance (DI)	0 0 0 0 0 0 1 1
UNI status	0 0 0 1 0 0 0 1
EVC status	0 0 1 0 0 0 0 1
CE-VLAN ID/EVC map	0 0 1 0 0 0 1 0

Table 10.4 Subinformation Element Identifiers

Subinformation element	Identifier Bits 8 7 6 5 4 3 2 1
UNI identifier	0 1 0 1 0 0 0 1
EVC parameters	0 1 1 0 0 0 0 1
EVC identifier	0 1 1 0 0 0 1 0
EVC map entry	0 1 1 0 0 0 1 1
Bandwidth profile	0 1 1 1 0 0 0 1

10.2 E-LMI Messages

STATUS message is sent by the UNI-N to the UNI-C in response to a STATUS ENQUIRY message to indicate the status of EVCs or for the exchange of sequence numbers. It may be sent without a STATUS ENQUIRY to indicate the status of a single EVC.

The STATUS message described in Table 10.2 can include EVC service attributes and Parameters that enable automatic configuration of CE devices based upon the network configuration.

EVC status information elements are arranged in the message in ascending order of EVC reference IDs; the EVC with the lowest EVC reference ID is first, the second-lowest EVC reference ID is second, and so on. If all information elements cannot be sent in a single Ethernet frame, more STATUS messages are sent with type of report full status continued. The asynchronous STATUS message must contain a single EVC status information element, which precedes the CE-VLAN ID/EVC information element (IE).

The information elements carried in the STATUS message are sequence number, DI, UNI status, EVC status, and CE-VLAN ID/EVC map. The STATUS message can be FULL or asynchronous. In a full STATUS message, the UNI-C uses the EVC Status information element to detect a change in status of configured EVCs. When the UNI-C sends a STATUS ENQUIRY message with a report type of full status, the UNI-N responds with a STATUS message containing an EVC status information element for each EVC configured at that UNI. Each EVC status information element contains an active bit and a partially active bit indicating the availability or unavailability of that EVC. For a point-to-point EVC, the EVC status is active if and only if the active bit is set to 1. For a multipoint-to-multipoint EVC, the status of the EVC is defined as in Table 10.5.

If the UNI-C receives an EVC status information element indicating that the EVC is not active, the CE stops transmitting frames on the EVC until the UNI-C receives an EVC status information element for that EVC indicating a status of active or partially active.

Since there is a delay between the time the MEN makes an EVC available and the time that the UNI-N transmits an EVC status information element notifying the UNI-C, there is a possibility of the CE receiving frames on an EVC marked as not active.

Similarly, since there is a delay between the time the MEN detects that an EVC has become not active or partially active and the time UNI-N transmits an EVC status information element notifying the UNI-C, there is a possibility of the MEN receiving frames on a not active or partially active EVC and drop of frames.

Asynchronous status is used to notify the UNI-C that the EVC has changed status without waiting for a request from UNI-C. The UNI-N uses the EVC status information element to inform the UNI-C about a change in status of a configured EVC. This STATUS message may be sent when the UNI-N detects the EVC status change, and the report type is set to single EVC asynchronous status.

STATUS ENQUIRY is sent by the UNI-N to request status or to verify sequence numbers. The UNI-C must send a STATUS message in response to a STATUS ENQUIRY message. The STATUS ENQUIRY message consists of protocol version, message type, report type, sequence numbers, and DI. Brief descriptions of each message element are given in Section 10.3.

Table 10.5 Status for a Multipoint-to-Multipoint EVC

Active Bit	Partially Active Bit	Status
1	0	Active
0	0	Not active
0	1	Partially active
1	1	Not defined

10.3 E-LMI Message Elements

Protocol version, which is a 1-octet field, indicates the version supported by the sending entity (UNI-C or UNI-N).

Message type, which is the second part of every message, identifies the function of the E-LMI message being sent. Its coding is in coded as in Table 10.6. Bit 8 is reserved for possible future use as an extension bit.

Report type information element indicates the type of enquiry requested when included in a STATUS ENQUIRY message or the contents of the STATUS message. The length of this information element is 3 octets (Table 10.7).

Sequence numbers information element exchanges sequence numbers between the UNI-N and the UNI-C on a periodic basis. This allows each protocol entity to detect if it has not received messages and to acknowledge receipt of messages to the other entity. The length of this information element is 4 octets (Tabled 10.8).

CE-VLAN ID/EVC map information element conveys how CE VLAN IDs are mapped to specific EVCs (Table 10.9). The maximum number of bytes needed to carry this information element depends on the number of VLAN IDs mapped to an EVC. When the number of octets needed exceeds the maximum length that can be specified in the TLV length octet (255), this information element can be repeated for the same EVC.

The EVC reference ID allows the UNI-C to correlate information received in the CE-VLAN ID/EVC map IE and the EVC status information element to the same EVC. It is a binary encoded number in the range 0 to 65,535. The EVC reference ID is locally significant, which means that a given EVC can have a different value of EVC reference ID at each of the UNIs in the EVC. If the sequence number exceeds the 6-bit counter, it rolls over to zero.

UNI status information element conveys the status and other relevant UNI service attributes of the UNI as defined in MEF 10.3 [2]. This information element cannot be repeated in a STATUS message. The length of this information element depends on the number and size of UNI identifier subinformation element (Tables 10.10 and 10.11).

Table 10.6 Message Type Coding

Bits								
8	7	6	5	4	3	2	1	*Message Type*
0	1	1	1	1	1	0	1	STATUS
0	1	1	1	0	1	0	1	STATUS ENQUIRY

Table 10.7 Report Type Coding

Bits								
8	7	6	5	4	3	2	1	*Report Type*
0	0	0	0	0	0	0	0	Full status
0	0	0	0	0	0	0	1	E-LMI check
0	0	0	0	0	0	1	0	Single EVC asynchronous status
0	0	0	0	0	0	1	1	Full status continued

10.3 E-LMI Message Elements

Table 10.8 Sequence Numbers Information Element

8 7	6 5	4 3 2 1	Octet					
8	7	6	5	4	3	2	1	Octet
Sequence numbers information element identifier per Table 10.3								1
Length of sequence numbers contents (= 00000010)								2
Send sequence number								3
Receive sequence number								4

Table 10.9 CE-VLAN ID/EVC Map Information Element

8	7	6	5	4	3	2	1	Octet
CE-VLAN ID/EVC map information element identifier per Table 10.3								1
Length of CE-VLAN ID/EVC Map information element								2
EVC Reference ID								3
EVC Reference ID–continue								4
Reserve 0	Last IE	CE-VLAN ID/EVC Map Sequence #						5
Reserve 0						Untagged/ Priority Tagged	Default EVC	6
EVC Map Entry Sub-Information element								7–10

EVC status information element (Tables 10.12 and 10.13) conveys the status and attributes of a specific EVC on the UNI. This information element can be repeated, as necessary to indicate the status of all configured EVCs on the UNI.

DI information element (Table 10.14) reflects the current state of UNI and EVC information that is active on the UNI-N and UNI-C. Whenever there is mismatch in DI, it is time to exchange UNI and EVC information between UNI-N and UNI-C. The format is shown in Table 10.14.

Table 10.10 UNI Status Information Element

8	7	6	5	4	3	2	1
UNI status information element identifier per Table 10.3							
Length of UNI status information element contents							
CE-VLAN ID/EVC map type							
Bandwidth profile subinformation element per Section 0							
UNI identifier subinformation element per Section 0							

Table 10.11 CE-VLAN ID/EVC Map Type Coding

Bits 8 7 6 5 4 3 2 1	CE-VLAN ID/EVC Map Type
0 0 0 0 0 0 0 1	All to one bundling
0 0 0 0 0 0 1 0	Service multiplexing with no bundling
0 0 0 0 0 0 1 1	Bundling

Table 10.12 EVC Status Information Element

8	7	6	5	4	3	2	1	Octet
EVC Status information element identifier per Information element							Identifier bits 8 7 6 5 4 3 2 1	1
Protocol version							Not applicable	
Message type							Not applicable	
Report Type							0 0 0 0 0 0 0 1	
Sequence Numbers							0 0 0 0 0 0 1 0	
Data Instance (DI)							0 0 0 0 0 0 1 1	
UNI Status							0 0 0 1 0 0 0 1	
EVC Status							0 0 1 0 0 0 0 1	
CE-VLAN ID/EVC Map Table							0 0 1 0 0 0 1 0	
Length of EVC Status information element								2
EVC Reference ID								3
EVC Reference ID (Continue)								4
EVC Status Type								5
Reserve 0	Reserve 0	Partially active	Active	New				
EVC Parameters subinformation element								6
EVC ID subinformation element								7
Bandwidth profile subinformation element								8

Table 10.13 EVC Status Coding

Bits 3 2 1	EVC Status
0 0 0	Not active
0 0 1	New and not active
0 1 1	New and active
0 1 0	Active
1 0 0	Partially active
1 0 1	New and partially active

Bandwidth profile subinformation element (Table 10.15) conveys the characterization of the length and arrival for a sequence of the service frames at a reference point UNI. This subinformation element is included in the UNI status and EVC status information elements. It can be repeated, up to eight times in an EVC status information element, when there are eight per CoS identifier bandwidth profiles. The bandwidth profile information elements can be repeated up to 8 times in the EVC status IE and appear one time in the UNI IE. When this subinformation element appears in the UNI IE, this bit is set to 0. When this subinformation appears in the EVC Status IE, if this bit is set to zero, then the bandwidth profile is a per-EVC bandwidth profile.

10.3 E-LMI Message Elements

Table 10.14 Data Instance Information Element Format

8	7	6	5	4	3	2	1	Octet
\multicolumn{8}{l}{Data Instance information element identifier per table}	1
Length of DI information element contents (00000101)								2
Reserved 0								3
Data Instance								4
Data Instance – continue								5
Data Instance – continue								6
Data Instance – continue								7

Data instance: Any integer value packed in 4 bytes. Value "0x00000000" is unique and will be used by UNI-C to send its first message to UNI-N. Value "0x00000000" is never sent by the UNI-N.

Table 10.15 Bandwidth Profile Subinformation Element

8	7	6	5	4	3	2	1	Octet
Bandwidth profile subinformation element								1
Length of bandwidth profile subinformation element contents (= 00001100)								2
Reserve 0					CM	CF	Per CoS bit	3
CIR Magnitude								4
CIR Multiplier in binary. CIR = (CIR multiplier) * $10^{(CIR\ magnitude)}$ [Kbps]								4.1 / 4.2
CBS magnitude								5
CBS multiplier CBS = (CBS multiplier) * $10^{(CBS\ magnitude)}$ [Kb].								5.1
EIR magnitude								6
EIR multiplier in binary. EIR = (EIR multiplier) * $10^{(EIR\ magnitude)}$ (Kbps)								6.1 / 6.2
EBS magnitude								7
EBS multiplier EBS = (EBS multiplier) * $10^{(EBS\ magnitude)}$ (Kb)								7.1
user_priority bits 111	user_priority bits 110	user_priority bits 101	user_priority bits 100	user_priority bits 011	user_priority bits 010	user_priority bits 001	user_priority bits 000	8

The coding of the various fields in the bandwidth profile subinformation element is shown in Table 10.16.

EVC Map Entry subinformation element (Table 10.17) is to specify one or more CE-VLAN ID values of 1, 2, ..., 4,095.

The UNI identifier subinformation element (Table 10.18) conveys the value of UNI identifier.

The EVC identifier subinformation element (Table 10.19) conveys the value of EVC identifier.

The EVC parameters subinformation element conveys the service attributes of an existing EVC on the UNI. This subinformation element can be repeated, as necessary, in a STATUS message to indicate the service attributes of all configured EVCs on the UNI. The format and coding of this information element are shown in Tables 10.20 and 10.21.

Table 10.16 Coding in Bandwidth Profile Subinformation Element

Field Name	Value	Meaning
Per CoS bit (octet 3, bit 1)	0	user_priority bit values are ignored and not processed
	1	user_priority bit values are significant
Coupling flag (CF) (octet 3, bit 2)	0	Coupling flag not set
	1	Coupling flag set
Color mode flag (CM) (octet 3, bit 3)	0	Color mode flag is not set
	1	Color mode flag is set
user_priority bits 000 (octet 8, bit 1)	0	Bandwidth profile does not apply to frames with user_priority = 000
	1	Bandwidth profile applies to frames with user_priority = 000
user_priority bits 001 (octet 8, bit 2)	0	Bandwidth profile does not apply to frames with user_priority = 001
	1	Bandwidth profile applies to frames with user_priority = 001
user_priority bits 010 (octet 8, bit 3)	0	Bandwidth profile does not apply to frames with user_priority = 010
	1	Bandwidth profile applies to frames with user_priority = 010
user_priority bits 011 (octet 8, bit 4)	0	Bandwidth profile does not apply to frames with user_priority = 011
	1	Bandwidth profile applies to frames with user_priority = 011
user_priority bits 100 (octet 8, bit 5)	0	Bandwidth profile does not apply to frames with user_priority = 100
	1	Bandwidth profile applies to frames with user_priority = 100
user_priority bits 101 (octet 8, bit 6)	0	Bandwidth profile does not apply to frames with user_priority = 101
	1	Bandwidth profile applies to frames with user_priority = 101
user_priority bits 110 (octet 8, bit 7)	0	Bandwidth profile does not apply to frames with user_priority = 110
	1	Bandwidth profile applies to frames with user_priority = 110
user_priority bits 111 (octet 8, bit 8)	0	Bandwidth profile does not apply to frames with user_priority = 111
	1	Bandwidth profile applies to frames with user_priority = 111

Table 10.17 EVC Map Entry Subinformation Element

8	7	6	5	4	3	2	1	Octet
\multicolumn{8}{l}{EVC Map Entry sub-information element identifier per}	
Subinformation element				Identifier bits 8 7 6 5 4 3 2 1				1
UNI identifier				0 1 0 1 0 0 0 1				
EVC parameters				0 1 1 0 0 0 0 1				
EVC identifier				0 1 1 0 0 0 1 0				
EVC map entry				0 1 1 0 0 0 1 1				
Bandwidth profile				0 1 1 1 0 0 0 1				
Table								
Length of EVC map entry contents								2
CE-VLAN ID								3
								4
CE-VLAN ID								5
								6
...								...

10.4 E-LMI System Parameters and Procedures

The behavior of the E-LMI protocol is defined by a set of procedures that need to be carried out based on the events at the CE and the CEN, and the received E-LMI

10.4 E-LMI System Parameters and Procedures

Table 10.18 UNI Identifier Subinformation Element

8	7	6	5	4	3	2	1	Octet
UNI Identifier subinformation element identifier per								1
Length of UNI identifier contents								2
ASCII octet								3
ASCII octet								4
...								...
...								...

Table 10.19 EVC Identifier Subinformation Element

8	7	6	5	4	3	2	1	Octet
EVC Identifier subinformation element identifier per								

Subinformation element	Identifier bits 8 7 6 5 4 3 2 1	1
UNI identifier	0 1 0 1 0 0 0 1	
EVC parameters	0 1 1 0 0 0 0 1	
EVC identifier	0 1 1 0 0 0 1 0	
EVC map entry	0 1 1 0 0 0 1 1	
Bandwidth profile	0 1 1 1 0 0 0 1	

Table								
Length of EVC identifier contents								2
ASCII octet								3
ASCII octet								4
...								...

Table 10.20 EVC Parameters Subinformation Element

8	7	6	5	4	3	2	1	Octet
EVC parameters subinformation element identifier								1
Length of EVC parameters subinformation element contents								2
Reserve 0				EVC Type				3

Table 10.21 EVC Type Coding

Bits			
3	2	1	EVC Type
0	0	0	Point-to-point EVC
0	0	1	Multipoint-to-multipoint EVC

messages or protocol data units (PDUs) by the UNI-C and the UNI-N.

The E-LMI procedures are characterized by a set of E-LMI messages that will be exchanged at the UNI. These message exchanges can be asynchronous or periodic.

Periodic message exchanges are governed by timers, status counters, and sequence numbers.

The polling timer (T391) is the timer for UNI-C to poll status of UNI and EVCs (full status) via STATUS ENQUIRY messages. The timer starts with transmission of STATUS ENQUIRY message. If STATUS message is not received, an error is recorded.

T391 can take values from 5 seconds to 30 seconds; 10 seconds is the recommended default value for T391.

The polling verification timer (T392) is the timer for UNI-N to respond STATUS ENQUIRY messages. T392 starts with transmission of STATUS message and stops with receiving STATUS ENQUIRY. It is greater than T391 and takes values of 5 to 30 seconds with a recommended default of 15 seconds. If a STATUS ENQUIRY message is not received, an error is recorded.

The polling counter (N391) is a counter of UNI-C for full STATUS polling counts. It takes values in the range of 1 to 65,000 with a recommended default of 360.

The status counter (N393) is a counter for both UNI-C and UNI-N for consecutive error counts.

Sequence numbers allow the UNI-N and the UNI-C to determine the status of the E-LMI process including correlating STATUS ENQUIRY messages with STATUS messages.

The UNI-C and the UNI-N maintain the send and receive internal counters. The send sequence counter maintains the value of the send sequence number field of the last sequence numbers information element sent. The receive sequence counter maintains the value of the last received send sequence number field in the sequence numbers information element and maintains the value to be placed in the next transmitted received sequence number field.

The value zero in the receive sequence number indicates that the receive sequence number field contents are undefined; this value is normally used after initialization. Figure 10.5 shows an example of the use of the send and receive sequence numbers.

10.4.1 Periodic Polling Process

The periodic polling process can be summarized as:

1. At least every T391 seconds, the UNI-C sends a STATUS ENQUIRY message to the UNI-N.
2. At least every N391 polling cycles, the UNI-C sends a full status STATUS ENQUIRY.
3. The UNI-N responds to each STATUS ENQUIRY message with a STATUS message and resets the polling verification timer, which is used by the UNI-N to detect errors. If the UNI-C sends a STATUS ENQUIRY requesting full status, the UNI-N responds with a STATUS message with the report type specifying full status. The STATUS message sent in response to a STATUS ENQUIRY contains the sequence numbers and report type information elements.

If the UNI-N cannot fit EVC status information elements and service attributes and parameters for all EVCs into a single full status STATUS message, the UNI-N

10.5 UNI-C and N Procedures

```
                STATUS ENQUIRY(SND Sq Cntr, RCV Sq Cntr)

         CE                                                      CEN
          |         STATUS ENQUIRY(5,4)                            |
          |─────────────────────────────────────────────────────→ |
   PT     |         STATUS (5,5)                                   |
 expires  | ←───────────────────────────────────────────────────── |
          |         STATUS ENQUIRY(6,5)                            |
          |─────────────────────────────────────────────────────→ |
          |         STATUS (6,6)                                   |
          | ←───────────────────────────────────────────────────── |
          |         STATUS ENQUIRY(m, n)                           |
   PT     |─────────────────────────────────────────────────────→ |
 expires  |         STATUS (n+1, m)                                |
          | ←───────────────────────────────────────────────────── |
                              PT= PollTimer
```

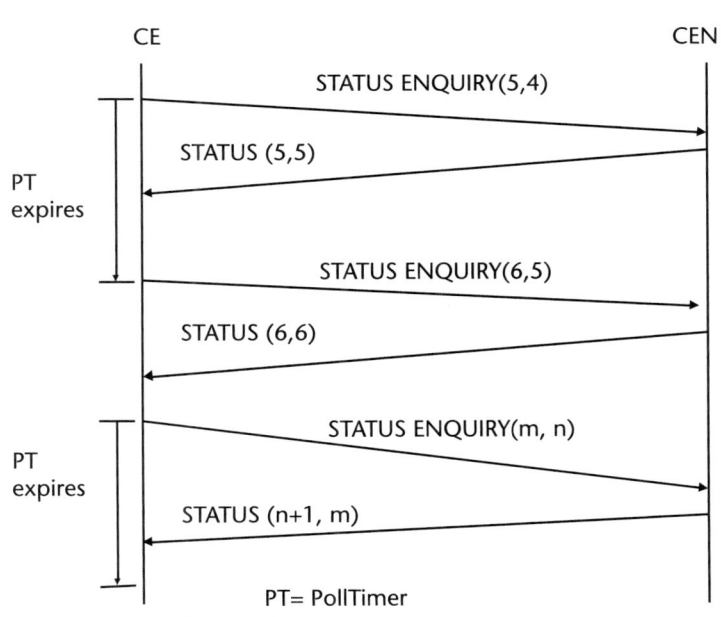

Figure 10.5 E-LMI status check example.

responds with a full status continued STATUS message, containing as many EVC status information elements as allowed by the Ethernet frame size.

If the STATUS message is a full status or full status continued STATUS message, the UNI-C updates its configuration according to the status of the UNI and the status and service attributes of each configured EVC.

Upon receipt of a full status continued STATUS message, the UNI-C continues to request EVC status by sending a full status continued STATUS ENQUIRY message (without waiting for the polling timer to expire). The UNI-C S restarts polling timer with value T391 each time it transmits a full status continued STATUS ENQUIRY message. When the UNI-N responds with a full status STATUS message, it is an indication that all information has been sent.

The UNI-C compares the EVC reference ID sent in the full status with the previously reported EVC reference ID.

In addition to triggering full status and full status continued reports every N391 polling cycles, the data instance is used to trigger such reports each time there is a change in EVC or UNI information.

10.5 UNI-C and N Procedures

When the UNI-C comes up for first time or is restarted, the UNI-C sets its DI to 0 and sends the ELMI STATUS ENQUIRY with report type full status.

The UNI-C will then receive full status or full status continued reports, including the latest UNI and EVC information and update its local database. After the reports, the UNI-C's local DI is set to the UNI-N DI that is received in the full status report. The DI value will stay the same until the status procedure is complete.

When the UNI-N first comes up, it sets its DI value to a nonzero value that is different from the DI value received in the first message received from the UNI-C. Any change in information related to UNI or EVC including status change results in incrementing DI value to reflect the change in data.

For ELMI check, on receipt of ELMI STATUS ENQUIRY, with ELMI check, UNI-N responds with ELMI STATUS and includes the current value of DI.

E-LMI notifies the UNI-C of a newly added EVC using a full status STATUS message.

In addition, the UNI-N and the UNI-C use the information provided by periodic polling for error monitoring and detect reliability errors (i.e., nonreceipt of STATUS/STATUS ENQUIRY messages or invalid sequence numbers in a sequence numbers information element) and protocol errors. The UNI-N and the UNI-C ignore messages (including their sequence numbers) containing these errors.

Unrecognized information and subinformation elements are ignored by both the UNI-C and UNI-N. When an error is detected, the appropriate management entity is notified.

10.6 Conclusion

LMI is a protocol terminated by the UNI-C of the CE and by the UNI-N of the CEN. It provides the status of an EVC and UNI and enables the CE to configure itself to access Carrier Ethernet services. E-LMI messages, STATUS and STATUS ENQUIRY, are encapsulated inside Ethernet frames. This chapter described E-LMI message elements and procedures between UNI-C and UNI-N. E-LMI is not widely used in the industry today. Demarcation devices supporting E-LMI are few. Service providers are not eager to provide E-LMI services either. This might change as operation support systems that can manage E-LMI become available and user demand for E-LMI picks up.

References

[1] MEF 16, Ethernet Local Management Interface, January 2006.
[2] MEF 10.3, Ethernet Services Attributes Phase 3, October 2013.

CHAPTER 11

Provider Bridges, Provider Backbone Bridges, and Provider Backbone Transport

11.1 Introduction

Carrier Ethernet has proven itself as an alternative technology for metro area networks (MANs) and wide area networks (WANs) in addition to enterprise networks. In order for Carrier Ethernet to be an alternative for backbone, reliability, and scalability needed to be supported in addition to its QoS and ENNI capabilities. In addition to simplicity compared to alternatives, more features have been added to Carrier Ethernet, while keeping the cost low.

Core networks connect networks of different cities, regions, countries, or continents. The complexity of these technologies imposes substantial financial burdens on network operators, both in the area of capital expenditures (CAPEX) and operational expenditures (OPEX). Low-cost Ethernet is very attractive. New extensions to the Ethernet protocols [1–4] have been developed to transform Ethernet to a technology ready for use in MANs/WANs.

Services supported in the Carrier Ethernet network are end-to-end. This results in no changes to the customer's LAN equipment, providing end-to-end usage of the technology, contributing to wider interoperability and low cost. SLAs provide end-to-end performance, based on rate, frame loss, delay, and jitter, and enable traffic engineering (TE) to fine-tune the network flows.

A high degree of scalability is needed for handling different traffic types and for user separation inside the network, especially, the scalability in terms of address space, maximum transmission speed, and maximum transmission distance. High bandwidth and various tagging and tunneling capabilities help Ethernet to support scalability.

Network errors needed to be detected and repaired, without the users noticing it. It had to be possible to monitor, diagnose, and centrally manage the network with standard, vendor-independent implementations and to provide services rapidly. Carrier-class OAM needed to be implemented. A 50-ms recovery time during failures that has been inherited from SONET is also considered one of the parameters of reliability for Ethernet networks.

PBT is a group of enhancements to the Ethernet that are defined in the IEEE's Provider Backbone Bridging Traffic Engineering (PBBTE). PBBTE separates the

Ethernet service layer from the network layer thus enabling the development carrier-grade public Ethernet services. Its capabilities may be summarized as:

- Enabling traffic engineering in Ethernet networks. PBT tunnels reserve appropriate bandwidth and support the provisioned QoS metrics that guarantee that SLAs will be met without having to overprovision network capacity.
- Multiplexing of any service inside PBT tunnels, including Ethernet and MPLS services. This flexibility allows service providers to deliver native Ethernet initially and MPLS-based services (VPWS, VPLS) if and when they require.
- Protecting a point-to-point Ethernet tunnel by allowing to provision an additional backup tunnel to provide resiliency.
- Scalability by turning off MAC learning features, the undesirable broadcast functionality that creates MAC flooding and limits the size of the network is removed. Additionally, PBT offers a full 60-bit addressing scheme that enables virtually limitless numbers of tunnels to be set up in the service provider network.
- Better service management by having more effective alarm correlation, service-fault correlation, and service-performance correlation, as a result of the fact that the network knows both the source and destination address in addition to the route since each packet is self-identifying.

The underlying protocols for PBT are as follows:

- 802.1AB [5] Link Layer Discovery Protocol, which is used to discover the network layout and forwarding this information to the control plane or management layer;
- The 802.1ag [1] protocol to monitor the links and trunks in the PBT layout;
- PBT, which is an expansion of the PBB protocol, IEEE 802.1ah [2];
- 802.1Qay [3], provider backbone bridges with traffic engineering or PBT protocol.

In the following sections, we will describe 802.1AB and 802.1ah protocols first and then PBT-TE.

11.2 IEEE 802.1AB

The 802.1AB standards (Station and Media Access Control Connectivity Discovery) define the Link Layer Discovery Protocol (LLDP) that is designed for advertising information to stations attached to the same 802 LAN, to populate physical topology and device discovery management information databases. The protocol facilitates the identification of stations connected by IEEE 802 LANs/MANs, their points of interconnection, and access points for management protocols.

The protocol can operate with all IEEE 802 access protocols and network media. Network management information schema and object definitions of LLDP are suitable for storing connection information about adjacent stations and compatible with the IETF PTOPO Management Information Base (MIB) [6].

An IEEE 802.1AB enabled device advertises its system information to its neighbors. The neighborssave the system information in a MIB, which can then be used by a management protocol such as SNMP. With the MIB, the network management system knows the network topology, which systems are connected to it, and port status.

Normally a network management system would discover the network status. The purpose of allowing the network to discover itself and then tell it to a management station is to:

- Facilitate multivendor interoperability and use of standard management tools to discover and make available physical topology information for network management;
- Make it possible for network management to discover certain configuration inconsistencies or malfunctions that can result in impaired communication at higher layers;
- Provide information to assist network management in making resource changes and/or reconfigurations that correct configuration inconsistencies or malfunctions.

In the next sections, we will describe 802.1AB architecture, LLDP protocol, and MIB.

11.2.1 Architecture

LLDP is a link layer protocol and optional part of 802 LAN protocol stack. It is a media independent protocol intended to be run on all IEEE 802 devices, allowing a LLDP agent to learn higher layer management reachability and connection endpoint information from adjacent devices. It runs on all IEEE 802 media and over the datalink layer only, allowing two systems running different network layer protocols to learn about each other.

Architecturally, LLDP runs on top of the uncontrolled port of an IEEE 802 MAC client. It may be run over an aggregated MAC client as described in [7]. It may be desirable for stations to prohibit the transmission of LLDP PDUs over the uncontrolled port until the controlled port has been authorized. The spanning tree state of a port does not affect the transmission of LLDP PDUs.

Current applications use ports at transport (TCP/UDP) protocol layer to identify points of access in a machine and to define connections using endpoint service identifiers (pair protocol, port). At LLC over Ethernet, this function is performed using link service access points (LSAP), that provide interface ports for users above logical link control (LLC) sublayer.

Each LLDP agent (Figure 11.1) is responsible for causing the following tasks to be performed:

- Maintaining current information in the LLDP local system MIB.

- Extracting and formatting LLDP local system MIB information for transmission to the remote port
- Provide LLDP agent(s) status at regular intervals or whenever there is a change in the system condition or status.
- Recognizing and processing received LLDP frames.
- Maintaining current values in the LLDP remote system MIB.
- Using the procedure somethingChangedLocal() and the procedure somethingChangedRemote() to notify the PTOPO MIB manager and MIB managers of other optional MIBs whenever a value status change has occurred in one or more objects in the MIBs depicted in Figure 11.1.

MIB objects are generally accessed through SNMP. Objects in the MIB are defined using the mechanisms of the structure of management information (SMI). The LLDP MIB manager is responsible for updating and maintaining each LLDP remote systems MIB associated with the LLDP agent.

11.2.2 Principles of Operation

LLDP runs over the datalink layer; therefore, devices that use different network layer protocols can still identify each other. An 802-based LAN with LLDP contains agents that use the services of the LLC and the MAC address to transmit and receive information from and to other LLDP agents. It uses a special multicast MAC address to send the information to other agents.

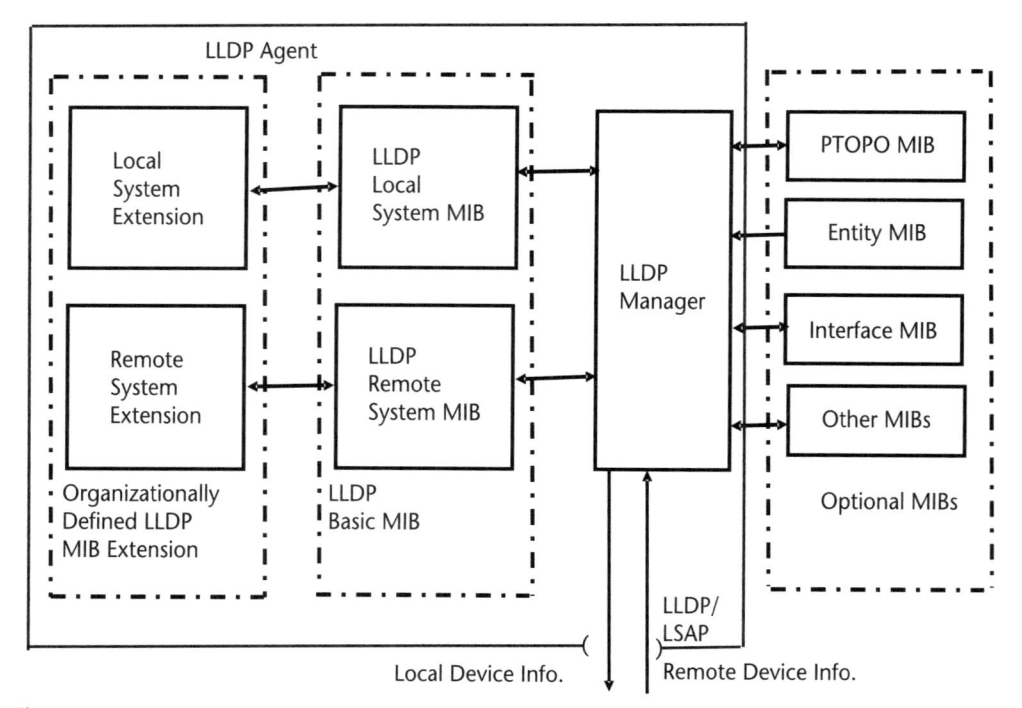

Figure 11.1 LLDP components.

11.2 IEEE 802.1AB

On Ethernet networks, LLDP header and messages are encapsulated in an 802 Slow Protocol frame, which includes a multicast destination address and the source MAC address of the transmitting interface. Although LLDP is defined as an Ethernet protocol, any Layer-2 protocol can use it.

The transmission of the Link Layer Discovery Protocol Data Unit (LLDPDU) can be triggered by two factors:

- The expiration of a timer that is the transmit countdown timing counter, and status or value changes in one of the information elements in the local system.
- The LLDP agent can transmit information as well as receive information about the remote systems. It is possible to turn either the receiving or the sending functions of the agent off. This makes it possible to configure an implementation, which restricts a local LLDP agent either to transmit or receive only or to do both.

When the transmission of the LLDPDU has been started, the local system will put its new information from its MIB in a TLV (type, length, and value). The TLVs are inserted into an LLDPDU, which will be transmitted using the special multicast address.

When an LLDP receive module receives the incoming LLDPDU, it will recognize LLC entity and MAC address. It will then use the information from different TLVs to fill the LLDP remote systems MIB.

The LLDP protocol is essentially a one-way protocol. Each device configured with an active LLDP agent sends periodic messages to the slow protocols multicast MAC address. The device sends the periodic messages on all physical interfaces enabled for LLDP transmission, and listens for LLDP messages on the same set on interfaces. Each LLDP message contains information identifying the source port as a connection end-point identifier. It also contains at least one network address that can be used by an NMS to reach a management agent on the device via the indicated source port. Each LLDP message contains a configurable time-to-live value, which tells the recipient LLDP agent when to discard each element of learned topology information. Additional optional information may be contained in LLDP PDUs to assist in the detection of configuration inconsistencies.

The LLDP protocol is designed to advertise information useful for discovering pertinent information about a remote peer and to populate topology management information as defined in [6]. It is not intended to act as a configuration protocol for remote devices or as a mechanism to signal control information between peers.

During the operation of LLDP it may be possible to discover configuration inconsistencies between devices on the same physical LAN. This protocol does not provide a mechanism to resolve those inconsistencies, rather a means to report discovered information to higher-layer management entities.

11.2.3 802.1AB Frame Format

The frame format of the 802.1AB standard is derived from a normal Ethernet frame and contains:

- Destination address is the special LLDP multicast address 01-80-C2-00-00-0E,
- Source address contains the MAC-address of the sending port.
- EtherType is the LLDP ethertype, 88-CC.

The information fields in each LLDP frame are contained in an LLDPDU as a sequence of short, variable length, information elements known as TLVs that each include type, length, and value fields where:

- Type identifies what kind of information is being sent.
- Length indicates the length of the information string in octets.
- Value is the actual information that needs to be sent (for example, a binary bit map or an alphanumeric string that can contain one or more fields).

Each LLDPDU includes four mandatory TLVs and optional TLVs as selected by network management:

- Chassis ID TLV identifies the chassis that contains the LAN station associated with the transmitting LLDP agent. It is also used for the MAC service access point (MSAP), which is used to identify a system.
- Port ID TLV contains the port component of the MSAP identifier associated with the transmitting LLDP agent.
- Time-to-live TLV contains an integer value in the range 0 to 65,535 seconds, which is used to indicate the time to live of the LLDPDU. When this time reached zero, the LLDPDU should be thrown away. Normally, the LLDPDU is renewed before this exceeding of time.
- From zero to n optional TLVs, as allowed by the space limitation of the LLDPDU, the optional TLVs can be inserted in any order.
- The end of LLDPDU TLV is the last TLV in the LLDPDU and is used to mark the end of the LLDPDU.

The chassis ID and the port ID values are concatenated to form a logical MSAP identifier that is used by the recipient to identify the sending LLDP agent/port. Both the chassis ID and port ID values can be defined in a number of convenient forms such as port 1 (portID) and IP-Phone (ChassisID), as defined in LLDP MIB [8].

A zero value in the TTL field of time-to-live TLV tells the receiving LLDP agent how long all information pertaining to this LLDPDU's MSAP identifier will be valid so that all the associated information can later be automatically discarded by the receiving LLDP agent if the sender fails to update it in a timely manner.

The currently defined optional TLVs may be used to describe the system and/or to assist in the detection of configuration inconsistencies associated the MSAP identifier: IEEE 802.1 Organizationally Specific TLV set contains Port VLAN ID TLV, Port and Protocol VLAN ID TLV, VLAN Name TLV, and Protocol Identity TLV. An IEEE 802.3 organizationally specific TLV set contains:

- MAC/PHY configuration/status TLV (indicates the auto-negotiation capability and the duplex.

- Speed of IEEE 802.3 MAC (PHYs).
- Power via MDI TLV indicating the capabilities and current status of IEEE 802.3 PMDs that either require or are able to provide power over twisted-pair copper links.
- Link aggregation TLV indicating the current link aggregation status of IEEE 802.3 MACs.
- Maximum frame size TLV indicating the maximum supported IEEE 802.3 frame size.

Organizationally specific TLVs (see Figure 11.2) can be defined by either the professional organizations or the individual vendors that are involved with the particular functionality being implemented within a system. The basic format and procedures for defining organizationally specific TLVs are provided below.

Information for constructing the various TLVs to be sent is stored in the LLDP local system MIB. The selection of which particular TLVs to send is under control of the network manager. Information received from remote LLDP agents is stored in the LLDP remote systems MIB. The LLDP remote system's MIB is updated after all TLVs have been validated.

The LLDPDU is checked to ensure that it contains the correct sequence of mandatory TLVs and then each optional TLV is validated in succession. LLDPDUs and TLVs that contain detectable errors are discarded.

LLDP is designed to operate in conjunction with MIBs defined by IETF, IEEE 802, and others. LLDP agents automatically notify the managers of these MIBs whenever there is a value or status change in an LLDP MIB object.

LLDP-managed objects for the local system are stored in the LLDP local system MIB. Information received from a remote LLDP agent is stored in the local LLDP agent's LLDP remote system MIB. LLDP MIBs are the following:

- IETF physical topology in [6] allows an LLDP agent to expose learned physical topology information, using a standard MIB. LLDP is intended to support the PTOPO MIB.
- IETF entity MIB in [9] allows the physical component inventory and hierarchy to be identified. Chassis IDs passed in the LLDPDU may identify entPhysicalTable entries. SNMP agents that implement the LLDP MIB should implement the entPhysicalAlias object from the Entity MIB version 2 or higher.

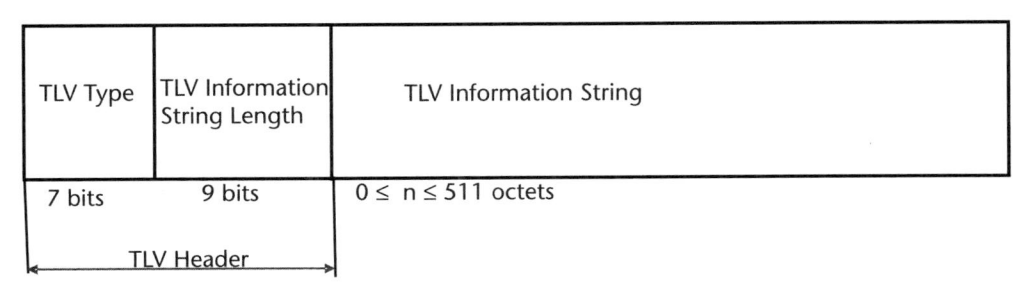

TLV Type	TLV Information String Length	TLV Information String
7 bits	9 bits	$0 \leq n \leq 511$ octets

TLV Header

Figure 11.2 Organizationally specific TLV format.

- IETF interfaces MIB in [8] provides a standard mechanism for managing network ports. Port IDs passed in the LLDPDU may identify ifTable (or entPhysicalTable) entries. SNMP agents that implement the LLDP MIB should also implement the ifTable and ifXTable for the ports that are represented in the Interfaces MIB.

11.3 Provider Backbone Bridges

Provider backbone bridges (PBBs), namely MAC-in-MAC encapsulation, is based on IEEE 802.1ah [2]. PBB supports complete isolation of individual client-addressing fields as well as isolation from address fields used in the operator backbone. Client provider bridge (PB) frames are encapsulated and forwarded in the backbone network based on new B-DA (backbone destination address), B-SA (backbone source address) and B-VID (backbone VLAN-ID).

Although Q-in-Q supports a tiered hierarchy (i.e., no tag, C-Tag/C-VLAN ID and S-Tag/S-VLAN ID), the service provider can create 4,094 customer VLANs, which is insufficient for large metropolitan and regional networks. The 802.1ah introduces a new 24-bit tag field (I-SID), service instance identifier, to overcome 12-bit S-VID (S-VLAN ID) defined in PB. This 24-bit tag field is proposed as a solution to the scalability limitations encountered with the 12 bit S-VID defined in PBs.

PBB service is connectionless. The Multiple Spanning Tree Protocol (MSTP) is used to prevent loops. MSTP differs from STP in that it is capable to allow frames assigned to different VLANs to follow separate paths through the network. This is based on different MSTP instances, within various regions of LANs. These regions are all connected into a single spanning tree, the common spanning tree (CST). VLAN tags are reserved on a network, rather than a per-port basis. Flooding is used when destination MAC addresses are not recognized.

PBBs operate the same way as traditional Ethernet bridges. IEEE 802.1ag connectivity fault management (CFM) addresses the end-to-end OAM such as loopback at specific MAC, link trace, and continuity check.

PBB implementation uses the MAC addresses of the Ethernet UNIs (ingress ports), rather than customer MACs in the switch forwarding tables. This eliminates the MAC address explosion by greatly reducing the number of MAC addresses that must be learned and maintained by switches in the service provider's core infrastructure.

Keeping the number of MAC addresses to a minimum reduces the aging out and relearning of MAC addresses, thus enhancing end-to-end performance and making network forwarding far more stable.

The MAC learning process observes the MAC addresses of frames received at each port and updates filtering database conditionally on the state of receiving port.

PBB networks are managed via SNMP. Separation of addresses of both customer and service providers creates a secure environment. When one side needs to make changes to its network, this side should not have to worry about the overlapping MAC addresses or VLANs. This leads to a simpler operation. There are no forwarding loops and broadcast storms between customer and service providers

11.3 Provider Backbone Bridges

that lead to robustness. By not trading addresses between different tiers would reduce memory and processing power, eventually reduces capital expenditure.

11.3.1 802.1ah Frame Format

Evolution of headers from 802.1 to 802.1ah is depicted in Figure 11.3.

The upper side of the frame (payload to customer DA) is the part that is used in a provider bridges-enabled network [10]. The lower part is the part that is added by a PBEB (PB edge bridge) when the frame travels through a backbone network.

Extra MAC header added by the 802.1ah protocol can operate as a normal Ethernet frame. Therefore, it can be used by Ethernet devices, which do not have PBB-enabled.

Backbone source and destination address have the same function as they would have in normal Ethernet, just as the EtherType and backbone-VLAN ID. The next field contains the instance-tag (I-TAG) that comes in two flavors, which are used by the PBEBs: short service instance TAG and long service instance TAG.

Short service instance TAG (32 bits) consists of the 24-bit I-SID, which identifies the service in the provider's network, and 4 control fields:

- Priority indicates the customer frame's priority.
- Drop eligible is a 1-bit field that indicates the customer drop eligibility.
- Res1 is a 2-bit field that is used for any future format variations.
- Res2 is a 2-bit field that is reserved for future format variations.

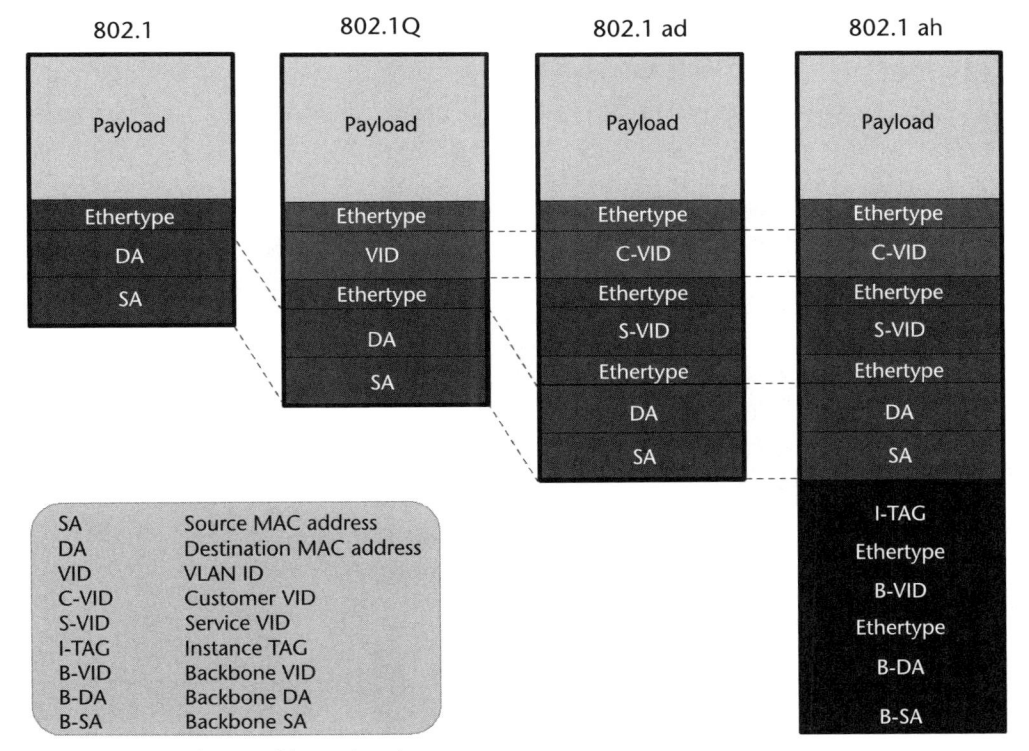

Figure 11.3 Evolution of frame headers.

Long service instance TAG (128 bits) consists of the short tag, including the customer source and destination address. The purpose of the long service instance TAG is to indicate that an Ethernet frame is encapsulated inside the PBB header, while the short service instance TAG is intended for multiprotocol encapsulation. The fields are distinguished by a different EtherType. The EtherType can be used by PBEBs to known what kind of frame is encapsulated (DIX, 802.3, and so forth)

The frame format is depicted in Figure 11.4. Each field is described here:

- Payload;
- EtherType (802.1Q= 81-00)(16 bits);
- Customer VLAN ID (16 bits);
- EtherType (802.1ad= 88-a8)(16 bits);
- Service VLAN ID (16 bits);
- EtherType (standard EtherType, for example, EtherType IPv4 (08-00) or IPv6 (86-DD)) (ts)16 b;
- Customer-SA (MAC) (48 bits);
- Customer-DA (MAC) (48 bits);
- I-TAG (Short I-TAG: 32 bits, Long I-TAG: 128 bits);
- EtherType (can be used to specify what is encapsulated) (16 bits);
- Backbone VLAN ID (16 bits);
- EtherType (specifies whether Long or Short I-TAG) (16 bits);
- Backbone-SA (MAC) (48 bits);
- Backbone-DA (MAC) (48 bits);
- B-TAG-backbone tag, which is identical to S-TAG and optional in the frame;
- I-TAG-Service Instance Tag, which is optional in frame and encapsulates customer addresses, introduces service instance identifier (I-SID) that allows each backbone edge bridges (BEBs) to support a number of backbone service instances and permits the unambiguous identification up to 224backbone service instances within a single provider backbone bridge network (PBBN).

Long service instance tag indicates that an Ethernet frame is encapsulated inside the PBB header, while the short service instance tag is intended for multiprotocol encapsulation. The fields are distinguished by different EtherType. The EtherType (before the I-Tag field) can be used by PBEBs to know the type of frame that is encapsulated such as DIX and 802.3. The I-TAG frame consists of the following fields:

- *Priority code point (I-PCP):* This 3-bit field encodes the priority and drop-eligible parameters of the service request primitive associated with this frame

| B-DA | B-SA | B-TAG | I-TAG | L/T | User Data | FCS |

Figure 11.4 PBB frame format.

using the same encoding as specified for VLAN tags in Chapter 6. The PBBN operates on the priority associated with the B-TAG.

- *Drop eligible indicator (I-DEI):* This 1-bit field carries the drop-eligible parameter of the service request primitive associated with this frame. The PBBN operates on the drop eligibility associated with the B-TAG.
- *No customer addresses (NCA):* This 1-bit field indicates whether the C-DA and C-SA fields of the tag contain valid addresses. A value of 0 indicates that the C-DA and C-SA fields contain valid addresses. A value of 1 indicates that the C-DA and C-SA fields do not contain valid addresses.
- *Reserved 1 (Res1):* This 1-bit field is used for any future format variations. The Res1 field contains a value of 0 when the tag is encoded and is ignored when the tag is decoded.
- *Reserved 2 (Res2):* This 2-bit field is used for any future format variations. The Res2 field contains a value of 0 when the tag is encoded. The frame will be discarded if this field contains a nonzero value when the tag is decoded.
- *Service instance identifier (I-SID):* This 24-bit field carries the service instance identifier of the backbone service instance.
- *Encapsulated customer destination address (C-DA):* If the NCA bit is 0, this 48-bit field carries the address in the destination address parameter of the service request primitive associated with this frame. If the NCA bit is 1, this field has a value of 0 when the tag is encoded and is not used when the tag is decoded.
- *Encapsulated customer source address (C-SA):* If the NCA bit is 0, this 48-bit field carries the address in the source address parameter of the service request primitive associated with this frame. If the NCA bit is 1, this field has a value of 0 when the tag is encoded and is not used when the tag is decoded. The service instance identifier is encoded in a 24-bit field. A BEB may not support the full range of I-SID values, but supports the use of any I-SID values in the range 0 through a maximum N, where N is a power of 2 and specified for that implementation.

11.3.2 PBB Principles of Operation

The PBB located at the backbone of the PBT network is called backbone core bridge (BCB). The bridge located at the edge of PBT network is the BEB. BCB is an S-VLAN bridge used within the core of a PBBN. BEB is a system that encapsulates customer frames for transmission across a PBBN.

A provider backbone edge bridge (PBEB) connects the customers' 802.1ad provider bridge-enabled networks to the provider network. A customer Ethernet frame coming from the customer network arrives at a PBEB to be transmitted to the backbone. PBEB adds a service provider MAC header to the customer Ethernet frame, thus allowing customer (C-MAC) addresses and VLANs to be independent of the backbone address (B-MAC) and backbone VLANs administered by the PBBN operator to relay those frames across the backbone. The S-VLANs used to encapsulate customer frames are known as backbone VLANs (B-VLANs). The resources supporting those VLANs are considered to be part of the PBBN.

PBEB checks the service provider MAC address against its forwarding tables and forwards Ethernet frame as if it is an ordinary frame. As a result, in the core of the network, the switches do not have to be PBB-enabled. The frame travels through the network and arrives on the other edge at another edge bridge. This PBEB will de-encapsulate the extra MAC header, so the customer's frame is again present on the other side of the network. The PBEB forwards the Ethernet frame by using the internal forwarding table and the frame arrives at its destination.

Spanning trees can be employed in PB and BB networks. Customer spanning trees may extend over the provider network. However, PB network and BB network spanning trees must be decoupled to scale the provider network (Figure 11.5).

11.3.3 Provider Backbone Bridge Network

The BEB is of three types: I-type backbone edge bridge (I-BEB), B-type backbone edge bridge (B-BEB), and IB-type backbone edge bridge (IB-BEB).

I-component is responsible for encapsulating frames received from customers, assigning each to a backbone service instance and destination identified by a backbone destination address, backbone source address, and a service instance identifier (I-SID) (Figure 11.6).

I-component allows each BEB to support more than one such virtual medium for each combination of backbone address and VLAN identifier. If the I-component does not know which of the other BEBs provides connectivity to a given customer address, it uses a provisioned default encapsulating B-MAC address that reaches all the other BEBs logically attached to that virtual medium.

Each I-component learns the association between customer source addresses received (encapsulated) from the backbone and the parameters that identify the

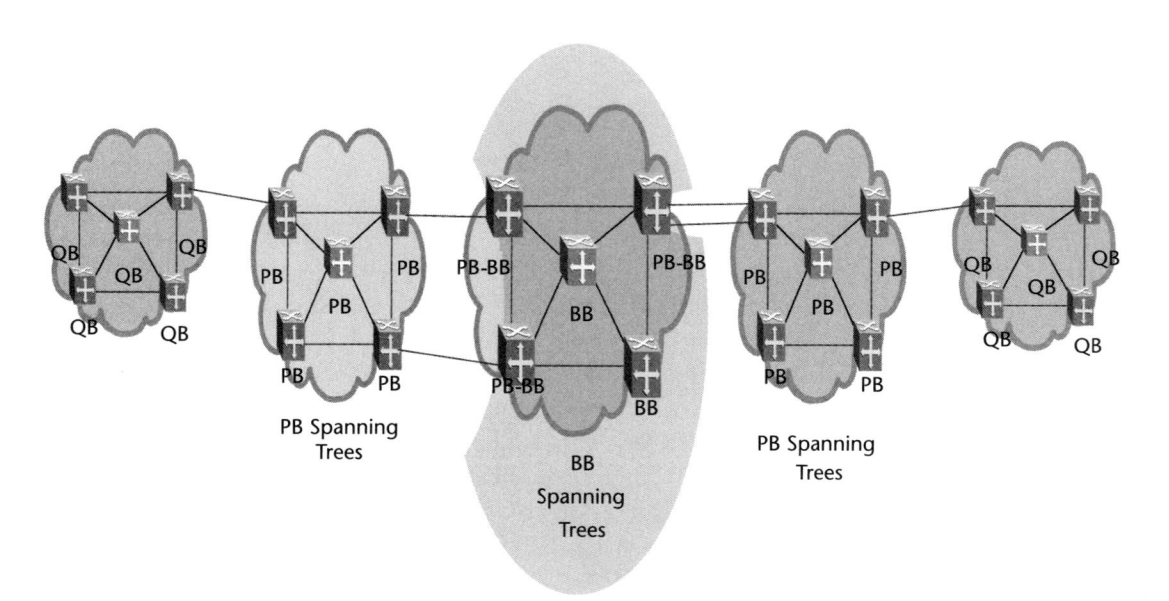

Figure 11.5 Spanning trees in PB and BB networks.

11.3 Provider Backbone Bridges

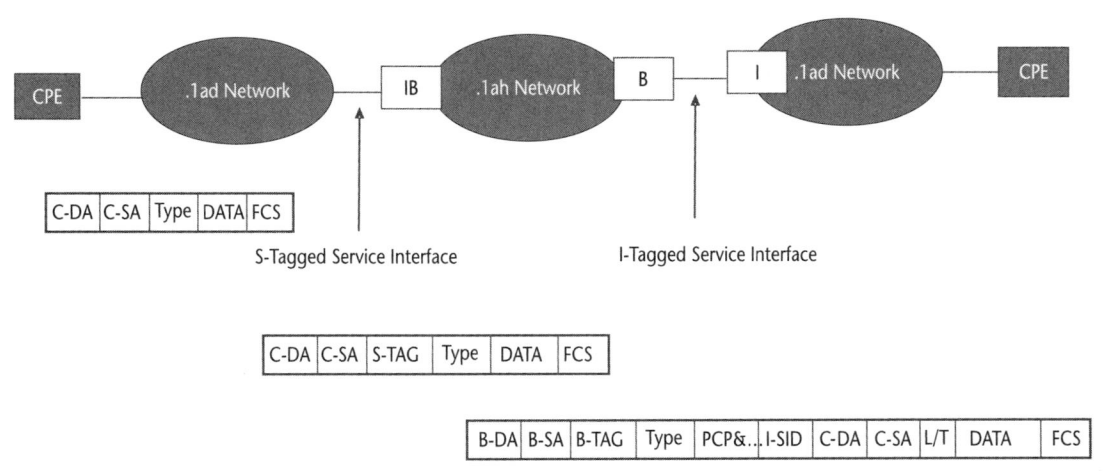

Figure 11.6 Multiprovider Ethernet service delivery.

backbone connection, so subsequent frames to that address can be transmitted to the correct BEB.

B-component is an S-VLAN component with one or more customer backbone ports (CBPs). B-component is responsible for the following:

- Relaying encapsulated customer frames to and from I-components or other B-components when multiple domains interconnect, either within the same BEB or externally connected;
- Checking that ingress/egress is permitted for frames with that I-SID;
- Translating the I-SID (if necessary) and using it to assign the supporting connection parameters (backbone addresses if necessary and VLAN identifiers) for the provider backbone bridge network (PBBN);
- Relaying the frame to and from the provider network port(s) that provide connectivity to the other bridges within and attached to the backbone.

The number of I-components (0 or more) and B-components (0 or 1) within a given BEB is a direct consequence of the type and number of external interfaces supported by that BEB. The I-component and B-component may be in the same bridge or may be in different bridges. The types of BEBs may be classified as I-BEB, B-BEB, and IB-BEB.

PBBN provides multipoint tunnels between provider bridged networks (PBNs) (Figure 11.7) where each B-VLAN carries many S-VLANs. S-VLANs may be carried on a subset of a B-VLAN. For example, all point-to-point S-VLANs could be carried on a single-multipoint B-VLAN providing connection to all end points (Figure 11.8). An I-SID uniquely identifies an S-VLAN within the backbone. When a new service instance is created, I-SID to/from S-VID mapping is provisioned.

Regardless of the I-SID address size, the map tables only have 4,096 entries since only one I-SID exists per S-VLAN and only 4,096 S-VLANs exist per provider bridge. A different S-VID in each PBN maps to the I-SID.

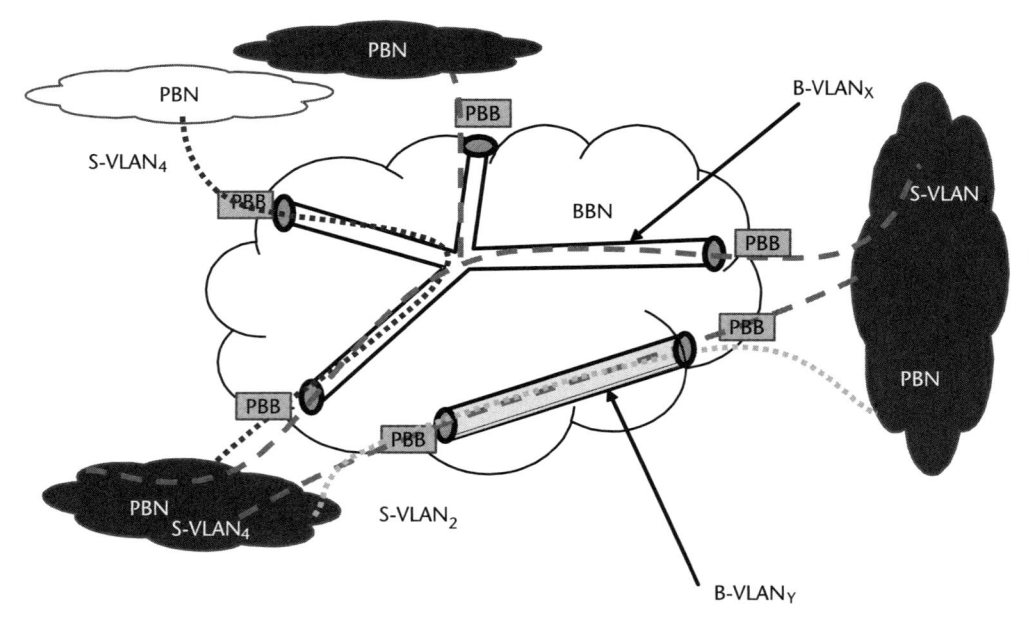

Figure 11.7 PBBN consisting of multiple tunnels between PBNs.

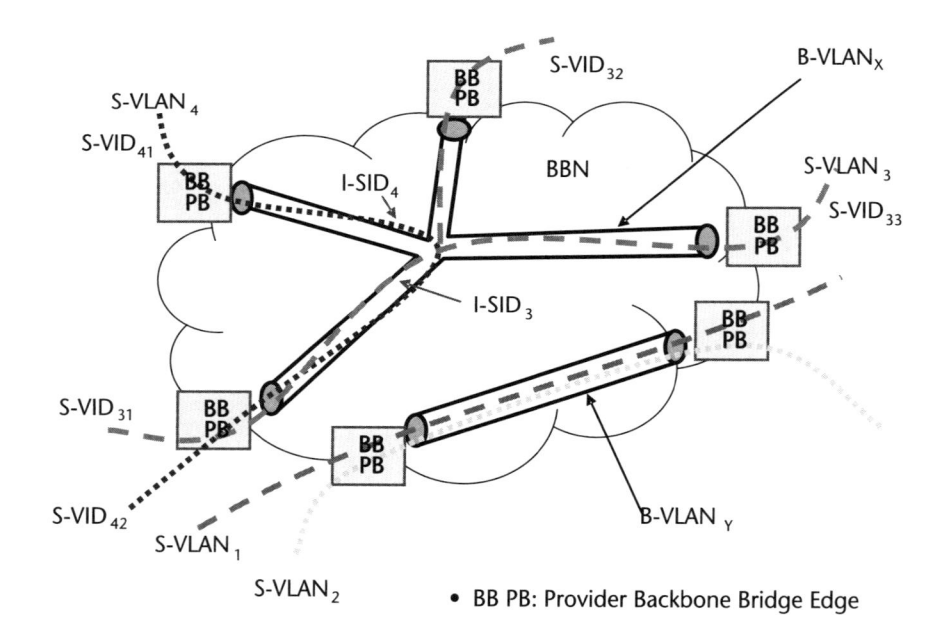

- BB PB: Provider Backbone Bridge Edge

-An I-SID uniquely identifies a S-VLAN within the Backbone•
-The MAP Shim translates between S-VID and I-SID•
-The I-SID to (from) S-VID mapping is provisioned when a new service instance is created

Figure 11.8 Single I-SID per S-VLAN.

11.3 Provider Backbone Bridges

Backbone POP MAC address, B-MAC addresses, identify the edge provider backbone bridges (BB PB) (Figure 11.9). B-MAC addresses are learned by other edge backbone edge bridges. The backbone edge MAC address determines which edge on the B-VLAN will receive the frame. Frames may be flooded by sending with broadcast or multicasts DA B-MACs to the B-VLAN.

The PBB allows customer's MAC addresses to overlap with the provider's MAC addresses, because the customers' service frames are tunneled by provider backbone bridges and are not used when switching frames inside the provider's network. As a result, customers are free to assign identifier and class of service values to their VLANs, without concern that the service provider will alter them. Meanwhile, the service provider does not need to worry about coordinating VLAN administration with its customers.

Service provider's switches only use the provider MAC header. There is no need for them to maintain visibility of customers' MAC addresses, reducing the burden on the forwarding tables in the provider's network. This also ensures that changes to the customers' networks do not impact the provider network, improving the stability of the service provider's network.

Customer security is improved, because the service provider switches are no longer inspecting the customer MAC header.

The provider backbone bridge tunnels the customers' service frames and control frames, like STP frames through the provider's network. This allows control protocols, like STP, to be used separately by the customers' networks and the service providers' network. However, the STP that is used in the customers' networks, should not interact with the STP used in the service provider's part of the network, which is PBB-enabled.

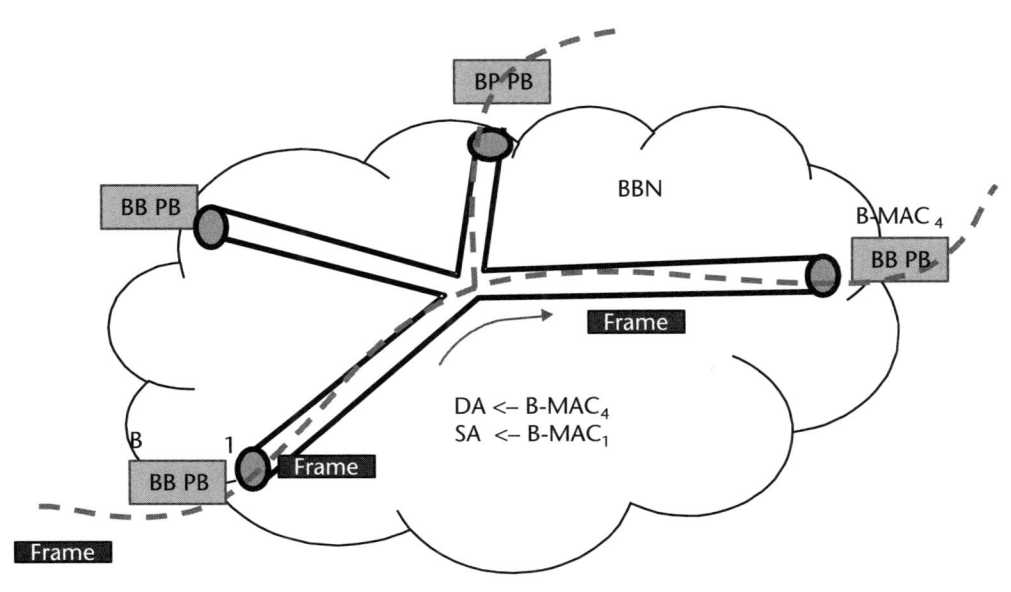

Figure 11.9 BB PB.

11.4 Provider Backbone Transport

Traffic-engineered provider backbone bridging (PBB-TE), PBT, is an Ethernet derivative intended to bring connection-oriented characteristics and deterministic behavior to Ethernet. It turns off Ethernet's spanning tree and media-access-control address-flooding and learning characteristics. That lets the Ethernet behave more like a traditional carrier transport technology.

The frame format of PBT is exactly as the format used for implementing PBB. The difference is in the meaning of the frame's fields. The VID and the B-DA fields together form a 60-bit globally unique identifier.

The control plane is used to manage the forwarding tables of the switches; to create PBT tunnels, all switches need to be controlled from one (PBT) domain. This is necessary for the control plane to fill the forwarding tables of the switches and thus set up a path. Since the control plane fills the forwarding tables, learning and flooding/broadcasting and spanning tree have become obsolete for PBT. This technique enables circuit-switching on an Ethernet.

There are a total of 4,094 VLANs, which can be set up according to the 802.1Q standard. When using, for example, 64 VIDs for PBT, the other 4,030 can be used for ordinary connectionless Ethernet.

PBT is a combination of Ethernet extensions, created to establish Carrier Ethernet. IEEE 1Qay [3] was based on extensions of [1, 2, 5, 10]. It was developed with scalability and QoS in mind. To use Ethernet as a transport technique in MAN/WAN, OAM functionality is required. PBT is able to use standards in [1, 4, 5, 11] and the developing MEF specification in [12]. PBT can take full advantage of the MEF work on an external network-to-network interface (ENNI) specification [13] to allow peering of Ethernet services.

When using PBT, it becomes possible to set up a path with QoS and a backup path over an Ethernet-based network (Figure 11.10). PBT tunnels are created, and appropriate bandwidth is reserved to guarantee SLA without having an overprovisioned network capacity. This provisioning needs to be handled by the network management layer. The precise implementation of the provisioning is not described

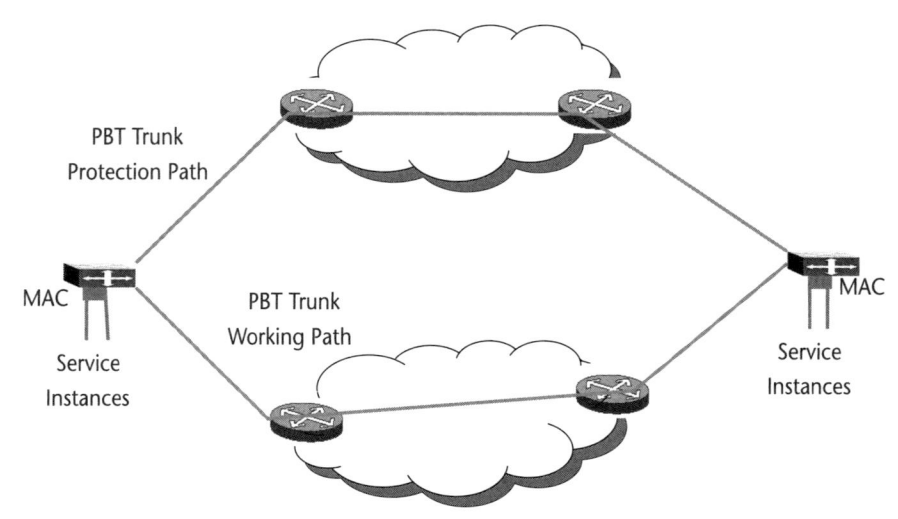

Figure 11.10 PBT-TE protection.

yet in [3]. A single physical part of the network can have PBT, PBB, and normal Ethernet frames on different VLANs.

PBT packets will be switched on VID and destination MAC address. There are two VLANs when path protection is enabled: one VLAN provides the current work path and the other provides the backup path. Fast reroute uses reserved set of B-VID tags.

PBB-TE reuses existing implementations of VLANs and double-tagging and combines them with the network separation and layering principles of PBB.

In the PBB model, the B-VID identifies a packet flooding domain that interconnects different PB networks. In the PBB-TE model, the B-VID identifies a specific path through the network, in combination with the B-DA address.

Forwarding model PBB-TE is intended to be deployed as a connection-oriented, packet-switched network layer. PBB-TE runs on independent VLAN learning (IVL) capable switches, which allow packets to be forwarded based on the 60-bit concatenation of the B-VID and Destination MAC address. Broadcast frames are discarded. Packets with unknown VID + MAC entries are discarded before being flooded.

End-to-end OAM and protection mechanism are based on [1, 4]. A 12-bit VID field is used to identify alternate paths to the associated destination MAC address, for protection switching. One VID is assigned for the working path and another VID is assigned for the protection path. Multiple VIDs could support k-shortest path routing or be assigned to protection paths assigned to different failure scenarios.

Any service with a defined EtherType can be transported over PBB-TE. For example, E-Line and E-LAN, pseudowires, and OAM packets can run over PBB-TE.

PBT can use 802.1AB for network management auto-discovery of network nodes. Bridge-forwarding tables are populated using a network management system or GMPLS control plane. Control plane management system directly provision packet forwarding into the forwarding tables. GMPLS can be used to populate forwarding tables by GMPLS-controlled Ethernet label switching (GELS) [14]. Provider link state bridging (PLSB) using IS-IS protocol helps PBB-TE support multipoint, multicast and connectionless services. Protocols used by PBT are depicted in Figure 11.11.

11.4.1 Principles of Operation

PBT ignores spanning tree states; however, STP can be used on the non-PBT-VLANs. 802.1AB tells the management how the network layout looks like. In turn, the network management fills the forwarding tables of the Layer-2 devices. This makes provisioning and adding nodes much simpler.

Forwarding of the packets is based on explicit routes through the network. When the continuity check messages are lost in the network, resulting in a fault condition, the VLAN-ID (VID) at the sending host will be changed to the backup VID. This process should take place within 50 milliseconds.

Let's consider the example in Figure 11.12 where customer A and B networks are LANs. LAN A is connected to the provider via a PBEB A. The provider network contains several other provider backbone core bridges (PBCB) C and D. Customer B is connected to the provider network through PBEB B.

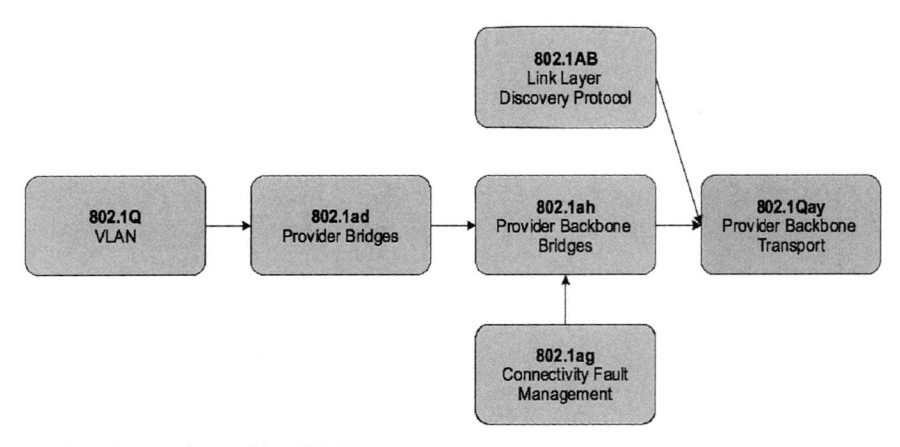

Figure 11.11 Protocols used by PBB-TE.

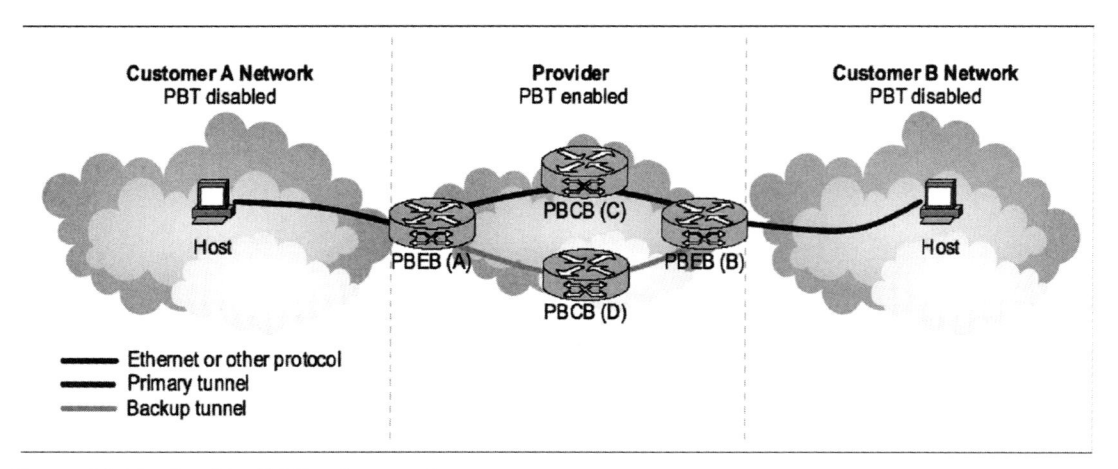

Figure 11.12 Provider PBT implementation.

A management layer or a suitable control plane manages the core network with the PBEBs and PBCBs. This control plane fills the forwarding database in the PBCBs with the correct entries, so they are able to forward traffic through the path/tunnel. For example, PBCB C has an entry for PBEB B in combination with a certain backbone VLAN-ID, the 60-bit identifier, to forward traffic on a certain port in the direction of PBEB D. These tunnels are created by the operator through the management layer.

When a customer from site A sends traffic to a customer at site B, its destination will be reached as following:

- The traffic originating from customer A is sent through the customer's network to the PBEB A.
- The PBEB A is configured to add the 24-bit I-SID of a certain value to the I-TAG field, based on the S-VID of the customer's frame.

11.4 Provider Backbone Transport

- With the use of this value, a tunnel and its backup tunnel can be selected. These tunnels, identified by the 60-bit identifier, are reserved before the tunnel becomes operational.
- The PBEB A also adds the backbone destination address of the PBEB B and its own backbone source address and the backbone VLAN-ID.
- The encapsulated frame will be sent through the tunnel and will arrive at the PBCB C.

The PBCB C has a forwarding entry that it selects based on the B-VID. It will forward the frame to the PBEB B. The PBEB B will de-encapsulate the frame and forward the frame based on the I-SID, which is associated with a certain VLAN-ID. The frame will be delivered to customer B.

Paths are being monitored by the 802.1ag protocol, so when an error occurs, the backup tunnel are automatically switched.

11.4.2 End-to-End Carrier Ethernet with Multiple PBT Domains and Single Domain

Implementing end-to-end carrier Ethernet connection with QoS, protection, and scalability requires implementing PBT inside the customer's network as well as in provider's network. Each network can have its own domain.

This network has multiple PBT domains (see Figure 11.13), three control planes exist, in order to ensure the PBT tunnels to be created and ended correctly. Every time a frame travels from one domain to another the frame has to be de-encapsulated and encapsulated.

PBEBs create single points of failure in the connections. This could be solved by installing additional PBEBs between the domains. Each domain is monitored by CFM per domain. When a link error occurs on one domain, the other domains will not know the problem has occurred. To solve this issue, the management layer should be handled on another (OSI) layer.

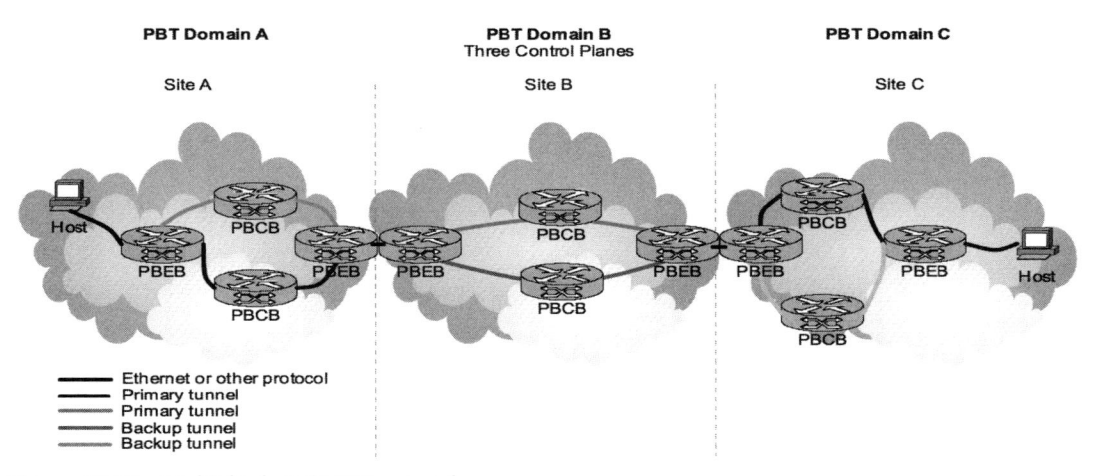

Figure 11.13 Multiple-domain PBT network.

Customer and provider networks can be under one PBT domain (Figure 11.14). This example is more applicable to corporate networks connecting offices of multiple locations.

A single-domain solution also has its specific advantages and disadvantages. For instance, one of the advantages is that only one control plane is required, but a disadvantage is that this solution has less scalability for the different partners.

Another example of multiple PBT domain networks is where access networks implementing PBT while backbone connecting them with non-PBT network (Figure 11.15). This example is also very much applicable to enterprise networks.

11.4.3 PBB-TE Network

The PBB-TE provides a method for full traffic engineering of Ethernet switched path (ESP) paths in a PBBN network. To do this, PBBTE replaces the MSTP control plane with either a management plane or an external control plane and then

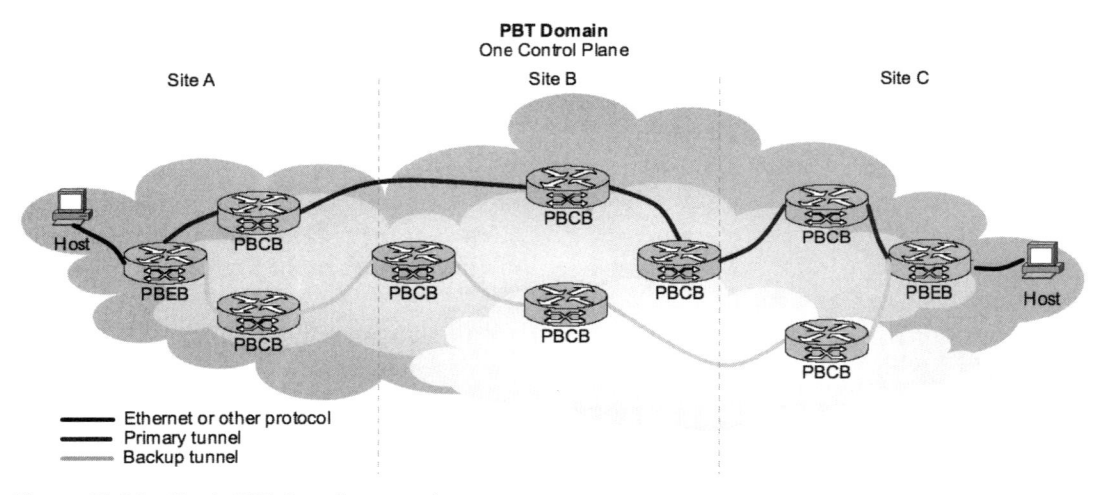

Figure 11.14 Single PBT domain network.

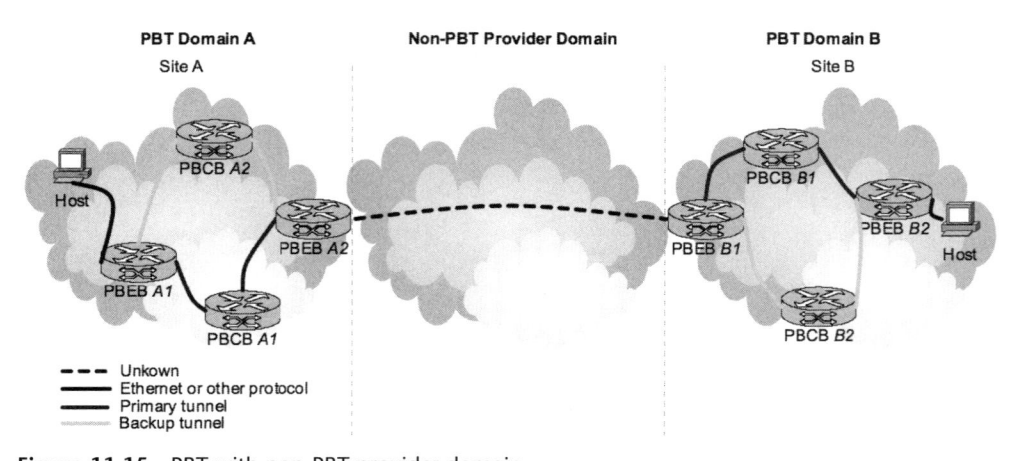

Figure 11.15 PBT with non-PBT provider domain.

populates the bridge filtering tables of the corresponding B-components and BCBs by creating static filtering table entries for the ESP-MACs and ESP-VIDs.

The external PBB-TE management/control plane is responsible for maintaining and controlling all the topology information to support point-to-point or point-to-multipoint unidirectional ESPs over the PBBN.

The PBB-TE topology can coexist with the existing active MSTP or with the shortest path bridging (SPB) topology by allocating B-VID spaces to PBB-TE, MSTP, or SPB, or PBB-TE can be stand-alone.

PBB-TE takes control of the ESP-VID range from the BCB and BEB of the PBBN. It will maintain routing tree(s) which provides corouted reverse path(s) from the B-DA to the B-SA. The ESP-VID(s) used in this reverse ESP does not have to be the same one used for the forward ESP. Each of the provisioned ESPs are identified by a 3-tuple: <ESP-MAC DA, ESP-MAC SA, ESP-VID>.

PBB-TE can provide point-to-point and point-to-multipoint services. A point-to-point PBB-TE service instance, which is called a PBB-TE trunk, is provided by a pair of corouted unidirectional point-to-point ESPs and by a pair of 3-tuples: < DA1, SA1,VID1>, < SA1, DA1, VID2>. The VIDs can be the same or different.

A point-to-multipoint PBB-TE service is provided by one multipoint ESP and n unidirectional point-to-point ESPs, routed along the leaves of the multicast ESP and correspondingly by n+1 3-tuples: <DA, SA, VID>, <SA, SA1, VID1>,...,<SA, SAn, VIDn>.

In a PBB-TE region, BEBs mark the demarcation between the PBBN and the networks attached to it. The protection domain is defined to be the area between the customer backbone ports (CBPs) of BEBs.

ESPs are provisioned from one CBP to the other, each one identified by the tuple <B-DA,B-SA, B-VID>. Each ESP represents a unidirectional path and the ESP pairs that form the bidirectional path are defining a PBB-TE trunk. The ESPs belonging to the same PBB-TE trunk are corouted, but can also be identified by different B-VIDs.

Two PBB-TE trunks can be provisioned, working trunk and protecting trunk, each identified one 3-tuple: <CBP-B, CBP-A, VID-1> and <CBP-A, CBP-B, VID-2>. The working trunk consists of two ESPs identified by the 3-tuples <CBP-B, CBP-A, VID-3> and <CBP-A, CBP-B, VID-4>

On the terminating BEBs, the VLAN membership of each participating port has to be configured for the B-components.

Each of the PBB-TE trunks is monitored by an independent maintenance association (MA). One MA is set to monitor PBB-TE trunk-1 and one is set to monitor PBB-TE trunk-2. Each of these two MAs is associated with a pair of ESPs and identified by their corresponding 3-tuples. The MA that monitors the PBB-TE trunk-1 contains VID-1 and VID-2 in its VID list. Two Up MEPs, associated with this MA, are configured on the CBPs that terminate each PBB-TE trunk.

Each of these MEPs has its own primary VID, VID-1 for the MEP on the West B and VID-2 for the MEP on the East B. Each MEP receives frames that are tagged with any of the VIDs in the MA list, but sends frames that are tagged only with that MEP's primary VID.

The MEP for the working trunk on the West B and the corresponding MEP on the East B exchange CCM frames over primary VIDs. If a fault occurs at any of the ESPs, the MEP on the receiving end will be notified. In particular, if a fault on the working ESP occurs, the MEP on the East B-component will declare a remote MEP defect. A bridge does not set remote MEP CCM defect within 3.25 * CCMinterval seconds of the receipt of a CCM, and sets remote MEP CCM defect within 3.5*CCMinterval seconds after the receipt of the last CCM.

All subsequent CCMs sent by MEP will have the RDI field set for as long as proper CCMs are not received by the MEP. A reception of a CCM frame with the RDI field set will cause a change of the B-VID parameter of the CBP on West B-component, to the pre-configured value of the protection ESP. The resulting behavior will be to move the specific service instance to the protection PBB-TE trunk.

11.4.4 PBT: MPLS Internetworking

MPLS is the primary technology in the core of IP-based converged networks. MPLS uses pseudowire tunnels for the transport of services across a core network. At the edge, PBB-TE can support access to the core and provide Carrier Ethernet services, including E-Line, E-LAN, and E-Tree.

11.5 Conclusion

PBT is the technology to make end-to-end Ethernet feasible. With scalable addressing, QoS, service OAM, G.8031 [11] protection methods, and MEF-defined services, PBT is a low-cost alternative for carrier class networking.

In fact, some service providers are in the process of implementing PBT. We expect to see more deployment in the coming years.

References

[1] IEEE 802.1ag -2007, Virtual Bridged Local Area Networks – Amendment 5: Connectivity Fault Management, December 2007.
[2] IEEE 802.1ah 2008 - IEEE Standard for Local and Metropolitan Area Networks -- Virtual Bridged Local Area Networks Amendment 7: Provider Backbone Bridges.
[3] IEEE 802.1Qay-2009, Provider Backbone Bridge Traffic Engineering, RFC 2922, Physical Topology MIB, September 2000.
[4] ITU-T Y.1731, OAM Functions and Mechanisms for Ethernet Based Networks, May 2006.
[5] IEEE 802.1AB, Station and Media Access Control Connectivity Discovery, 2005.
[6] RFC 2922, Physical Topology MIB, September 2000.
[7] IEEE Std. 802.3, 2000.
[8] IETF RFC 2863, The Interfaces Group MIB, June 2000.
[9] IETF RFC 2737 Entity MIB, Version 2, December 1999.
[10] IEE 802.1ad, OAM Functions and Mechanisms for Ethernet Based Networks, May 2006.
[11] G.8031, Ethernet Linear Protection Switching, 2011.
[12] MEF 35 Service OAM Performance Monitoring Implementation Agreement, April 2012.

[13] MEF 26, External Network-Network Interface, January 2010.
[14] RFC 5828, Generalized Multiprotocol Label Switching (GMPLS) Ethernet Label Switching Architecture and Framework, March 2010.

CHAPTER 12
Transport MPLS

12.1 Introduction

MPLS was originally developed by IETF in order to address core IP router performance issues. It became a dominant technology for carriers' converged IP/MPLS core networks and a platform for data services such as IP-VPN. With increasing packet networking, the ITU-T became interested in adapting MPLS to make it a carrier-class network.

MPLS first appeared on the transport networking scene in the form of Generalized Multiprotocol Label Switching (GMPLS), a set of protocols or protocol extensions offered up by the IETF that could be implemented as a distributed control plane for the setup and teardown of connections based on circuits, wavelengths, or whole fibers in a transport network.

GMPLS's focus is solely on improving the management of core optical resources, not on packet transport efficiency. Some carriers have taken advantage of GMPLS to support topology and neighbor discovery in order to improve provisioning of wavelengths or circuits in their networks. However, this is limited to typically core optical switches and large multiservice provisioning platform (MSPPs).

Transport-MPLS (T-MPLS) offers packet-based alternatives to SONET circuits and promises operators much greater flexibility in how packet traffic is transported through their metro and core optical networks. In T-MPLS, a new profile for MPLS is created so that MPLS label-switched paths (LSPs) and pseudowires can be engineered to behave like TDM circuits or Layer-2 virtual connections, replacing SONET section, line, and path. SONET will remain as a physical interface (PHY) along with gigabit and 10-Gb Ethernet.

The T-MPLS is a connection-oriented, packet-switched transport layer network technology based on the MPLS bearer plane modeled in [1]. Its architectural principles are:

- A T-MPLS layer network can operate independently of its clients and its associated control networks.
- Transport connections can have very long holding times. Therefore, T-MPLS includes features traditionally associated with transport networks, such as protection switching and operation and maintenance (OAM) functions.

- A T-MPLS network will not peer directly with an IP/MPLS network. An LSP initiated from an IP/MPLS network element will be encapsulated before it transits a T-MPLS network. Similarly, if IP/MPLS is used as a server layer for T-MPLS, then an LSP initiated from a T-MPLS network element will be encapsulated before it transits an IP/MPLS network. This results in control plane independence between T-MPLS and IP/MPLS.

T-MPLS is intended to be a separate layer network with respect to MPLS. However, T-MPLS will use the same data-link protocol ID (e.g., EtherType), frame format and forwarding semantics as defined for MPLS frames. The semantics used within a label space are defined in [2].

T-MPLS may be described as a transport network profile of IETF RFCs and [1]. Reference [3] defined architecture of T-MPLS layer network while [4] defined data plane and UNI/NNI. Equipment functionalities and management are based on [5, 6].

OAM, which is based on [7, 8], is specific to the transport network and functionality is referenced from [9]. This provides the same OAM concepts and methods (e.g., connectivity verification, alarm suppression, remote defect indication) already available in other transport networks, without requiring complex IP data plane capabilities.

The result is T-MPLS, a connection-oriented packet transport network based on MPLS that provides managed point-to-point connections to different client layer networks such as Ethernet. However, unlike MPLS, it does not support a connectionless mode and is intended to be simpler in scope, less complex in operation, and more easily managed. Layer-3 features have been eliminated and the control plane uses a minimum of IP to lead to low-cost equipment implementations.

As an MPLS subset, T-MPLS abandons the control protocol family that IETF defines for MPLS. It simplifies the data plane, removes unnecessary forwarding processes, and adds ITU-T transport style protection switching and OAM functions.

The forwarding behavior of T-MPLS is a subset of IETF-defined MPLS. This common data/forwarding plane retains the essential nature of MPLS and ensures that interoperability and interworking will be readily achievable. The features are end-to-end LSP protection, survivability methods mimicking the current linear [10], ring [11], and shared mesh options, per-layer monitoring, fault management and operations that leverage existing SONET OAM mechanisms.

T-MPLS has strict data and control plane separation. It does not use Layer-3 and IP-related features, such as LSP merge, penultimate hop popping, and equal-cost multiple path (ECMP), eliminating the need for packet transport network elements to process IP packets, while also improving OAM. The changes allow establishment of bidirectional end-to-end LSPs, configured and monitored via a central network management system. Control plane traffic via three types of signaling communication channels, namely, in-fiber/in-band via native IP packets, in-fiber/in-band via dedicated LSP, in or out-fiber/out-band, is supported.

The management plane will be used for manual/automated provisioning, in the same way as SDH and OTN/WDM networks. However, the control plane for T-MPLS is envisaged to be ASON/GMPLS and will thus enable more dynamic and intelligent operation.

MPLS fast reroute (FRR) capability requires the use of LSP merge that is excluded from T-MPLS. Because no control plane is involved, protection switching performance can, in principle, be fast.

Bidirectional T-MPLS LSPs are supported by pairing the forward and backward directions to follow the same path (i.e., the same nodes and links). The pairing relationship between the forward and the backward directions is known in each node traversed by the bidirectional LSP. Both LSP and E-LSP are supported.

Both TTL and EXP as defined in [12] use either the pipe or the short-pipe model. (Three tunnel modes are defined in [12]: uniform, short pipe, and pipe. In uniform mode, any changes made to the EXP value of the top label on a label stack are propagated both upward as new labels are added and downward as labels are removed. In the short-pipe mode, the IP precedence bits on an IP packet are propagated upward into the label stack as labels are added. When labels are swapped, the existing EXP value is kept. If the topmost EXP value is changed, this change is propagated downward only within the label stack, not to the IP packet. The pipe mode is just like the short-pipe mode, except the PHB on the MPLS to IP link is selected based on the removed EXP value rather than the recently exposed DSCP value. QoS within each T-MPLS sublayer is decoupled from any other one and from client QoS. Diff-Serv as defined in [12] is supported for only pipe and short-pipe models.

Both per-platform and per-interface label space are supported. T-MPLS will not reserve labels for its own use independently of MPLS. This helps to ensure that interoperability and interworking will be readily achievable.

Transport MPLS networks can be used to implement Metro Ethernet E-Line services and E-LAN services.

12.2 Differences from MPLS

In order to define a subset of MPLS that is connection-oriented and can use the established transport OAM model, several MPLS protocol features are excluded from T-MPLS. The key differences of T-MPLS compared with MPLS include:

- Use of bidirectional LSPs: T-MPLS pairs the forward and backward LSPs that follow the same nodes and link to provide bidirectional connections while MPLS LSPs are unidirectional.
- No penultimate hop popping (PHP) option: This mechanism pops the MPLS label at the penultimate node, sending IP packets to the last node. This eases the processing at the final node, but also means that MPLS OAM packets cannot reach this node.
- No LSP merging option: LSP merge means that all traffic forwarded along the same path to the same destination may use the same MPLS label. Although this improves scalability, it makes OAM and performance monitoring (PM) difficult because the traffic source becomes ambiguous and unknown. In fast reroute MPLS link protection, it must be possible to merge two LSPs into one at a node. However, this can create problems in maintaining OAM integrity.

- No ECMP option: ECMP allows traffic within one LSP to be routed along multiple network paths. Not only does this require additional IP header processing, as well as MPLS label processing, but it makes OAM more complex as continuity check (CC) and PM flows may follow different paths. ECMP allows MPLS packets to be sent over multiple LSPs to the same end point.

T-MPLS and MPLS system connectivity requires interworking between IP/MPLS pseudowires (PWs) and T-MPLS PW at a very preliminary stage. The interworking requirements have been defined in [7, 8]. There is a working assumption that T-MPLS OAM should re-use where possible the recently developed OAM standards for Ethernet [13, 14].

12.3 Architecture

Transport MPLS is a connection-oriented packet transport technology, based on MPLS frame formats (Figure 12.1). It profiles MPLS, avoiding the complexity and need for IP routing capability and deeper packet inspection. T-MPLS allows for guaranteed SLAs and defines protection switching and restoration. Fault localization and multi-operator service offerings are possible.

A T-MPLS layer network is operated with network management and/or by a control plane. The control plane inside the T-MPLS network is an automatically switched optical network (ASON) [15, 16]/GMPLS [17].

T-MPLS leverages MPLS as a service layer, as depicted in Figures 12.2 and 12.3. Its characteristic information, client/server associations, the topology, and partitioning of T-MPLS transport networks [3] use a subset of the MPLS architecture [1].

The T-MPLS architecture defines one layer network that is a path layer network. The MPLS layer network characteristic information can be transported through MPLS links supported by trails in other path layer networks such as Ethernet MAC layer network.

T-MPLS is intended to be a separate layer network with respect to MPLS. However, T-MPLS will use the same data-link protocol ID such as EtherType, frame format, and forwarding semantics as defined for MPLS frames. The semantics used within a label space are defined in [2].

T-MPLS is conceived within the ITU-T's layered network architecture, such that layer networks (i.e., servers) operate independently of their clients (networks or services) and their associated management and signaling networks. As a result, T-MPLS networks will be able to carry customer traffic completely transparently

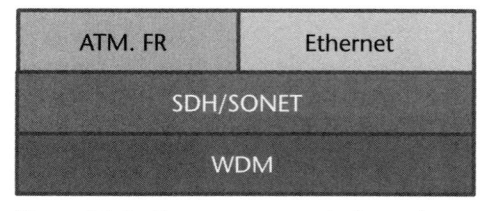

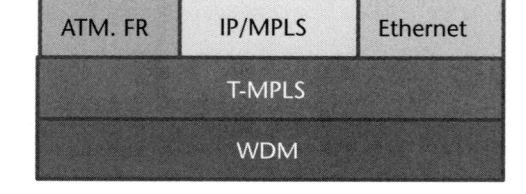

ATM. FR	Ethernet
SDH/SONET	
WDM	

ATM. FR	IP/MPLS	Ethernet
T-MPLS		
WDM		

Figure 12.1 Transport network alternatives.

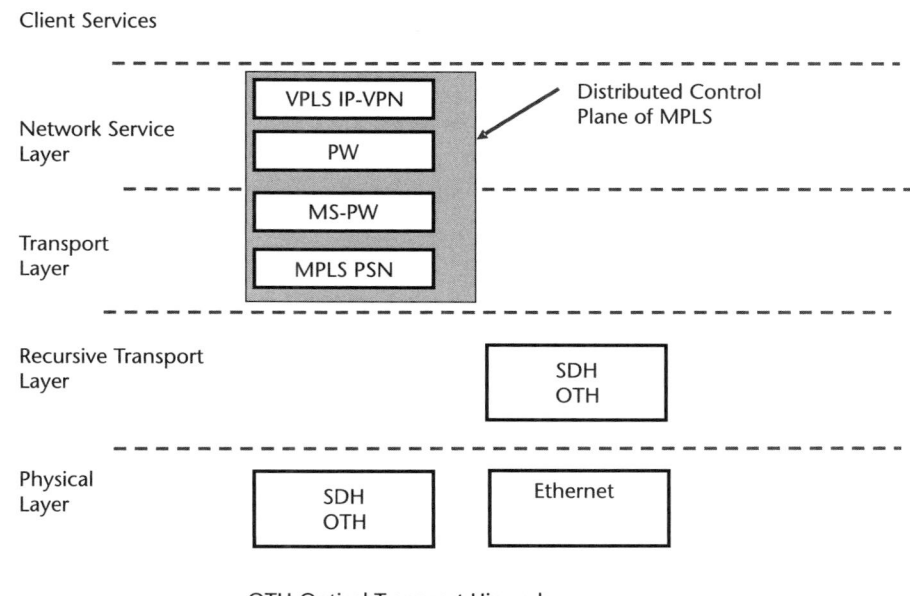

12.3 Architecture

Figure 12.2 Layers of MPLS networks.

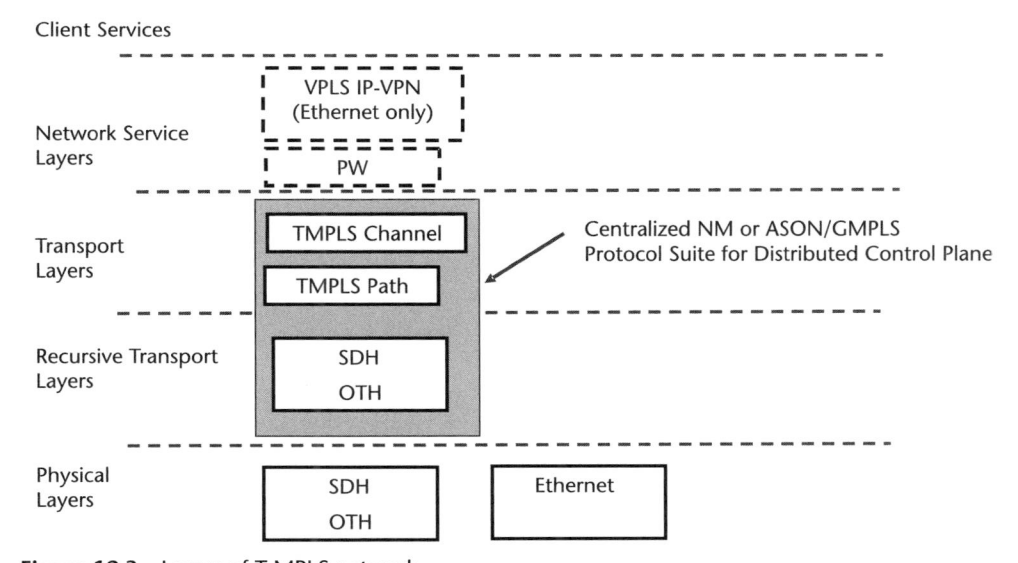

Figure 12.3 Layers of T-MPLS network.

and securely, without unwanted interaction (Figure 12.4). In other words, T-MPLS can transport any type of packet service such as Ethernet, IP, and MPLS. The precise adaptation or client handling needs to be defined on a service-by-service basis.

The OAM methodology for T-MPLS is based on [9]. However, [9] did not include the complete set of tools that transport networks conventionally provide such as tools for PM or on-demand OAM.

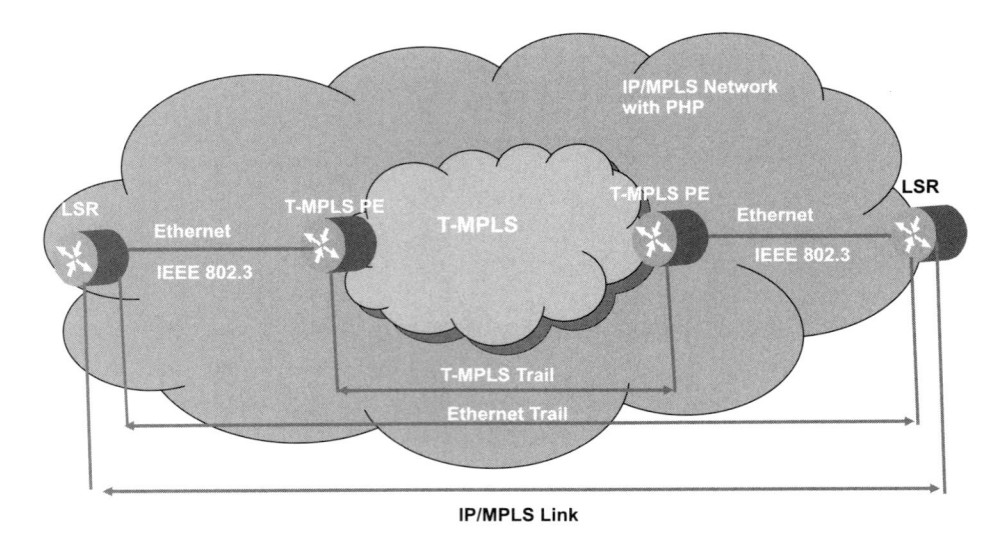

Figure 12.4 IP/MPLS via Ethernet over T-MPLS network.

The control plane for T-MPLS has not been defined. It is expected that the ITU-T's ASON control model will be applied, using GMPLS protocols.

12.3.1 T-MPLS Interfaces

T-MPLS, similar to MPLS, defines the UNI, which is the interface between a client and service node and NNI, which is between two service nodes (Figure 12.5).

As depicted in Figure 12.6, T-MPLS NNI may carry informational elements of data plane that may include a data communication network (DCN) supporting management plane and control plane communications, control plane providing signaling and routing, and management plane. The ENNI itself may consist of informational elements of three planes as data plane including OAM, control plane, and the management plane.

T-MPLS NNIs for T-MPLS-over-ETH (MoE), T-MPLS-over-Synchronous Digital Hierarchy (SDH) (MoS), T-MPLS-over-optical transport hierarchy (OTH) (MoO), T-MPLS-over-plesiosynchronous digital hierarchy (PDH) (MoP), and T-MPLS-over-resilient packet ring (RPR) (MoR) are defined in [4]. The payload bandwidths of PDH, SDH, and OTH for MOR are given in Tables 12.1, 12.2, and 12.3.

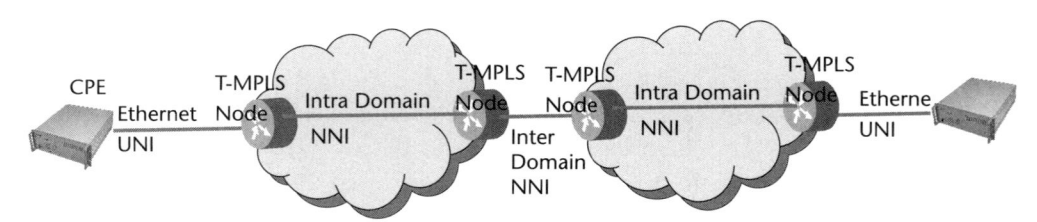

Figure 12.5 Multi-operator T-MPLS network.

12.3 Architecture

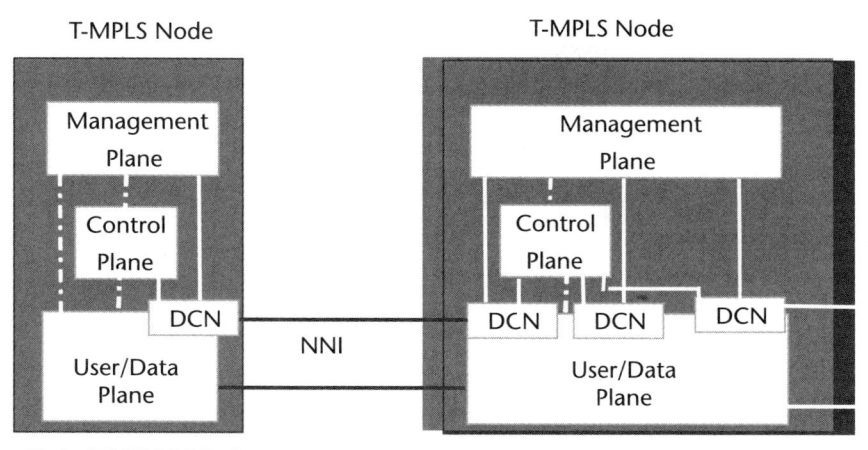

Figure 12.6 T-MPLS NNI planes.

Table 12.1 The Bandwidth of the Payload of PDH Path Signals

PDH Type	PDH Payload (Kbps)
P11s	1,536 − (64/24) ≈ H1,533
P12s	1,980
P31s	33,856
P32e	4,696/4,760 * 44,736 = 44,134
P11s-Xv, X = 1 to 16	=1,533 to =24,528
P12s-Xv, X = 1 to 16	1,980 to 31,680
P31s-Xv, X = 1 to 8	33,856 to 270,848
P32e-Xv, X = 1 to 8	=44,134 to =353,072

Table 12.2 The Bandwidth of the Payload of SDH VCs

Virtual Container (VC) Type	VC Payload (Kbps)
VC-11	1,600
VC-12	2,176
VC-2	6,784
VC-3	48,384
VC-4	149,760
VC-4-4c	599,040
VC-4-16c	2,396,160
VC-4-64c	9,584,640
VC-4-256c	38,338,560
VC-11-Xv, X = 1 to 64	1,600 to 102,400
VC-12-Xv, X = 1 to 64	2,176 to 139,264
VC-2-Xv, X = 1 to 64	6,784 to 434,176
VC-3-Xv, X = 1 to 256	48,384 to 12,386,304
VC-4-Xv, X = 1 to 256	149,760 to 38,338,560

Table 12.3 The Bandwidth of OTH Optical Channel Data Units (ODUs)

ODU Type	Optical Payload Unit (OPU) Payload (Kbps)
ODU1	2,488,320
ODU2	238/237 9,953,280 = 9,995,277
ODU3	238/236 39,813,120 = 40,150,519
ODU1-Xv, X = 1 to 256	2,488,320 to 637,009,920
ODU2-Xv, X = 1 to 256	=9,995,277 to =2,558,709,902
ODU3-Xv, X = 1 to 256	=40,150,519 to H 10,278,532,946

12.4 T-MPLS Frame Structure

T-MPLS pushes an additional header structure to packet. Figure 12.7 depicts the frame format of a T-MPLS packet for Ethernet physical medium while Figure 12.8 illustrates the relationship between network layers and T-MPLS header.

12.5 T-MPLS Networks

A typical T-MPLS network configuration is given in Figure 12.9 where the T-MPLS network provides a primary LSP and backup LSP. The switching between the primary and secondary LSP tunnels can take place within 50 ms. These T-MPLS tunnels can support both Layer-3 IP/MPLS traffic flows and Layer-2 traffic flows via pseudowires.

In Figure 12.9, T-MPLS tunnel is transported over MPLS LSPs. Various services can be transported over these T-MPLS tunnels as depicted in Figure 12.10.

12.5.1 T-MPLS Protection

The T-MPLS protection can be linear or ring. The linear protection switching for T-MPLS as defined in [10] can be 1 + 1 or 1:1. The ring protection switching is defined in [11].

The 1 + 1 architecture operates with unidirectional switching while the 1:1 architecture operates with bidirectional switching. In the 1 + 1 architecture, a protection transport entity is dedicated to each working transport entity. The normal traffic is copied and fed to both working and protection transport entities with a permanent bridge at the source of the protected domain. The traffic on working and protection transport entities is transmitted simultaneously to the sink of the

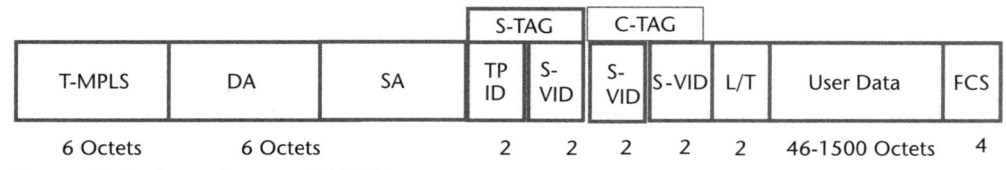

T-MPLS	DA	SA	TP ID	S-VID	S-VID	S-VID	L/T	User Data	FCS
6 Octets	6 Octets		2	2	2	2	2	46-1500 Octets	4

(S-TAG: TP ID, S-VID; C-TAG: S-VID, S-VID)

Figure 12.7 Frame format of T-MPLS.

12.5 T-MPLS Networks

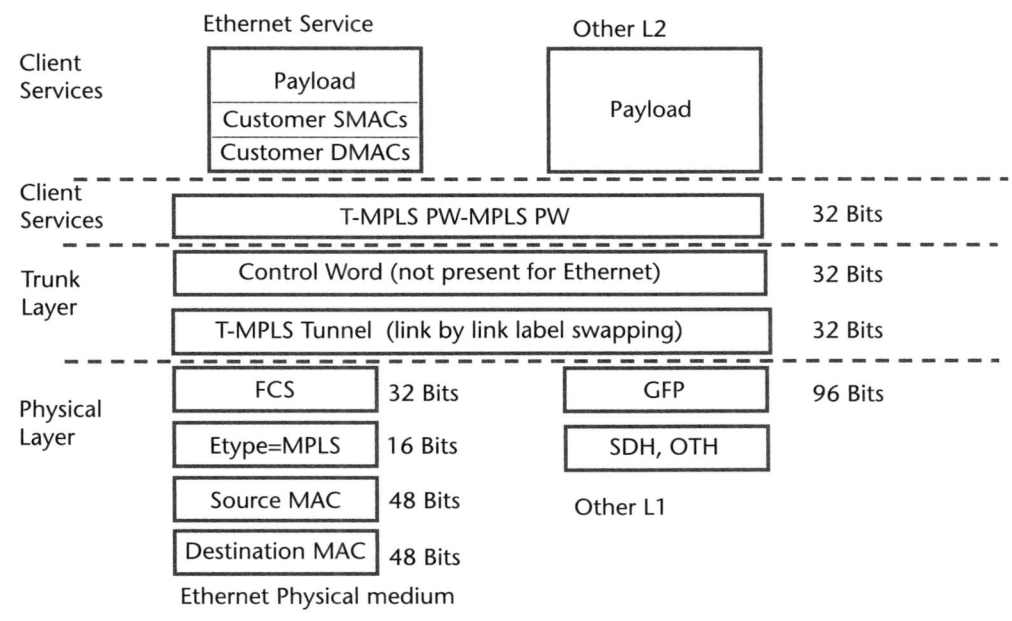

Figure 12.8 Mapping of T-MPLS frame structure into network layers.

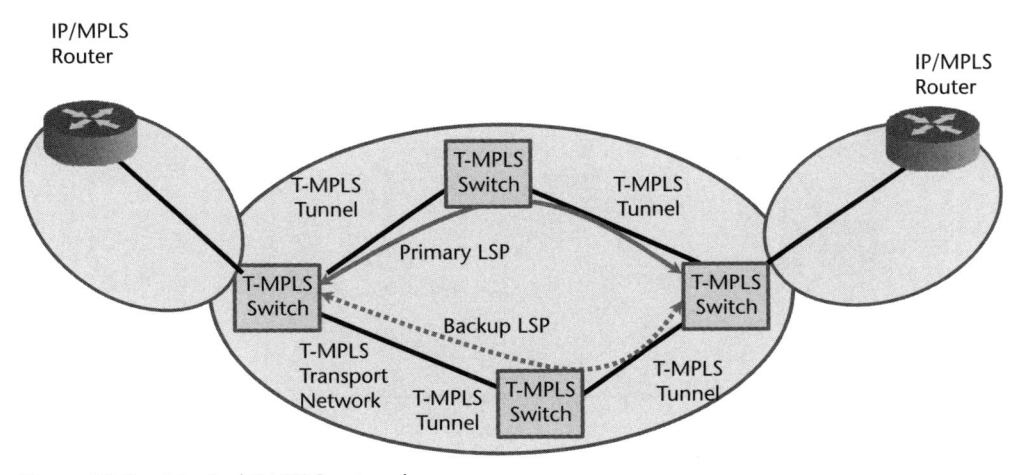

Figure 12.9 A typical T-MPLS network.

protected domain, where a selection between the working and protection transport entities is made based on some predetermined criteria, such as server defect indication.

Figure 12.11 depicts a unidirectional 1 + 1 trail protection scheme where protection switching is performed by the selector at the sink side of the protection domain based on purely local protection sink information. The working (protected) traffic is permanently bridged to the working and protection connection at the source side of the protection domain. If connectivity check packets are used to detect defects of the working and protection connection, they are inserted at the source of the protection domain of both working and protection side and detected and extracted at the sink side of the protection domain.

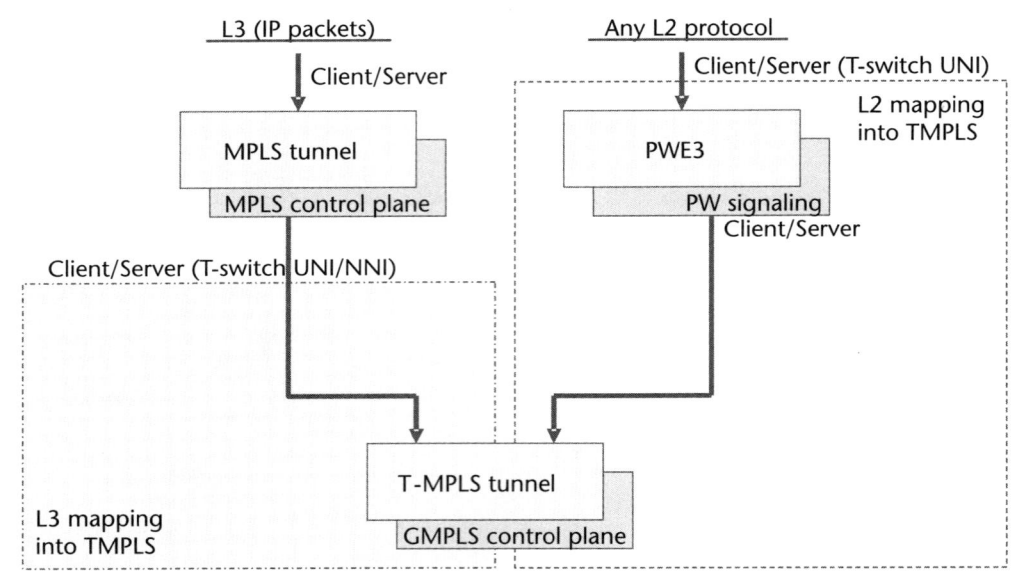

Figure 12.10 Services over T-MPLS tunnels.

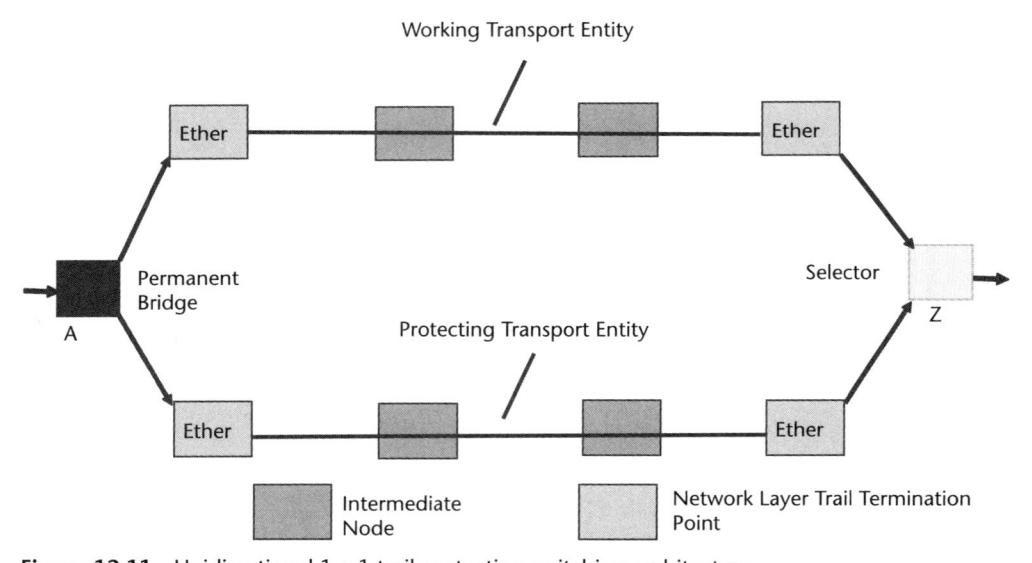

Figure 12.11 Unidirectional 1 + 1 trail protection switching architecture.

For example, if a unidirectional defect in the direction of transmission from node A to node Z occurs for the working connection as in Figure 12.11, this defect will be detected at the sink of the protection domain at node Z and the selector at node Z will switch to the protection connection.

In the 1:1 architecture, the protection transport entity is dedicated to the working transport entity. However, the normal traffic is transported either on the working transport entity or on the protection transport entity using a selector bridge at the source of the protected domain. The selector at the sink of the protected domain selects the entity that carries the normal traffic. Since source and sink need to be coordinated to ensure that the selector bridge at the source and the selector at the sink select the same entity, an APS protocol is necessary.

In 1:1 bidirectional protection switching operation, the protection switching is performed by both the selector bridge at the source side and the selector at the sink side of the protection domain based on local or near-end information and the APS protocol information from the other side or far end.

If connectivity check packets are used to detect defects of the working and protection connection, they are inserted at the both working and protection side.

The protection operation mode can be nonrevertive or revertive. In nonrevertive types, the service will not be switched back to the working connection if the switch requests are terminated. In nonrevertive mode of operation, when the failed connection is no longer in an SF (signal fail) or SD (signal degrade) condition, and no other externally initiated commands are present, a no request state is entered. During this state, switching does not occur.

In revertive types, the service will always return to the working connection if the switch requests are terminated. In revertive mode of operation, under conditions where working traffic is being transmitted via the protection connection and when the working connection is restored, if local protection switching requests have been previously active and now become inactive, a local wait to restore state is entered. This state normally times out and becomes a no request state after the wait-to-restore timer has expired. Then reversion back to select the working connection occurs. The wait-to-restore timer deactivates earlier if any local request of higher priority preempts this state.

Except for the case of 1 + 1 unidirectional switching, an APS signal is used to synchronize the action at the A and Z ends of the protected domain. Request/state type, requested signal, bridged signal, and protection configuration are communicated between A and Z.

The APS payload structure in a T-MPLS OAM frame is undefined. The field values for the APS octets are given in Table 12.4.

APS protocol must prevent misconnections and minimize the number of communication cycles between A and Z ends of the protected domain, in order to minimize the protection switching time. The communication may be once Z → A, twice Z → A and A Z, or three times (Z → A, A → Z, and Z → A). This is referred to as 1-phase, 2-phase, and 3-phase protocols.

For example, in a bidirectional 1:1 trail protection configuration, if a defect in the direction of transmission from node Z to node A occurs for the working connection Z-to-A, this defect will be detected at the node A. The APS protocol initiates the protection switching. The protocol is as follows:

- Node A detects the defect.
- The selector bridge at node A is switched to protection connection A to Z and the merging selector at node A switches to protection connection Z to A.
- The APS command sent from node A to node Z requests a protection switch.
- After node Z validates the priority of the protection switch request, the merging selector at node Z is switched to protection connection A to Z and the selector bridge at the node Z is switched to protection connection Z to

Table 12.4 APS Channels Field Value Description

1111 Lockout of Protection (LP)	1110 Signal Fail for Protection (SF-P)	1101 Forced Switch (FS)
1100 Signal Fail (SF)	1010 Signal Degrade (SD)	1000 Manual Switch (MS)
0110 Wait to Restore (WTR)	Request/State	0100 Exercise (EXER)
0010 Reverse Request (RR)	0001 Do Not Revert (DNR)	0000 No Request (NR)
Others reserved for future international standardization	0 No APS Channel	A
1 APS channel	0 1+1 (Permanent Bridge)	B
1 (1:1)$_n$ (Selector Bridge) ($n \geq 1$)	0 Unidirectional switching	D
1 bidirectional switching	0 Nonrevertive operation	Protection Type
R	1 Revertive operation	0 Null Signal
Requested Signal 1–254, Normal Traffic Signal 1–254	255 Unprotected Traffic Signal	0 Null Signal
Bridged Signal 1–254 Normal Traffic Signal 1–254	255 Unprotected Traffic Signal	

A (i.e., in the Z to A direction the working traffic is sent on both working connection Z to A and protection connection Z to A).

- Then the APS command sent from node Z to node A is used to inform node A about the switching.
- Now the traffic flows on the protection connection.

T-MPLS subnetwork connection protection is used to protect a section of a connection where two separate routes are available. The section can be within an operator's network or multiple operators' networks. Two independent subnetwork connections exist, which act as working and protection transport entities for the protected traffic.

The T-MPLS sublayer trail termination functions provide OAM functions to determine the status of the working and protection T-MPLS sublayer trails.

The protection switching types can be a unidirectional switching type or a bidirectional switching type. In unidirectional switching, only the affected direction of the connection is switched to protection, the selectors at each end are independent. Unidirectional 1+1 SNC protection can be either revertive or non-revertive. This type is applicable for 1 + 1 T-MPLS trail and SNC/S protection.

In bidirectional switching, both directions of the connection, including the affected direction and the unaffected direction, are switched to protection. In this case, APS needs to coordinate the two endpoints. Bidirectional 1:1 SNC/S protection should be revertive. This type is applicable for 1:1 T-MPLS trail and SNC/S protection.

Protection switching can be triggered by:

- Manually via clear, lockout of protection (LP), forced switch, or manual switch commands. Clear command clears all of the externally initiated switch commands. LP fixes the selector position to the working connection, prevents the selector from switching to the protection connection when it is selecting the working connection, and switches the selector from the protection to the working connection when it is selecting the protection connection.

Forced switch switches the selector from the working connection to the protection connection, unless a higher-priority switch request, LP, is in effect. Manual switch (MS) switches the selector from the working connection to the protection connection, unless an equal or higher-priority switch request (i.e., LP, FS, SF, or MS) is in effect.

- Automatically via signal fail or signal degrade associated with a protection domain.
- Protection switching function of wait to restore, reverse request, do not revert, and no request. The wait-to-restore state is only applicable for the revertive mode and applies to a working connection. This state is entered by the local protection switching function in conditions where working traffic is being received via the protection connection when the working connection is restored, if local protection switching requests have been previously active and now become inactive. It prevents reversion back to select the working connection until the wait-to-restore timer has expired. The wait-to-restore timer may be configured by the operator in 1-minute steps between 5 and 12 minutes. An SF or SD condition will override the WTR. The no request state is entered by the local protection switching function under all conditions where no local protection switching requests (including wait to restore) are active.

The target switching time for trail and SNC protection is 50 ms. The switching time excludes the detection time necessary to initiate the protection switch and the hold-off time. A hold-off timer is started when a defect condition is declared and runs for a nonresettable period, which is provisionable from 0 to 10 seconds in steps of 100 ms. When the timer expires, protection switching is initiated if a defect condition is still present at this point.

12.6 Conclusion

T-MPLS is a connection oriented carrier-grade packet transport technology. With OAM&P based on traditional transport concepts, T-MPLS aimed at achieving full OAM capabilities. T-MPLS is able to converge any L2 and L3 protocol and a common packet-based networking technology and exploit a common distributed control plane with lambda and/or TDM-based transport layer. Its OPEX and CAPEX are expected to be lower than MPLS.

ITU-T Recommendations specify the architectural approach of T-MPLS, but are focused on the data plane aspects of T-MPLS. Many other areas require further work in order to develop a fully fledged suite of T-MPLS standards that can be applied to large scale network deployments.

For T-MPLS (or its successor, MPLS-TP) to succeed, it must ultimately address operator concerns about its interoperability with IP/MPLS, its support for multipoint-to-multipoint connections, its potential alignment with recent OAM standards work for Ethernet.

References

[1] ITU-T G.8110, MPLS Layer Network Architecture, 2005.
[2] IETF RFC 3031, Multiprotocol Label Switching Architecture, 2001.
[3] G.8110.1: Architecture of Transport MPLS (T-MPLS) Layer Network, 2006.
[4] G.8112: Interfaces for the Transport MPLS (T-MPLS) Hierarchy (TMH), 2006.
[5] G.8121: Characteristics of Multi-Protocol Label Switched (MPLS) Equipment Functional Blocks, 2006.
[6] ITU-T G.8151, Management Aspects of the T-MPLS Network Element, 2007.
[7] G.8113, Operations, Administration and Maintenance Mechanism for MPLS-TP in Packet Transport Network (PTN), 2011.
[8] ITU-T G.8114 (draft), Operation & Maintenance Mechanisms for T-MPLS Layer Networks, 2008.
[9] ITU-T Y.1711, Operation & Maintenance Mechanism for MPLS Networks, 2004.
[10] ITU-T G.8131, Linear Protection Switching for Transport MPLS (T-MPLS) Networks, 2007.
[11] ITU-T, Draft recommendation G.8132, T-MPLS Shared Protection Ring, 2008.
[12] RFC 3270, Multi-Protocol Label Switching (MPLS) Support of Differentiated Services, May 2002.
[13] ITU-T Y.1731-OAM Functions and Mechanisms for Ethernet Based Networks, May 2006.
[14] IEEE 802.1ag-2007, Virtual Bridged Local Area Networks – Amendment 5: Connectivity Fault Management, December 2007.
[15] ITU-T G.807, Requirements for an Automatic Switched Transport Networks (ASTN), 2004.
[16] ITU-T G.808, (808.1) Generic Protection Switching: Linear Trail and Subnetwork Protection, 2010.
[17] RFC3471, Generalized Multi-Protocol Label Switching (GMPLS) Signaling Functional Description, January 2003.

CHAPTER 13

MPLS Transport Profile

13.1 Introduction

MPLS transport profile (MPLS-TP) has been defined by the IET to be used as a network layer technology in transport networks. It is a continuation of T-MPLS. The objective is to achieve the transport characteristics of Synchronous Optical Network/Synchronous Digital Hierarchy (SONET/SDH) that are connection-oriented and have a high level of availability, quality of service, and extensive operations, administration, and maintenance (OAM) capabilities. MPLS-TP is expected to enable multiprotocol label switching (MPLS) to be deployed in a transport network and operated in a similar manner to existing transport technologies and support packet transport services with a similar degree of predictability to that found in existing transport networks.

The ITU Study Group 15 took MPLS as a starting point and has labored to develop transport MPLS (T-MPLS), a stripped-down version suitable for connection-oriented transport. It is a packet-based transport network that will provide an evolution path for next-generation networks reusing a profile of existing MPLS and complementing it with transport-oriented OAM and protection capabilities. T-MPLS promises multiservice provisioning, multilayer OAM, and protection resulting in optimized circuit and packet resource utilization. The objective is to have an infrastructure for any client traffic type, in any scale, at the lowest cost per bit.

ITU-T approved the first version of T-MPLS architecture in 2006. By 2008, some vendors started supporting T-MPLS in their optical transport products. At the same time, the IETF was working on a new mechanism called pseudowire emulation edge-to-edge (PWE3) that emulates the essential attributes of a service such as ATM, TDM, frame relay, or Ethernet over a packet-switched network (PSN), which can be an MPLS network [1].

A Joint Working Group (JWT) was formed between the IETF and the ITU-T to achieve mutual alignment of requirements and protocols and come up with another approach. The T-MPLS is renamed to MPLS-TP to produce a converged set of standards for MPLS-TP. The MPLS-TP is a packet-based transport technology based on the MPLS traffic engineering (MPLS-TE) and pseudowire (PW) data plane architectures defined in [2–4].

MPLS-TP is client-agnostic that can carry L3, L2, L1 services and physical layer agnostic that can run over IEEE Ethernet PHYs, SONET/SDH and optical transport network (OTN), wavelength division multiplexing (WDM), and so forth. Its OAM functions are similar to those available in traditional optical transport networks such as SONET/SDH and OTN. Several protection schemes are available at the data plane similar to those available in traditional optical transport networks.

With MPLS-TP, network provisioning can be achieved via a centralized network management system (NMS) and/or a distributed control plane. The generalized multiprotocol label switching (GMPLS) can be used as a control plane that provides a common approach for management and control of multilayer transport networks.

Networks are typically operated from a network operation center (NOC) using an NMS that communicates with the network elements (NEs). The NMS provides FCAPS management functions (i.e., fault, configuration, accounting, performance, and security management) as defined in [5].

For MPLS-TP, NMS can be used for static provisioning while the GMPLS can be used dynamic provisioning of transport paths. The control plane is mainly used to provide restoration functions for improved network survivability in the presence of failures and facilitates end-to-end path provisioning across network or operator domains. The operator has the choice to enable the control plane or to operate the network in a traditional way without control plane by means of an NMS. The NMS also needs to configure the control plane and interact with the control plane for connection management purposes.

Static and dynamic provisioning models are possible. The static provisioning model is the simplified version commonly known as static MPLS-TP. This version does not implement even the basic MPLS functions, such as Label Distribution Protocol (LDP) and Resource Reservation Protocol-Traffic Extension (RSVP-TE), since the signaling is static. However, it does implement support for generic associated channel label (GAL) and generic associated channel (G-ACh), which is used in supporting OAM functions. There are several proposals for a dynamic provisioning version, one of which includes GMPLS as a signaling mechanism.

MPLS-TP uses a subset of IP/MPLS standards where features that are not required in transport networks such as IP forwarding, penultimate hop popping (PHP) and equal cost multiple path (ECMP) are not supported (as in T-MPLS) or made optional. However, MPLS-TP defines extensions to existing IP/MPLS standards and introduces established requirements from transport networks. Among the key new features are comprehensive OAM capable of fast detection, localization, troubleshooting and end-to-end SLA verification, linear and ring protection with sub-50-ms recovery, separation of control and data plane, and fully automated operation without control plane using NMS.

MPLS-TP allows for operation in the networks without requiring the nodes to implement IP forwarding. This means that the OAM operation has to be able to operate in IP and non-IP modes. The control plane is optional and the protocols such as OAM and protection are structured to be able to operate fully without a control plane.

The essential features of MPLS-TP are [6]:

13.1 Introduction

- A connection-oriented packet switching technology its architecture based on [7].
- No PHP.
- There is no modification to MPLS forwarding architecture. Existing PW and label switched path (LSP) constructs will be used.
- Bidirectional T-MPLS LSPs are supported by pairing the forward and backward directions to follow the same path (i.e., the same nodes and links).
- Time to live (TTL) is supported according to [8] for only the pipe and short-pipe models (as discussed in Chapter 13).
- Both E-LSP (EXP inferred label-switched path) and L-LSP (label only inferred) are supported.
- Support QoS-guaranteed multiservice by PWE3.
- EXP is supported for only the pipe and short-pipe models.
- Packet loss probability comparable to TDM transport with a single drop precedence (i.e., green color frames) and statistical multiplexing gain with two drop precedence values (i.e., green and yellow frames) are supported. PHB scheduling class (PSC) and drop precedence are inferred directly from the EXP field in the MPLS shim header.
- The model used for TTL and EXP is consistent, either both use the pipe or both use the short-pipe model.
- Both per-platform and per-interface label spaces are supported.
- OAM tools for operation and fault management are based on [9] where the OAM function is responsible for monitoring the LSP/PWE and initiating path recovery actions.
- High reliability, equipment protection, and network protection less than 50 ms where protection switching and survivability are based on [26].
- LSP merging is not supported (e.g., no use of LDP multipoint-to-point signaling in order to avoid losing LSP head-end information).
- Point-to-multipoint multicast (i.e., not multipoint to multipoint) is supported. Similarly, only point-to-multipoint connection is supported while multipoint-to-multipoint connection is not.
- Equal Cost Multipath Protocol is not supported.
- IP forwarding is not required to support of OAM or data packets.
- Different options are supported for signaling communication network (SCN) links:
  - Shared trail SCN links;
  - Shared hop SCN links;
  - Independent SCN links.
- It can be used with static provisioning systems or with control plane. With static provisioning, there will be no dependency on routing or signaling (e.g. GMPLS or, IGP, RSVP, BGP, LDP).

- Be able to interoperate with existing MPLS and PWE control and forwarding planes.
- Support high accuracy timing and clock.

13.2 Frame Format

MPLS-TP uses the MPLS frame format and MPLS label for packet forwarding. In addition, MPLS-TP defines G-ACh to support OAM.

Figure 13.1 illustrates the structure of the MPLS label. The MPLS label is a fixed 4-byte identifier added to the packet to switch the packet to its destination without the need for any routing table (Layer-3) look-ups. MPLS header is called the shim header. One or more labels are pushed on the packet at the ingress router forming a label stack. The first label is called the top label or the transport label, and other labels are used by different MPLS applications if needed.

MPLS label is a short fixed length physically contiguous identifier that is used to identify a forwarding equivalence class (FEC), which aggregates individual flows.

In conventional IP forwarding, a particular router will typically consider two packets to be in the same FEC if there is some address prefix X in that router's routing tables such that X is the longest match for each packet's destination address. As the packet traverses the network, each hop in turn reexamines the packet and assigns it to an FEC.

The 3 bits of EXP (experimental bits) field represent eight packet priorities, called traffic class.

The stack (S) field (1-bit) supports a hierarchical label stack. A packet with no label can be thought of as a packet whose label stack is empty (i.e., whose label stack has depth 0) (see Figure 13.2). If a packet's label stack is of depth m, the label at the bottom of the stack is called the level-1 label and the label at the top is called the level-m label. The label stack is used for routing packets through LSP tunnels.

The TTL field (8 bits) provides conventional IP TTL functionality.

To support FCAPS (fault, configuration, accounting, performance, and security), G-ACh is created that runs on PWE3 and carries OAM messages.

For PWs, the MPLS G-ACh uses the first 4 bits of the PW control word to provide the initial discrimination between data packets and packets belonging to the associated channel [10] (Figure 13.3). The first 32 bits following the bottom of

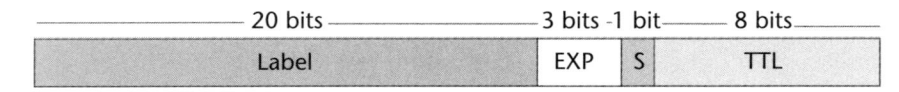

20 bits	3 bits	1 bit	8 bits
Label	EXP	S	TTL

Figure 13.1 MPLS packet format.

Label	EXP	S	TTL	Label	EXP	S	TTL	Optional Control word	Ethernet Service Payload

Figure 13.2 MPLS-TP packet format for PWs.

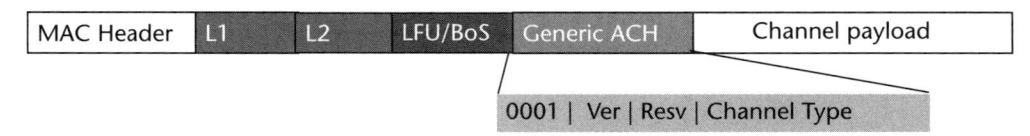

| MAC Header | L1 | L2 | LFU/BoS | Generic ACH | Channel payload |

0001 | Ver | Resv | Channel Type

Figure 13.3 Generic exception label and generic associated channel for LSP monitoring and alarming.

stack label then have a defined format called an associated channel header (ACH), which further defines the content of the packet. The ACH is both a demultiplexer for G-ACh traffic on the PW, and a discriminator for the type of G-ACh traffic.

When the control message is carried over a section or an LSP, rather than over a PW, a reserved label, Transport Alert Label as a Label For yoU (LFU), with a value of 13 at the bottom of the label stack (Figure 13.3), indicates that the packet payload is something other than a client data packet. This reserved label is referred to as the GAL as defined in [11]. When a GAL is found, it indicates that the payload begins with an ACH. The GAL is a demultiplexer for G-ACh traffic on the section or the LSP, and the ACH is a discriminator for the type of traffic carried on the G-ACh.

MPLS-TP forwarding follows the normal MPLS model, and thus a GAL is invisible to an LSR unless it is the top label in the label stack. The only other circumstance under which the label stack may be inspected for a GAL is when the TTL has expired.

13.3 Architecture

MPLS-TP architectural requirements are defined in [6]. They are briefly:

- MPLS-TP does not modify the MPLS forwarding architecture and is based on existing PW and LSP constructs.
- Point-to-point LSPs may be unidirectional or bidirectional. Congruent bidirectional LSPs can be constructed.
- MPLS-TP LSPs do not merge with other LSPs at an MPLS-TP label switch router (LSR). If a merged LSP is created, it will be detected.
- Packets are forwarded solely based on switching the MPLS or PW label. LSPs and/or PWs can be established in the absence or presence of a dynamic control plane (i.e., dynamic routing or signaling).
- OAM and protection mechanisms and forwarding of data packets operate without IP forwarding support.
- LSPs and PWs are monitored through the use of OAM in the absence of control-plane or routing functions.

The MPLS-TP layered architecture supporting the requirements above is depicted in Figure 13.4. The LSP (PSN tunnel) can transport multiple PWs, each demultiplexed by a unique PW MPLS label. A 4-octet control word may be added

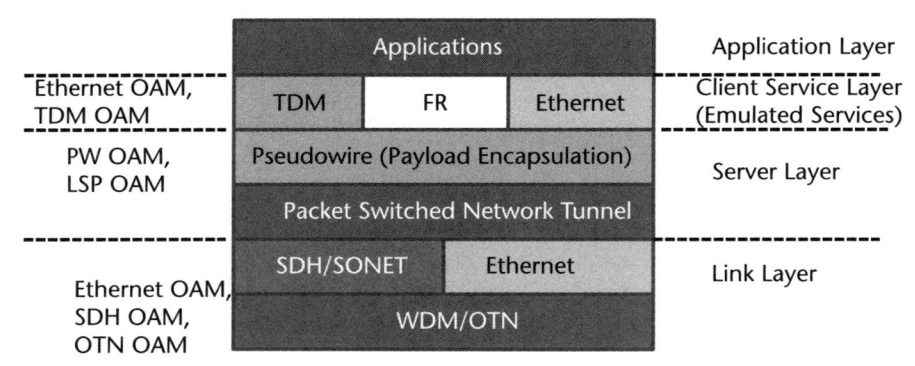

Figure 13.4 MPLS-TP network layering.

to the MPLS payload field to identify different payload types encapsulated in PW such as Ethernet, frame relay, ATM, PPP, PDH, and SDH.

For a PW server layer, the client signal is an attachment circuit (AC) (Figure 13.5 and 13.6). An AC may be an Ethernet port, an Ethernet VLAN, a frame relay DLCI, or a frame relay port. A PW forwarder binds an AC to a particular PW. The PW encapsulates service-specific PDUs or circuit data received from attachment circuits, carries encapsulated data over a PSN tunnel, and manages the signaling, timing, order or other aspects of the service at the boundaries of the PW. From an OAM perspective, the PW provides status and alarm management for each MPLS-TP service instance.

The relationship between the client layer network and the MPLS-TP server layer network is defined by the MPLS-TP network boundary and the label context (Figure 13.7). It is not explicitly indicated in the packet. In terms of the MPLS label stack, when the native service traffic type is itself MPLS-labeled, then the S bits of all the labels in the MPLS-TP label stack carrying that client traffic are 0. Otherwise, the bottom label of the MPLS-TP label stack has the S bit set to 1.

The S bit is used to identify when MPLS label processing stops and network layer processing starts. Only one label stack entry (LSE) contains the S bit (bottom of stack bit) set to 1.

13.3.1 Data Plane

MPLS-TP employs a standard MPLS data plane as defined in [2]. Sections, LSPs, and PWs (Figure 13.8) are the data transport entities providing a packet transport service. LSPs and PWs abstract the data plane of the client layer network from the

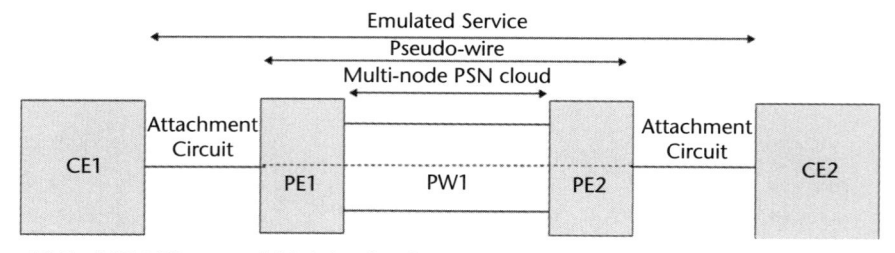

Figure 13.5 MPLS-TP network high level architecture.

13.3 Architecture

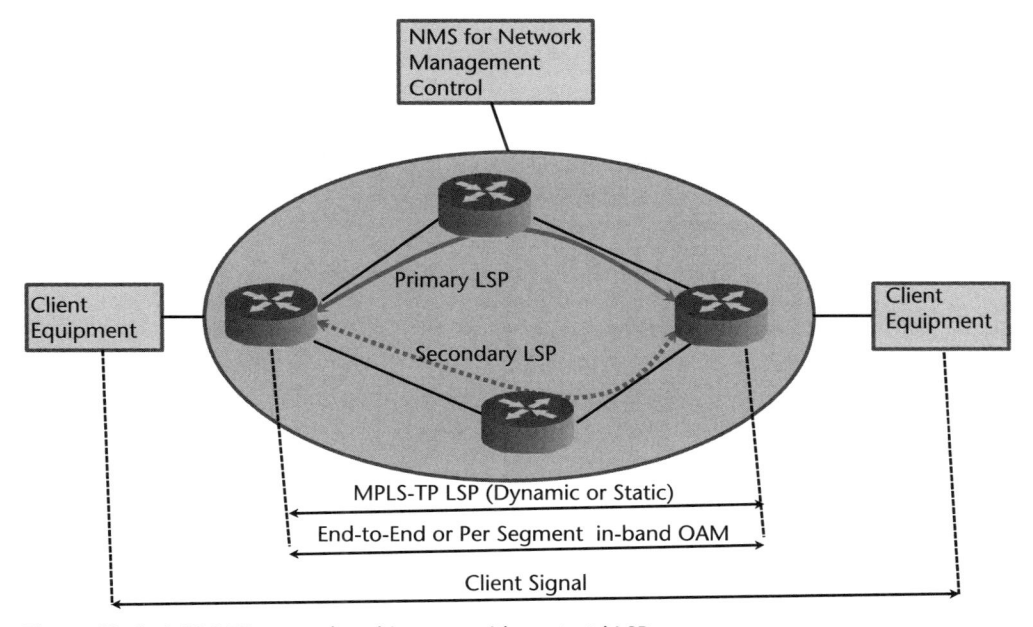

Figure 13.6 MPLS-TP network architecture with protectd LSPs.

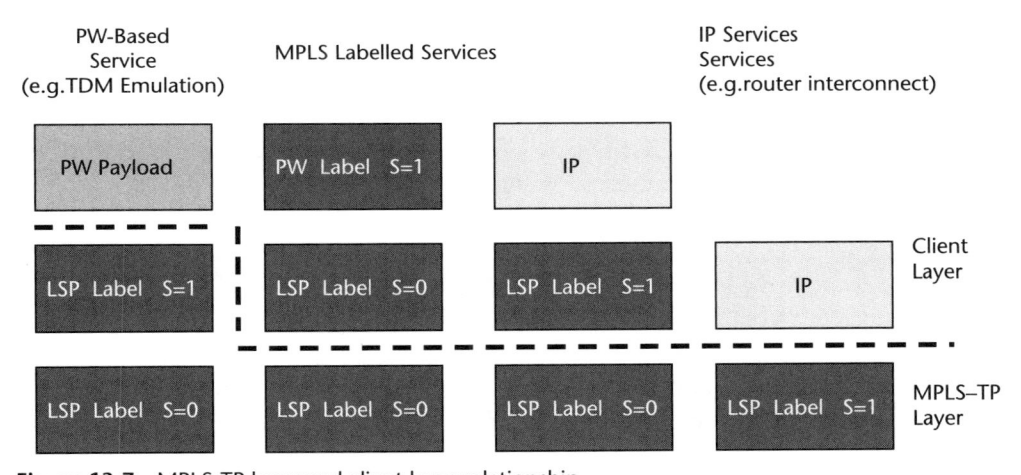

Figure 13.7 MPLS-TP layer and client layer relationship.

MPLS-TP data plane by encapsulating payload. The PWs can emulate services such as Ethernet, frame relay, or PPP/high-level data link control (HDLC). The adaptation supports IP packets and MPLS-labeled packets services such as PWs, Layer-2 VPNs, and Layer-3 VPNs.

In addition, the data plane provides MPLS-TP forwarding function based on the label that identifies the transport path (i.e., LSP or PW). The label value specifies the processing operation to be performed by the next hop at that level of encapsulation.

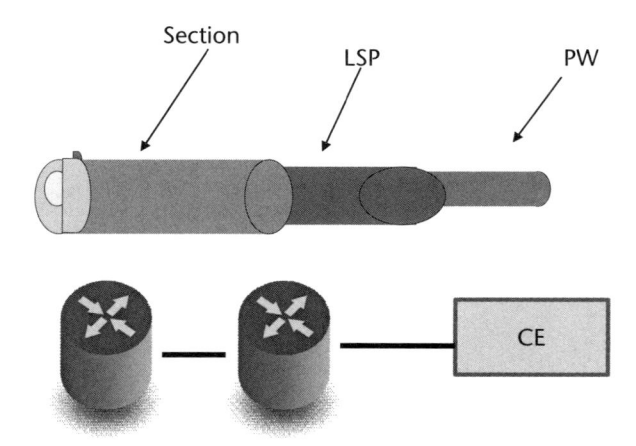

Figure 13.8 LSP, section, PW.

The lowest server layer provided by MPLS-TP is an MPLS-TP LSP. The client layers of an MPLS-TP LSP may be network-layer protocols, MPLS LSPs, or PWs. Therefore, the LSP payload can be network-layer protocol packets and PW packets.

The links traversed by a layer N+1 MPLS-TP LSP are Layer-N MPLS-TP sections. Such an LSP is referred to as a client of the section layer, and the section layer is referred to as the server layer with respect to its clients.

The MPLS label stack associated with an MPLS-TP section at layer N consists of N labels, in the absence of stack optimization mechanisms. In order for two LSRs to exchange non-IP MPLS-TP control packets over a section, an additional label, the GAL must be at the bottom of the label stack.

A section provides a means of identifying the type of payload it carries. If the section is a data link, link-specific mechanisms such as a protocol type indication in the data-link header may be used. If the section is an LSP, the payload is implied by the LSP label. If the LSP payload is MPLS-labeled, the payload is indicated by the setting of the S bit (Figure 13.7).

LSP can be point-to-point unidirectional, point-to-point associated bidirectional, point-to-point co-routed bidirectional, and point-to-multipoint unidirectional. Point-to-point unidirectional LSPs are supported by the basic MPLS architecture. Multipoint-to-point and multipoint-to-multipoint LSPs are not supported.

A point-to-point associated bidirectional LSP between LSR A and LSR B consists of two unidirectional point-to-point LSPs, one from A to B and the other from B to A, which are regarded as a pair providing a single logical bidirectional transport path.

A point-to-point co-routed bidirectional LSP is a point-to-point associated bidirectional LSP with the additional constraint that its two unidirectional component LSPs in each direction follow the same nodes and links (i.e., same path).

In point-to-multipoint unidirectional LSP functions, an LSR may have more than one pair (egress interface, outgoing label) associated with the LSP, and any packet on the LSP is transmitted out to all associated egress interfaces.

MPLS-TP LSPs use the MPLS label-switching operations and TTL processing procedures defined in [2, 17, 18]. In addition, MPLS-TP PWs use the single-segment PWs (SS-PW) and optionally the multisegment PWs (MS-PW) forwarding operations defined in [4, 19].

MPLS-TP supports quality of service capabilities via the MPLS differentiated services (DiffServ) architecture [20]. Both E-LSP (Exp-Inferred-LSP) and L-LSP (label-only-inferred LSP) MPLS DiffServ modes are supported. Data-plane quality-of-service capabilities are included in the MPLS-TP in the form of traffic-engineered (TE) LSPs [21]. The uniform, pipe, and short pipe DiffServ tunneling and TTL processing in [18, 20] may be used for MPLS-TP LSPs.

ECMP load-balancing is not performed on an MPLS-TP LSP. However, MPLS-TP LSPs may operate over a server layer that supports load-balancing, which is transparent to MPLS-TP.

PHP is disabled by default on MPLS-TP LSPs.

13.3.2 MPLS-TP Router Types

An MPLS-TP LSR is either an MPLS-TP provider edge (PE) router or an MPLS-TP provider (P) router for a given LSP. An MPLS-TP PE router is an MPLS-TP LSR that adapts client traffic and encapsulates it to be transported over an MPLS-TP LSP by pushing a label or using a PW. An MPLS-TP PE exists at the interface between a pair of layer networks. For an MS-PW, an MPLS-TP PE may be either a switching PE (S-PE) or a terminating PE (T-PE).

An MPLS-TP label edge router (LER) is an LSR that exists at the endpoints of an LSP and therefore pushes or pops the LSP label.

A PE, which pushes or pops an LSP label, resides at the edge of a given MPLS-TP network domain of a service provider's network. It has links to another MPLS-TP network domain or to a customer edge (CE), except for the case of a PW S-PE router, which is not restricted to the edge of an MPLS-TP network domain.

An MPLS-TP provider router is an MPLS-TP LSR that switches LSPs that carry client traffic, but does not adapt client traffic and encapsulate it to be carried over an MPLS-TP LSP. A provider router does not have links to other MPLS-TP network domains.

A PW S-PE router is a PE capable of switching the control and data planes of the preceding and succeeding PW segments in an MS-PW. It terminates the PSN tunnels of the preceding and succeeding segments of the MS-PW and includes a PW switching point for an MS-PW. Furthermore, an S-PE can exist anywhere a PW must be processed or policy applied.

A PW T-PE router is a PE where ACs are bound to a PW forwarder. A terminating PE is present in the first and last segments of an MS-PW.

CE equipment is on either side of the MPLS-TP network are peers and view the MPLS-TP network as a single link.

13.3.3 Service Interfaces

An MPLS-TP network consists of two layers, the transport service layer and the transport path layer. The transport service layer provides the interface between CE nodes and the MPLS-TP network. Each packet transmitted by a CE node for transport over the MPLS-TP network is associated at the receiving MPLS-TP PE node with a single logical point-to-point connection at the transport service layer between the ingress PE and the corresponding egress PE to which the peer CE is attached. Such a connection is called an MPLS-TP transport service instance, and the set of client packets belonging to the native service associated with such an instance on a particular CE-PE link is called a client flow.

The transport path layer provides aggregation of transport service instances over LSPs as well as aggregation of transport paths.

An MPLS-TP PE node can provide two types of interface to the transport service layer, user-network interface (UNI) providing the interface between a CE and the MPLS-TP network (Figure 13.9) and network-network interface (NNI) providing the interface between two MPLS-TP PEs in different administrative domains (Figure 13.10). A UNI service interface may carry only network layer clients by MPLS-TP LSPs or both network-layer and non-network-layer clients where a PW is required to adapt the client traffic received over the service interface. An NNI service interface may be to an MPLS LSP or a PW.

The UNI for a particular client flow may or may not involve signaling between the CE and PE, and if signaling is used, it may or may not traverse the same

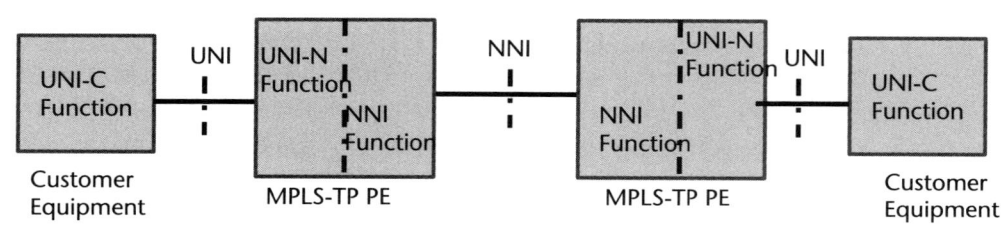

Figure 13.9 Transport service layer interfaces.

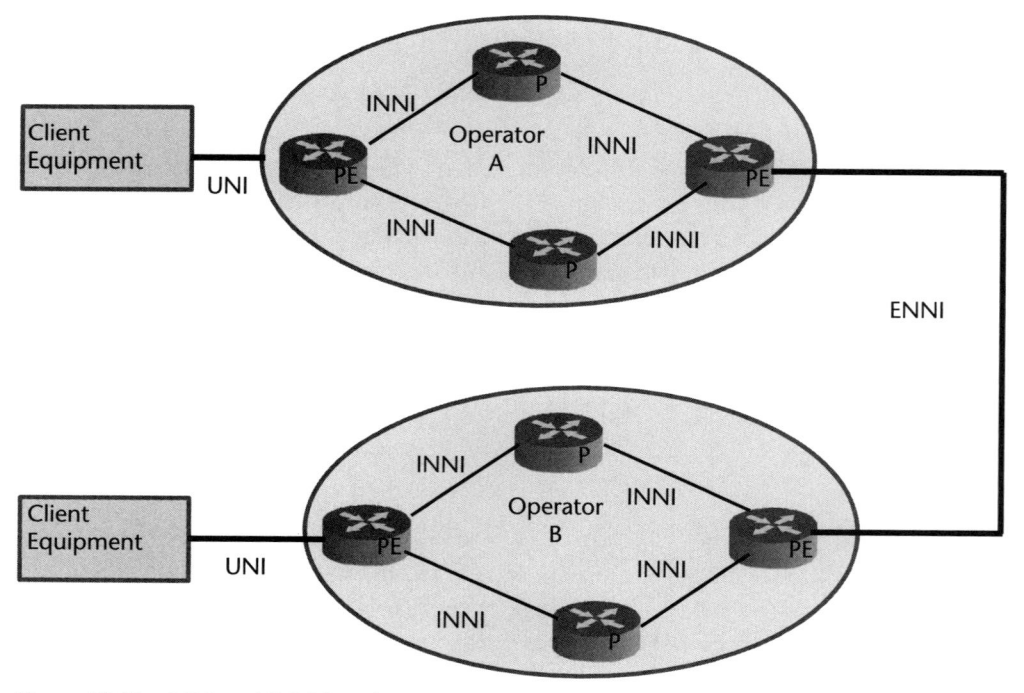

Figure 13.10 INNI and ENNI interfaces.

attachment circuit that supports the client flow. Similarly, the NNI for a particular transport service instance may or may not involve signaling between the two PEs, and if signaling is used, it may or may not traverse the same data-link that supports the service instance.

13.3.4 IP Transport Service

MPLS-TP network can provide a point-to-point IP transport service between attached CE nodes. The IP packet is extracted from the link-layer frame and associated with a service LSP based on the source MAC address and the IP protocol version. The packet formed after encapsulation by adding the service LSP label to the frame is mapped to a tunnel LSP where the tunnel LSP label is pushed, and the packet is transmitted over the outbound interface associated with the LSP.

For packets received over a tunnel LSP carrying the service LSP label, the steps are performed in the reverse order.

MPLS-TP uses PWs to provide a virtual private wire service (VPWS), a virtual private local area network service (VPLS), a virtual private multicast service (VPMS), and an Internet Protocol local area network service (IPLS) [22–24]. PWs and their associated labels may be configured or signaled.

If the MPLS-TP network provides a Layer-2 interface as a service interface to carry both network-layer and non-network-layer traffic, then a PW is required to support the service interface. The PW is a client of the MPLS-TP LSP server layer. Multisegment PWs may be used to provide a packet transport service.

Figure 13.11 depicts an MPLS-TP network with single segment PWs while Figure 13.12 shows the architecture for an MPLS-TP network when multisegment PWs are used. The corresponding MPLS-TP protocol stacks including PWs are shown in Figure 13.13. In this figure, the transport service [25] is identified by the PW demultiplexer (demux) label, and the transport path layer is identified by the LSP demux label [25].

With network-layer adaptation, the MPLS-TP domain provides either a unidirectional or bidirectional point-to-point connection between two PEs in order to deliver a packet transport service to attached CE nodes. For example, a CE may be an IP, MPLS, or MPLS-TP node. As shown in Figures 13.14 and 13.15, there is an attachment circuit between the CE1 and PE1 providing the service interface, a bidirectional LSP across the MPLS-TP network to the corresponding PE node on

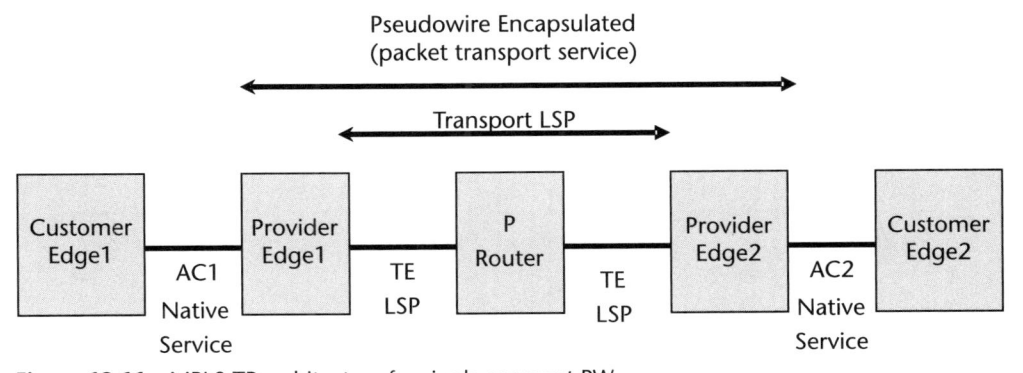

Figure 13.11 MPLS-TP architecture for single-segment PW.

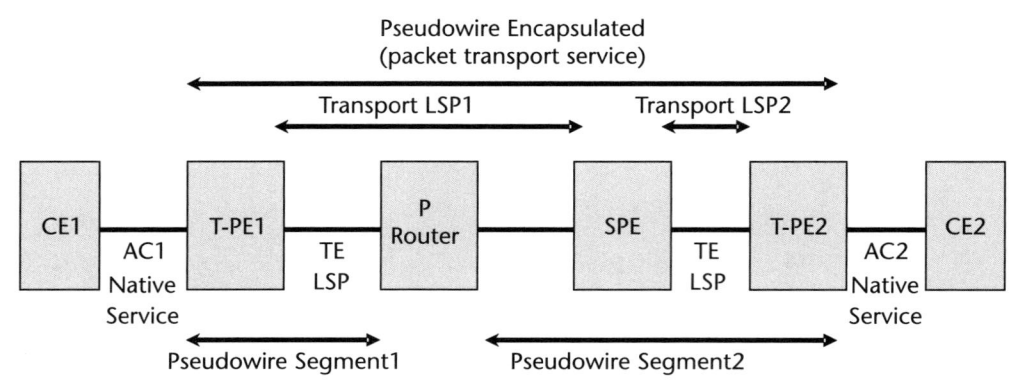

Figure 13.12 MPLS-TP architecture (multisegment PW).

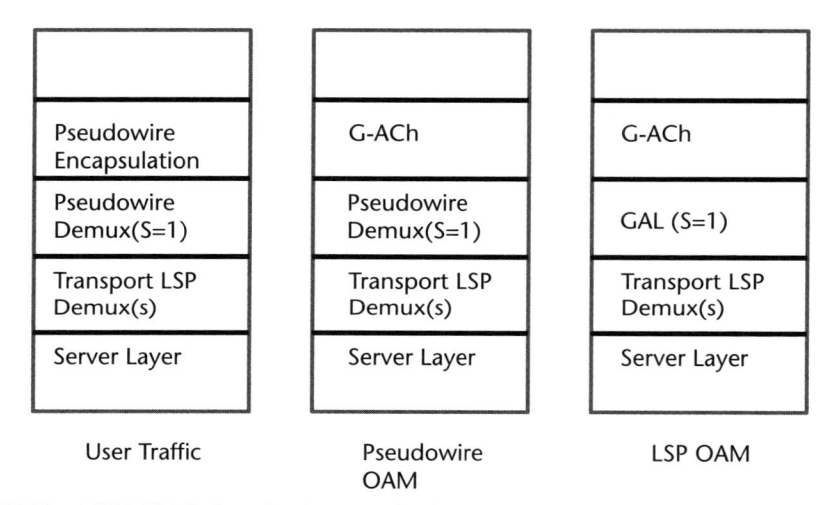

Figure 13.13 MPLS-TP label stack using pseudowires.

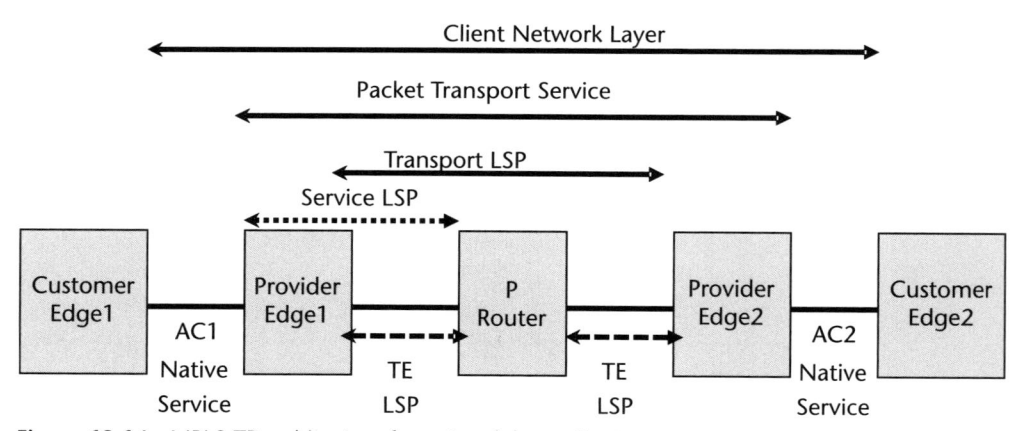

Figure 13.14 MPLS-TP architecture for network-layer clients.

13.3 Architecture

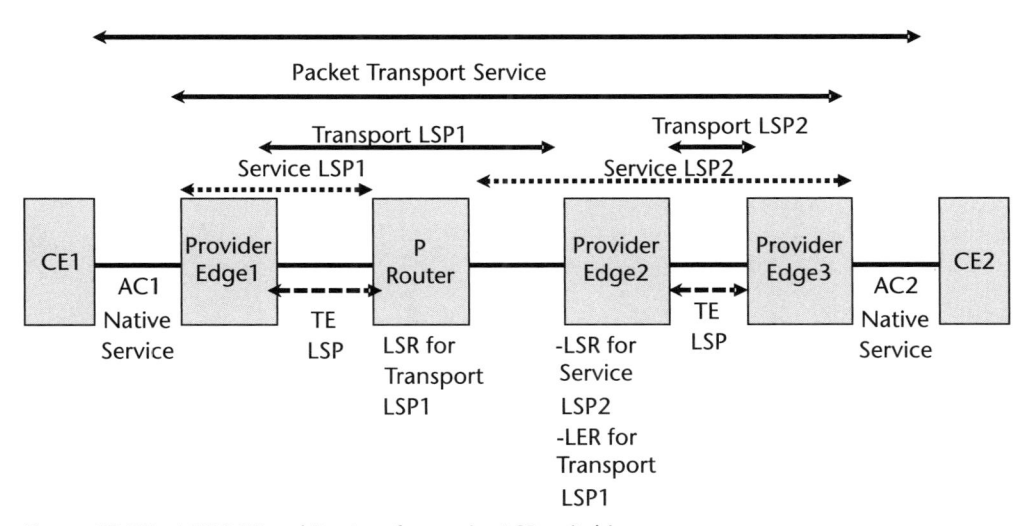

Figure 13.15 MPLS-TP architecture for service LSP switching.

the right, and an attachment circuit between that PE node and the corresponding CE node for this service.

Client packets are received at the ingress service interface of the PE. The PE pushes one or more labels onto the client packets that are then label switched over the transport network. Correspondingly, the egress PE pops any labels added by the MPLS-TP networks and transmits the packet for delivery to the attached CE via the egress service interface.

In Figure 13.16, the Transport Service layer [25] is identified by the service LSP (SvcLSP) demultiplexer (demux) label, and the transport path layer [25] is identified by the transport (Trans) LSP demux label. The functions of the encapsulation label and the service label (SvcLSP demux) shown above may alternatively be represented by a single label stack entry. The S bit is always zero when the client layer

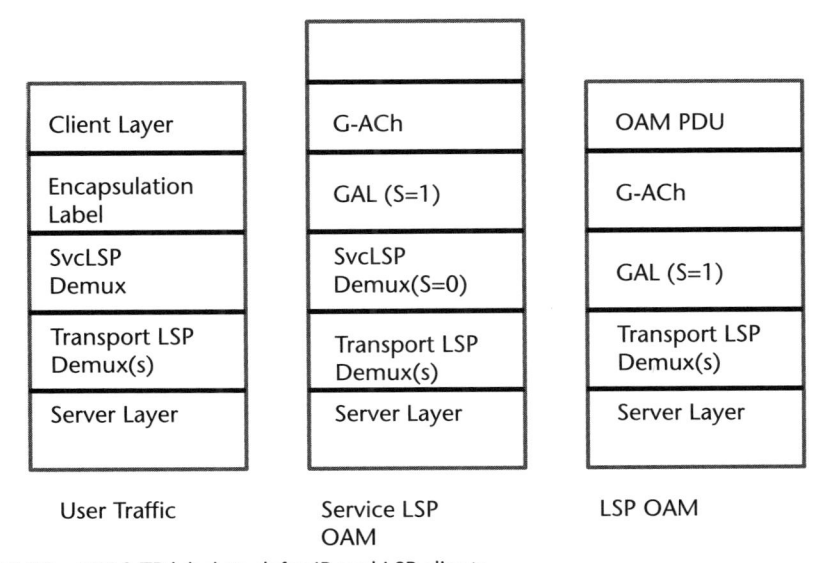

Figure 13.16 MPLS-TP label stack for IP and LSP clients.

is MPLS-labeled. It may be necessary to swap a service LSP label at an intermediate node.

Within the MPLS-TP transport network, the network-layer protocols are carried over the MPLS-TP network using a logically separate MPLS label stack, which is the server stack. The server stack is entirely under the control of the nodes within the MPLS-TP transport network and not visible outside that network. Figure 13.16 shows how a client network protocol stack (which may be an MPLS label stack and payload) is carried over a network layer client service over an MPLS-TP transport network.

A label may be used to identify the network-layer protocol payload type. Therefore, when multiple protocol payload types are to be carried over a single-service LSP, a unique label stack entry needs to be present for each payload type. Such labels are referred to as encapsulation labels (Figure 13.16). An encapsulation label may be either configured or signaled.

Both an encapsulation label and a service label should be present in the label stack when a particular packet transport service is supporting more than one network-layer protocol payload type.

Service labels are typically carried over an MPLS-TP transport LSP edge-to-edge. An MPLS-TP transport LSP is represented as an LSP transport demux label (Figure 13.16). Transport LSP is commonly used when more than one service exists between two PEs.

13.3.5 Generic Associated Channel

The MPLS generic associated channel (G-ACh) [11] is an auxiliary logical data channel supporting control, management, and OAM traffic associated with MPLS-TP transport entities, MPLS-TP sections, LSPs, and PWs in the data plane.

For correct operation of OAM mechanisms, OAM packets follow the same path that data frames use, which is called fate-sharing. In addition, in MPLS-TP it is necessary to discriminate between user data payloads and other types of payload such as packets associated with a signaling communication channel (SCC) or a channel used for a protocol to coordinate path protection state. This is achieved by either a generic control channel associated to the LSP, PW, or section with no IP encapsulation or an IP encapsulation.

MPLS-TP makes use of G-ACh to support FCAPS functions by carrying packets related to OAM, a protocol used to coordinate path protection state, SCC, management communication channel (MCC), or other packet types in-band over LSPs, PWs, or sections. The G-ACh is similar to the PW-associated channel [10], which is used to carry OAM packets over PWs. The G-ACh is indicated by an ACH, presenting for all sections, LSPs, and PWs that make use of FCAPS functions supported by the G-ACh.

Figure 13.17 shows the reference model depicting how the control channel is associated with the pseudowire protocol stack. This is based on the reference model for virtual circuit connection verification (VCCV) in [12].

PW associated channel messages that control the PW PDUs are encapsulated using the PWE3 encapsulation such that they are handled and processed in the same manner as the PW PDUs.

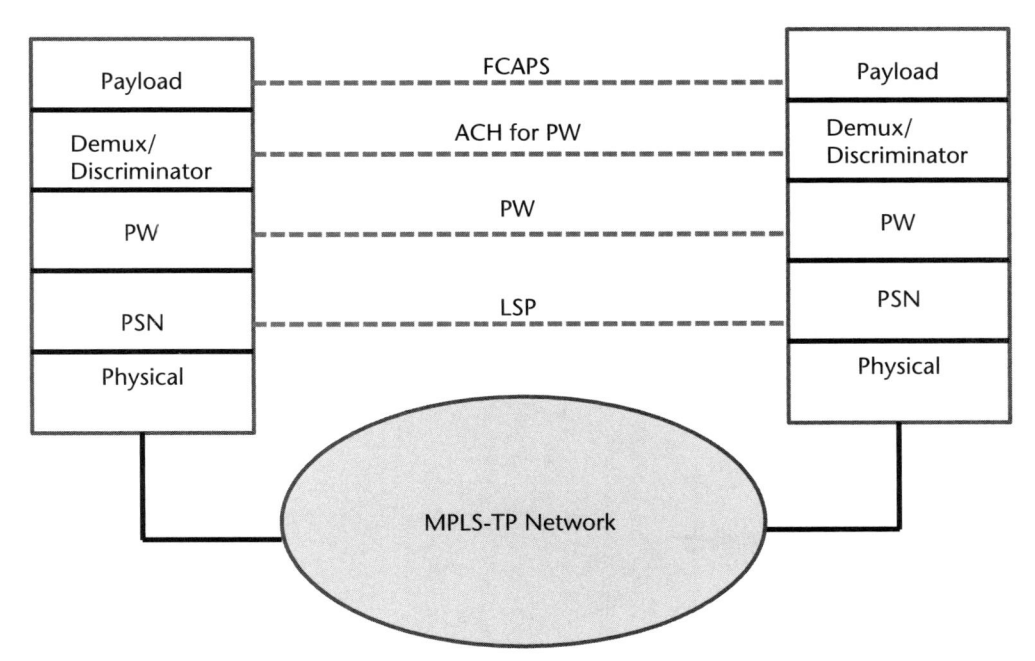

Figure 13.17 PWE3 Protocol stack reference model showing the G-ACh [13].

Figure 13.18 shows the reference model depicting how the control channel is associated with the LSP protocol stack.

13.3.6 Control Plane

The control plane is responsible for end-to-end, segment LSPs and PWE-3 application labels, determining and defining primary and backup paths, and configuring the OAM function along the path. OAM is responsible for monitoring and driving switches between primary and backup paths for the end-to-end path and path segments.

Figure 13.19 illustrates the interactions between the MPLS-TP control plane, the forwarding plane, the management plane, and OAM for point-to-point MPLS-TP LSPs or PWs.

The use of a control plane protocol is optional in MPLS-TP. LSPs and PWs can be provisioned by a network management system similar to provisioning a SONET network. However, they can be configured by the network using generalized (G)-MPLS and Targeted Label Distribution Protocol (T-LDP), respectively. G-MPLS is based on the TE extensions to MPLS (MPLS-TE). It may also be used to set up the OAM function and define recovery mechanisms. T-LDP is part of the PW architecture and is widely used today to signal PWs and their status.

A distributed dynamic control plane may be used to enable dynamic service provisioning. NMS may be centralized or distributed while control plane is distributed. The control plane may be transported over the server layer, an LSP, or a G-Ach, and capable of activating MPLS-TP OAM functions such as fault detection and localization.

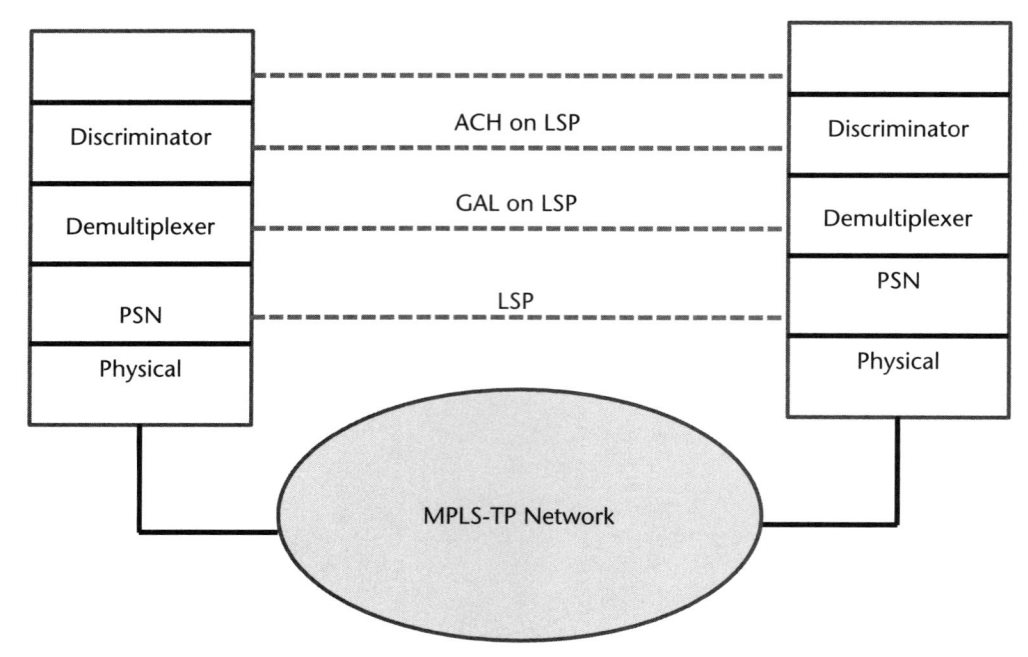

Figure 13.18 MPLS Protocol stack reference model showing the LSP-associated control channel [13]

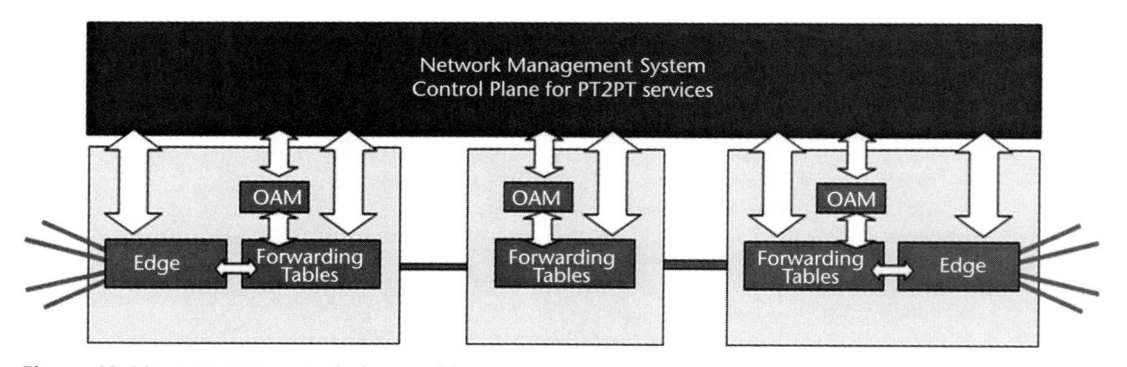

Figure 13.19 MPLS-TP control plane architecture context.

The distributed MPLS-TP control plane may provide signaling, routing, traffic engineering, and constraint-based path computation. It is based on existing MPLS and PW control plane protocols, GMPLS signaling for LSPs, and targeted LDP (T-LDP) signaling for PWs. However, PW control and maintenance take place separately from LSP tunnel signaling.

In a multidomain environment, the MPLS-TP control plane supports different types of interfaces at domain boundaries or within the domains such as UNI, internal network-network interface (INNI), and external network-network interface (ENNI).

The MPLS-TP control provides functions to ensure its own survivability and to enable it to recover gracefully from failures and degradations. These include graceful restart and hot redundant configurations. The control plane is logically decoupled from the data plane such that a control-plane failure does not imply a failure of the existing transport paths.

A PW or LSP may be statically configured without the support of a dynamic control plane. Static operation is independent for a specific PW or LSP instance; therefore, it is possible to configure a PW statically while setting up the LSP by a dynamic control plane. When static configuration mechanisms are used, care must be taken to ensure that loops are not created.

In order to monitor, protect, and manage a segment or multisegments of an LSP, a hierarchical LSP [2], called a subpath maintenance element (SPME), which does not carry user traffic is instantiated. For example, in Figure 13.20, two SPMEs are configured to allow monitoring, protection, and management of the LSP concatenated segments. One SPME is defined between LER2 and LER3, and a second SPME is set up between LER4 and LER5. Each of these SPMEs may be monitored, protected, or managed independently [22].

13.3.7 Network Management

The network management architecture and requirements for MPLS-TP are based on ITU-T G.7710/Y.1701 [14].

Element management system (EMS) can be used to manage MPLS-TP NEs such as LSR, LER, PE, S-PE, or T-PE. The management communication channel (MCC), realized by the G-ACh, provides a logical operations channel between NEs. Management at the network level can be provided by a network management system (NMS). Between EMS and NMS, Netconf [15], SNMP, or CORBA

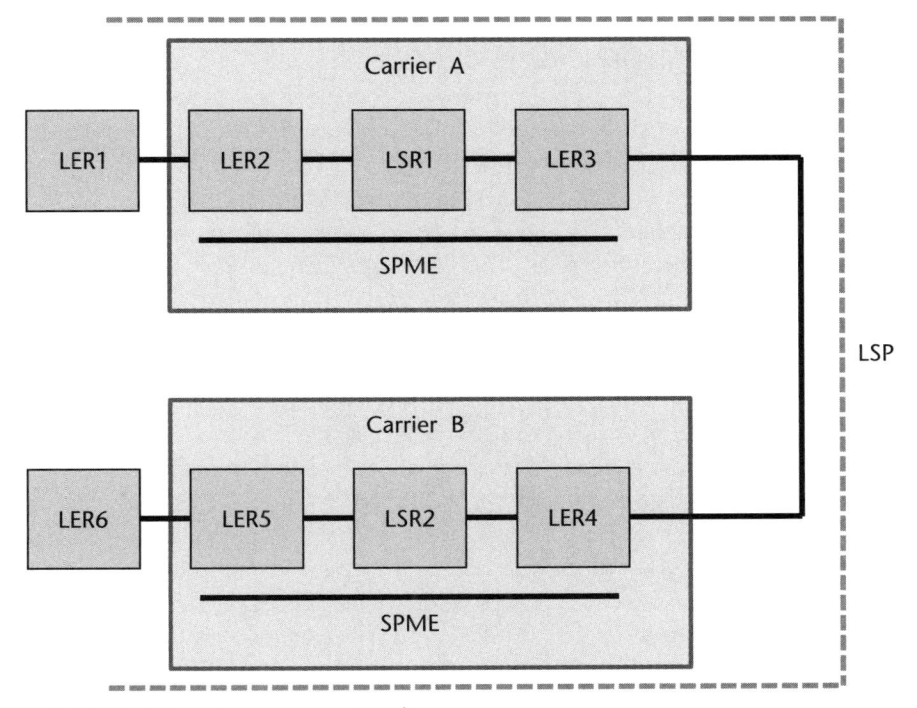

Figure 13.20 SPMEs in intercarrier network.

(Common Object Request Broker Architecture) protocol may be used. With this architecture, FCAPS can be realized.

13.4 OAM

The MPLS-TP OAM architecture supports a wide range of OAM functions to verify connectivity, monitor path performance, and to generate, filter, and manage local and remote defect alarms. These functions are applicable to any layers, MPLS-TP sections, LSPs, and PWs.

The MPLS-TP OAM functions are able to perform without relying on a dynamic control plane or IP functionality in the data path. In an MPLS-TP network where IP functionality is available, all existing IP/MPLS OAM functions such as LSP Ping, Bidirectional Forward Detection (BFD), Connectivity Check and Connectivity Verification may be used. The default mechanism for OAM demultiplexing in MPLS-TP LSPs and PWs is the generic associated channel [11]. This uses a combination of an ACH and a GAL to create a control channel associated to an LSP, section, or PW. Forwarding based on IP addresses for OAM or user data packets is not required for MPLS-TP.

References [27] and BFD for MPLS LSPs [28] defined mechanisms that enable an MPLS LSR to identify and process MPLS OAM packets when the OAM packets are encapsulated in an IP header. These mechanisms are based on TTL expiration and/or use an IP destination address in the range 127/8 for IPv4.

OAM and monitoring in MPLS-TP are based on the concept of maintenance entities, where a maintenance entity (ME) can be viewed as the association of two maintenance entity group endpoints (MEPs). A maintenance entity group (MEG) is a collection of one or more MEs that belongs to the same transport path and that are maintained and monitored as a group. The MEPs limit OAM responsibilities of an OAM flow to the domain of a transport path or segment, in the layer network being monitored and managed. A MEG may also include a set of maintenance entity group intermediate points (MIPs). MEP and MIP support continuity checks and connectivity verification messages between the two endpoints, allowing the detection of lost connectivity and connection misconfiguration. The continuity check (CC)/continuity verification (CV) messages can be exchanged with a period that can be set from 3.3 ms to 10 minutes, thus allowing for very fast failure detection. In summary, service OAM concepts described in Chapter 7 is pretty much carried over to MPLS-TP.

MPLS-TP OAM also supports performance monitoring, including delay and loss measurements defined in [29] to detect performance degradations.

OAM packets follow the same path as traffic data using the G-ACh. A G-ACh packet may be directed to an individual MIP along the path of an LSP or MS-PW by setting the appropriate TTL in the label stack entry for the G-ACh packet. When the location of MIPs along the LSP or PW path is not known by the MEP, the LSP path tracing may be used to determine the appropriate setting for the TTL to reach a specific MIP. Any node on an LSP and PW can send an OAM packet on that LSP. An OAM packet can only be received to be processed at an LSP endpoint, a PW endpoint, or on the expiry of the TTL in the LSP or PW label stack entry.

The MPLS-TP OAM framework is applicable to sections, LSPs, MS-PWs and SPMEs. It supports co-routed and associated bidirectional or unidirectional point-to-point transport paths and point-to-multipoint transport paths. Management, control, and OAM protocol functions may need response packets to be delivered from the receiver back to the originator of a message exchange.

Let's assume U is the upstream LER (Figure 13.21), and D is the downstream LER and Y is an intermediate LSR along the LSP. Return path packet transmission can be from D to U, or from Y to U, or from D to Y. In-band or out-of-band return paths can be used for the traffic from a downstream node D to an upstream node U. The LSP between U and D is bidirectional, and therefore D has a path via the MPLS-TP LSP to return traffic back to U, or D has an out-of-band mechanism for directing traffic back to U.

An arbitrary part of a transport path that can be monitored via OAM independent of the end-to-end monitoring (OAM) is called a tandem connection [7]. Tandem connection monitoring (TCM) allows monitoring of a subset or a segment of LSP in terms of connectivity, fault, quality, and alarms. In MPLS-TP, TCM can be implemented with path segment tunnels (PST), enabling a subset of the segments of LSP or MS-PWs to be monitored independently of any end-to-end OAM.

OAM packets are subject to the same forwarding treatment as the user data packets; therefore, they have the same per-hop behavior (PHB) scheduling class (PSC) in both E-LSP and L-LSP.

13.4.1 OAM Hierarchy

OAM hierarchy for MPLS-TP maybe expressed as (Figure 13.22):

- OAM for physical layer involved in power level and signal quality, including alarms such as loss of light, loss of signal (LoS), loss of frame (LoF), errored second (ES), and severely errored second (SES);
- OAM at link layer such as 802.3ah functions (i.e., facility loopback, AIS, RDI and Dying Gasp);
- OAM at the LSP layer that is the OAM for end-to-end and segment LSPs;
- OAM for PW layer.

The MPLS-TP OAM framework builds upon the concept of an MEG and its associated MEPs and MIPs, as described above.

MEs define a relationship between two points (MEPs) of a transport path to which maintenance and monitoring operations apply. The collection of one or

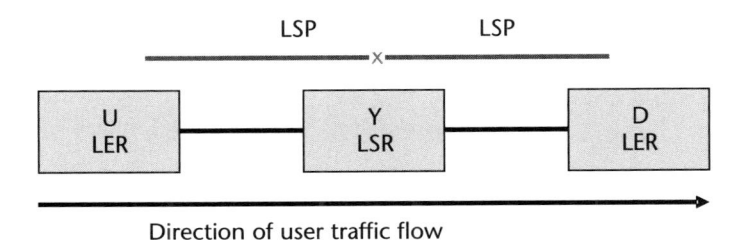

Figure 13.21 Return path reference model.

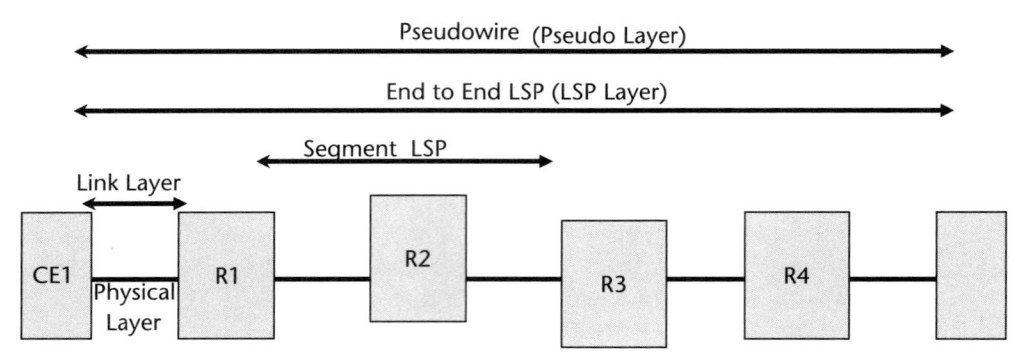

Figure 13.22 OAM hierarchy.

more MEs that belongs to the same transport path and that are maintained and monitored as a group are known as an MEG. Intermediate points in between MEPs are called MIPs. MEPs and MIPs are associated with the MEG and can be shared by more than one ME in a MEG.

In case of unidirectional point-to-point transport paths, a single unidirectional ME is defined to monitor it. In case of associated bidirectional point-to-point transport paths, two independent unidirectional MEs are defined to independently monitor each direction. In case of co-routed bidirectional point-to-point transport paths, a single bidirectional ME is defined to monitor both directions congruently. In case of unidirectional point-to-multipoint transport paths, a single unidirectional ME for each leaf is defined to monitor the transport path from the root to that leaf.

OAM functionalities are available not only on a transport path granularity (e.g., LSP or MS-PW), but also on arbitrary parts of transport paths, defined as tandem connections, between any two arbitrary points along a transport path. SPMEs are hierarchical LSPs instantiated to provide monitoring of a portion of a set of transport paths (LSPs or MS-PWs) that follow the same path between the ingress and the egress of the SPME. In the TCM, there is a direct correlation between all fault management and performance monitoring information gathered for the SPME and the monitored path segment of the end-to-end transport path.

MEPs are responsible for originating almost all of the proactive and on-demand monitoring OAM functionality for the MEG. A MEP is capable of originating and terminating OAM packets for fault management and performance monitoring. These OAM packets are carried within the G-ACh with the proper encapsulation and an appropriate channel type as defined in [11].

A server MEP is a MEP of a MEG that encapsulates and transports the MPLS-TP layer network or the sublayer being referenced. It can run appropriate OAM functions for fault detection within the server sublayer network, and provides a fault indication to its client MPLS-TP layer network such as lock report (LKR) and alarm indication signal (AIS) that are originated by intermediate nodes and triggered by server layer events.

A policing function is normally co-located with a MEP at UNI or NNI.

In the context of an MPLS-TP LSP, only LERs can implement MEPs. In the context of an SPME, any LSR of the MPLS-TP LSP can be an LER of SPMEs, monitoring infrastructure of the transport path. Regarding PWs, only terminating PW endpoints (T-PEs) can implement MEPs while for SPMEs supporting one or more PWs both T-PEs and S-PEs can implement SPME MEPs. Any MPLS-TP LSR can implement a MEP for an MPLS-TP section. A node hosting a MEP can either support per-node MEP or per-interface MEPs.

An MIP, which is a point between the MEPs of a MEG for a PW, LSP, or SPME, reacts some OAM packets and forwards all the other OAM packets. It can generate OAM packets only in response to OAM packets that it receives from the MEG to which it belongs. The OAM packets generated by the MIP are sent to the originating MEP. A MIP can be a per-node MIP in an unspecified location within the node or a per-interface MIP where two or more MIPs per node on both sides of the forwarding engine.

The following MPLS-TP MEGs are specified in [16]:

- A section maintenance entity group (SMEG), allowing monitoring and management of MPLS-TP sections between MPLS LSRs;
- An LSP maintenance entity group (LMEG), allowing monitoring and management of an end-to-end LSP between LERs;
- A PW maintenance entity group (PMEG), allowing monitoring and management of an end-to-end single-segment (SS)/MS-PWs (between T-PEs). A PMEG can be configured on any SS-PW or MS-PW. It is intended to be deployed in scenarios where it is desirable to monitor an entire PW between a pair of MPLS-TP enabled T-PEs rather than monitoring the LSP aggregating multiple PWs between PEs.
- An LSP SPME ME group (LSMEG), allowing monitoring and management of an SPME between a given pair of LERs and/or LSRs along an LSP. An LSMEG is an MPLS-TP SPME with an associated maintenance entity group intended to monitor an arbitrary part of an LSP between the MEPs instantiated for the SPME. An LSMEG can monitor an LSP path segment and may support the forwarding engine(s) of the node(s) at the edge(s) of the path segment.
- A PW SPME ME group (PSMEG), allowing monitoring and management of an SPME between a given pair of T-PEs and/or S-PEs along an MS-PW, to monitor an arbitrary part of an MS-PW between the MEPs instantiated for the SPME independently of the end-to-end monitoring (PMEG).

A PSMEG may include the forwarding engine(s) of the node(s) at the edge(s) of the path segment. A PSMEG is no different than an SPME; it is simply named as such to discuss SPMEs specifically in a PW context.

Hierarchical LSPs are also supported in the form of SPMEs. In this case, each LSP in the hierarchy is a different sublayer network that can be monitored, independently from higher and lower level LSPs in the hierarchy, on an end-to-end basis (from LER to LER) by a SPME. It is possible to monitor a portion of a hierarchical LSP by instantiating a hierarchical SPME between any LERs/LSRs along the hierarchical LSP.

Figure 13.23 depicts these MEGs in a network path between two customer equipment (i.e., CE1 and CE2). Domain 1 may represent the section of the path within Carrier 1 while Domain 2 may represent the section of the path within Carrier 2.

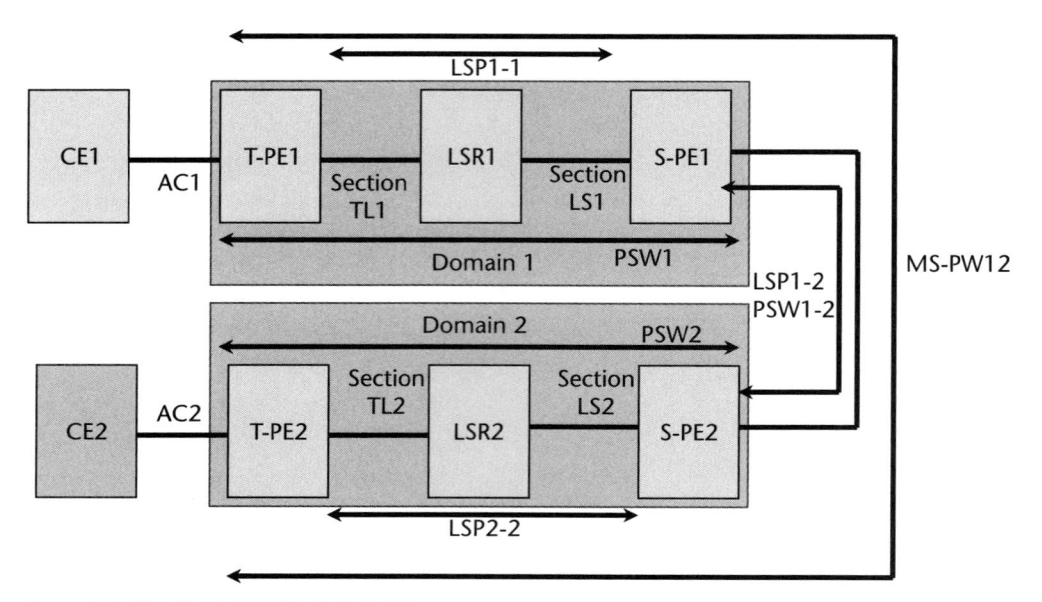

Figure 13.23 The MPLS-TP OAM MEGs.

T-PE1 (terminating provider edge 1) is adjacent to LSR1 via the MPLS-TP section TL1 and label-switched router 1 (LSR1) is adjacent to switching provider edge 1 (S-PE1) via the MPLS-TP section LS1. Similarly, in domain 2, T-PE2 is adjacent to LSR2 via the MPLS-TP section TL2 and LSR2 is adjacent to S-PE2 via the MPLS-TP Section LS2.

Figure 13.23 also shows a bidirectional MS-PW (MS-PW12) between AC1and AC2. The MS-PW consists of three bidirectional PW path segments: PW1 path segment between T-PE1 and S-PE1 via the bidirectional LSP1-1, PW1-2 path segment between S-PE1 and S-PE2, via the bidirectional LSP1-2, and PW2 path segment between S-PE2 and T-PE2 via the bidirectional LSP2-2.

Let's consider an end-to-end LSP traversing two carrier networks in Figure 13.24. OAM is end to end or per segment where a segment is between MEPs. The OAM in each segment is independent of any other segment. Similarly, protection or restoration is always between MEPs that are per segment or end-to-end basis.

OAM monitoring for a PW with switching and without switching is depicted in Figures 13.25 and 13.26 and the related packet format is depicted in Figure 13.3. At the endpoints or PW stitch point, the process verifies the operational status of the PW and works with the native attachment circuit technology by transporting and acting upon native attachment circuit OAM.

In Figure 13.25, PW over LSP monitoring is depicted where the PW has no switching. End-to-end LSP OAM is used since PW OAM cannot create MIPs at the intercarrier boundary without a PW switching function.

In Figure 13.26, PW over LSP monitoring is depicted where the PW has switching. End-to-end LSP OAM is not required since the PW switching points can support a MIP.

13.4 OAM

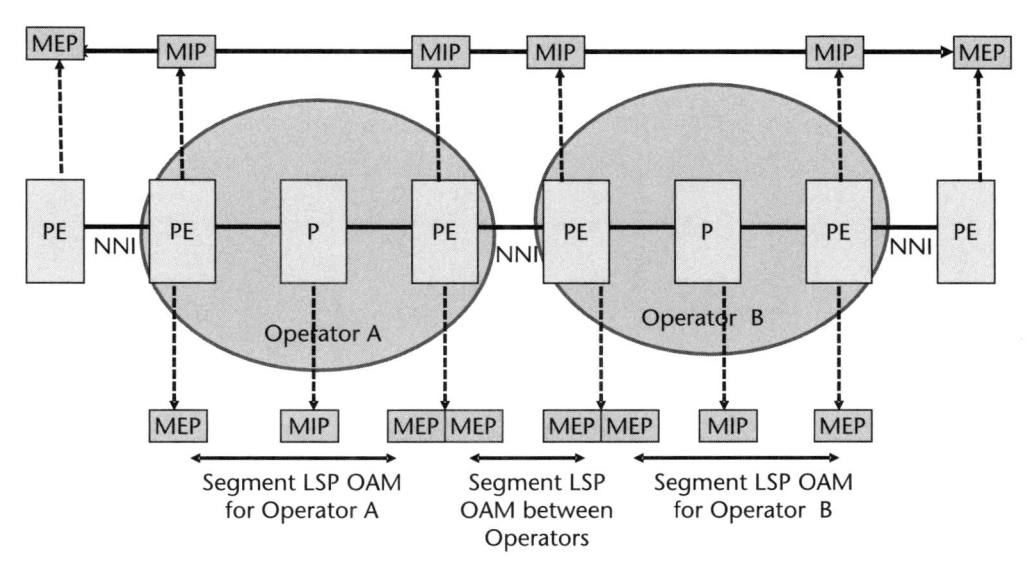

Figure 13.24 An LSP monitoring.

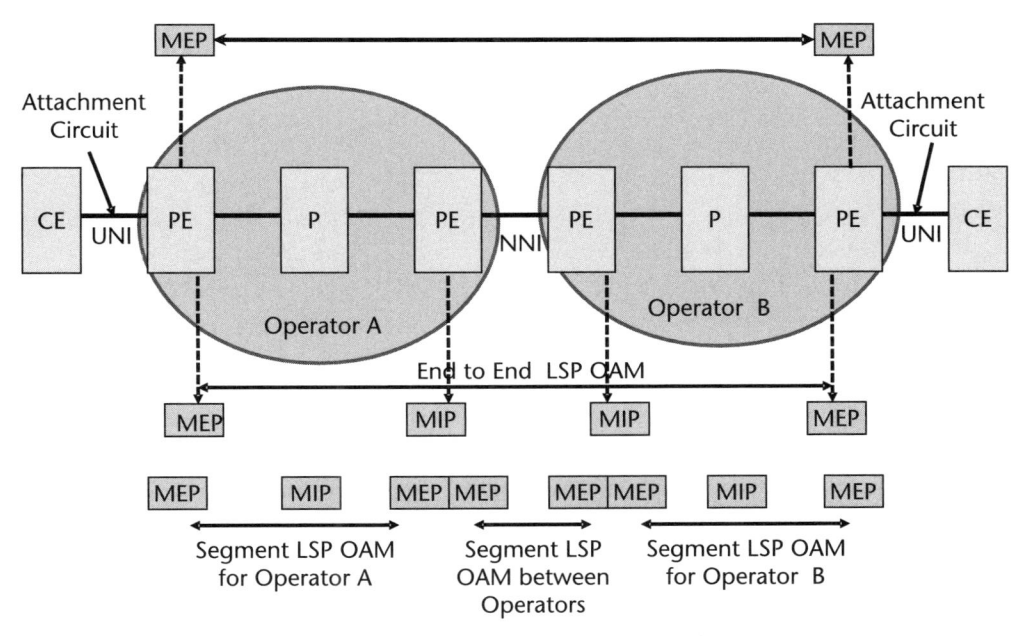

Figure 13.25 Monitoring of a PW over LSP where PW has no switching.

13.4.2 OAM Functions for Proactive Monitoring

Proactive monitoring for an MEG is usually configured at transport path creation time and performed in-service. Such transactions are universally MEP to MEP in operation while notifications can be node to node such as MS-PW transactions or node to MEPs such as AIS. They are connectivity verification (CV), loss measurement (LM), and delay measurements configured at the MEPs and typically reported outside to a management system.

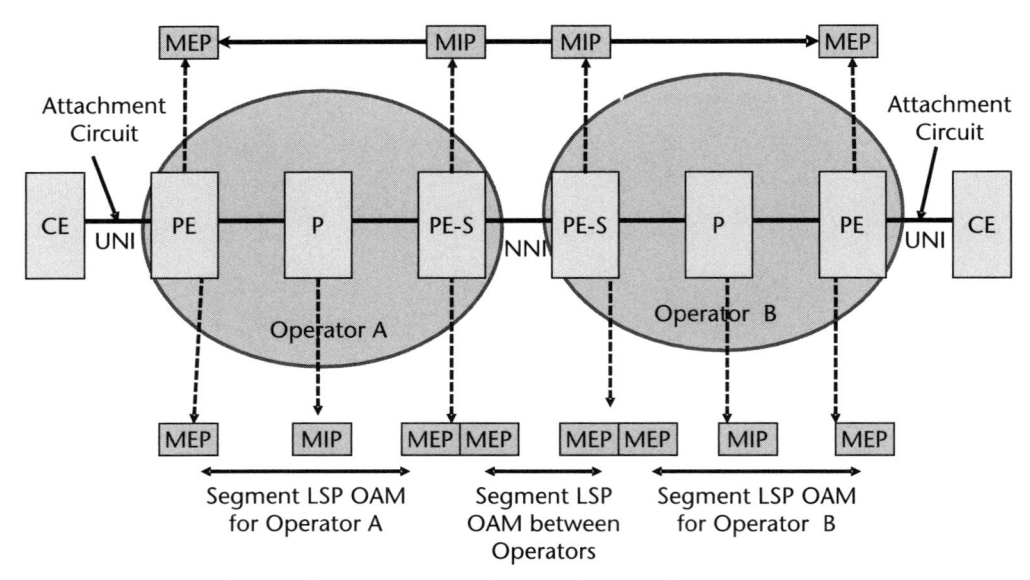

Figure 13.26 Monitoring of a PW over LSP where PW has switching.

Continuity check and connectivity verification functions are used to detect a loss of continuity defect (LOC) between two MEPs in a MEG to detect an unexpected connectivity defect between two MEGs such as mismerging or misconnection, as well as unexpected connectivity within the MEG with an unexpected MEP. Each CC-V OAM packet also includes a globally unique source MEP identifier, whose value needs to be configured on the source and peer sink MEP(s). The packet is transmitted at a regular, operator configurable, rate. Default transmission periods for fault management, performance management, and protection switching are 1 second, 100 ms, and 3.3 ms, respectively.

If protection switching with CC-V defect entry criteria of 12 ms is required in conjunction with the requirement to support 50-ms recovery time, then a transmission period of 3.33 ms is necessary. Sometimes, the requirement of 50 ms recovery time is associated with the requirement for a CC-V defect entry criteria period of 35 ms: in these cases, a transmission period of 10 ms can be used.

When proactive CC-V is enabled, a sink MEP detects a loss of continuity (LOC) defect when it fails to receive proactive CC-V OAM packets from the source MEP. A sink MEP identifies a misconnectivity defect (e.g., mismerge, misconnection, or unintended looping) when the received packet carries an unexpected globally unique source MEP identifier.

If proactive CC-V OAM packets are received with the expected source MEP identifier and a transmission period different than the locally configured reception period, then a CC-V period misconfiguration defect is detected.

If proactive CC-V OAM packets are received with the expected globally unique source MEP identifier but with an unexpected encapsulation, then a CC-V unexpected encapsulation defect is detected. The sink MEP does not receive any CC-V OAM packets with the expected MEP identifier and an unexpected encapsulation for an interval equal at least to 3.5 times the longest transmission period of the proactive CC-V OAM packets, a misconnectivity defect is raised. It blocks all the traffic (including also the user data packets) that it receives from the misconnected transport path.

Loss measurement (LM) is used to exchange periodically (i.e., proactive LM) counter values for ingress and egress packets transmitted and received by the transport path.

Each MEP counts its transmitted and received user data packets. These counts are then correlated in real time with the peer MEP in the ME to calculate packet loss.

For a MEP, near-end packet loss refers to packet loss associated with incoming data packets from the far-end MEP while far-end packet loss refers to packet loss associated with egress data packets (towards the far-end MEP.

Proactive LM can be operated in one-way and two-way. One-way LM is applicable to both unidirectional and bidirectional (co-routed or associated) transport paths while two-way LM is applicable only to bidirectional (co-routed or associated) transport paths.

MIPs, as well as intermediate nodes, do not process the LM information and forward these proactive LM OAM packets as regular data packets.

Proactive DM is used to measure the long-term packet delay and packet delay variation in the transport path monitored by a pair of MEPs. Periodic DM OAM packets from a MEP to a peer MEP are sent. Proactive DM can be operated in one-way or two-way.

MIPs, as well as intermediate nodes, do not process the DM information and forward these proactive DM OAM packets as regular data packets.

Out-of-service throughput testing can be performed on-demand verifying the bandwidth/throughput of an LSP or PW before it is put in service, in one-way or two-way modes between MEPs and between MEP and MIP. OAM test packets are at increasing rate (up to the theoretical maximum), computing the percentage of OAM test packets received and reporting the rate at which OAM test packets begin to drop. This rate is dependent on the OAM test packet size.

When configured to perform such tests, a source MEP inserts OAM test packets with a specified packet size and transmission pattern at a rate to exercise the throughput.

13.4.3 Data Plane Loopback, RDI, AIS

Data plane loopback is an out-of-service function that places a transport path, at either an intermediate or terminating node, into a data plane loopback state such that all traffic received on the looped back interface is sent on the reverse direction of the transport path without modification.

The MEG is locked such that user data traffic is prevented from entering/exiting that MEG, and test traffic is inserted at the ingress of the MEG. This test traffic can be generated from an internal process residing within the ingress node or injected by external test equipment connected to the ingress node. The MEG should be put into a locked state before the diagnostic test is started.

OAM packets are sent to the MIPs or MEPs in the data plane loopback mode via TTL expiry. MIPs can be addressed with more than one TTL value on a co-routed bidirectional path set into data plane loopback. If the loopback function is to be performed at an intermediate node it is only applicable to co-routed bidirectional paths. If the loopback is to be performed end to end, it is applicable to both co-routed bidirectional or associated bidirectional paths.

When there is a signal fail condition, remote defect indication (RDI) is transmitted by a sink MEP to communicate the condition to its source MEP. In case of co-routed and associated bidirectional transport paths, the RDI can be piggybacked onto the CC-V packet. In case of unidirectional transport paths, the RDI can be sent only using an out-of-band return path if it exists and its usage is enabled by policy actions.

When the signal fail condition clears, the MEP should stop transmitting the RDI to its peer MEP. When RDI piggybacked onto CC-V, the RDI will be cleared from subsequent transmission of CC-V packets. A MEP should clear the RDI defect upon reception of an RDI indicator cleared.

The alarm reporting relies upon an alarm indication signal (AIS) packet to suppress alarms following detection of defect conditions at the server layer. When a server MEP has a signal fail condition, it notifies that to the co-located MPLS-TP client/server adaptation function which then generates OAM packets with AIS information in the downstream direction to allow the suppression of secondary alarms at the MPLS-TP MEP in the client layer. MIPs and intermediate nodes forward these AIS OAM packets as regular data packets.

AIS condition is cleared if no AIS packet has been received in 3.5 times the AIS transmission period which is traditionally one per second.

Similarly, the lock reporting function upon a locked report (LKR) packet is used to suppress alarms following administrative locking action in the server layer.

When a server MEP is locked, the MPLS-TP client layer adaptation function generates packets with LKR information to allow the suppression of secondary alarms at the MEPs in the client layer. MIPs, as well as intermediate nodes, do not process the LKR information and forward these LKR OAM packets as regular data packets.

Locked condition is cleared if no LKR packet has been received for 3.5 times the transmission period. The LKR transmission period is traditionally one per second.

Lock instruct (LKI) command instructs the peer MEP(s) to put the MPLS-TP transport path into a locked condition. It is single-side provisioning for administratively locking (and unlocking) an MPLS-TP transport path. It is also possible to administratively lock (and unlock) an MPLS-TP transport path using two-side provisioning, where the NMS administratively puts both MEPs into an administrative lock condition. In this case, the LKI function is not required/used. MIPs, as well as intermediate nodes treat LKI OAM packets as regular data packets.

An MEP, upon receiving a single-side administrative unlock command from NMS, sends an LKI removal request OAM packet to its peer MEP(s).

13.4.4 Client Failure Indication, Route Tracing, and Lock Instruct

The client failure indication (CFI) is used to help process client defects and propagate a client signal defect condition from the local attachment circuit, where the defect was detected, to the far-end attachment circuit, in case the client of the transport path does not support a native defect/alarm indication mechanism such as AIS. MIPs, as well as intermediate nodes, do not process the CFI information and forward these proactive CFI OAM packets as regular data packets.

It is often necessary to trace a route covered by a MEG from a source MEP to its peer MEP(s) including all the MIPs in-between for trouble shooting purposes.

Route tracing discovers the full list of MIPs and peer MEPs. In case a defect exists, the route trace function will only be able to trace up to the defect and needs to be able to return the incomplete list of OAM entities, allowing one to localize the fault.

Route tracing is a basic means to perform connectivity verification. For this function to work properly, a return path must be present.

13.5 Protection Switching

MPLS-TP also supports sub-50-ms fully automated protection switching without any control plane and by relying on fast OAM messages and the protection state coordination (PSC) mechanism to synchronize the both ends in case of failure. The OAM message can be exchanged at rates up to 3.3 ms intervals to achieve protection switching in sub-50-ms time.

The fast, accurate, and coordinated protection switching time is achieved for LSP and PW by linear protection switching that uses a PSC protocol very similar to that in ITU-T G.8131 [31]. When deployed in ring topologies, a ring protection mechanism similar to that in ITU-T G.8132 [32] can be used. MPLS-TP supports various modes of protection (i.e., 1+1, 1:1, N:1) and also provides protection on every layer (i.e., PW, LSP, and section). All these protection-switching mechanisms are very similar to the current mechanisms being used by SONET/SDH and OTN networks

MPLS-TP uses existing GMPLS and PW control plane-based restoration mechanisms applicable to bidirectional paths. Traditional PW redundancy can be used for PE/AC failure protection. In addition, 1+1 and 1:1 LSP protection are supported, as is full LSP reroute mechanism (as is common in MPLS networks).

Different protection schemes of two concurrent traffic paths (1+1), one active and one standby path with guaranteed bandwidth on both paths (1:1), and one active path and a standby path the resources of which are shared by one or more other active paths (shared protection) can be provided.

MPLS-TP recovery schemes are applicable to all levels in the MPLS-TP domain (i.e., section, LSP, and PW) providing segment and end-to-end recovery. These mechanisms, which can be data-plane, control-plane, or management-plane based, support the coordination of protection switching at multiple levels to prevent race conditions occurring between a client and its server layer.

13.6 Security Considerations

Protection against possible attacks on data plane and control plane are necessary. G-ACh itself and its signaling need to be secured. MPLS-TP control plane security is discussed in [32].

A peer MPLS-TP node can be flooded with G-Ach messages to deny services to others. G-Ach packets can be intercepted as well. To protect against these potential attacks, G-ACh message throttling mechanisms can be used. Messages between line cards and the control processor will be limited.

The contents of G-ACh messages need to be protected. OAM traffic can reveal sensitive information such as performance data and details about the current state of the network. Therefore, authentication, authorization, and encryption should be in place. This will prevent unauthorized access to vital equipment and it will prevent third parties from learning about sensitive information about the transport network.

In case of misconfiguration, some nodes can receive OAM packets that they cannot recognize. In such a case, these OAM packets should be silently discarded in order to avoid malfunctions whose effect may be similar to malicious attacks (e.g., degraded performance or even failure). Further considerations about data plane attacks via G-ACh.

13.7 Conclusion

MPLS-TP is a continuation of T-MPLS. It is client-agnostic that can carry L3, L2, L1 services and physical-layer agnostic that can run over IEEE Ethernet PHYs, SONET/SDH and optical transport network (OTN), wavelength division multiplexing (WDM), and so forth. Its OAM functions are similar to those available in traditional optical transport networks and Metro Ethernet networks. However, there are substantial differences between Metro Ethernet OAM PDUs and MPL-TP OAM PDUs. In order for them to interoperate, mapping between them is necessary.

MPLS-TP uses a subset of IP/MPLS standards. PHP and ECMP are not supported or made optional. Control and data plane are separated. The operation is fully automated without control plane by using NMS.

Alignment between Carrier Ethernet OAM and MPLS-TP OAM is very helpful in providing end-to-end OAM where access is formed by Carrier Ethernet networks while the backbone is formed by MPLS-TP networks.

References

[1] RFC 3916, Requirements for Pseudo-Wire Emulation Edge-to-Edge, September 2004.
[2] RFC 3031, Multiprotocol Label Switching Architecture, January 2001.
[3] RFC 3985, Pseudowire Emulation Edge-to-Edge (PWE3) Architecture, March 2005.
[4] RFC 5659, An Architecture for Multi-Segment Pseudowire Emulation Edge-to-Edge, October 2009.
[5] ITU-T Recommendation M.3400, TMN Management Functions, April 1997.
[6] RFC 5654, Requirements of an MPLS Transport Profile, September 2009.
[7] ITU-T G.805, Generic Functional Architecture of Transport Networks, March 2000.
[8] IETF RFC 3443, Time to Live (TTL) processing in Multi-Protocol Label Switching (MPLS) Networks, January 2003.
[9] ITU-T Y.1711, Operation & Maintenance Mechanism for MPLS Networks, February 2004.
[10] RFC 4385, Pseudowire Emulation Edge-to-Edge (PWE3) Control Word for Use over an MPLS PSN, February 2006.

[11] RFC 5586, MPLS Generic Associated Channel, June 2009.
[12] RFC 5085, Pseudowire Virtual Circuit Connectivity Verification (VCCV): A Control Channel for Pseudowires, December 2007.
[13] RFC 5921, A Framework for MPLS in Transport Networks, July 2010.
[14] G.7710-Common Equipment Management Function Requirements, July 2007.
[15] RFC 4741, NETCONF Configuration Protocol, December 2006.
[16] Operations, Administration and Maintenance Framework for MPLS-Based Transport Networks, draft-ietf-mpls-tp-oam-framework-11.txt, February 2011.
[17] RFC 3032, MPLS Label Stack Encoding, January 2001.
[18] RFC 3443, Time to Live (TTL) Processing in Multi-Protocol Label Switching (MPLS) Networks, 2003.
[19] RFC 3985, Pseudo Wire Emulation Edge-to-Edge (PWE3) Architecture, March 2005.
[20] RFC 3270, Multi-Protocol Label Switching (MPLS) Support of Differentiated Services, May 2002.
[21] RFC 3209, RSVP-TE: Extensions to RSVP for LSP Tunnels, December 2001.
[22] RFC 5921, A Framework for MPLS in Transport Networks, July 2010.
[23] RFC 4664, Framework for Layer 2 Virtual Private Networks (L2VPN), September 2006.
[24] BGP Based Virtual Private Multicast Service Auto-Discovery and Signaling, draft-raggarwa-l2vpn-p2mp-pw-02.txt, July 2009.
[25] RFC 5654, Requirements of an MPLS Transport Profile, September 2009.
[26] ITU-T Y.1720, Protection Switching for MPLS Networks, 2002.
[27] RFC 4379, Detecting Multi-Protocol Label Switched (MPLS) Data Plane Failures, 2006.
[28] RFC 5884, Bidirectional Forwarding Detection (BFD) for MPLS Label Switched Paths (LSPs), 2010.
[29] Y.1731, OAM Functions and Mechanisms for Ethernet Based Networks, May 2006.
[30] ITU-T G.8131, Linear Protection Switching for Transport MPLS (T-MPLS) Networks, 2007.
[31] ITU-T Draft recommendation G.8132, T-MPLS Shared Protection Ring, 2008.
[32] RFC 5920, Security Framework for MPLS and GMPLS Networks, July 2010.

CHAPTER 14

Virtual Private LAN Services

14.1 Introduction

Multiprotocol Label Switching (MPLS) is generally accepted as the common convergence technology and facilitates the deployment and management of virtual private networks (VPNs). MPLS-based VPN can be classified as either a Layer-2 [1, 2] or a Layer-3 point-to-point service or multipoint service:

- Layer-3 multipoint VPNs or Internet Protocol (IP) VPNs that are often referred to as virtual private routed networks (VPRN);
- Layer-2 point-to-point VPNs, which basically consist of a collection of separate virtual leased lines (VLL) or pseudowires (PW);
- Layer-2 multipoint VPNs, or virtual private LAN services (VPLS).

VPLS is a multipoint service, but unlike IP VPNs, it can transport non-IP traffic and leverages advantages of Ethernet. VPLS is also used within a service provider's network to aggregate services for delivery to residential and enterprise customers.

Regardless of how the VPN service is utilized, most enterprise customers use routers at the LAN/WAN boundary. However, VPLS is a Layer-2 VPN service and allows the use of Layer-2 switches as the customer edge (CE) device.

VPLS is introduced by RFC 2764 [1], as a VPN service, which emulates a LAN segment using IP-based facilities (Figure 14.1) for a given users. In other words, VPLS creates a Layer-2 broadcast domain that is fully capable of learning and forwarding on Ethernet MAC addresses. Multiple CEs such as bridges, router, Layer-2 switches across a shared Ethernet and IP/MPLS service provider network infrastructure are interconnected via transparent LAN Service (TLS).

Two VPLS solutions are proposed (Table 14.1):

- RFC 4761 [3]: VPLS using BGP that uses BGP for signaling and discovery. The signaling is similar to RFC 4360 [4]. the encapsulation is similar to Martini encapsulation. BGP signaling mechanism is used in BGP VPNs; therefore, operators can use the same signaling for VPLS.

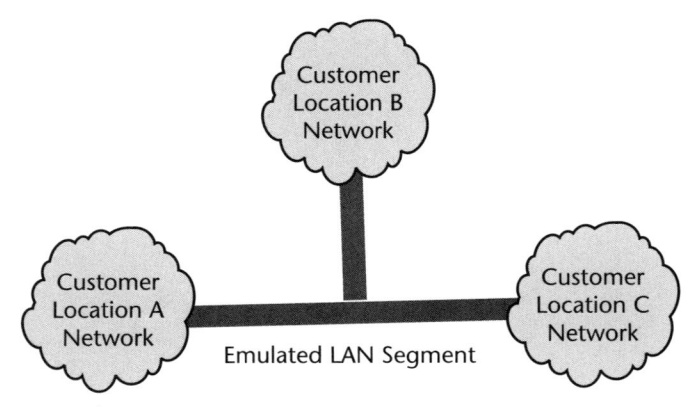

Figure 14.1 VPLS emulating a LAN segment.

Table 14.1 VPLS Solutions

VPLS Implementation Model	Discovery	Signaling
RC 4761 (BGP-based VPLS)	BGP	BGP
RFC 4762 (LDP-based VPLS)	None	LDP

- RFC4762 [5]: VPLS using label distribution that uses LDP signaling and basically an extension to the Martini draft [6, 7]. Broadcast mode is used.

Both approaches assume tunnel LSPs between PEs (Figure 14.2). Pseudowires (PWE3s) are set up over tunnel LSPs [i.e., virtual connection (VC) LSPs]. IDs for a PWE3 is VC-ID, which is the VPN ID. RFC 4762 [5] is more widely implemented.

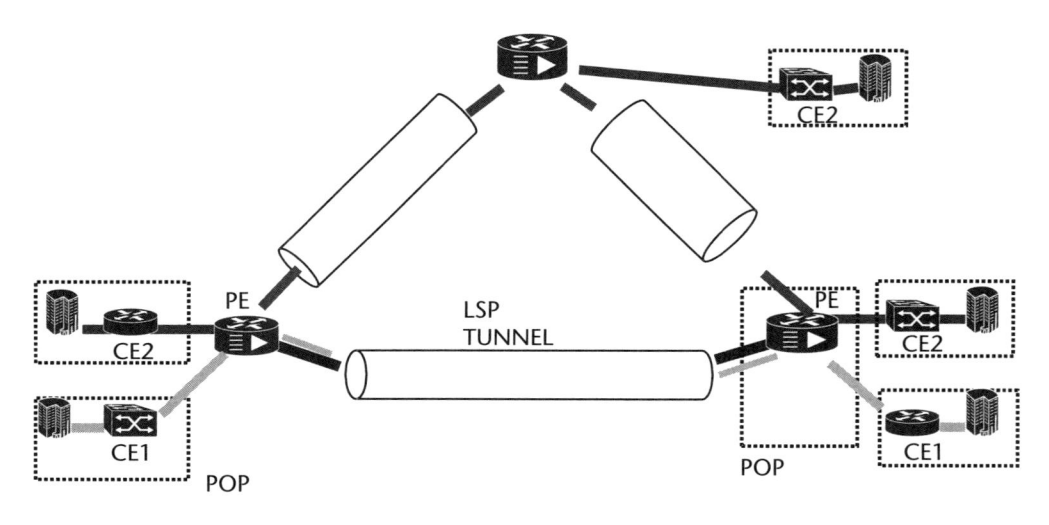

Figure 14.2 Tunnel LSPs between PEs for VPLS.

14.1 Introduction

VPLS as described in [5] is a class of VPN that allows the connection of multiple sites in a single bridged domain over a provider managed IP/MPLS network (Figure 14.3).

All customer sites that are in one or more metro areas in a VPLS instance appear to be on the same LAN, regardless of their location. VPLS uses an Ethernet interface as the customer handoff, simplifying the LAN/WAN boundary and allowing for rapid and flexible service provisioning.

Use of IP/MPLS routing protocols and procedures instead of the Spanning Tree Protocol, and MPLS labels instead of VLAN IDs, improves the scalability of the VPLS service.

In VPLS, customers maintain complete control over their routing, and since all the customer routers in the VPLS are part of the same subnet (LAN). The result is a simplified IP addressing plan, especially when compared to a mesh constructed from many separate point-to-point connections. The service provider also benefits from reduced complexity to manage the VPLS service since it has no awareness or participation in the customer's IP addressing space and routing.

A VPLS-capable network consists of CEs, Provider Edges (PE), and a core MPLS network:

- The CE device is a router or switch located at the customer's premises. It can be owned and managed by the customer or owned and managed by the service provider. It is connected to the PE via an attachment circuit (AC). The Ethernet is the interface between the CE and the PE.
- The PE device is where all the VPN intelligence resides, the VPLS originates and terminates, and all the necessary tunnels are set up to connect to all the other PEs. As VPLS is an Ethernet Layer-2 service, the PE must be capable of media access control (MAC) learning, bridging, and replication on a per-VPLS basis.
- The IP/MPLS core network interconnects the PEs. It does not really participate in the VPN functionality. Traffic is simply switched based on the MPLS labels.

The basis of any multipoint VPN service (IP VPN or VPLS) is the full mesh of MPLS tunnels that are label-switched paths (LSPs) set up between all the PEs

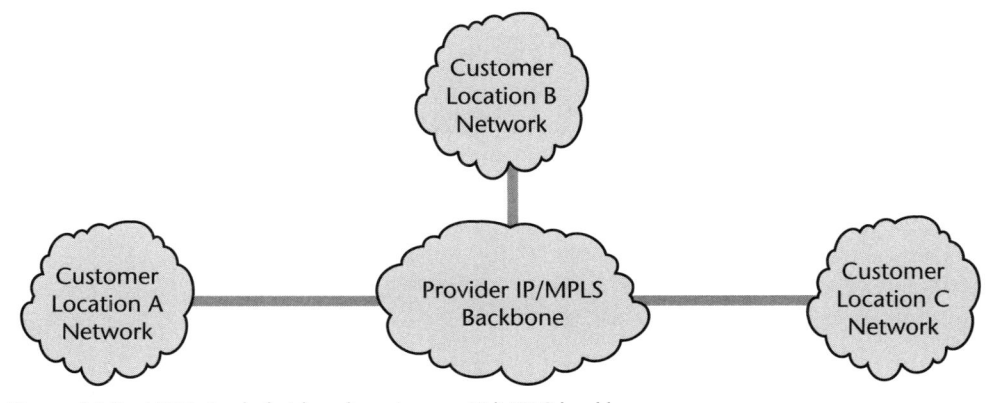

Figure 14.3 VPLS single-bridge domain over IP/MPLS backbone.

participating in the VPN service. For each VPN instance, an Ethernet PW runs inside tunnels. CEs appear to be connected by a single LAN (Figure 14.4). An Ethernet bridge determines where PE needs to send a frame. As a result, VPLS can be defined as a group of virtual switch instances (VSIs) that are interconnected using Ethernet over MPLS circuits in a full mesh topology to form a single, logical bridge.

In addition to its MPLS functions, a VPLS-enabled PE has a VPLS code module based on IETF RFCs and a bridging module based on an IEEE 802.1D learning bridge, as depicted in Figure 14.5. A service provider network inside rectangle looks like a single Ethernet bridge. If a CE is a router, then a PE only sees one MAC per customer location.

Depending on the exact VPLS implementation, when a new PE or VPLS instance is added, the amount of effort to establish this mesh of LSPs can vary dramatically. Once the LSP mesh is built, the VPLS instance on a particular PE is now able to receive Ethernet frames from the customer site and, based on the MAC address, to switch those frames into the appropriate LSP. This is possible because VPLS enables the PE router to act as a learning bridge with one MAC table per VPLS instance on each PE. In other words, the VPLS instance on the PE router has a MAC table that is populated by learning the MAC addresses as Ethernet frames enter on specific physical or logical ports, exactly the same way that an Ethernet switch does.

Once an Ethernet frame enters via a customer-facing ingress port, the destination MAC address is looked up in the MAC table and the frame is sent unaltered as long as the MAC table contains the MAC address into the LSP that will deliver

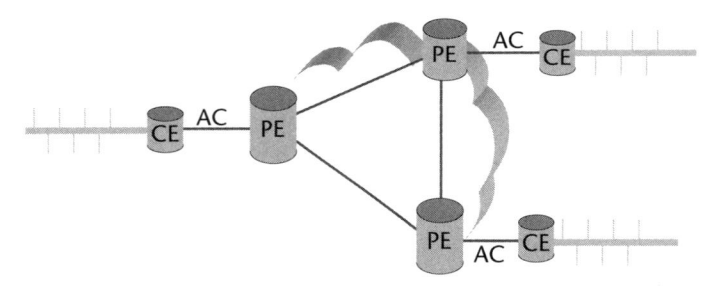

Figure 14.4 Full mesh of PEs.

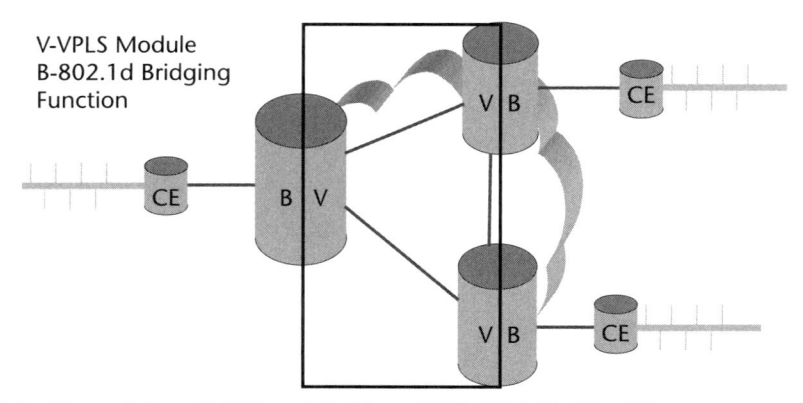

Figure 14.5 PE consisting of VPLS module (V) and IEEE 802.1d bridge (B).

it to the correct PE attached to the remote site. If the MAC address is not in the MAC address table, the Ethernet frame is replicated and flooded to all logical ports associated with that VPLS instance, except the ingress port where it just entered. Once the PE hears back from the host that owns that MAC address on a specific port, the MAC table is updated in the PE. Similar to a switch, the MAC addresses that have not been used for a certain amount of time are aged out.

Each MPLS LSP consists of two unidirectional LSPs, providing bidirectional transport for each bidirectional PW. A VPLS network is established by first defining the MPLS LSPs that will support the VPLS PW tunnels. The paths are determined using the open path-shortest first (OSPF) link-state protocol, which determines the shortest path for the LSP to the target destination. A full bidirectional mesh of LSPs needs to be established between all participating VPLS PEs. The Label Distribution Protocol (LDP) or Resource Reservation Protocol (RSVP) is then used to distribute the label information to allow label swapping at each node. After that, the PWE3 tunnels are established over the existing LSPs. Either LDP or BGP is used to exchange PWE3 labels.

In order to establish MPLS LSPs, OSPF-TE, and RSVP-TE can be used as well where OSPF-TE will take bandwidth availability into account when calculating the shortest path, while RSVP-TE allows reservation of bandwidth.

The MPLS mechanisms and tunnels are merely used as transport. Pseudowires are generic and designed to be implemented on any suitable packet-switched network. It is therefore possible to envisage a VPLS-like approach based on other tunnel mechanisms, such as MPLS-TP and PBB-TE.

There are two key components of VPLS, PE discovery, and signaling. PE discovery can be via provisioning application, BGP, and RADIUS. Signaling can be via targeted LDP and BGP.

The advantages of VPLS may be summarized as:

- Complete customer control over their routing where there is a clear demarcation of functionality between service provider and customer that makes troubleshooting easier;
- Ability to add a new site without configuration of the service provider's equipment or the customer equipment at existing sites;
- Faster provisioning, with potential for customer provisioned bandwidth on-demand;
- Minimize MAC address exposure, improving scaling by having one MAC address per site (i.e., one MAC per router) or per service;
- Improve customer separation by having CE router to block unnecessary broadcast or multicast traffic from customer LANs;
- MPLS core network emulates a flat LAN segment that overcomes distance limitations of Ethernet-switched networks and extends Ethernet broadcast capability across WAN;
- Point-to-multipoint connectivity connecting each customer site to many customer sites;
- A single CE-PE link transmits Ethernet packets to multiple remote CE routers;

- Fewer connections required to get full connectivity among customer sites;
- Adding, removing, or relocating a CE router requires configuring only the directly attached PE router, resulting in substantial OpEx savings.

14.2 Data Plane

Data plane of VPLS deals with frame encapsulation, classification, MAC address learning and aging, and forwarding. They are described next.

14.2.1 VPLS Encapsulation

In a VPLS, a customer Ethernet frame without preamble is encapsulated with a header as defined in RFC 4448 [8]:

- If the frame, as it arrives at the PE, has an encapsulation that is used by the local PE as a service delimiter, to identify the customer and/or the particular service of that customer, then that encapsulation may be stripped before the frame is sent into the VPLS. As the frame exits the VPLS, the frame may have a service-delimiting encapsulation inserted.
- If the frame, as it arrives at the PE, has an encapsulation that is not service delimiting, then its encapsulation should not be modified by the VPLS.

According to these rules, If a customer frame arrives at a customer-facing port with a VLAN tag that identifies the customer's VPLS instance, the tag would be stripped before it is encapsulated in the VPLS. At egress, the frame may be tagged again, if a service-delimiting tag is used, or it may be untagged if none is used. At both ingress and egress, dealing with service delimiters is a local action that neither PE has to signal to the other.

If a customer frame arrives at a customer-facing port with a VLAN tag that identifies a VLAN domain in the customer L2 network, then the tag is not modified or stripped.

14.2.2 Classification and Forwarding

VPLS packets are classified as belonging to a given service instance and associated forwarding table based on the interface over which the packet is received. After that they are forwarded based on the destination MAC address. The former mapping is determined by configuration.

In order to offer different classes of service within a VPLS, 802.1p bits in a customer Ethernet frame with a VLAN tag are mapped to EXP bits in the PW and/or tunnel label. This mapping can be different for each VPLS, as each VPLS customer may have its own view of the required behavior for a given setting of 802.1p bits.

When a bridge receives a packet to a destination that is not in its forwarding information base (FIB), a PE will flood packets to an unknown destination to all other PEs in the VPLS. An Ethernet frame whose destination MAC address is the broadcast MAC address is sent to all stations in that VPLS.

In forwarding a broadcast Ethernet frame or one with an unknown destination MAC address, the split horizon rule is used to prevent loops. When a PE receives a broadcast Ethernet frame or one with an unknown destination MAC address from an attached CE, the PE must send a copy of the frame to every other attached CE, as well as to all other PEs participating in the VPLS. When the PE receives the frame from another PE, this PE must send a copy of the packet only to attached CEs, and not send the frame to other PEs, since the other PE would have already done so.

14.2.3 MAC Address Learning and Aging

VPLS is a multipoint service; therefore, the entire service provider network appears as a single logical learning bridge for each VPLS. The logical ports of this service provider (SP) bridge are the customer ports as well as the PWs on a VPLS edge (VE), which is the designated multihomed PE router acting as the end point for the VPLS PW from the remote PE router. The SP bridge learns MAC addresses at its VEs while a learning bridge learns MAC addresses on its ports. Source MAC addresses of packets with the logical ports on which they arrive are associated in the FIB to forward packets. If a VE learns a source MAC address S-MAC on logical port P and later sees S-MAC on a different port R, then the VE updates its FIB to reflect the new port R.

The Ethernet MAC learning based on the 6-octet MAC address is called unqualified learning. If the key for learning includes the VLAN tag when present, it is called qualified learning.

Choosing between qualified and unqualified learning mainly involves whether one wants a single global broadcast domain (unqualified) or a broadcast domain per VLAN (qualified). The latter makes flooding and broadcasting more efficient, but requires larger MAC tables.

In unqualified learning, MAC addresses need to be unique and nonoverlapping among customer VLANs, or else they cannot be differentiated within the VPLS instance. An application of unqualified learning is port-based VPLS service for a given customer where all the traffic on a physical port, which may include multiple customer VLANs, is mapped to a single VPLS instance.

In qualified learning, each customer VLAN is assigned to its own VPLS instance. MAC addresses among customer VLANs may overlap with each other, but they will be handled correctly because each customer VLAN has its own FIB. As a result, the qualified learning can result in large FIB table sizes, as the logical MAC address is now a VLAN tag + MAC address.

For STP to work in a qualified learning mode, a VPLS PE must be able to forward STP BPDUs over the proper VPLS instance.

VPLS PEs have an aging mechanism to remove a MAC address associated with a logical port, similar to learning bridges. This is needed so that a MAC address can be relearned if it moves from a logical port to another logical port. In addition, aging reduces the size of a VPLS MAC table to only active MAC addresses.

The age of a source MAC address S-MAC on a logical port P is the time since it was last seen as a source MAC on port P. If the age exceeds the aging time T,

S-MAC is flushed from the FIB, which keeps track of the mapping of customer Ethernet frame addressing and the appropriate PW to use. Every time S is seen as a source MAC address on port P, the S-MAC's age is reset.

PEs that learn remote MAC addresses should be able to remove unused entries associated with a PW label via an aging mechanism such as an aging timer. The aging mechanism conserves memory and increases efficiency of operation. The aging timer for the MAC address S-MAC should be reset when a packet with the source MAC address S-MAC is received.

14.3 LDP-Based VPLS

An interface participating in a VPLS must be able to flood, forward, and filter Ethernet frames. Each PE will form remote MAC address to PW associations and associate directly attached MAC addresses to local customer facing ports.

Connectivity between PEs can be via MPLS transport tunnels as well as other tunnels over PWs such as GRE, L2TP and IPSec. The PE runs the LDP signaling protocol and/or routing protocols to set up PWs, setting up transport tunnels to other PEs and delivering traffic over PWs.

The AC providing access to a customer site could be a physical Ethernet port, a VLAN, an ATM PVC carrying Ethernet frames, an Ethernet PW, and so forth.

14.3.1 Flooding, Forwarding, and Address Learning

One of the capabilities of an Ethernet service is that frames sent to broadcast addresses and to unknown destination MAC addresses are flooded to all ports. To achieve flooding within the service provider network, all unknown unicast, broadcast, and multicast frames are flooded over the corresponding PWs to all PE nodes participating in the VPLS, as well as to all ACs.

To forward a frame, a PE associates a destination MAC address with a PW. MAC addresses on both ACs and PWs are dynamically learned. Packets are forwarded and replicated across both ACs and PWs.

Unlike BGP VPNs [9], reachability information is not advertised and distributed via a control plane. Reachability is obtained by standard learning bridge functions in the data plane. Standard learning, filtering, and forwarding actions [10] are performed when a PW or AC state changes.

It may be desirable to remove or unlearn MAC addresses that have been dynamically learned for faster convergence. This is accomplished by sending an LDP address withdraw message with the list of MAC addresses to be removed to all other PEs over the corresponding LDP sessions. An LDP address withdraw message contains a new TLV, the MAC List TLV, specifying a list of MAC addresses that can be removed or unlearned using the LDP address withdraw message. Its format is depicted in Figure 14.6.

The address withdraw message with MAC list TLVs expedites the removal of MAC addresses as the result of a topology change (e.g., failure of the primary link for a dual-homed VPLS-capable switch). PEs that do not understand the message can continue to participate in the VPLS.

14.3 LDP-Based VPLS

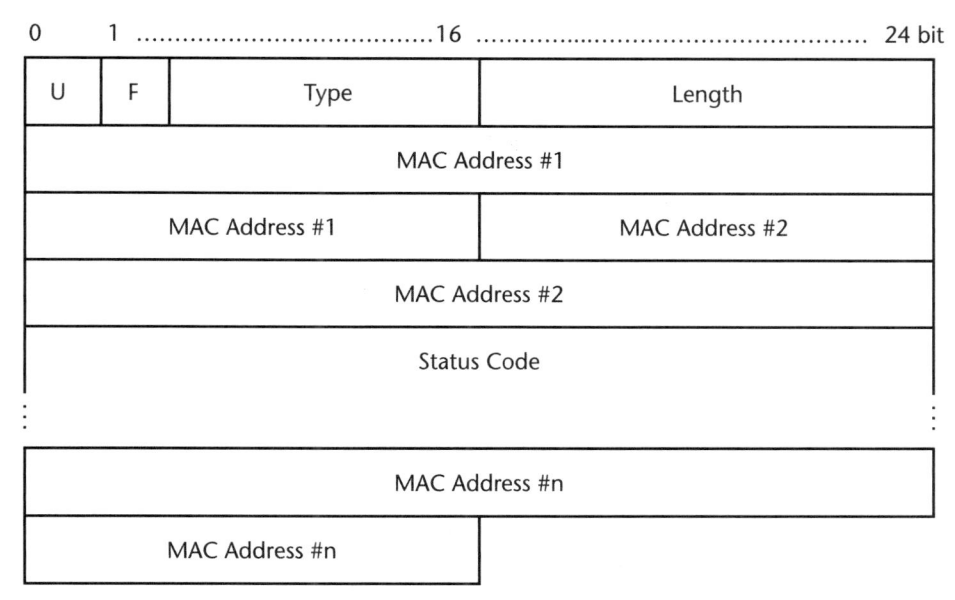

```
 0     1 ..........................16 ........................................ 24 bit
┌───┬───┬──────────────┬──────────────────────┐
│ U │ F │     Type     │       Length         │
├───┴───┴──────────────┴──────────────────────┤
│              MAC Address #1                 │
├──────────────────────┬──────────────────────┤
│    MAC Address #1    │    MAC Address #2    │
├──────────────────────┴──────────────────────┤
│              MAC Address #2                 │
├─────────────────────────────────────────────┤
│                Status Code                  │
├─────────────────────────────────────────────┤
│                     ⋮                       │
├─────────────────────────────────────────────┤
│              MAC Address #n                 │
├──────────────────────┬──────────────────────┤
│    MAC Address #n    │                      │
└──────────────────────┴──────────────────────┘
```

Figure 14.6 MAC list TLV address.

- U bit: Unknown bit. This bit must be set to 1. If the MAC address format is not understood, then the TLV is not understood and must be ignored.
- F bit: Forward bit. This bit must be set to 0. Since the LDP mechanism used here is targeted, the TLV must not be forwarded.
- Type: Type field. This field must be set to 0x0404. This identifies the TLV type as MAC list TLV.
- Length: Length field. This field specifies the total length in octets of the MAC addresses in the TLV. The length must be a multiple of 6.
- MAC address: The MAC address(es) being removed.

14.3.2 Tunnel Topology

Tunnels are set up between PEs to aggregate traffic. PWs are signaled to demultiplex encapsulated Ethernet frames from multiple VPLS instances that traverse the transport tunnels.

The topology of a VPLS must be loop-free topology such as full mesh of PWs. In a full mesh topology of PWs established between PEs, there is no need to relay packets between PEs. A simpler loop-breaking rule is the split horizon rule, whereby a PE must not forward traffic from one PW to another in the same VPLS mesh is instantiated.

To prevent forwarding loops, the split horizon rule is used where a PE must not forward traffic from one PW to another in the same VPLS mesh. The fact that there is always a full mesh of PWs between the PE devices ensures that every destination within the VPLS will be reached by a broadcast packet.

VPLS and PW relationships are depicted in Figures 14.7 and 14.8.

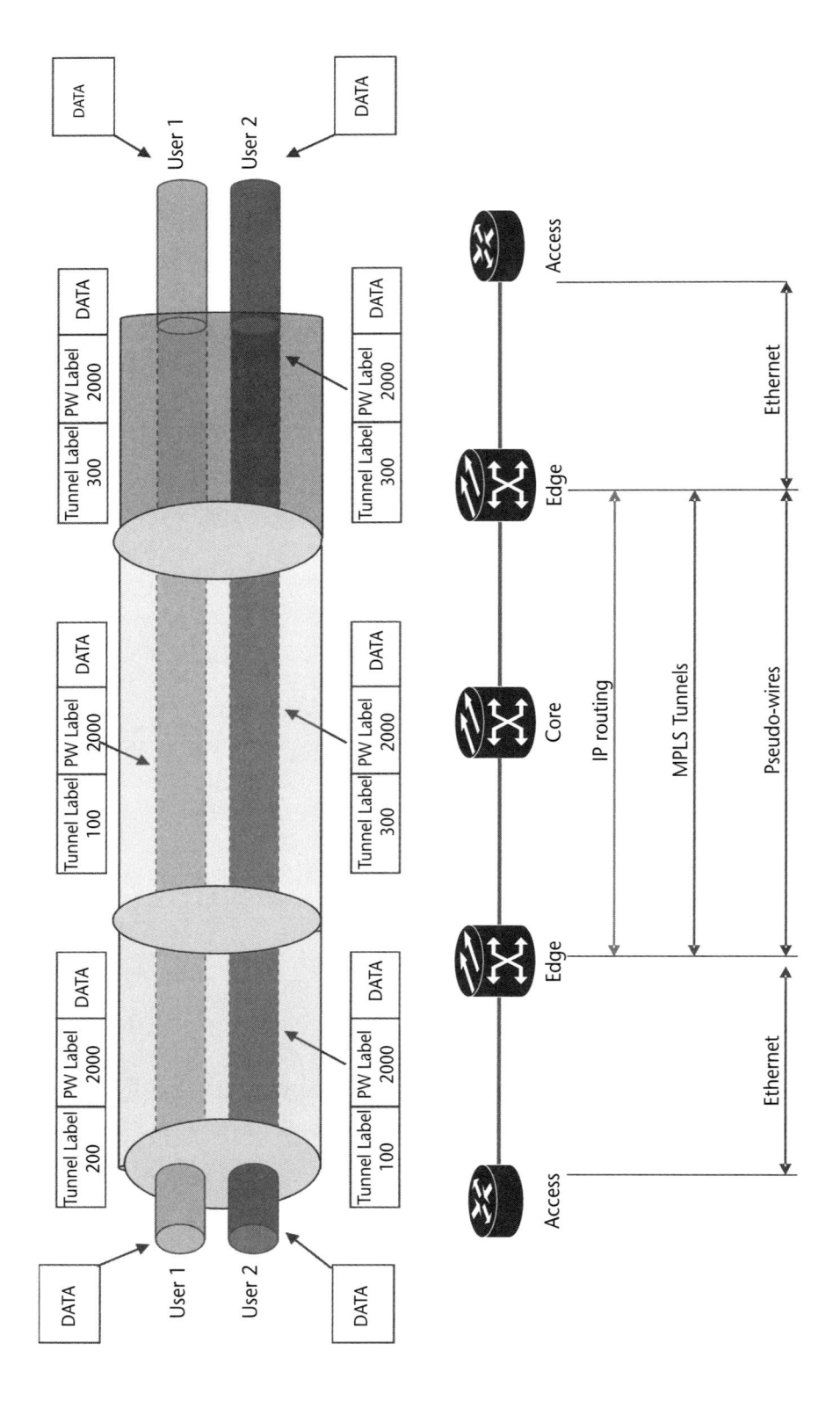

Figure 14.7 Usage of PWs in VPLS: (a) tunnels and PWs running between edge routers and (b) data carried over PWs in LSP tunnels.

| Outer Label | Inner Label | Control Word | Payload |

PW Packet

| 0000 | Reserved | Sequence Number (166) |

Control Word

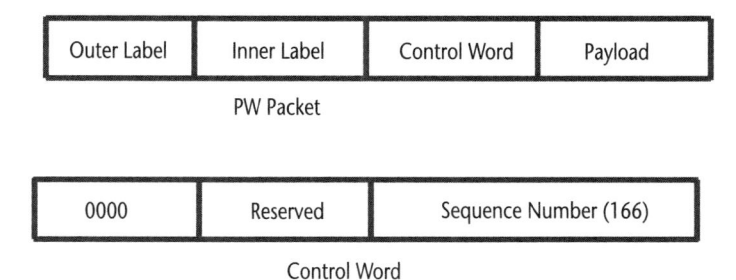

Figure 14.8 Format of PW packet and its control word.

14.3.3 Discovery

A PE either can be configured with the identities of all the other PEs in a given VPLS or can use some protocol to discover the other PEs. The latter is called auto-discovery.

The former approach is fairly configuration-intensive, since the PEs participating in a given VPLS are fully meshed. When a PE is added to or removed from the VPLS, the VPLS configuration on all PEs in that VPLS must be changed. Therefore, auto-discovery is critical to keep operational costs low, in automatically creating the LSP mesh. PEs can be manually configured when needed.

Multiple VPLS services can be offered over the same set of LSP tunnels. Signaling specified in [4, 5] is used to negotiate a set of ingress and egress virtual VC labels on a per service basis. The VC labels are used by the PE routers for demultiplexing traffic arriving from different VPLS services over the same set of LSP tunnels.

In RFC 4762 [5], the LDP is used to set up these tunnels. Alternatively the Resource Reservation Protocol–Traffic Engineering (RSVP-TE) or a combination of LDP and RSVP-TE can be used. Multipoint VPNs can be created on top of this full mesh, hiding the complexity of the VPN from the backbone routers.

For every VPLS instance, a full mesh of inner tunnels that are called PWs is created between all the PEs that participate in the VPLS instance. An auto-discovery mechanism locates all the PEs participating in a given VPLS instance. The service provider can either configure the PE with the identities of all the other PEs in a given VPLS, or can select its preferred auto-discovery mechanism, such as the Remote Authentication Dial-In User Service (RADIUS).

A PW consists of a pair of point-to-point, single-hop unidirectional LSPs in opposite directions, each identified by a PW label, also called a VC label. PW labels are exchanged between a pair of PEs using the targeted LDP signaling protocol. The VPLS identifier is exchanged with the labels, so that both PWs can be linked and be associated with a particular VPLS instance. The PW labels only have local significance between each pair of PEs. Each PE will form a MAC address to PW associations.

In VPLS, unlike BGP VPNs as defined in RFC 4364 [9], reachability information is not advertised and distributed via a control plane. Reachability is obtained by standard learning bridge functions in the data plane. PE routers need to implement a virtual bridge (Figure 14.5) functionality to support MAC learning, packet replication and forwarding. The creation of PWs with a pair of LSPs enables a PE to participate in MAC learning. When the PE receives an Ethernet frame with an unknown source MAC address, the PE knows on which VC it was sent.

When a packet arrives on a PW, if the source MAC address is unknown, it needs to be associated with the PW, so that outbound packets to that MAC address can be delivered over the associated PW. Similarly, when a packet arrives on an access port, if the source MAC address is unknown, it needs to be associated with the access port, so that outbound packets to that MAC address can be delivered over the associated access port.

Unknown packets (i.e., the destination MAC address has not been learned) are replicated and forwarded on all LSPs to the PE routers participating in that service until the target station responds and the MAC address is learned by the PE routers associated with that service.

Since LDP-based VPLS [5] does not specify auto-discovery, the service provider must know explicitly which PEs are part of the VPLS instance. For every VPLS instance present on a PE, the service provider will have to configure that PE with the addresses of all other PEs that are part of that VPLS instance. This process is operationally intensive and subject to human error.

Every time a PE joins a VPLS domain, the service provider must manually look up the other PEs that are part of that VPLS domain. Once this information is attained, a full mesh of LDP sessions between that PE and every other PE that is part of the VPLS domain is built. This overhead of a full mesh of LDP sessions is required because LDP does not have the advantage of BGP's route reflector architecture. For a VPLS instance with a small number of sites, the burden of LDP may not be that noticeable. However, it becomes more significant with the growth of the service, especially when LDP signaling sessions are authenticated via MD5. With a full LDP mesh, MD5 keys need to be configured on either end of every LDP session.

If the VPLS instance spans multiple autonomous systems (ASs), the globally significant 32-bit VCID used by LDP signaling requires manual coordination between ASs. For a VPLS instance spanned two ASs, the providers would need to use the same LDP VCID for that VPLS.

14.3.4 LDP-Based Signaling

A full mesh of LDP sessions is used to establish the mesh of PWs. Once an LDP session has been formed between two PEs, all PWs between these two PEs are signaled over this session. A hierarchical topology can be used in order to minimize the size of the VPLS full mesh when there is a large number of PWs.

Label mapping messages are exchanged between participating PEs to create tunnels while label withdrawal messages are used to tear down the tunnels. The message has FEC TLV (Figure 14.9) with PWid FEC element or generalized ID FEC element.

Both ends of a PW need to be configured with the PWid FEC, which is to be unique only within the targeted LDP session between the two PEs, and IP addresses. If pseudowires are members of a Layer-2 VPN and a discovery mechanism is used, or where switched on-demand PWs are used, or when distributed VPLS is used, it is desirable that only one end of a PW is configured with the configuration information. That information can be combined with the information carried in

14.3 LDP-Based VPLS

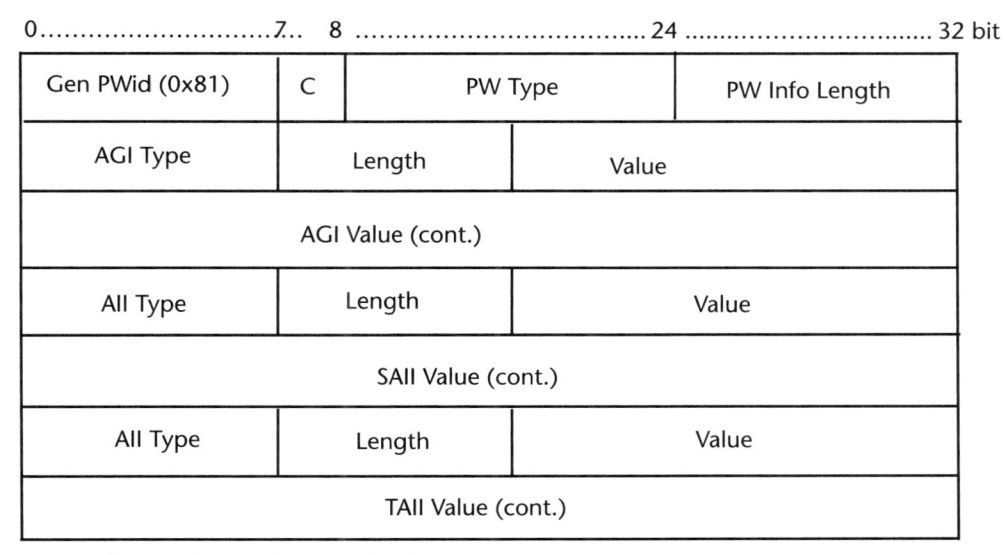

0........................7..	8	24	...............32 bit
Gen PWid (0x81)	C	PW Type	PW Info Length
AGI Type	Length	Value	
AGI Value (cont.)			
AII Type	Length	Value	
SAII Value (cont.)			
AII Type	Length	Value	
TAII Value (cont.)			

Attachment Group Identifier (AGI)
Source Attachment Individual Identifier (SAII)
Target Attachment Individual Identifier (TAII)

Figure 14.9 Generalized ID FEC element [7].

the discovery mechanism to successfully complete the signaling procedures. The PWid FEC needs to be globally unique.

The generalized ID FEC (Figure 14.9) tries to address all these cases. One can view the generalized ID FEC as one signaling scheme that covers the superset of cases, including the PWid provisioning model.

The PWId FEC contains PW type, control bit (indicates the presence of the control word), group ID, PW ID, and interface parameters sub-TLV. Generalized PW FEC [11] contains only PW type, control bit, and attachment group identifier (AGI), source attachment individual identifier (SAII), and target attachment individual identifier (TAII) that replace the PW ID.

The group ID and the interface parameters are contained in separate TLVs, called the PW grouping TLV and the interface parameters TLV.

Either of these types of PW FEC may be used for the setup of TDM PWs with the appropriate selection of PW types and interface parameters. LDP signaling is designed for setting up point-to-point connections and used in Martini PW services. After the label exchange, traffic parameters and OAM messages need to be negotiated for efficient signaling of each PW information.

The PWid FEC element can be used if a unique 32-bit value has been assigned to the PW, and if each end point has been provisioned with that value. The generalized PWid FEC element requires that the PW end points be uniquely identified. In addition, the end-point identifiers are structured to support applications where the identity of the remote end points needs to be auto-discovered rather than statically configured.

The generalized PWid FEC element is FEC type 0x81. Its fields are depicted in Figure 14.9:

- Control bit (C): This bit is used to signal the use of the control word as specified in [7].
- PW type: The allowed PW types are Ethernet (0x0005) and Ethernet tagged mode (0x004), as specified in [12].
- PW information length as specified in [7].
- AGI, length, value: The unique name of this VPLS. The AGI identifies a type of name, and length denotes the length of value, which is the name of the VPLS. We use the term AGI interchangeably with VPLS identifier.
- TAII, SAII: These are null because the mesh of PWs in a VPLS terminates on MAC learning tables, rather than on individual attachment circuits. The use of non-null TAII and SAII is reserved for future enhancements.

14.3.5 Data Forwarding on an Ethernet PW

Let's consider a customer with three sites (Figure 14.10), through CE1, CE2, and CE3, respectively. Assume that this configuration was determined using an unspecified auto-discovery mechanism. CE1, CE2, and CE3 are end stations at different customer sites and their ACs to their respective PE devices have been configured in the PEs. A VPLS has been set up among PE1, PE2, and PE3 such that the PEs belong to a particular VPLS identifier (AGI), 100.

Three PWs among PE1, PE2 and PE3 need to be created, each consisting of a pair of unidirectional LSPs or VCs. For VC label signaling between PEs, each PE initiates a targeted LDP session to the peer PE and communicates to the peer PE what VC label to use when sending packets for the considered VPLS. In this example, PE1 signals PW label 102 to PE2 and 103 to PE3, and PE2 signals PW label

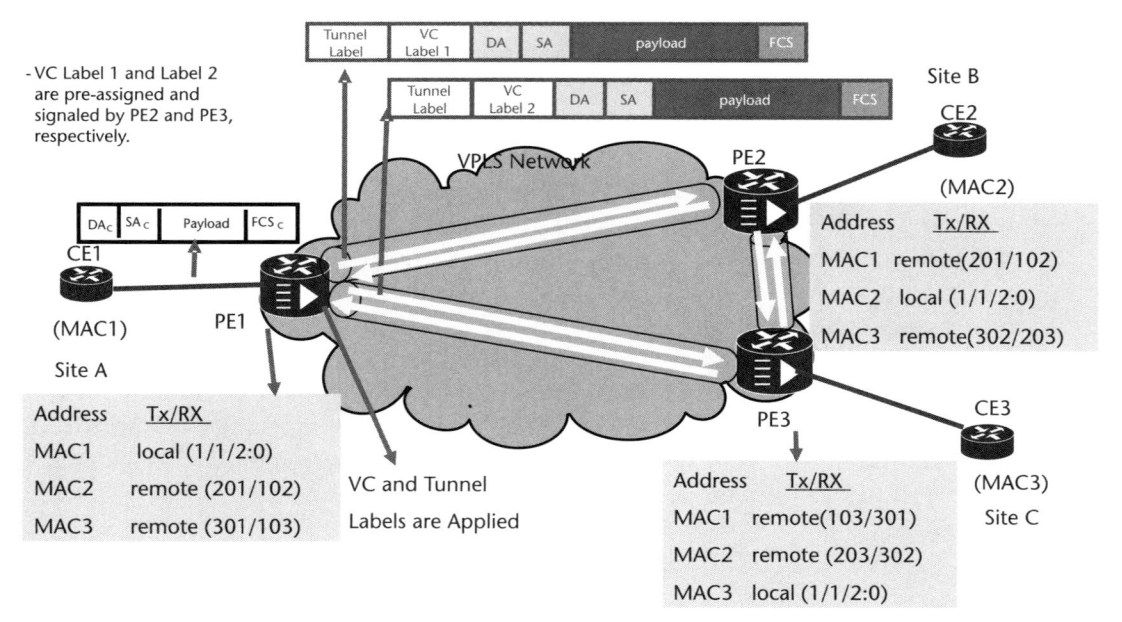

Figure 14.10 A VPLS example.

201 to PE1 and 203 to PE3. The specific VPLS instance is identified in the signaling exchange using a service identifier.

Assume a packet from CE1 is bound for CE2. When it leaves CE1, say it has a source MAC address of MAC1 and a destination MAC of MAC2. If PE1 does not know where MAC2 is, it will flood the packet, by sending it to PE2 and PE3. When PE2 receives the packet, it will have a PW label of 201. PE2 can conclude that the source MAC address MAC1 is behind PE1, since it distributed the label 201 to PE1. It can therefore associate the MAC address MAC1 with PW label 102.

Once the VPLS instance with Svc-id 100 has been created, the first packets can be sent and the MAC learning process starts. Assume CE2 is sending a packet to PE2 destined for CE1:

- PE2 receives the packet and learns from the source MAC address that CE2 can be reached on local port 1/1/2:0, stores this information in the FIB for Svc-id 100.
- PE2 does not yet know the destination MAC address CE1, so it floods the packet to PE1 with VC label 201 on the corresponding MPLS outer tunnel and to PE3 with VC label 203 on the corresponding MPLS outer tunnel.
- PE1 learns from VC label 201 that CE2 is behind PE2; it stores this information in the FIB for Svc-id 100.
- PE3 learns from VC label 203 that CE2 is behind PE2; it stores this information in the FIB for Svc-id 100.
- PE1 strips off label 201, does not know the destination CE1 and floods the packet on ports 1/1/2:0. PE1 does not flood the packet to PE3 because of the split horizon rule.
- PE3 strips off label 203. PE3 does not know the destination M1 and sends the packet on port 1/1/2:0, therefore, does not flood the packet to PE1 because of the split horizon rule.
- CE1 receives the packet. When CE1 receives the packet from CE2, it replies with a packet to CE2.
- PE1 receives the packet from CE1 and learns that CE1 is on local port 1/1/2:0. It stores this information in the FIB for Svc-id 100.
- PE1 already knows that CE2 can be reached via PE2 and therefore only sends the packet to PE2 using VC label 102
- PE2 receives the packet for CE2. It knows that CE2 is reachable on port 1/1/2:0.
- CE2 receives the packet.

14.3.6 Hierarchical VPLS

Hierarchical VPLS (H-VPLS) was designed to address scalability issues in VPLS. In VPLS, all PE nodes are interconnected in a full mesh to ensure that all destinations can be reached. In H-VPLS, a new type of node is introduced called the multitenant unit (MTU) (Figure 14.11), which aggregates multiple CE connections into a single PE, to reduce the number of PE-to-PE connections.

The MTU is connected to the nearest PE using a spoke PW. PE-to-PE PWs are now referred to as hub PWs. Full connectivity is only required between PEs, which allows more linear scaling of the network.

Spoke PWs are created between the MTUs and the PE routers. Connection for the spoke PW implementation can be either an IEEE 802.1Q tagged connection or an MPLS LSP with LDP signaling. Spoke connections are generally created between Layer-2 switches placed at the MTU and the PE routers placed at the service provider's point of presence (PoP). This considerably reduces both the signaling and replication overhead on all devices.

It is often beneficial to extend the VPLS service tunneling techniques into the access switch domain. This can be accomplished by treating the access device as a PE and provisioning PWs between this device and other edge devices, as a basic VPLS.

In H-VPLS, a CE is attached to an MTU via an AC. An AC from a specific customer is associated with a virtual bridge which is dedicated to that customer within the considered MTU.

An AC may be a physical or a virtual LAN (VLAN) tagged logical port. In the basic scenario, an MTU has one uplink to a PE. This uplink contains one spoke PW for each VPLS served by the MTU. The end points of this spoke PW are an MTU and a PE. Spoke PWs can be implemented using LDP-signaled MPLS PWs, if the MTU is MPLS enabled. Alternatively, they can be implemented using service provider VLANs (S-VLAN), whereby every VLAN on the MTU-PE uplink of an Ethernet aggregation network identifies a spoke PW.

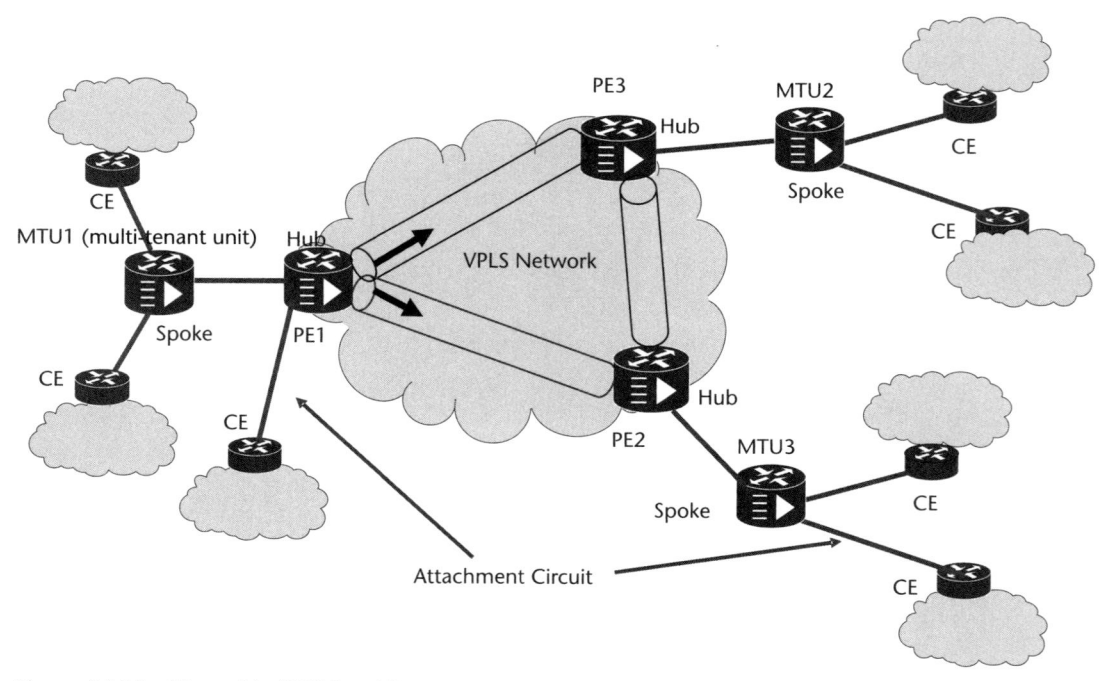

Figure 14.11 Hierarchical VPLS architecture.

14.3 LDP-Based VPLS

Let's consider the operation in Figure 14.11:

- Ethernet frames with known MAC addresses are switched accordingly within the VPLS.
- Frames with unknown or broadcast MAC addresses that are received from the PW are replicated and sent to all attached CE devices within the VPLS.
- Frames with unknown or broadcast MAC addresses that are received from a CE device are sent over the PW to the PE and to all other attached CE devices within the VPLS.
- Frames coming from the PW and CE devices with unknown MAC addresses are learned and aged within the VPLS.

The PE device needs to implement one VB for each VPLS served by the PE-attached MTUs. A particular spoke PW is associated with the PE VB dedicated to the considered VPLS instance. In the core network, the PE has a full mesh of PWs to all other PEs that serve the VPLS, as in the normal VPLS scenario.

H-VPLS also enables VPLS services to span multiple metro networks. A spoke connection that can be a simple tunnel is used to connect each VPLS service between the two metros. A set of ingress and egress VC labels is exchanged for each VPLS service instance to be transported over this LSP. The PE routers at each end treat this as a virtual spoke connection for the VPLS service in the same way as the PE-MTU connections. This architecture minimizes the signaling overhead and avoids a full mesh of VCs and LSPs between the two metro networks.

Spoke connectivity between MTU and PE depends on PE capabilities. PE that supports bridging functions, routing and MPLS encapsulation is called PE-rs. In Figure 14.12, each MTU-s has one PW to PE-rs. The PE-rs devices are connected in a basic VPLS full mesh. For each VPLS service, a single spoke PW is set up between the MTU-s and the PE-rs based on [7]. Unlike traditional PWs that terminate on a physical port or a VLAN, a spoke PW terminates on a virtual switch instance (VSI) on the MTU-s and the PE-rs devices. The VSI is a virtual MAC bridging instance with split horizon learning that is unknown MAC addresses received from a PW tunnel will not be forwarded to other PW tunnels to avoid learning loops.

The MTU-s and the PE-rs treat each spoke connection like an AC of the VPLS service. The PW label is used to associate the traffic from the spoke to a VPLS instance.

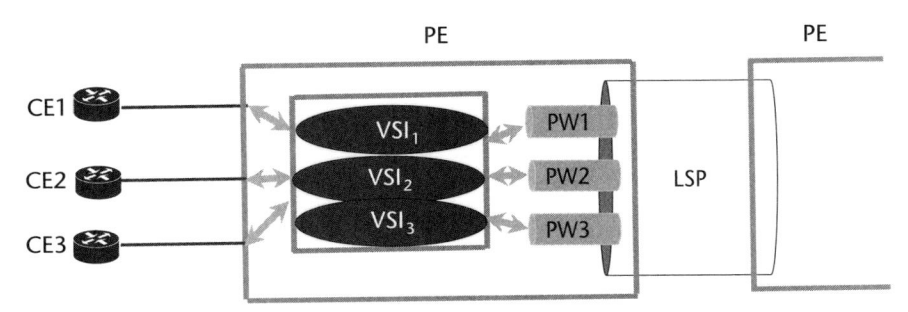

Figure 14.12 A diagram of MTU with VSI and PE-rs with VSI.

If PE does not support bridging function (i.e., PE-r), for every AC participating in the VPLS service, PE-r creates a point-to-point PW that terminates on the VSI of PE1-rs as shown in Figure 14.13.

For every port that is supported in the VPLS service, a PW is set up from the PE-r to the PE-rs. Once the PWs are set up, there is no learning or replication function required on the part of the PE-r. Traffic between CEs that are connected to PE-r is switched at PE1-rs and not at PE-r.

If PE-r devices use provider VLANs (P-VLANs) as demultiplexers instead of PWs, PE1-rs can treat them as such and map these circuits into a VPLS domain to provide bridging support between them.

This approach adds more overhead than the bridging-capable (MTU-s) spoke approach, since a PW is required for every AC that participates in the service versus a single PW required per service (regardless of ACs) when an MTU-s is used. However, this approach offers the advantage of offering a VPLS service in conjunction with a routed Internet service without requiring the addition of new MTU-s.

14.3.6.1 Virtual Private Wire Service (VPWS)

The VPWS-PE and VPLS-PE are functionally very similar in that they both use forwarders to map attachment circuits to PWs. The only difference is that while the forwarder in a VPWS-PE does a one-to-one mapping between the attachment circuit and PW, the forwarder in a VPLS-PE is a virtual switching instance (VSI) that maps multiple attachment circuits to multiple PWs.

In a VPWS, each CE device is presented with a set of point-to-point virtual circuits. The other end of each virtual circuit is another CE device. Frames transmitted without affecting their content by a CE on such a virtual circuit are received by the CE device at the other end point of the virtual circuit. The PE thus acts as a virtual circuit switch.

As mentioned before, an MTU-s supports layer-2 switching and bridging on all its ports including the spoke. It is treated as a virtual port. Packets to unknown

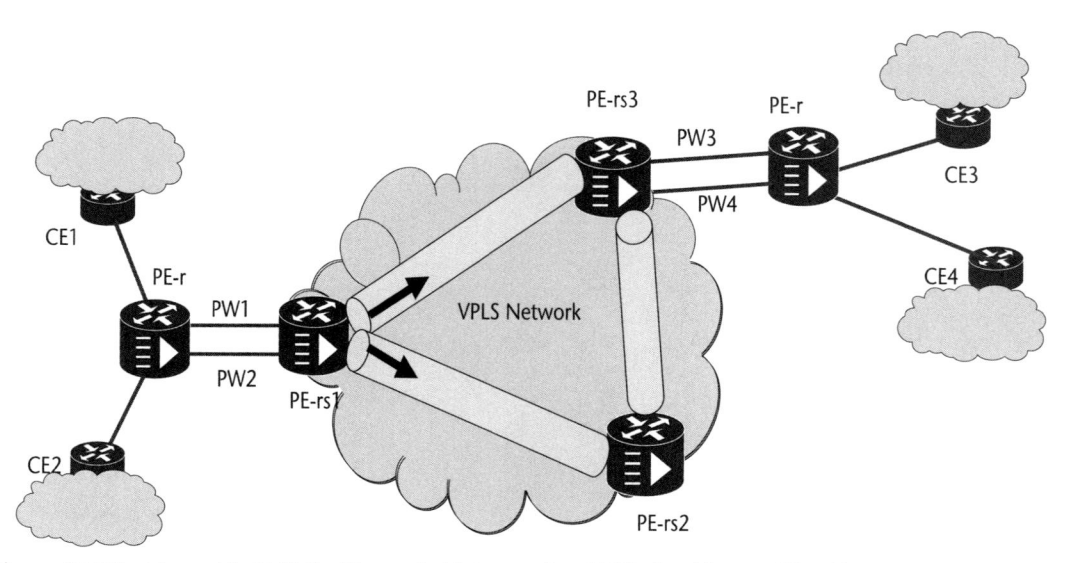

Figure 14.13 Hierarchical VPLS with non-bridging spokes (PW1 should go to PE_rs3)

destinations are replicated to all ports in the service including the spoke. Once the MAC address is learned, traffic between CEs that are connected to the spoke will be switched locally by the MTU-s, while traffic between a CE and any remote destination is switched directly onto the spoke and sent to the PE-rs over the point-to-point PW. Because the MTU-s is bridging-capable, only a single PW is required per VPLS instance for any number of access connections in the same VPLS service. This further reduces the signaling overhead between the MTU-s and PE-rs.

If the MTU-s is directly connected to the PE-rs, other encapsulation techniques, such as Q-in-Q, can be used for the spoke. At PE-rs, the spoke from the MTU-s is treated as a virtual port. The PE-rs will switch traffic between the spoke PW, hub PWs, and ACs once it has learned the MAC addresses.

14.3.6.2 Protection via Dual Homing

Protection against PW link failure and PE-rs equipment failure, MTU-s, or PE-r can be dual-homed into two PE-rs devices. These two PE-rs devices must be part of the same VPLS service instance. The MTU-s sets up two PWs connected to each PE-rs for each VPLS instance. If we use the 1 + 1 protection scheme, one of the two PWs can be designated as the primary one that carries active traffic while the other can be designated as the secondary one, which is a standby. The MTU-s negotiates the PW labels for both the primary and secondary PWs, but does not use the secondary PW unless the primary PW fails.

The MTU-s should control the usage of the spokes to the PE-rs devices. If the spokes are PWs, then LDP signaling is used to negotiate the PW labels. The hello messages of the LDP session could be used to detect failure of the primary PW.

Upon failure of the primary PW, MTU-s immediately switches to the secondary PW. At this point, the PE-rs3 that terminates the secondary PW starts learning MAC addresses on the spoke PW. All other PE-rs nodes in the network think that CE-1 and CE-2 are behind PE-rs1 and may continue to send traffic to PE1-rs until they learn that the devices are now behind PE-rs3. The unlearning process can take a long time and may adversely affect the connectivity of higher-level protocols from CE1 and CE2.

To enable faster convergence, the PE3-rs where the secondary PW got activated may send out a flush message, using the MAC List TLV, to all PE-rs nodes. Upon receiving the message, PE-rs nodes flush the MAC addresses associated with that VPLS instance.

14.3.6.3 Hierarchical VPLS Model Using Ethernet Access Network

Ethernet-based access networks are currently deployed by service providers to offer VPLS services to their customers. Tagged or untagged Ethernet traffic of customers can be tunneled via an additional VLAN tag (S-Tag) to the customer's data. Therefore, there is a one-to-one correspondence between an S-Tag and a VPLS instance.

The PE-rs needs to perform bridging functionality over the standard Ethernet ports toward the access network, as well as over the PWs toward the network core. In this model, the PE-rs may need to run STP towards the access network, in addition to split-horizon over the MPLS core. The PE-rs needs to map a S-VLAN to a VPLS-instance and its associated PWs, and vice versa.

Since each P-VLAN corresponds to a VPLS instance, the total number of VPLS instances supported is limited to 4K. This 4K limit applies only within an Ethernet access network and not to the entire network.

14.4 BGP Approach

The VPLS control plane functions of auto-discovery and provisioning of PWs are accomplished with a single BGP update advertisement. In the auto-discovery, each PE discovers other PEs that are part of a given VPLS instance via BGP. When a PE joins or leaves a VPLS instance, only the affected PE's configuration changes while other PEs automatically find out about the change and adapt.

The BGP route target (RT) community (or extended communities) [3, 12] is used to identify members of a VPLS. If VPLS is fully meshed, a single RT is adequate to identify a given VPLS instance. A PE announces usually via I-BGP that it belongs to a specific VPLS instance by annotating its network layer reachability information (NLRI) for that VPLS instance with RT, and acts on this by accepting NLRIs from other PEs that have RT. A PE announces that it no longer participates in that specific VPLS instance by withdrawing all NLRIs that it had advertised with RT. RT format is given in Figure 14.14. Detailed description of route target is given in RFC 4360 [4].

14.4.1 Auto-Discovery

In order to understand auto-discovery in BG-based VPLS, let's assume a new PE is added by the service provider. Per [3], a single BGP session is established between the new PE and a route reflector. The new PE then joins a VPLS domain when the VPLS instance is configured on that PE. Each VPLS instance is identified by a particular route target BGP extended community, which is configured as part of configuring a VPLS instance.

Once this occurs, the PE advertises that it is part of the VPLS domain via the route reflector to other PEs joined in that VPLS instance.

The advertisement carries the BGP route target extended community that is configured for that VPLS. This community identifies the advertisement with a particular VPLS. For the purpose of redundancy, the PE may establish BGP sessions with more than one route reflector.

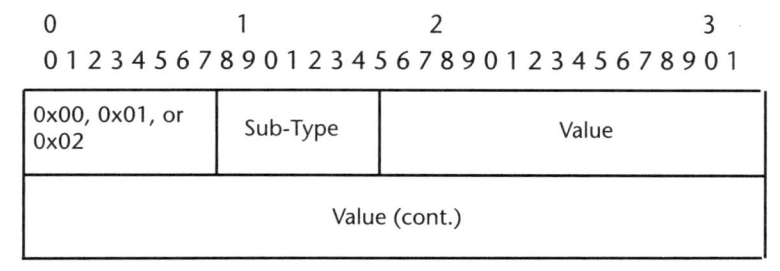

```
 0                   1                   2                   3
 0 1 2 3 4 5 6 7 8 9 0 1 2 3 4 5 6 7 8 9 0 1 2 3 4 5 6 7 8 9 0 1
```

0x00, 0x01, or 0x02	Sub-Type	Value
Value (cont.)		

Figure 14.14 RT format.

When a PE is added, only a BGP session between it and the route reflector is established. If the session is to be authenticated with MD5, then only keys for the two end points of that BGP session are configured. When a new VPLS instance is configured on that PE, it then advertises its availability via the route reflector, making all other relevant PEs aware of its presence. At the same time, BGP signaling automatically builds the mesh of LSPs for that VPLS instance.

If the VPLS instance needs to span multiple ASs, use of the route target for identifying a particular VPLS simplifies operations, as each AS can assign a particular route target to that VPLS on its own. This is possible because route target extended community embeds the autonomous system number, and these numbers are globally unique.

14.4.2 Signaling

Once discovery is done, each pair of PEs in a VPLS must be able to establish PWs to each other, transmit certain characteristics of the PWs that a PE sets up for a given VPLS, and tear down the PWs when they are no longer needed. This mechanism is called signaling.

Multiple streams are multiplexed in a tunnel where each stream is identified by an MPLS label if MPLS is used, called demultiplexer. The label determines to which VPLS instance a packet belongs and the ingress PE. Each stream may represent a service.

A distinct BGP update message can be used to send a demultiplexer to each remote PE. This would require the originating PE to send N such messages for N remote PEs. It is also possible PE to send one common update message that contains demultiplexers for all the remote PEs. Doing this reduces the control plane load both on the originating PE as well as on the BGP route reflectors (Figure 14.15) that may be involved in distributing this update to other PEs.

To accomplish this, label blocks are introduced. A label block, defined by a label base LB and a VE block size VBS, is a contiguous set of labels {LB, LB+1, ..., LB+VBS-1}. All PEs within a given VPLS are assigned unique VE IDs as part of their configuration. A PEx wishing to send a VPLS update sends the same label block information to all other PEs. Each receiving PE infers the label intended for

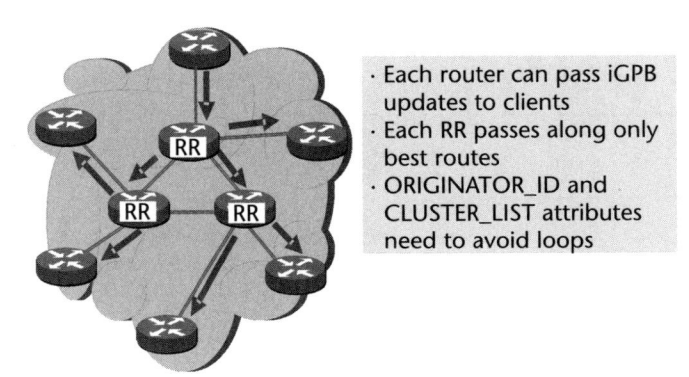

Figure 14.15 Route reflector.

PEx by adding its unique VE ID to the label base. In this manner, each receiving PE gets a unique demultiplexer for PEx for that VPLS.

14.4.2.1 VPLS BGP NLRI

A VPLS BGP NLRI (Figure 14.16) consists of a VE ID, a VE block offset, a VE block size, and a label base. The address family identifier (AFI) is the L2VPN AFI (25), and the subsequent address family identifier (SAFI) is the VPLS SAFI (65).

A PE participating in a VPLS must have at least one VPLS edge (VE) ID. If the PE is the VE, it typically has one VE ID. If the PE is connected to several u-PEs, it has a distinct VE ID for each u-PE. It may additionally have a VE ID for itself, if it itself acts as a VE for that VPLS.

VE IDs are typically assigned by the network administrator. A given VE ID should belong to only one PE, unless a CE is multihomed.

The Layer-2 info extended community (Figure 14.17) is used to signal control information about the PWs to be set up for a given VPLS. The extended community value, 0x800A, is allocated by IANA. This information includes the Encaps Type, which is the type of encapsulation on the PWs; control flags (Figure 14.18), which is the control information regarding the PWs, and the maximum transmission unit (MTU), which is used on the PWs.

14.4.3 BGP VPLS Operation

To create a new VPLS, VPLSnew, a network administrator must pick an RT for VPLSnew, RTnew. This will be used by all PEs that serve VPLSnew. To configure a given PE, PE1, to be part of VPLSnew, the network administrator only has to choose a VE ID, V, for PE1.

If PE1 is connected to u-PEs (Figure 14.19), PE1 may be configured with more than one VE ID. The PE1 may also be configured with a Route Distinguisher (RD),

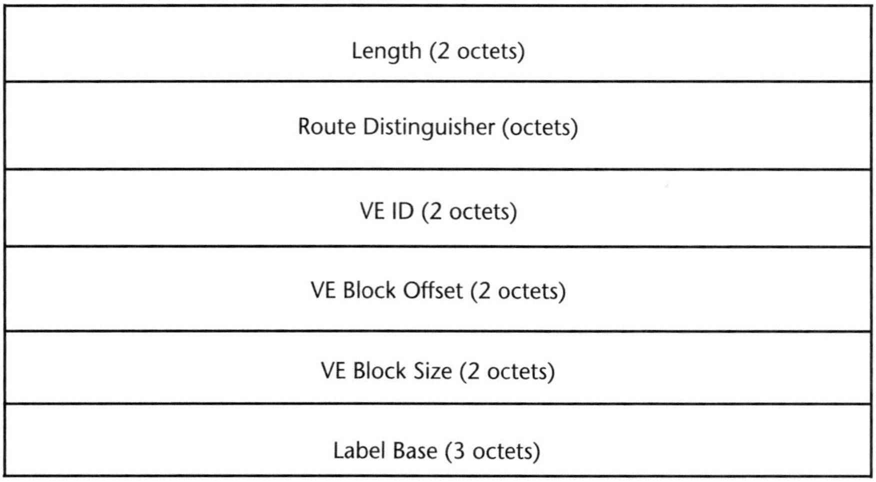

Length (2 octets)
Route Distinguisher (octets)
VE ID (2 octets)
VE Block Offset (2 octets)
VE Block Size (2 octets)
Label Base (3 octets)

Figure 14.16 BGP NLRI for VPLS information.

14.4 BGP Approach

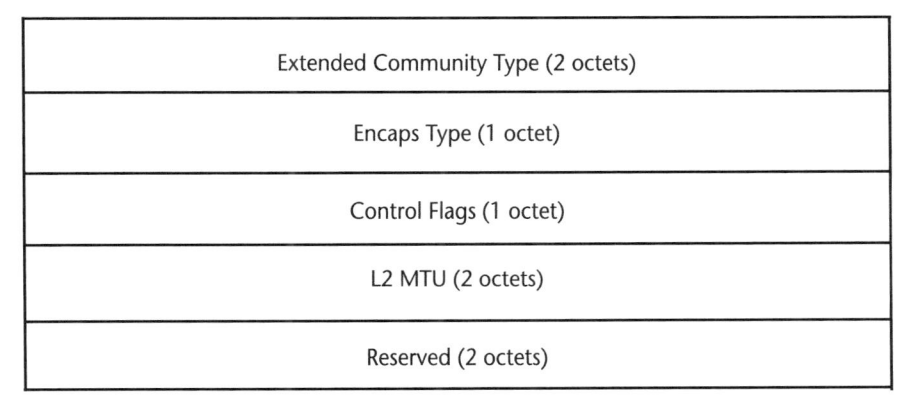

Extended Community Type (2 octets)
Encaps Type (1 octet)
Control Flags (1 octet)
L2 MTU (2 octets)
Reserved (2 octets)

Figure 14.17 Layer-2 info extended community.

```
0     1     2     3     4     5     6     7
|-----------------------------|-----|-----|
|            MBZ              |  C  |  S  |
|-----------------------------|-----|-----|
```

MBZ: Must be zero

C: A Control word [7] MUST or MUST NOT be present when sending VPLS packets to this PE, depending on whether C is 1 or 0, respectively

S: Sequenced delivery of frames MUST or MUST NOT be used when sending VPLS packets to this PE, depending on whether S is 1 or 0, respectively

Figure 14.18 Control flags bit vector.

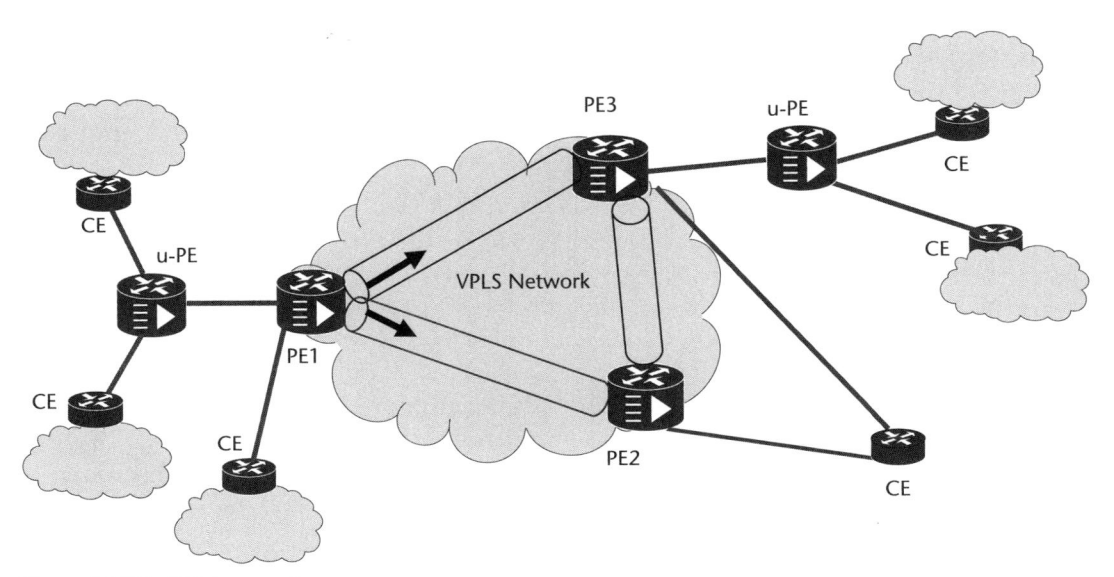

Figure 14.19 VPLS operation.

RDnew. Otherwise, it generates a unique RD for VPLSnew. PE1 then generates an initial label block and a remote VE set for V, defined by VE block offset VBO, VE block size VBS, and label base LB.

PE1 then creates a VPLS BGP NLRI with RD RDnew, VE ID V, VE block offset VBO, VE block size VBS and label base LB. It attaches a Layer-2 info extended community and an RT, RTnew. It sets the BGP next hop for this NLRI as itself, and announces this NLRI to its peers. The network layer protocol associated with the network address of the next hop for the combination <AFI=L2VPN AFI, SAFI=VPLS SAFI> is IP [13]. If the value of the length of the next hop field is 4, then the next hop contains an IPv4 address. If this value is 16, then the next hop contains an IPv6 address.

If PE1 hears from another PE, for example, PE2, a VPLS BGP announcement with RTnew and VE ID Z, then PE1 knows via auto-discovery that PE2 is a member of the same VPLS. PE1 then has to set up its part of a VPLS PW between PE1 and PE2. Similarly, PE2 will have discovered that PE1 is in the same VPLS, and PE1 must set up its part of the VPLS PW. Thus, signaling and PW setup is also achieved with the same update message.

If PE1's configuration is changed to remove VE ID V from VPLSnew, then PE1 withdraws all its announcements for VPLSnew that contain VE ID V. If all of PE1's links to its CEs in VPLS nw go down, then PE1 either withdraws all its NLRIs for VPLSnew or lets other PEs in the VPLSnew know that PE1 is no longer connected to its CEs.

14.4.4 Multi-AS VPLS

As in [3, 12], auto-discovery and signaling functions are typically announced via I-BGP. This assumes that all the sites in a VPLS are connected to PEs in a single AS. However, sites in a VPLS may connect to PEs in different ASs. In this case, I-BGP connection between PEs and PE-to-PE tunnels between the ASs must be established.

There are three methods for signaling inter-provider VPLS:

- VPLS-to-VPLS connections at the AS border routers (ASBRs): This requires an Ethernet interconnect between the ASes, and both VPLS control and data plane state on the ASBRs. This method is easy to deploy.

An AS border router (ASBR1) acts as a PE for all VPLSs that span AS1 and an AS to which ASBR1 is connected to, such as AS2 (Figure 14.20). The ASBR on the neighboring AS (ASBR2) is viewed by ASBR1 as a CE for the VPLSs that span AS1 and AS2. Similarly, ASBR2 acts as a PE for this VPLS for AS2 and views ASBR1 as a CE.

This method does not require MPLS protocol on the ASBR1-ASBR2 link, but requires L2 Ethernet. A VLAN ID is assigned to each VPLS traversing this link. Furthermore, ASBR1 performs the PE operations (discovery, signaling, MAC address learning, flooding, encapsulation, and so forth) for all VPLSs that traverse ASBR1. This imposes a significant load on ASBR1, both on the control plane and the data plane, which limits the number of multi-AS VPLSs.

- EBGP redistribution of VPLS information between ASBRs: This requires the VPLS control plane state on the ASBRs and MPLS on the AS-AS interconnect.

14.4 BGP Approach

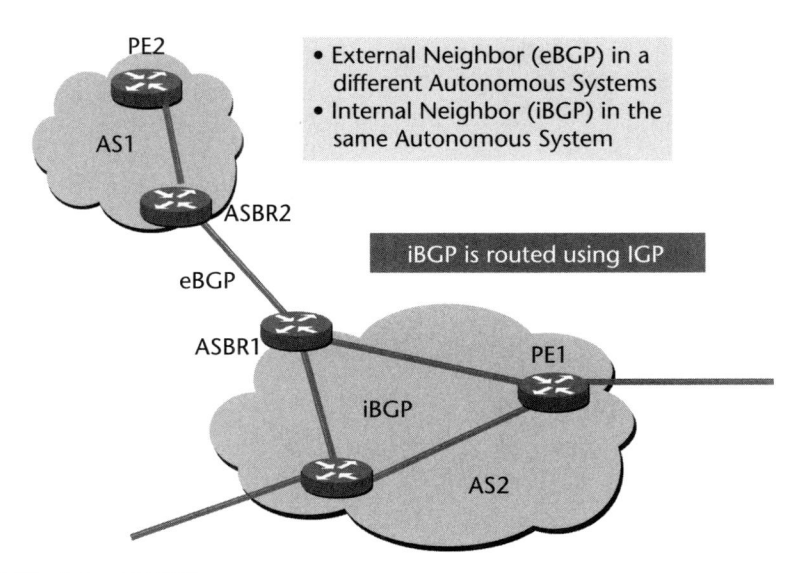

Figure 14.20 Inter-AS VPLS.

This method requires I-BGP peerings between the PEs in AS1 and ASBR1, an E-BGP peering between ASBR1 and ASBR2 in AS2, and I-BGP peerings between ASBR2 and the PEs in AS2.

- Multihop EBGP redistribution of VPLS information between ASs: This requires MPLS on the AS-AS interconnect, but no VPLS state on the ASBRs. In this method, there is a multihop E-BGP peering between the PEs (or preferably, a route reflector) in AS1 and the PEs (or route reflector) in AS2.

In the multi-AS configuration, a range of VE IDs identifying VPLS spanning multiple ASs is assigned for each AS. For example, AS1 uses VE IDs in the range 1 to 100, AS2 from 101 to 200, etc. If there are 10 sites attached to AS1 and 20 to AS2, the allocated VE IDs could be 1 to 10 and 101 to 120. This minimizes the number of VPLS NLRIs that are exchanged while ensuring that VE IDs are kept unique.

There will be no overlap between VE ID ranges among ASs, except when there is multihoming. When a VPLS site is connected to multiple PEs in the same AS or PEs in different ASs, the PEs connected to the same site can be configured either with the same VE ID or with different VE IDs. When PEs are in different ASs, it is mandatory to run STP on the CE device, and possibly on the PEs, to construct a loop-free VPLS topology.

14.4.5 Hierarchical BGP VPLS

The purpose of hierarchical BGP VPLS to scale the VPLS control plane when using BGP. The hierarchy:

1. Alleviates the full mesh connectivity requirement among VPLS BGP speakers;

2. Limits BGP VPLS message passing to just the interested speakers rather than all BGP speakers;
3. Simplifies the addition and deletion of BGP speakers.

The basic technique for the hierarchy is to use BGP route reflectors (RRs) [6], which help to achieve objectives in (1) and (3). A designated small set of route reflectors are fully meshed. A BGP session between each BGP speaker and one or more RRs is established.

In this approach, there is no need for a full mesh connectivity among all the BGP speakers. If a large number of RRs is needed, then this method will be used recursively.

The use of RRs introduces no data plane state and no data plane forwarding requirements on the RRs, and does not change the forwarding path of VPLS traffic. This is in contrast to the technique of hierarchical VPLS defined in [8].

One can define several sets of RRs, for example, a set to handle VPLS, another to handle IP VPNs, and another for Internet routing. Another partitioning could be to have some subset of VPLSs and IP VPNs handled by one set of RRs, and another subset of VPLSs and IP VPNs handled by another set of RRs. The use of route target filtering (RTF) [4] can make this simpler.

Limiting BGP VPLS message passing to just the interested BGP speakers is addressed by the use of RTF. RTF is also very effective in inter-AS VPLS. More details on how RTF works and its benefits are provided in [4].

14.5 Security

Data in a VPLS is only distributed to other nodes in that VPLS and not to any external agent or other VPLS. However, VPLS does not offer confidentiality, integrity, or authentication capabilities. VPLS packets are sent in the clear in the packet switched network. A device-in-the-middle can eavesdrop and may be able to inject packets into the data stream. If security is desired, the PE-to-PE tunnels can be IPsec tunnels. The end systems in the VPLS sites can use encryption to secure their data even before it enters the service provider network.

14.6 External Network-Network Interface

A service provider can choose VPLS to transport Ethernet service over its domain independent from hand-offs to other providers. The handoff which is an external network-network interface (ENNI) can be 802.1ad (Figure 14.21).

The 802.1ad packets can be mapped onto to the appropriate tunnel in the provider domain. Each provider network determines the best path within that network optimized within each domain. QoS is honored within each provider network. At interprovider boundaries, EXP- PCP mappings are done to preserve end-to-end QoS.

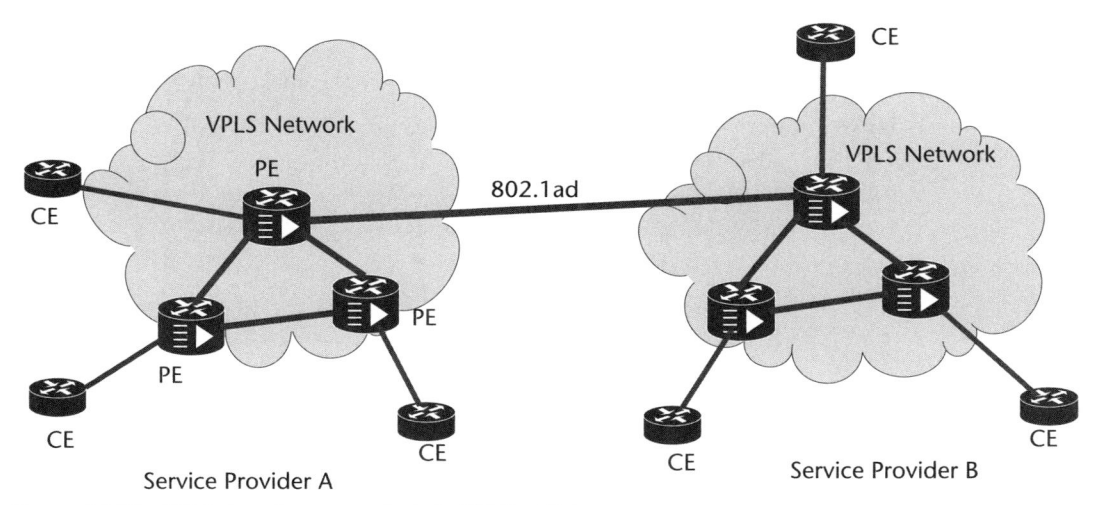

Figure 14.21 VPLS domains connected via 802.1ad links.

14.7 Conclusion

VPLS has received widespread industry support from both vendors and service providers and deployed widely. It offers flexible connectivity and a cost-effective solution for enterprise customers and carriers.

We have described two VPLS architectures, namely LDP and BGP approaches, to emulate LAN over an MPLS network. VPLS offers enterprise customers intersite connectivity, protocol transparency and scalability, and a simplified LAN/WAN boundary. VPLS simplifies the network management and reduces involvement of the service provider in the customer IP scheme. Instead of requiring customers to connect to an IP network, with the complexity of IP routing protocols, they connect with simple and low-cost Ethernet. All of this is provisioned using standards-based Ethernet and MPLS gear.

References

[1] RFC 2764, "A Framework for IP Based Virtual Private Networks," 2000.

[2] Andersson, L., and E. Rosen: "Framework for Layer-2 Virtual Private Networks," http://www.ietf.org/internet-drafts/draft-ietf-L2vpn-l2-framework-05.txt, 2004.

[3] RFC 4761, "Virtual Private LAN Service (VPLS) Using BGP for Auto-Discovery and Signaling," 2007.

[4] RFC 4360, "BGP Extended Communities Attribute," 2006.

[5] RFC 4762, "Virtual Private LAN Service (VPLS) Using Label Distribution Protocol (LDP) Signaling," 2007.

[6] Encapsulation Methods for Transport of Ethernet Over MPLS Networks, draft-ietf-pwe3-ethernet-encap-11.txt, 2005.

[7] RFC 4447, "Pseudowire Setup and Maintenance Using the Label Distribution Protocol (LDP)," 2006.

[8] RFC 4448, "Encapsulation Methods for Transport of Ethernet over MPLS Networks," 2006.

[9] RFC 4364, "BGP/MPLS IP Virtual Private Networks (VPNs)," 2006.

[10] IEEE 802.1ad, "Local and Metropolitan Area Networks Virtual Bridged Local Area Networks," 2005.

[11] RFC 2587, "Internet X.509 Public Key Infrastructure LDAPv2 Schema," 1999.

[12] RFC 4446, "IANA Allocations for Pseudowire Edge to Edge Emulation (PWE3)," 2006.

[13] Lasserre, M., and V. Kompella, "Virtual Private LAN Services over MPLS," http://www.ietf.org/internet-drafts/draft-ietf-l2vpn-vpls-ldp-07.txt, 2005.

CHAPTER 15
Information Modeling

15.1 Introduction

This chapter describes the information model to manage the Carrier Ethernet services described in Chapter 6.

A common management model is needed for Carrier Ethernet services to ensure that the information in operation support systems (OSSs), software-defined networking (SDN) orchestrators, SDN controllers, network management systems (NMSs), element management systems (EMSs), and network elements (NEs) from various vendors are logically consistent and allow service providers to readily integrate their capabilities into the management environment.

Three management information views are defined by ITU-T M.3100 [1]:

- The network element view is concerned with managing an NE, which includes the information required to manage the network element function and the physical aspects of the NE.
- Network view is concerned with the information representing the network, both physically and logically and how NE entities are related, topographically interconnected, and configured to provide and maintain end-to-end connectivity.
- The service view is concerned with how network views are utilized to provide a network service, the requirements of a network service such as availability, and how these requirements are met through the use of the network. Objects defined for a given view may be used in others, and any object may be used by any interface which requires it.

MEF is focused on the overall Carrier Ethernet management information model to manage the Carrier Ethernet services as defined by the MEF. The management functions supported by the information model are configuration management, performance management with performance monitoring and evaluation of performance, and fault management with alarm reporting and testing.

In G.8010 [2] two layer networks are defined in the Ethernet over Transport (EoT) network architecture: Ethernet MAC (ETH) layer network and Ethernet PHY (ETY) layer network.

The ETH layer network characteristic information can be transported through ETH links supported by trails in the ETY layer network or other path layer networks such as synchronous digital hierarchy (SDH) virtual container-n (VC-n), multiprotocol label switching (MPLS), and asynchronous transfer mode (ATM). MEF 7.2 [3] focuses on identifying the management objects for Carrier Ethernet network (CEN) and service management at the ETH layer.

The Ethernet services layer, also referred to as the ETH layer, is the specific layer network within a CEN responsible for the instantiation of Ethernet MAC oriented connectivity services and the delivery of Ethernet Protocol data units (PDUs) presented across well-defined internal and external interfaces. The ETH layer is responsible for all service-aware aspects associated with Ethernet MAC flows, including operations, administration, maintenance, and provisioning capabilities required to support Ethernet connectivity services. The service frame presented by the ETH layer external interfaces is an Ethernet unicast, multicast, or broadcast frame conforming to the IEEE 802.3-2012 [4] frame format.

The service and network views provide an abstraction of service and network resources allowing for flexibility in the management of the Ethernet services and the underlying network resources.

The service and network view abstraction resides at the service management layer (SML) and network management layer (NML) of telecommunications management network (TMN) respectively, as depicted in Figure 15.1 [3]. The service and network view abstraction provides service, flow, and connection-oriented information that may be reapplied at the element management layer (EML) and element layer (EL) nodal-oriented management information models. The business management layer (BML) provides functions for managing the overall business.

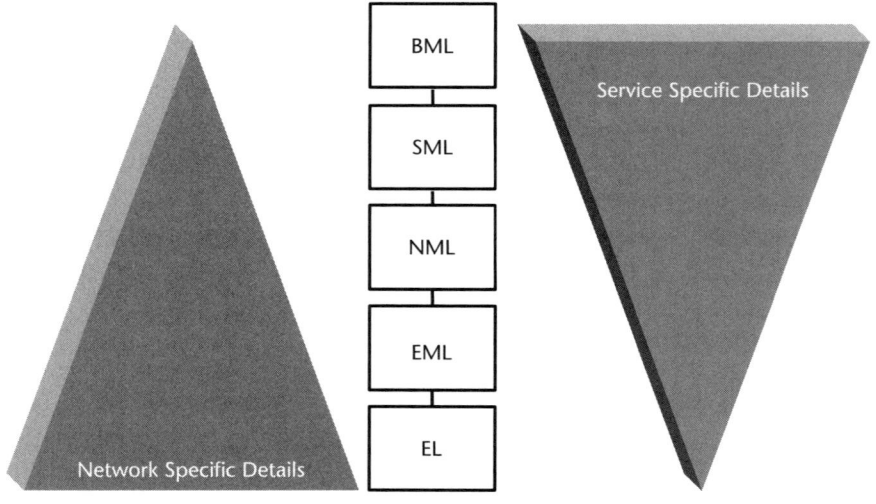

Figure 15.1 TMN functional layering [3].

As part of life-cycle service orchestration (LSO), MEF defined management views for carrier Ethernet network services as depicted in Figure 15.2. The following sections will describe MIBs, service view and resource view models for managing Carrier Ethernet Services with SNMP, and YANG models.

Current MEF modeling activities include MEF 7.3 service model [5], network resource provisioning data model [6], and API [7] and YANG [8] data models. We will summarize them next. For more details, we refer the readers to the references.

15.2 Information Modeling of Carrier Ethernet Services for EMS and NMS

A layer network domain (LND) represents an administration's view of the resources responsible for transporting a specific type of characteristic information [e.g., IP, ETH (layer 2), ETY (ETH PHY), MPLS, SONET/SDH]. Flows, connections, resources, and network topology can be managed and represented separately for each LND.

Flow domains (subnetworks) are composed of flow domains (subnetworks) and links. Furthermore, a flow domain (subnetwork) may be partitioned into subflow domains (subnetworks) and the links that connect them, as shown in Figure 15.3.

A carrier's network is usually partitioned along the lines of the network operations center (NOC) responsible for each flow domain or subnetwork. Within each flow domain (subnetwork) representing a NOC, flow domains (subnetworks)

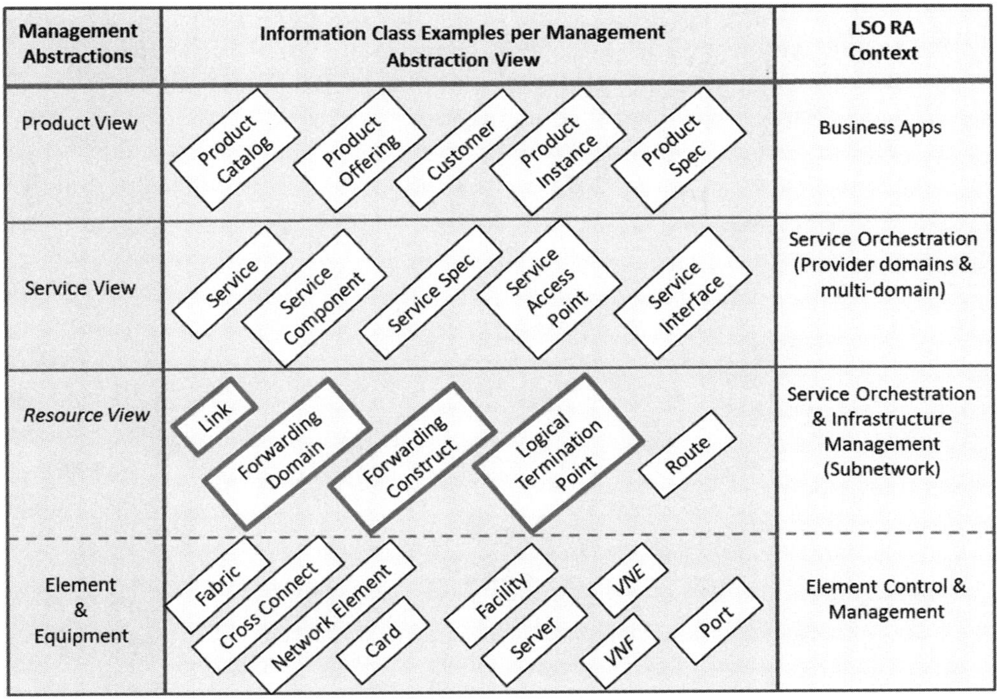

Figure 15.2 MEF management layering.

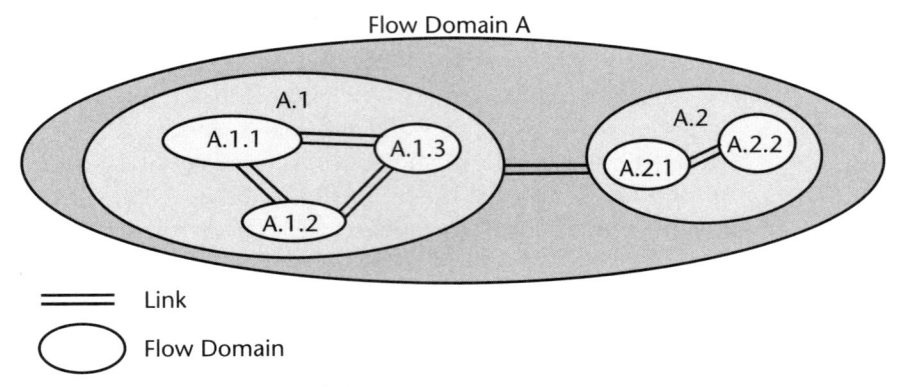

Figure 15.3 Partitioning example [3].

could be partitioned to describe the resources that are managed by a specific management system.

Topological elements represent the logical topology or structure of the flow domain (subnetworks) within an LND, including flow domains (subnetworks) and the links that connect them, as shown in Figure 15.4.

The flow domain (or subnetwork) provides capacity for carrying characteristic information within an LND. Flow domains (subnetworks) can be partitioned into a set of component flow domains (subnetworks) and links. In addition to representing flow domains in its own administration as component flow domains, a carrier can represent an external carrier network as a component flow domain, allowing the carrier to maintain a complete topology including connected external networks.

Flow domains can be used to represent a carrier's entire layer network, vendor-specific component flow domains, connected external carrier flow domains, and even "atomic" flow domains corresponding to individual bridges.

A link is a topological component that describes a fixed topological relationship between flow domains (subnetworks), along with the capacity supported by an underlying server LND trail. Links in the client LND are supported by trails in an underlying server LND.

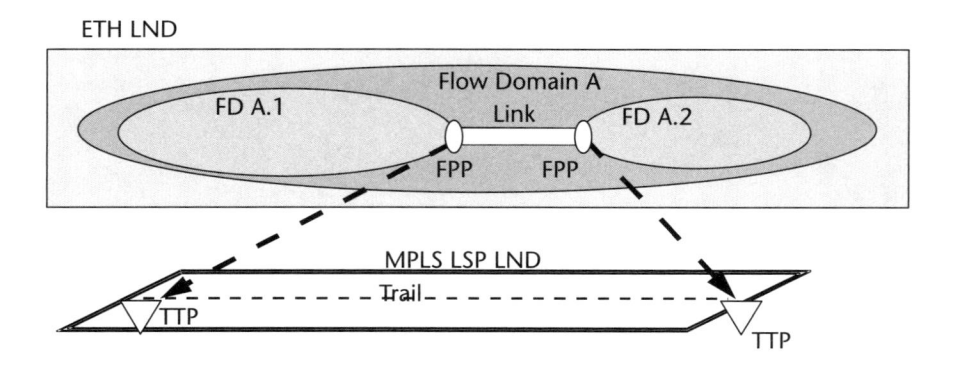

Figure 15.4 Topological elements [3].

15.2 Information Modeling of Carrier Ethernet Services for EMS and NMS

The termination of a link is called a flow point pool (FPP) or link end. The FPP or link end describes configuration information associated with an interface, such as a UNI or NNI. The FPP or link end is associated with the trail termination of the underlying server trail used to perform adaptation and transport of the characteristic information of the client LND.

Flow and connection are responsible for transporting characteristic information across the LND, across flow domains (subnetworks), and across links. A flow domain fragment (FDFr) or subnetwork connection (SNC) such as OVC is a connection responsible for transporting characteristic information across a flow domain or subnetwork. If the flow domain (subnetwork) that the FDFr (SNC) traverses is partitioned, the FDFr (SNC) may be partitioned into its component FDFrs (SNCs).

A subnetwork connection is terminated at connection termination points (CTPs). A flow domain fragment is terminated at flow points (FPs). Because subnetworks may be partitioned, several subnetwork connection endpoints may coincide at a single CTP. Likewise, a single FP may represent the termination of several FDFrs from the same partitioned FDFr. The CTP (FP) represents the actual point of termination of both SNCs (FDFrs) and link connections [3].

The Ethernet services are modeled from the point of view of the subscriber's equipment referred to as the customer edge (CE) that is used to access the service at the UNI into the provider edge (PE), as depicted in Figure 15.5. Service attributes represent the definition of service level specification (SLS), which is also called a service-level agreement (SLA).

In the service view there are no assumptions about the details of the CEN, which may consist of a single switch or a combination of networks based on various technologies as depicted in Figure 15.6.

In the following sections, we will describe objects involved in the management of Carrier Ethernet services.

15.2.1 Management of EVC, OVC and ELMI and Discovery of Ethernet

Figure 15.7 shows the configuration management functions involved in Carrier Ethernet services. They are mainly discovery of equipment supporting Carrier Ethernet; configuration of physical and logical ports, EVC; configuration of user profiles and SOAM components; and configuration of ELMI components.

The Carrier Ethernet manager establishes an Ethernet virtual connection (EVC) representing the UNI-to-UNI service across an Ethernet flow domain or a collection of Ethernet flow domains, as shown in Figure 15.8. During EVC establishment, the

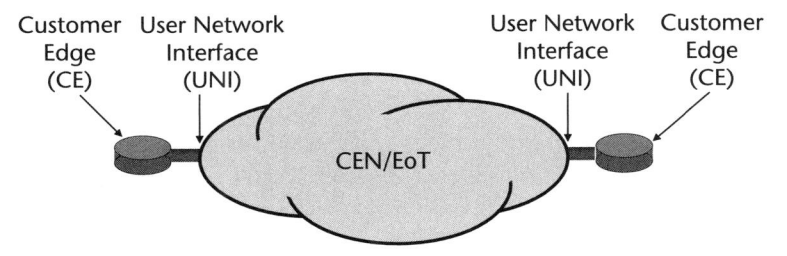

Figure 15.5 Ethernet services model [5].

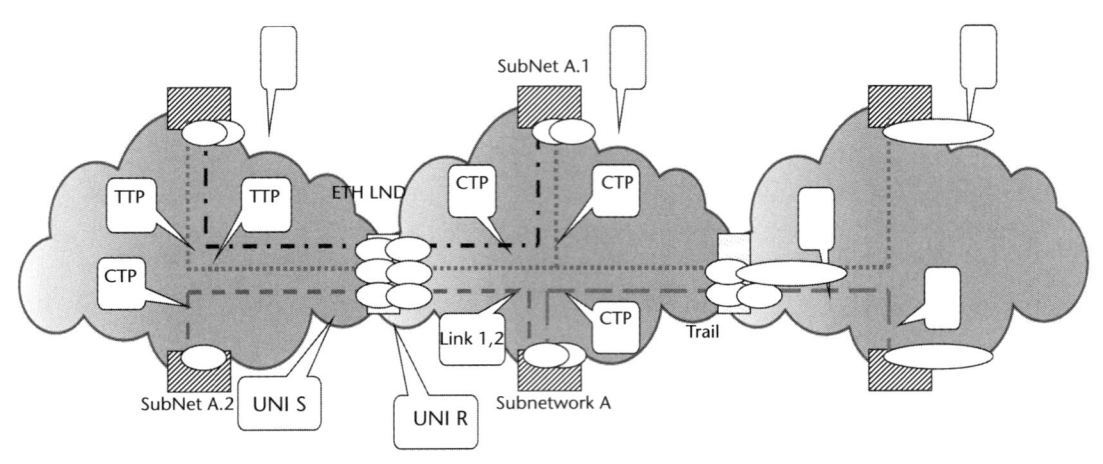

Figure 15.6 CEN consisting of multiple networks [3].

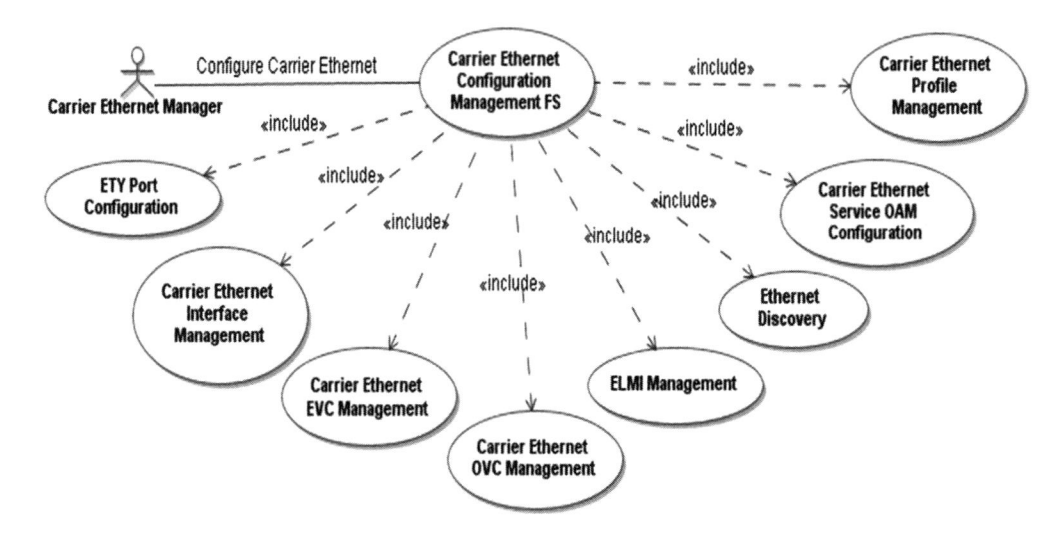

Figure 15.7 Carrier Ethernet configuration management functions [3].

managed system creates the EVC and associated flow points that are connected via the EVC.

The Carrier Ethernet manager establishes an Ethernet operator virtual connection (OVC) representing external interface-to-external interface (i.e., ENNI-to-ENNI or UNI-to-ENNI) connectivity across an Ethernet flow domain if the EVC crosses multiple operators (Figure 15.9). During OVC establishment, the managed system creates the OVC and associated OVC endpoints. An ETH OVC provides connectivity among the identified OVC endpoints.

Carrier Ethernet management of ELMI, as depicted in Figure 15.10, begins with requesting creation of ELMI profile on the managed system. Activation of ELMI, deactivation of ELMI, modification and querying of ELMI profile, and deletion of ELMI are part of the ELMI management.

15.2 Information Modeling of Carrier Ethernet Services for EMS and NMS

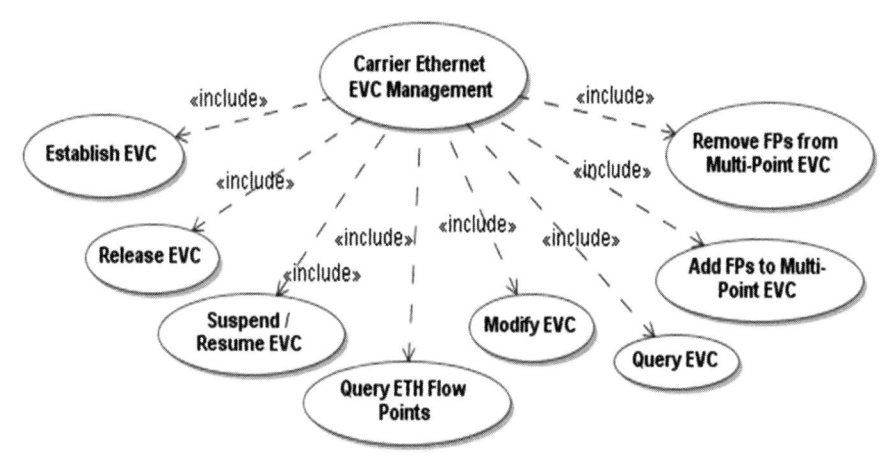

Figure 15.8 Carrier Ethernet EVC management [3].

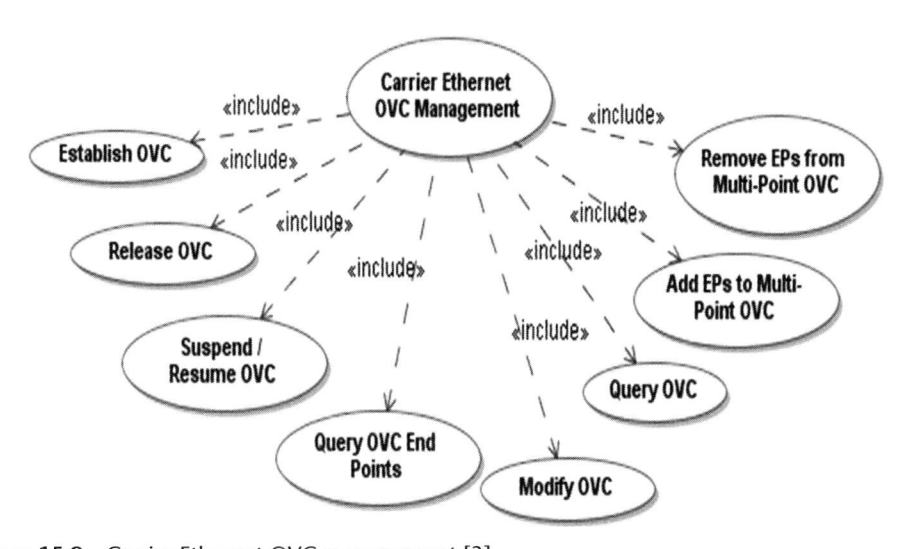

Figure 15.9 Carrier Ethernet OVC management [3].

For inventory, the management system needs to automatically discover physical ports, UNI and ENNI interfaces, connections, and flow domains (Figure 15.11).

15.2.2 Carrier Ethernet Service OAM Configuration

Carrier Ethernet SOAM configuration management involves in the management of maintenance entity (ME), maintenance domain (MD), maintenance entity group (MEG), MEG endpoint (MEP), and MEG intermediate point (MIP), as depicted in Figure 15.12.

MEG may belong to an EVC, OVC, UNI, ENNI, or LAG link service provider and operator. The configuration management deals with creation, deletion, modification, and querying of the components above. Furthermore, the configuration management deals with enabling and disabling of loopback, alarm indication signal (AIS), and connectivity verification.

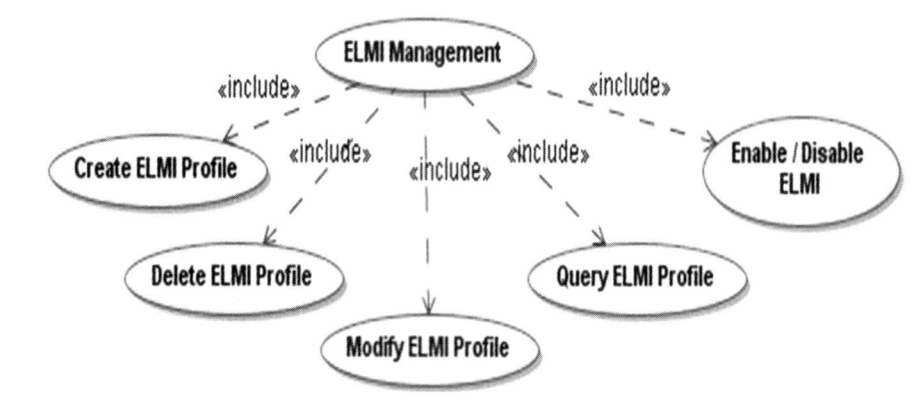

Figure 15.10 ELMI management [3].

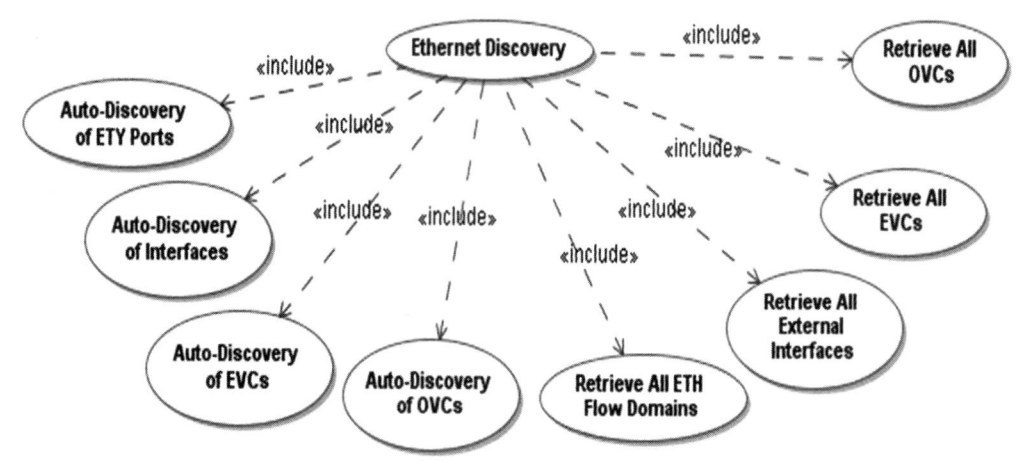

Figure 15.11 Ethernet discovery functions [3].

15.2.3 Carrier Ethernet Service Performance Management

Performance management of Carrier Ethernet services involves in periodic measurement of service performance parameters (i.e., delay, jitter, loss, and availability of connections), thresholds, and the transfer of collected performance data, as depicted in Figures 15.13 and 15.14.

In addition, measurements per physical and logical interface (i.e., frames received and transmitted) and associated thresholds are part of the performance management.

15.2.4 Carrier Ethernet Fault Management

Fault management of Carrier Ethernet services includes on-demand connectivity verification, port and connection-level loopbacks, link trace, port and connection-level event notifications, and testing, as depicted in Figures 15.15 and 15.16.

Severity of alarms (i.e., critical, major and minor) can be configured per physical port, logical port and connection.

15.2 Information Modeling of Carrier Ethernet Services for EMS and NMS

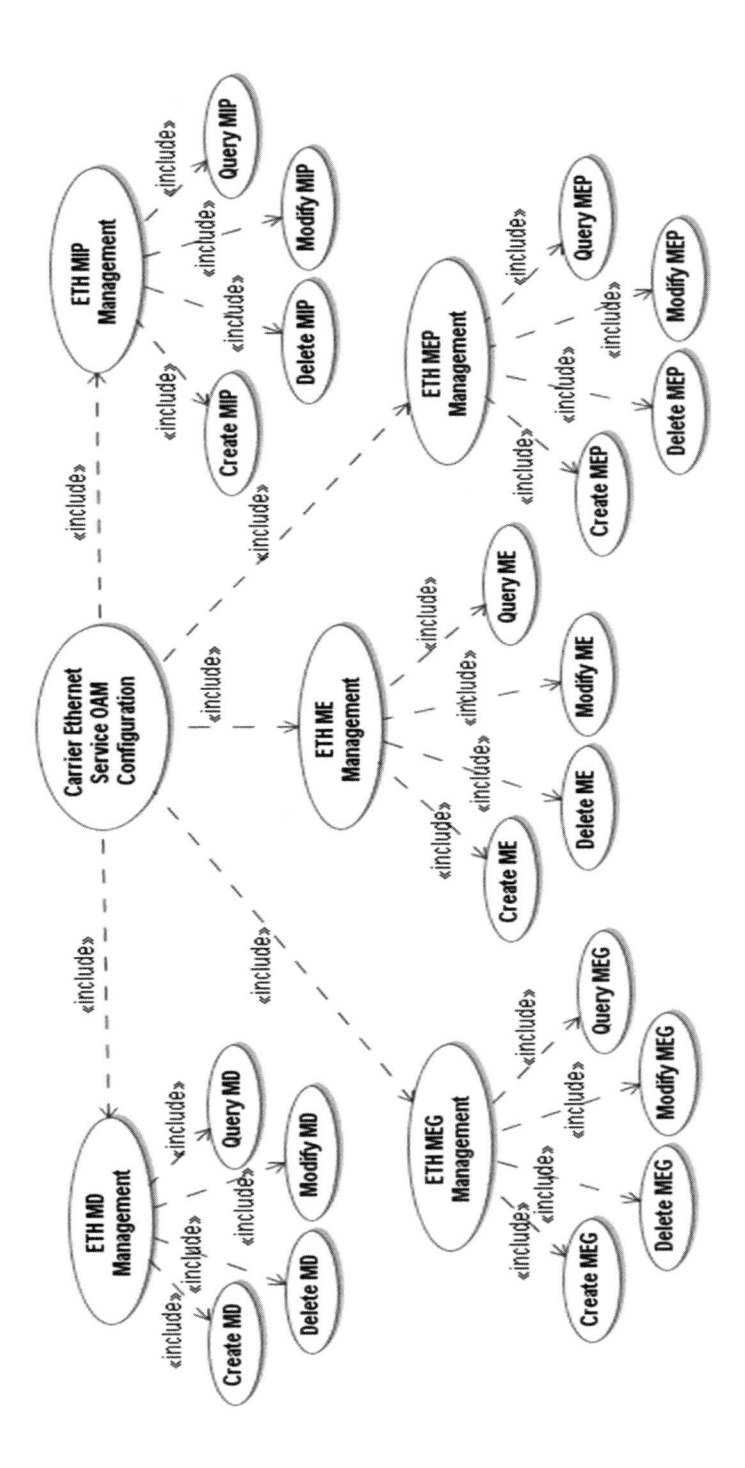

Figure 15.12 Carrier Ethernet service OAM configuration [3].

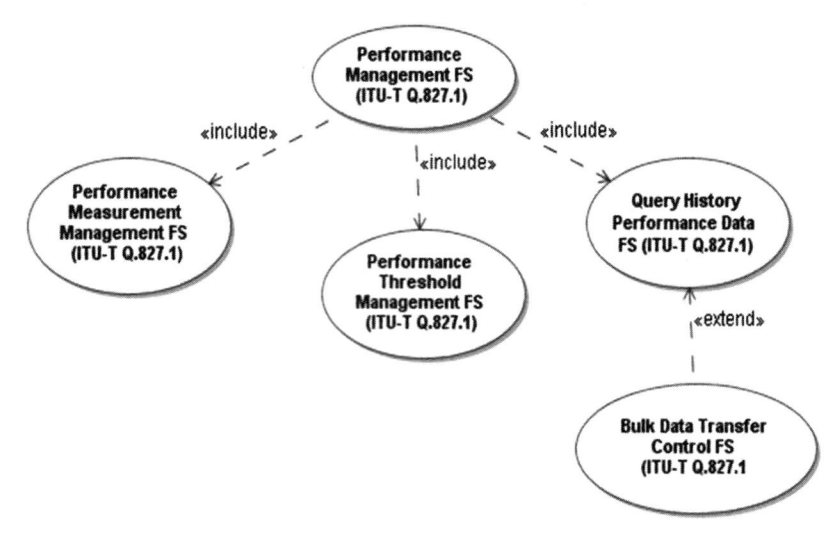

Figure 15.13 ITU-T Q.827.1 performance management functions [9].

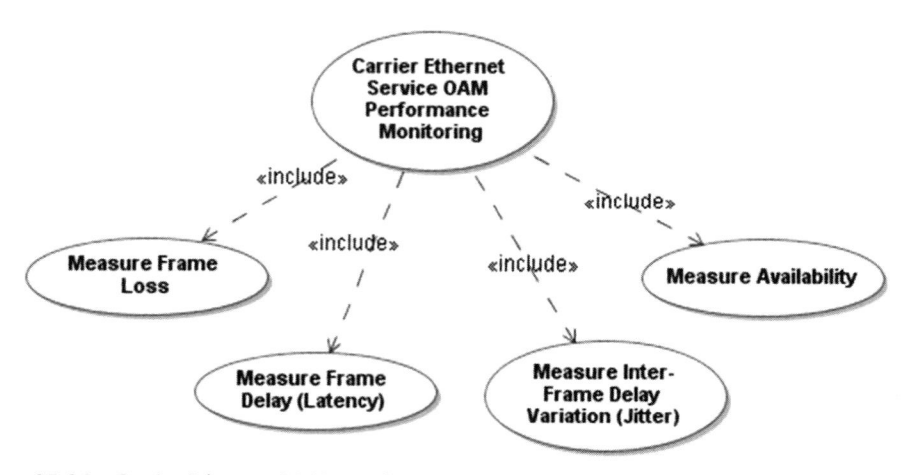

Figure 15.14 Carrier Ethernet SOAM performance monitoring.

15.2.5 Common Management Objects for Carrier Ethernet Services

Inheritance and containment diagrams for common management objects defined by ITU-T G.8010 [2] are given next. These objects are valid for the management of Carrier Ethernet services.

Objects such as network, eventLog, alarmRecord, and stateChange are common for all types of networks as depicted in Figure 15.17. Although objects such as log and alarmRecord are inherited from the top object, they are contained by the network object, as depicted in Figure 15.18.

15.2.6 Class Diagrams of ITU-T Q.840.1 Carrier Ethernet Management Entities

Figures 15.19, 15.20, and 15.21 are the inheritance diagram derived from ITU-T Q.840.1 [10] of the management entities providing the topology view, connectivity view, and reference data [3].

15.2 Information Modeling of Carrier Ethernet Services for EMS and NMS

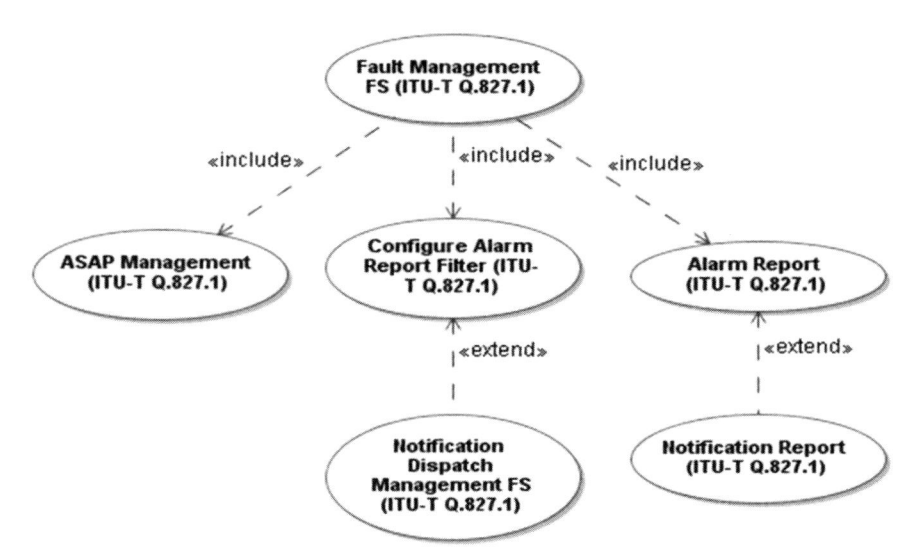

Figure 15.15 Fault management function set [3].

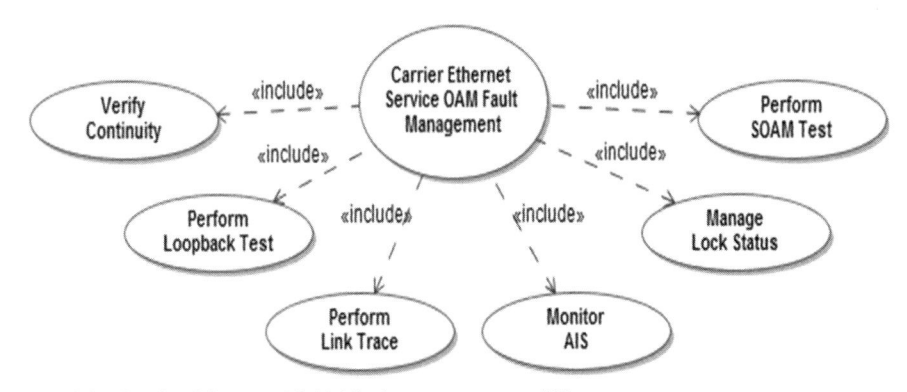

Figure 15.16 Carrier Ethernet SOAM fault management [3].

Flow domain, UNI, FPP, and associated attributes are listed in Figure 15.19, providing the topology information for the CEN.

FDFr_EVC, flow point, MAUTransport Port and associated attributes are listed in Figure 15.20, providing information for EVC.

Figures 15.22 and 15.23 are the relationship diagrams (including containment) derived from ITU-T Q.840.1 [10] of the network view, equipment view, and the service configuration view.

Figure 15.22 identifies relationships (i.e., supported by or contained) between network topology and connectivity objects, while Figure 15.23 defines relationships among Ethernet service configuration managed entities such as BandwidthProfile and cosBandwidthMapping.

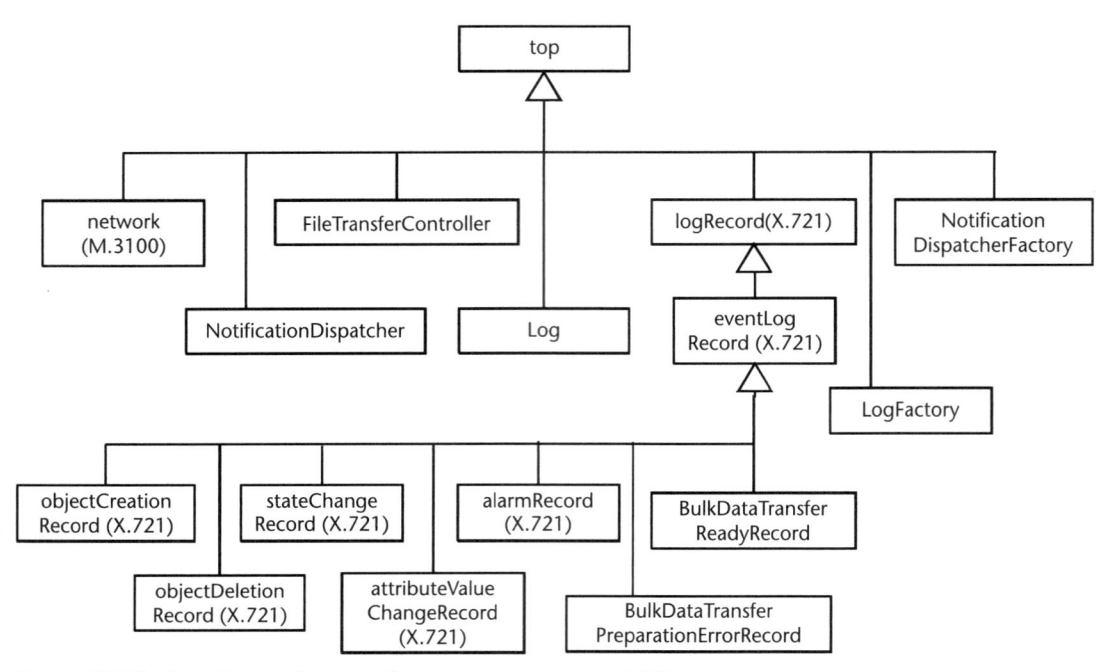

Figure 15.17 Inheritance diagram of common management [3].

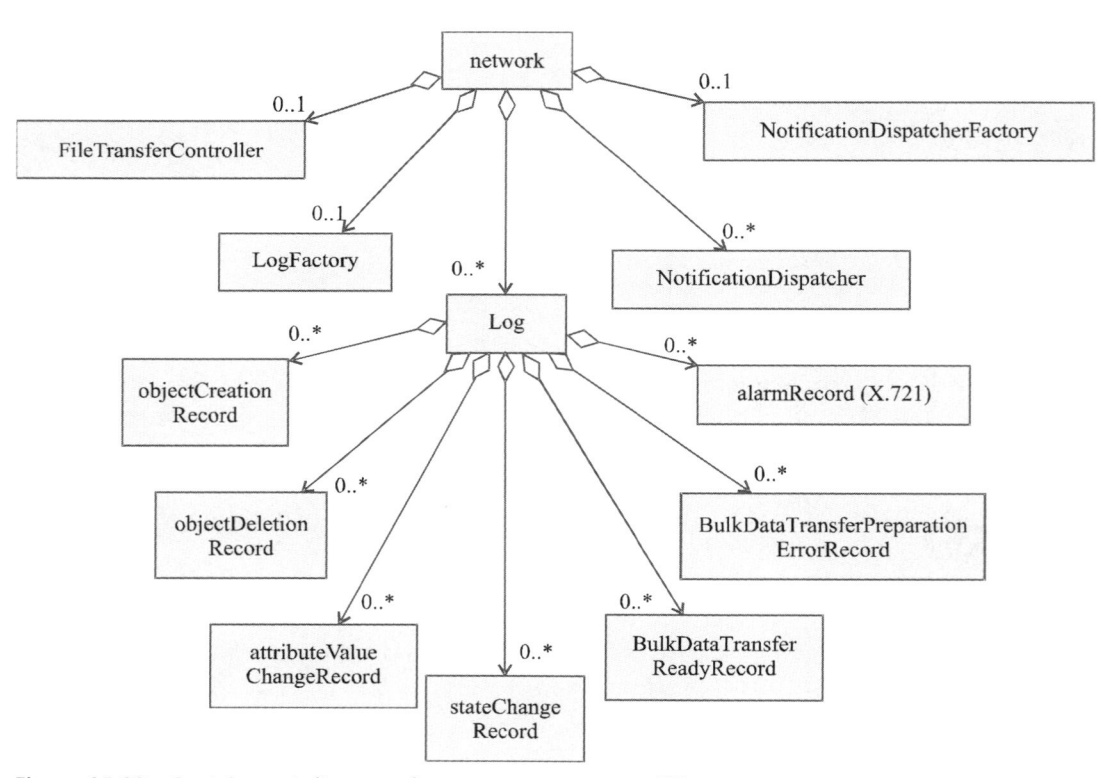

Figure 15.18 Containment diagram of common management [3].

15.2 Information Modeling of Carrier Ethernet Services for EMS and NMS

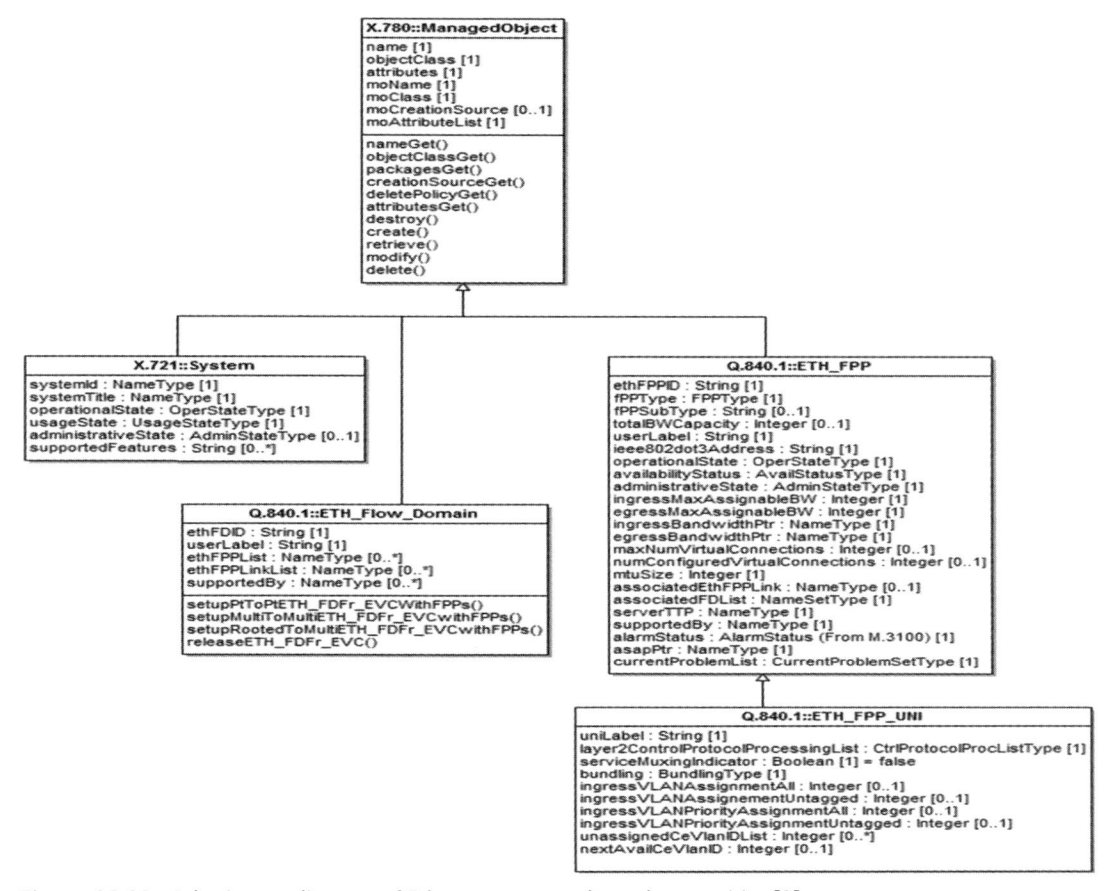

Figure 15.19 Inheritance diagram of Ethernet managed topology entities [3].

15.2.7 Carrier Ethernet Management Entities

Table 15.1 identifies the management entities necessary to manage Carrier Ethernet. The performance data sets for Carrier Ethernet are described in Table 15.2.

In Table 15.1, each management entity is described as being mandatory (denoted by M), optional (denoted by O), or conditional (denoted by C), with respect to applicability for Carrier Ethernet.

Applicable performance data sets for Carrier Ethernet services are described in Table 15.2 [9]. The performance data sets simply describe the category of the performance data set along with the individual counters associated with the set. For this logical model, the duration of the interval and amount of history to be stored are not specified; however, they can be specified within the management interface profiles and/or implementation agreements.

15.2.8 ENNI and Virtual UNI Related Objects

A containment tree for virtual UNI (VUNI), ENNI, FPP, FDFr_EVC, OVC, and OVC endpoint object classes and their associated attributes is depicted in Figure

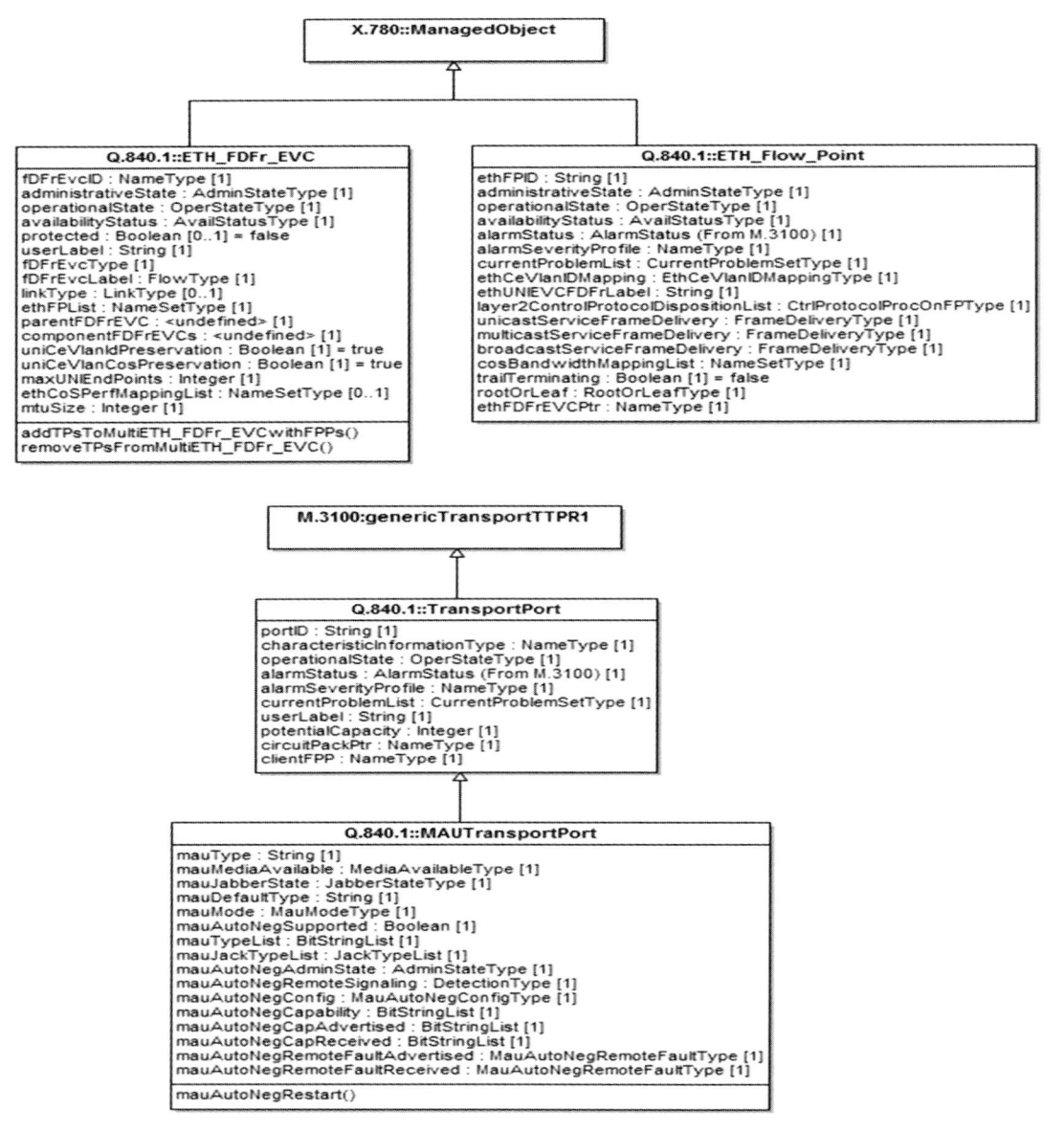

```
                    ┌─────────────────────────┐
                    │  X.780::ManagedObject   │
                    └─────────────────────────┘
                              △
                   ┌──────────┴──────────┐
┌──────────────────────────────────┐  ┌──────────────────────────────────────────────────┐
│      Q.840.1::ETH_FDFr_EVC       │  │           Q.840.1::ETH_Flow_Point                │
├──────────────────────────────────┤  ├──────────────────────────────────────────────────┤
│ fDFrEvcID : NameType [1]         │  │ ethFPID : String [1]                             │
│ administrativeState : AdminStateType [1] │ administrativeState : AdminStateType [1]    │
│ operationalState : OperStateType [1] │  operationalState : OperStateType [1]           │
│ availabilityStatus : AvailStatusType [1] │ availabilityStatus : AvailStatusType [1]    │
│ protected : Boolean [0..1] = false │ alarmStatus : AlarmStatus (From M.3100) [1]       │
│ userLabel : String [1]           │  │ alarmSeverityProfile : NameType [1]              │
│ fDFrEvcType [1]                  │  │ currentProblemList : CurrentProblemSetType [1]   │
│ fDFrEvcLabel : FlowType [1]      │  │ ethCeVlanIDMapping : EthCeVlanIDMappingType [1]  │
│ linkType : LinkType [0..1]       │  │ ethUNIEVCFDFrLabel : String [1]                  │
│ ethFPList : NameSetType [1]      │  │ layer2ControlProtocolDispositionList : CtrlProtocolProcOnFPType [1] │
│ parentFDFrEVC : <undefined> [1]  │  │ unicastServiceFrameDelivery : FrameDeliveryType [1] │
│ componentFDFrEVCs : <undefined> [1] │ multicastServiceFrameDelivery : FrameDeliveryType [1] │
│ uniCeVlanIdPreservation : Boolean [1] = true │ broadcastServiceFrameDelivery : FrameDeliveryType [1] │
│ uniCeVlanCosPreservation : Boolean [1] = true │ cosBandwidthMappingList : NameSetType [1] │
│ maxUNIEndPoints : Integer [1]    │  │ trailTerminating : Boolean [1] = false           │
│ ethCoSPerfMappingList : NameSetType [0..1] │ rootOrLeaf : RootOrLeafType [1]            │
│ mtuSize : Integer [1]            │  │ ethFDFrEVCPtr : NameType [1]                     │
├──────────────────────────────────┤  └──────────────────────────────────────────────────┘
│ addTPsToMultiETH_FDFr_EVCwithFPPs() │
│ removeTPsFromMultiETH_FDFr_EVC()    │
└──────────────────────────────────┘

              ┌──────────────────────────────┐
              │  M.3100:genericTransportTTPR1│
              └──────────────────────────────┘
                            △
              ┌──────────────────────────────┐
              │   Q.840.1::TransportPort     │
              ├──────────────────────────────┤
              │ portID : String [1]          │
              │ characteristicInformationType : NameType [1] │
              │ operationalState : OperStateType [1] │
              │ alarmStatus : AlarmStatus (From M.3100) [1] │
              │ alarmSeverityProfile : NameType [1] │
              │ currentProblemList : CurrentProblemSetType [1] │
              │ userLabel : String [1]       │
              │ potentialCapacity : Integer [1] │
              │ circuitPackPtr : NameType [1]│
              │ clientFPP : NameType [1]     │
              └──────────────────────────────┘
                            △
              ┌──────────────────────────────┐
              │  Q.840.1::MAUTransportPort   │
              ├──────────────────────────────┤
              │ mauType : String [1]         │
              │ mauMediaAvailable : MediaAvailableType [1] │
              │ mauJabberState : JabberStateType [1] │
              │ mauDefaultType : String [1]  │
              │ mauMode : MauModeType [1]    │
              │ mauAutoNegSupported : Boolean [1] │
              │ mauTypeList : BitStringList [1] │
              │ mauJackTypeList : JackTypeList [1] │
              │ mauAutoNegAdminState : AdminStateType [1] │
              │ mauAutoNegRemoteSignaling : DetectionType [1] │
              │ mauAutoNegConfig : MauAutoNegConfigType [1] │
              │ mauAutoNegCapability : BitStringList [1] │
              │ mauAutoNegCapAdvertised : BitStringList [1] │
              │ mauAutoNegCapReceived : BitStringList [1] │
              │ mauAutoNegRemoteFaultAdvertised : MauAutoNegRemoteFaultType [1] │
              │ mauAutoNegRemoteFaultReceived : MauAutoNegRemoteFaultType [1] │
              ├──────────────────────────────┤
              │ mauAutoNegRestart()          │
              └──────────────────────────────┘
```

Figure 15.20 Inheritance diagram of Ethernet managed connectivity entities [3].

15.24 while CoS profile and CoS bandwidth mapping and their associated attributes are depicted in Figure 15.25.

15.2.9 Fault Management Objects

Figure 15.26 illustrates the fault management objects (UML classes) and their relationships.

The fault management objects cover loopback, connectivity check, alarm indication signal (AIS), MEP, lock condition, link trace, and test scheduling.

15.2 Information Modeling of Carrier Ethernet Services for EMS and NMS

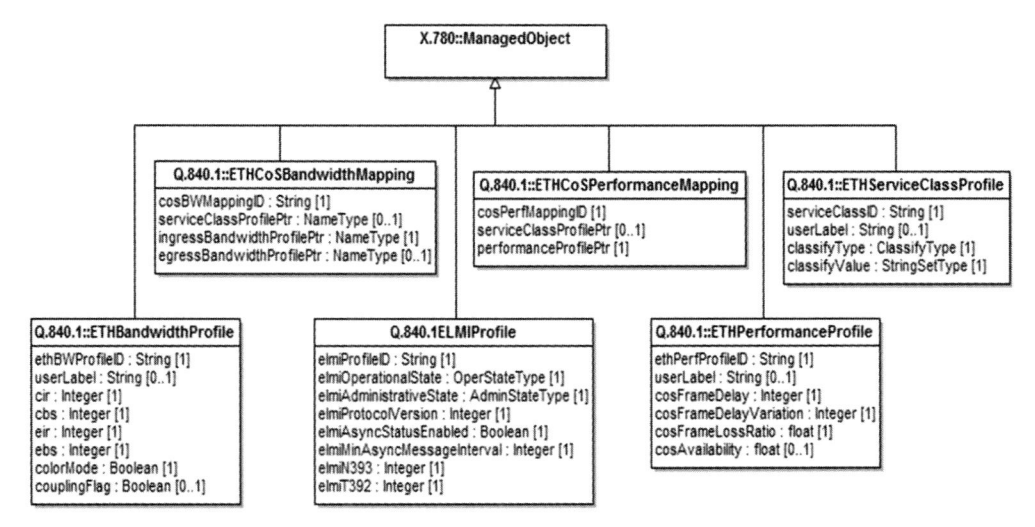

Figure 15.21 Inheritance diagram of Ethernet managed reference data entities [3].

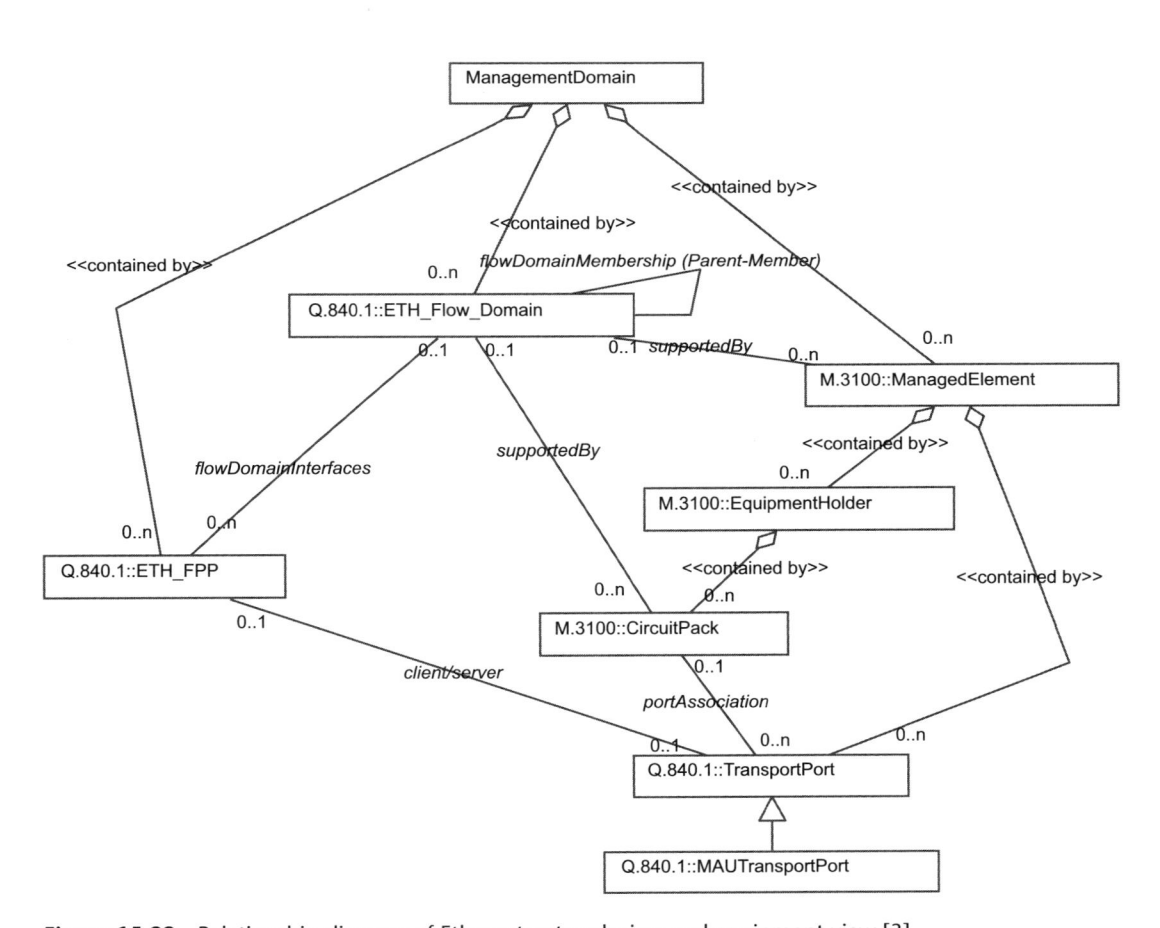

Figure 15.22 Relationship diagram of Ethernet network view and equipment view [3].

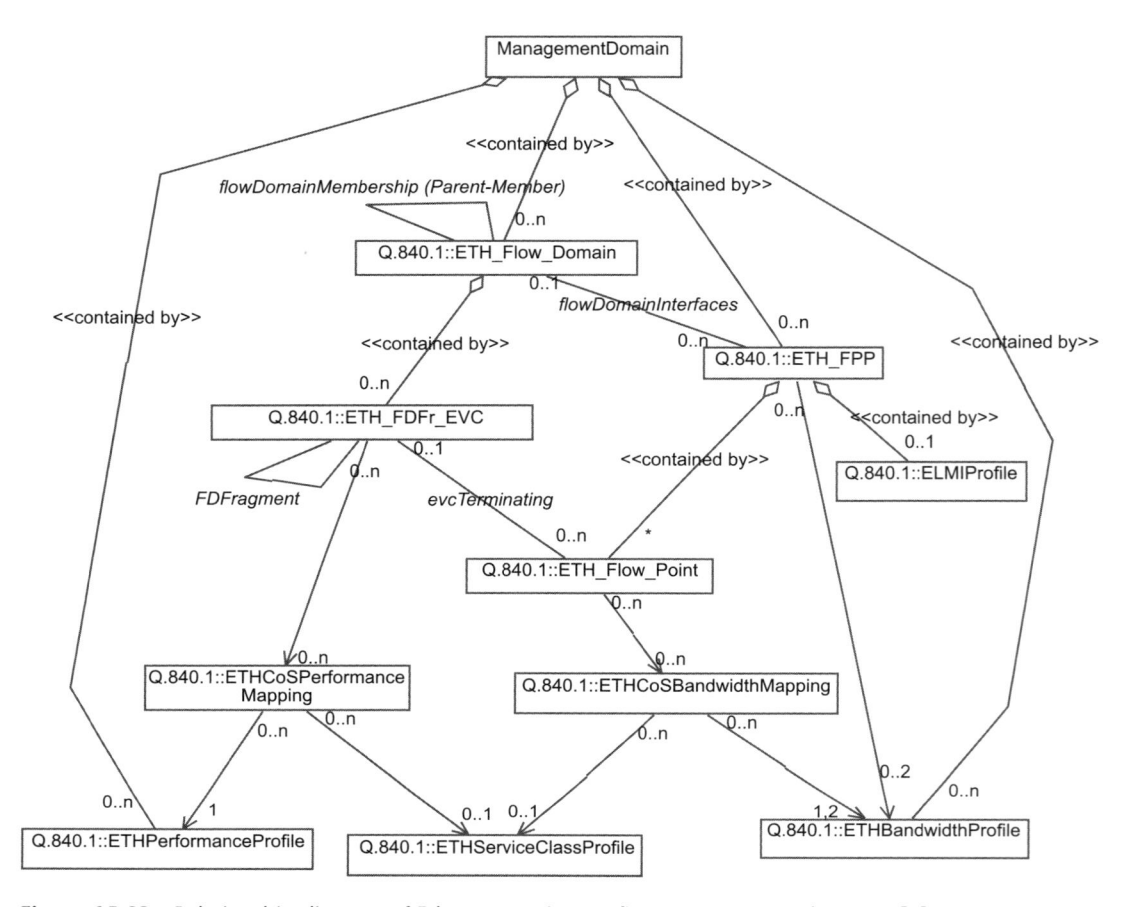

Figure 15.23 Relationship diagram of Ethernet service configuration managed entities [3].

15.2.10 Performance Monitoring Objects

Figure 15.27 illustrates the performance monitoring objects (UML classes) and their associations. The performance monitoring objects cover performance monitoring scheduling, delay, jitter, and loss.

15.2.11 ENNI and OVC MIBs

In addition to objects described in the previous sections for the management of Carrier Ethernet services, MEF defined MIB modules for ENNI, OVC, and SOAM PM and FM [14–17].

A MIB is a collection of managed objects that can be used to provision an entity, query an entity for status information, or define notifications that are sent to an NMS or an EMS. Collections of related objects are defined in MIB modules that are written using an adapted subset of Abstract Syntax One (ASN.1).

Figure 15.28 illustrates the relationship between the operational support system (OSS)/business support system (BSS), NMS, EMS, and NEs. The MIB modules are used in the interaction between the EMS (SNMP Manager) and the NE (SNMP Agent) via SNMP.

15.2 Information Modeling of Carrier Ethernet Services for EMS and NMS

Table 15.1 Carrier Ethernet Management Entities [3]

Management Entity	Reference (Where Defined)	Applicability to Carrier Ethernet
ManagedObject (superclass)	X.780	M (abstract superclass)
ManagementDomain	MTOSI Rel. 2.1, X.749	O
Network	M.3100 [1]	O
ManagedElement	M.3100	O
Equipment	M.3100	O
EquipmentHolder	M.3100	O
CircuitPack	M.3100	O
Alarm severity assignment profile	M.3100	O
GenericTransportTTP	M.3100	O
Log	X.721 [11]	O
LogRecord (superclass)	X.721	O
EventLogRecord (superclass)	X.721	O
AlarmRecord	X.721	O
System	X.721	O
CurrentData	Q.822 [12]	O
HistoryData	Q.822	O
ThresholdData	Q.822	O
FileTransferController	Q.827.1 [13]	O
LogFactory	Q.827.1	O
Log	Q.827.1	O
NotificationDispatcher	Q.827.1	O
NotificationDispatcherFactory	Q.827.1	O
ObjectCreationRecord	Q.827.1	O
ObjectDeletionRecord	Q.827.1	O
AttributeValueChangeRecord	Q.827.1	O
StateChangeRecord	Q.827.1	O
BulkDataTransferReadyRecord	Q.827.1	O
BulkDataTransferPreparationRecord	Q.827.1	O
EMS	Q.827.1 Amd. 1	C
ETH_Flow_Domain	Q.840.1	M
ETH_FPP (superclass)	Q.840.1	M (abstract superclass)
ETH_FPP_UNI	Q.840.1	M
ETH_FPP_Link	Q.840.1	O
ETH_FDFr_EVC	Q.840.1	M
ETH_Flow_Point	Q.840.1	M
ETHBandwidthProfile	Q.840.1	M
ETHServiceClassProfile	Q.840.1	M
ETHCoSBandwidthMapping	Q.840.1	M
ETHPerformanceProfile	Q.840.1	M
ETHCoSPerformanceMapping	Q.840.1	M
ELMIProfile	Q.840.1	C
TransportPort	Q.840.1	C
MAUTransportPort	Q.840.1	C

Table 15.1 (continued)

Management Entity	Reference (Where Defined)	Applicability to Carrier Ethernet
ETH_FPP_ENNI	MEF 7.2	M
ETH_FPP_VUNI	MEF 7.2	M
ETH_OVC	MEF 7.2	M
ETH_OVC_End_Point	MEF 7.2	M
EthMe	MEF 7.2	C
EthMeg	MEF 7.2	C
EthMp (superclass)	MEF 7.2	C
EthMep	MEF 7.2	C
EthMip	MEF 7.2	C
EthMd	MEF 7.2	C
EthMepPeerInfo	MEF 7.2	C
EthOamDmCfg	MEF 7.2	C
EthOamDmProactiveOneWayThreshold	MEF 7.2	C
EthOamDmProactiveTwoWayThreshold	MEF 7.2	C
EthOamLmCfg	MEF 7.2	C
EthOamLbCfg	MEF 7.2	C
EthOamLbStats	MEF 7.2	C
EthOamCcCfg	MEF 7.2	C
EthOamCcStats	MEF 7.2	C
EthOamAisCfg	MEF 7.2	C
EthOamAisStats	MEF 7.2	C
EthOamLtCfg	MEF 7.2	C
EthOamLtrStats	MEF 7.2	C
EthOamLckCfg	MEF 7.2	C
EthOamTestCfg	MEF 7.2	C
EthOamTestStats	MEF 7.2	C
EthVnidUniCfg	MEF 7.2	C
EthVnidOvcEndPointCfg	MEF 7.2	C
EthVnidRmi	MEF 7.2	C
EthVnidRpe	MEF 7.2	C
EthDiscardedDroppedFramesThreshold	MEF 7.2	C

ENNI-OVC MIB, UNI-EVC MIB, IF-MIB, and medium attachment unit (MAU)-MIB [14] are given in Tables 15.3 through 15.8.

15.2.12 SOAM FM MIB [16]

The SOAM Textual Convention (TC) MIB defines the textual conventions that are to be used with other MEF SOAM MIB modules. The SOAM TC MIB defines textual conventions for the following:

- MefSoamTcConnectivityStatusType: the connectivity status type of a MEG or MEP;

15.2 Information Modeling of Carrier Ethernet Services for EMS and NMS

Table 15.2 Performance Data Sets [3]

Performance Data Set	Reference	Applicability to Carrier Ethernet
ETH UNI anomalies performance data set	Q.840.1	M
ETH UNI traffic performance data set	Q.840.1	M
ETH ingress traffic management performance data set	Q.840.1	M
ETH egress traffic management performance data set	Q.840.1	M
ETH congestion discards performance data set	Q.840.1	M
ETH ELMI performance data set	Q.840.1	O
MAU termination performance data set	Q.840.1	C
ETH point-to-point EVC MEG performance data set (EthMegPerfDataSet)	MEF 7.2	C
ETH maintenance point performance data set (EthMpPerfDataSet)	MEF 7.2	C
ETH MEG endpoint loss measurement on-demand single-ended data set (EthOamLmOnDemandSingleEndedStats)	MEF 7.2	C
ETH MEG endpoint delay measurement on-demand two-way data set (EthOamDmOnDemandTwoWayStats)	MEF 7.2	C
ETH MEG endpoint delay measurement on-demand one-way data set (EthOamDmOnDemandOneWayStats)	MEF 7.2	C
ETH MEG endpoint delay measurement proactive one-way current data set (EthOamDmProactiveOneWayCurrentStats)	MEF 7.2	C
ETH MEG endpoint delay measurement proactive two-way current data set (EthOamDmProactiveTwoWayCurrentStats)	MEF 7.2	C
ETH MEG endpoint delay measurement proactive one-way history data set (EthOamDmProactiveOneWayHistoryStats)	MEF 7.2	C
ETH MEG endpoint delay measurement proactive two-way history data set (EthOamDmProactiveTwoWayHistoryStats)	MEF 7.2	C

M: mandatory; O: optional; C: conditional.

- MefSoamTcDataPatternType: defines the data pattern type used in data TLVs;
- MefSoamTcIntervalTypeAisLck: defines the interval for sending AIS and LCK PDUs;
- MefSoamTcMegIdType: defines the MEG ID type;
- MefSoamTcOperationTimeType: defines when an operation is initiated or stopped;
- MefSoamTcTestPatternType: defines the test pattern used in test TLVs.

The SOAM FM MIB is an extension to the connectivity fault management (CFM) MIBs as developed in IEEE 802.1ag [22] and IEEE 802.1ap [23], to support functionality defined by ITU-T Y.1731 [6] and by the SOAM-FM [24] specification.

The SOAM FM MIB is divided into the following groups:

- mefSoamNet that defines the objects necessary to support MEG unique functionality. This group augments the standard ieee8021CfmMaNetEntry row entry as found in 802.1ag [25].

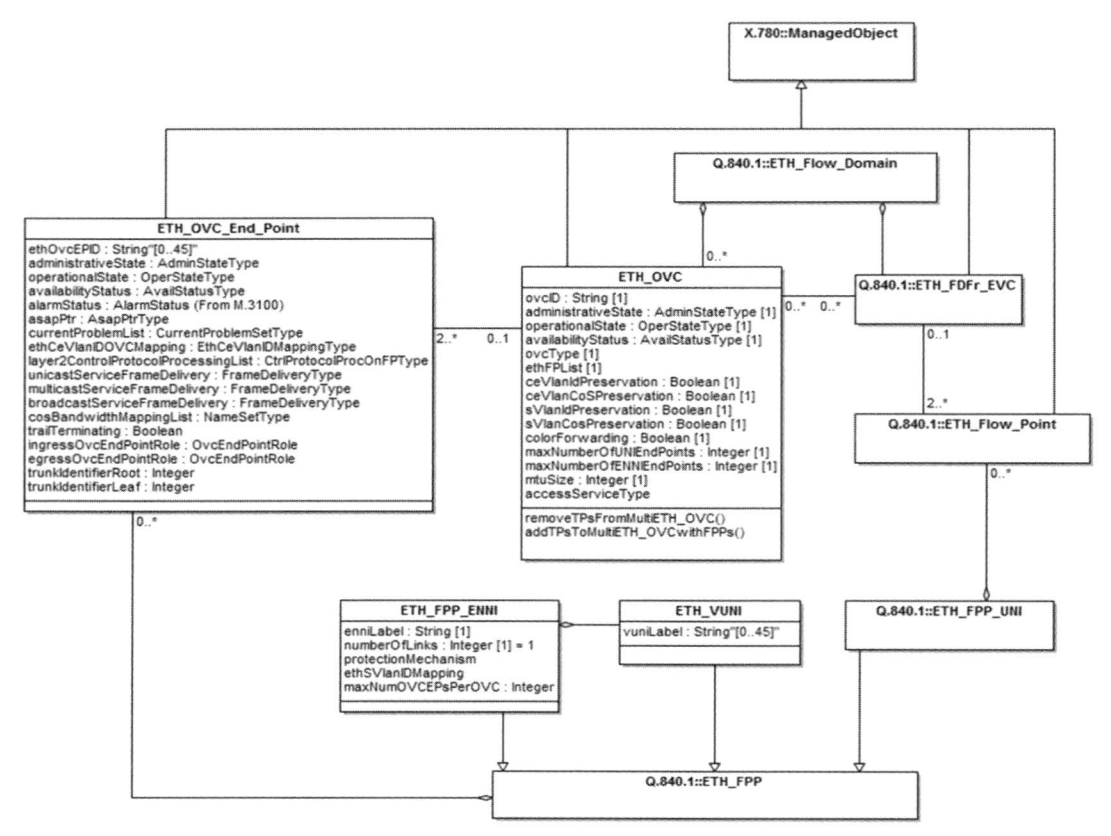

Figure 15.24 ENNI and VUNI related object classes [3].

- mefSoamMeg that defines the objects necessary to support the enhanced MEG/MA functionality. This group augments the standard ieee8021Cfm-MaCompEntry row entry as found in IEEE 802.1ap [23].
- mefSoamMep that defines the objects necessary to support the enhanced MEP functionality. This group augments the dot1agCfmMepEntry row entry as found in 802.1ag [22].
- mefSoamCc that defines the objects necessary to support the enhanced CCM functionality. This group augments the dot1agCfmMepEntry row entry as found in 802.1ag [22].
- mefSoamAis that defines the objects necessary to implement the ETH-AIS functionality. This group augments the dot1agCfmMepEntry row entry as found in 802.1ag [22].
- mefSoamLb that defines the objects necessary to support the enhanced CFM loopback functionality. This group augments the dot1agCfmMepEntry row entry as found in 802.1ag [22].
- mefSoamLt that defines the objects necessary to support the enhanced CFM Linktrace functionality. This group augments the dot1agCfmMepEntry row entry as found in 802.1ag [22].

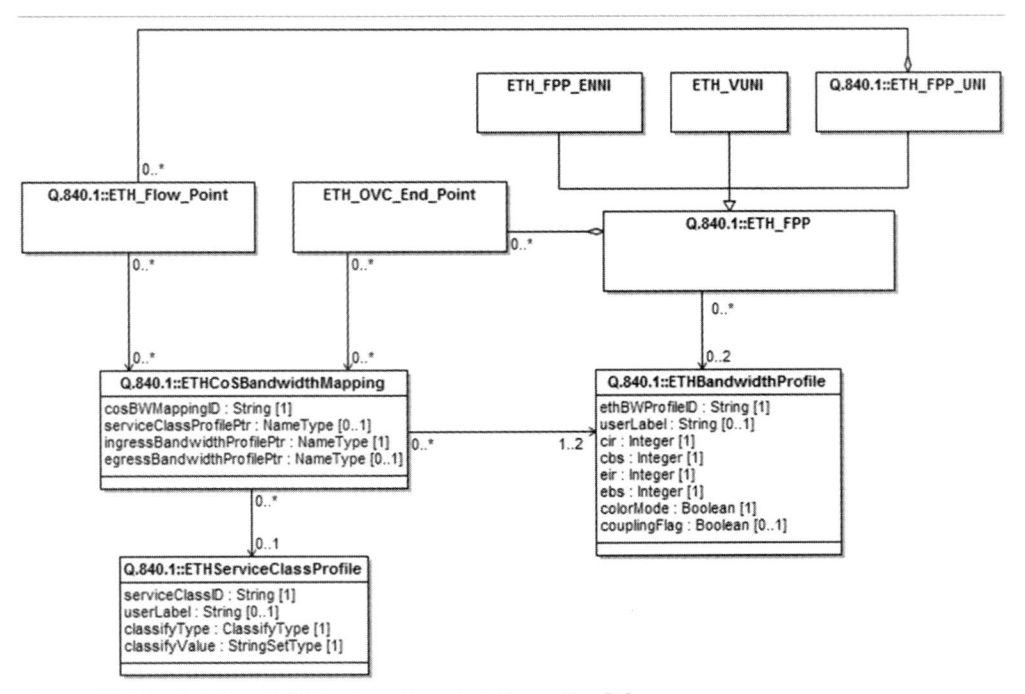

15.2 Information Modeling of Carrier Ethernet Services for EMS and NMS

Figure 15.25 ENNI and VUNI related bandwidth profiles [3].

- mefSoamLck that defines the objects necessary to implement the ETH-LCK functionality. This group augments the dot1agCfmMepEntry row entry as found in 802.1ag [22].
- mefSoamTest that defines the objects necessary to implement the ETH-Test functionality. This group augments the dot1agCfmMepEntry row entry as found in 802.1ag [22].
- mefSoamFmNotificationCfg that defines the objects necessary to configure the mefSoamFmNotifications.
- mefSoamFmNotifications that defines the notifications necessary to implement service OAM FM functionality.

15.2.13 SOAM PM MIB [24]

The SOAM PM objects in the MEF 36 MIB [26] extend the IEEE CFM MIB objects as well as providing new objects from ITU-T Y.1731 [6] and the SOAM PM IA.

The performance monitoring process is made up of PM sessions. A PM session can be initiated between two MEPs in a MEG and be defined as either a loss measurement (LM) PM session or delay measurement (DM) PM session.

The LM session can be used to determine the performance metrics FLR, availability, and resiliency. The DM session can be used to determine the performance metrics FD, IFDV, FDR, and MFD.

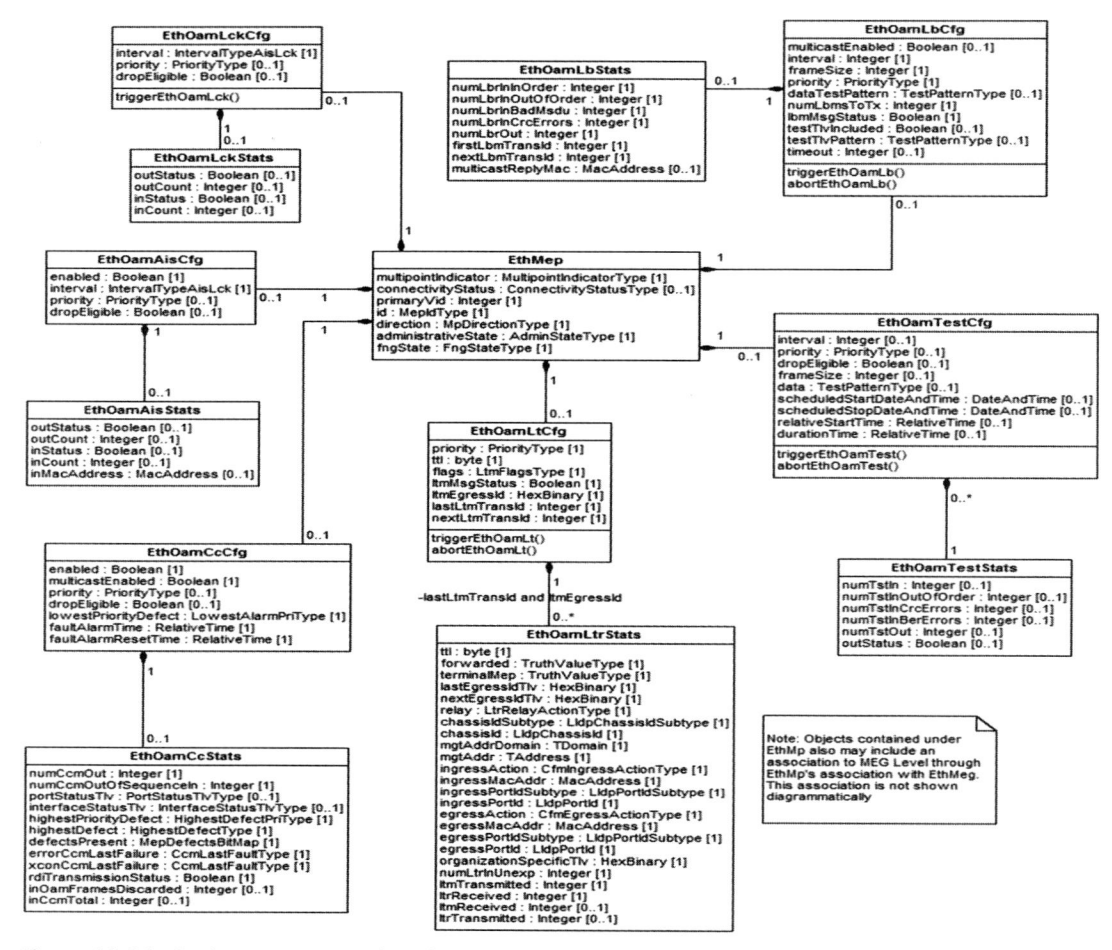

Figure 15.26 Fault management class diagram [3].

The PM session is defined by the specific PM function being run, start time, stop time, message period, measurement interval, and repetition time. The relationships of these different items are depicted in Figures 15.29 and 15.30.

- The start time is the time that the PM session begins and is applicable to on-demand PM sessions. For proactive PM sessions, the start time is not applicable as the PM session begins as soon as the PM session is configured and enabled.
- The stop time is the time that the PM session ends and is applicable to on-demand PM sessions. For proactive PM sessions, the stop time is not applicable as the PM session stops only when the PM session is disabled or deleted.
- The message period is the SOAM PM frame transmission frequency (the time between SOAM PM frame transmissions).
- The measurement intervals are discrete, nonoverlapping periods of time during which the PM session measurements are performed and results are gathered. The measurement interval can align with the PM session duration, but

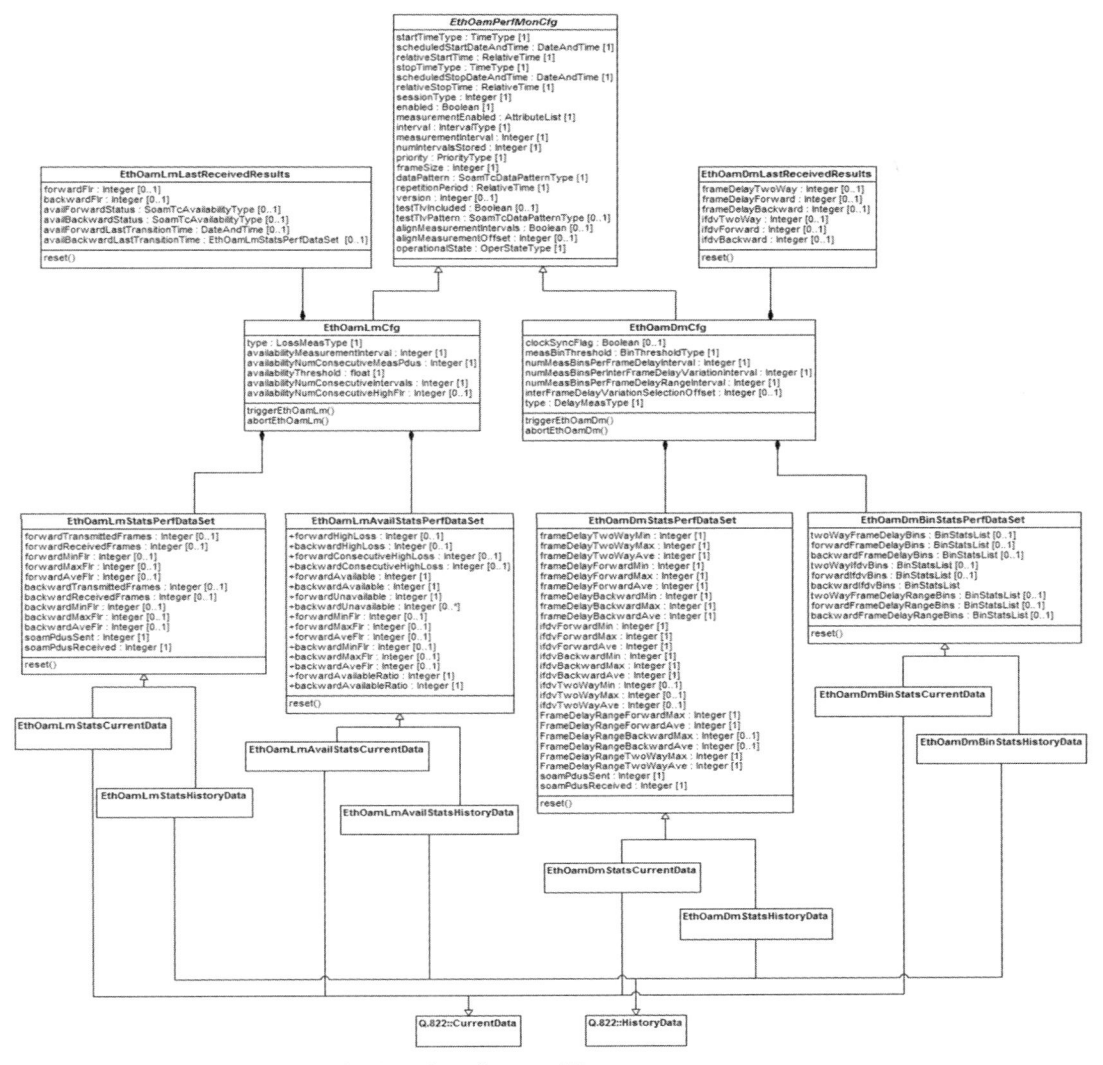

15.2 Information Modeling of Carrier Ethernet Services for EMS and NMS 415

Figure 15.27 Performance monitoring class diagram [3].

it does not need to. SOAM PM PDUs during a PM session are only transmitted during a measurement interval.

- The repetition time is the time between the start times of the measurement intervals.

The PM session can be configured to run forever or for a period of time. On-demand types have a definitive start and stop time that can be relative or absolute or can have a stop time of forever. Proactive types begin immediately when a PM session is configured and enabled and end when the PM session is deleted or disabled.

When a PM session is completed, either through the PM session being disabled or the stop time being reached, the current measurement interval is stopped, if it is in process at the time, and all the in process calculations are finalized.

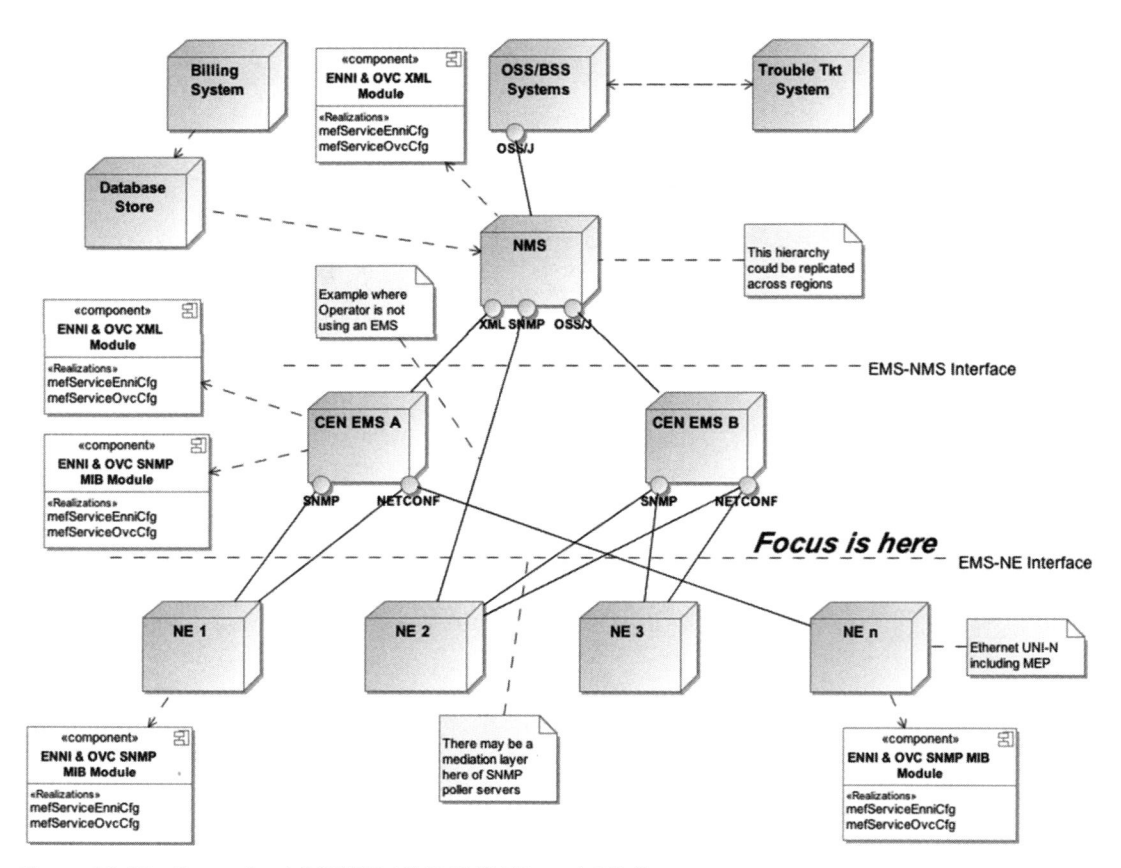

Figure 15.28 Generalized OSS/BSS-NMS-EMS-NE model [14].

Table 15.3 ENNI Service Attribute Alignment

MEF 26.1 [18] Attribute Name	MEF-UNI-EVC-MIB, MEF-ENNI-OVC-MIB, IF-MIB, MAU-MIB Objects
Operator ENNI identifier	mefServiceInterfaceCfgIdentifier [19], mefServiceEnniCfgIdentifier
Physical layer	mefServiceInterfaceCfgType [19], ifMauType (MAU-MIB)
Frame format	mefServiceInterfaceCfgFrameFormat [19]
Number of links	mefServiceEnniCfgNumberLinks
Protection mechanism	mefServiceEnniCfgProtection
ENNI maximum transmission unit size	ifMtu (IF-MIB)
Endpoint map	mefServiceOvcEndPtPerEnniCfgTable, mefServiceOvcEndPtPerVuniCfgTable
Maximum number of OVCs	mefServiceInterfaceStatusMaxVc [19]
Maximum number of OVC endpoints per OVC	mefServiceEnniCfgMaxNumberOvcEndPts

15.2 Information Modeling of Carrier Ethernet Services for EMS and NMS

Table 15.4 VUNI Service Attribute Alignment

MEF 28 [20] Attribute Name	MEF-ENNI-OVC-MIB Objects
VUNI identifier	mefServiceVuniCfgIdentifier
ENNI CE-VLAN ID value for ENNI frames with no C-tag or a C-tag whose VLAN ID value is 0	mefServiceVuniCfgCeVidUntagged
NA	mefServiceVuniCfgCePriorityUntagged
Maximum number of related OVC endpoints in the VUNI provider MEN	mefServiceVuniCfgMaxNumberOvcEndPoints
Ingress bandwidth profile per VUNI	mefServiceVuniCfgIngressBwpGrpIndex
Egress bandwidth profile per VUNI	mefServiceVuniCfgEgressBwpGrpIndex
NA	mefServiceVuniCfgL2cpGrpIndex

Table 15.5 OVC Service Attribute Alignment

MEF 26.1 [37] Attribute Name	MEF-ENNI-OVC-MIB Objects
OVC identifier	mefServiceOvcCfgIdentifier
OVC type	mefServiceOvcCfgServiceType
OVC endpoint list	mefServiceOvcEndPtPerEnniCfgTable, mefServiceOvcEndPtPerUniCfgTable, mefServiceOvcEndPtPerVuniCfgTable
Maximum number of UNI OVC endpoints	NA
Maximum number of ENNI OVC endpoints	mefServiceOvcStatusMaxNumEnniOvcEndPt
NA	mefServiceOvcStatusMaxNumVuniOvcEndPt
OVC maximum transmission unit size	mefServiceOvcStatusMaxMtuSize, mefServiceOvcCfgMtuSize
CE-VLAN ID preservation	mefServiceOvcCfgCevlanIdPreservation
CE-VLAN CoS preservation	mefServiceOvcCfgCevlanCosPreservation
S-VLAN ID preservation	mefServiceOvcCfgSvlanIdPreservation
S-VLAN CoS preservation	mefServiceOvcCfgSvlanCosPreservation
Color forwarding	mefServiceOvcCfgColorForwarding
	mefServiceOvcCfgColorIndicator
Service level specification	NA
Unicast service frame delivery	mefServiceOvcCfgUnicastDelivery
Multicast service frame delivery	mefServiceOvcCfgMulticastDelivery
Broadcast service frame delivery	mefServiceOvcCfgBroadcastDelivery
Layer 2 Control Protocol Tunneling	mefServiceOvcCfgL2cpGrpIndex
NA	mefServiceOvcCfgAdminState
NA	mefServiceOvcStatusOperationalState

A PM session can be dual-ended or single-ended. In a single-ended session, a controller MEP sends SOAM PM PDUs towards a responder MEP. The responder MEP sends SOAM PM PDUs towards the controller MEP in response to receiving SOAM PM PDUs from the controller MEP. All performance calculations are performed by the controller MEP, and results are only available on the controller MEP.

In a dual-ended session, a controller MEP sends SOAM PM PDUs towards a sink MEP. There are no responses sent towards the controller MEP. All performance calculations are performed by the sink MEP, and results are only available

Table 15.6 OVC per ENNI End Point Service Attribute Alignment

MEF 26.1 [37] Attribute Name	MEF-ENNI-OVC-MIB, MEF-UNI-EVC-MIB Objects
OVC endpoint identifier	mefServiceOvcEndPtPerEnniCfgIdentifier
NA	mefServiceOvcEndPtPerEnniCfgRole
Trunk identifiers	mefServiceOvcEndPtPerEnniCfgRootSvlanMap, mefServiceOvcEndPtPerEnniCfgLeafSvlanMap
Class of service identifiers	mefServiceCosCfgTable [36]
Ingress bandwidth profile per OVC endpoint	mefServiceOvcEndPtPerEnniCfgIngressBwpGrpIndex
Ingress bandwidth profile per ENNI class of service identifier	mefServiceOvcEndPtPerEnniCfgIngressBwpGrpIndex
Egress bandwidth profile per end point	mefServiceOvcEndPtPerEnniCfgEgressBwpGrpIndex
Egress bandwidth profile per ENNI class of service identifier	mefServiceOvcEndPtPerEnniCfgEgressBwpGrpIndex

Table 15.7 OVC per UNI End Point Service Attribute Alignment

MEF 26.1 [18] Attribute Name	MEF-ENNI-OVC-MIB, MEF-UNI-EVC-MIB Objects
UNI OVC identifier	mefServiceOvcEndPtPerUniCfgIdentifier (*)
NA	mefServiceOvcEndPtPerUniCfgRole (*)
OVC endpoint map	mefServiceOvcEndPtPerUniCfgCeVlanMap (*)
Class of service identifiers	mefServiceCosCfgTable [36]
Ingress bandwidth profile per OVC endpoint at a UNI	mefServiceOvcEndPtPerUniCfgIngressBwpGrpIndex (*)
Ingress bandwidth profile per class of service identifier at a UNI	mefServiceOvcEndPtPerUniCfgIngressBwpGrpIndex (*)
Egress bandwidth profile per OVC endpoint at a UNI	mefServiceOvcEndPtPerUniCfgEgressBwpGrpIndex (*)
Egress bandwidth profile per class of service identifier at a UNI	mefServiceOvcEndPtPerUniCfgEgressBwpGrpIndex (*)

Table 15.8 OVC per VUNI Endpoint Service Attribute Alignment

MEF 28 [21] Attribute Name	MEF-ENNI-OVC-MIB, MEF-UNI-EVC-MIB Objects
VUNI OVC identifier	mefServiceOvcEndPtPerVuniCfgIdentifier
NA	mefServiceOvcEndPtPerVuniCfgRole
OVC endpoint map	mefServiceOvcEndPtPerVuniCfgCeVlanMap
Class of service identifiers	mefServiceCosCfgTable [36]
Ingress bandwidth profile per OVC endpoint associated by a VUNI	mefServiceOvcEndPtPerVuniCfgIngressBwpGrpIndex
Ingress bandwidth profile per class of service identifier associated by a VUNI	mefServiceOvcEndPtPerVuniCfgIngressBwpGrpIndex
Egress bandwidth profile per OVC endpoint associated by a VUNI	mefServiceOvcEndPtPerVuniCfgEgressBwpGrpIndex
Egress bandwidth profile per class of service identifier associated by a VUNI	mefServiceOvcEndPtPerVuniCfgEgressBwpGrpIndex

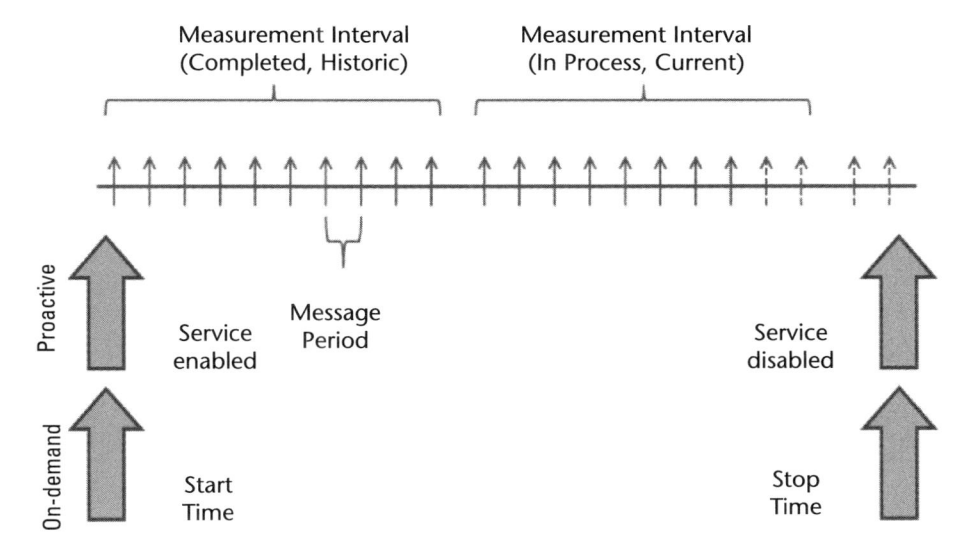

15.2 Information Modeling of Carrier Ethernet Services for EMS and NMS

Figure 15.29 Relationship between different timing parameters.

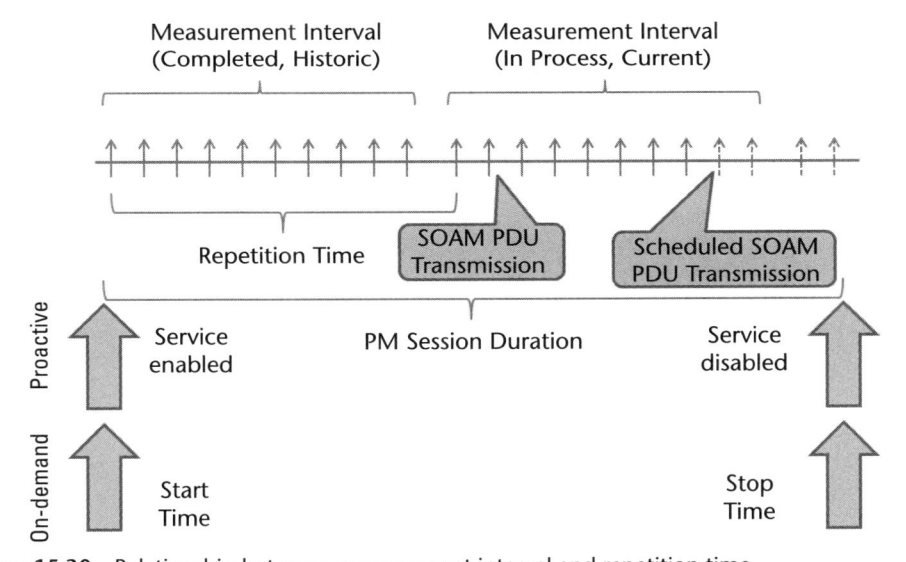

Figure 15.30 Relationship between measurement interval and repetition time.

on the sink MEP. Two dual-ended sessions can exist on an EVC or OVC where SOAM PDUs are sent in both directions.

PM sessions of type LMM/LMR, SLM/SLR, or DMM/DMR are single-ended types. PM sessions of type 1DM, 1SL, or CCM are dual-ended types. In summary:

- Controller MEPs send SOAM PM PDUs of type LMM, SLM, 1SL, DMM, CCM, or 1DM and receive SOAM PM PDUs of type LMR, SLR, and DMR.
- Responder MEPs send SOAM PM PDUs of type LMR, SLR, or DMR and receive SOAM PM PDUs of type LMM, SLM, or DMM.
- Sink MEPs receive SOAM PM PDUs of type 1SL, CCM, or 1DM.

The SOAM PM MIB is divided into a number of different object groupings: the PM MIB MEP Objects, PM MIB loss measurement objects, PM MIB delay measurement objects, TCA objects, and SOAM PM notifications.

The PM MIB per MEP objects are:

- mefSoamPmMepOperNextIndex: Indicates the next available index for row creation in the LM and DM configuration tables of a PM session on an MEP;
- mefSoamPmMepLmSingleEndedResponder: Indicates whether single-ended loss measurements (LMM) responders are enabled on an MEP;
- mefSoamPmMepSlmSingleEndedResponder: Indicates whether single-ended synthetic loss measurements (SLM) responders are enabled on an MEP.
- mefSoamPmMepDmSingleEndedResponder: Indicates whether single-ended delay measurements (DMM) responders are enabled on an MEP.

15.2.13.1 Loss Measurement Objects

The loss measurement (LM) objects are defined in six separate tables: mefSoamLmCfgTable, mefLmMeasuredStatsTable, mefSoamLmCurrentAvailStatsTable, mefSoamLmCurrentStatsTable, mefSoamLmHistoryAvailStatsTable, and mefSoamLmHistoryStatsTable.

For LM configuration, the *mefSoamLmCfgTable* includes configuration objects for the loss measurement PM session. A loss measurement session is created on an existing MEP by first accessing the *mefSoamPmMepOperNextIndex* object and using this value as the *mefSoamLmCfgIndex* during row creation.

A single loss measurement session can be used to measure frame loss, frame loss ratio, availability, and resiliency between a given pair of MEPs, for a given CoS frame set.

Configuration/status options are organized into eight general categories.

- LM session type, version, session enable, and counter enable:
    - mefSoamLmCfgType: LM PM session type;
    - mefSoamCfgVersion: G.8013/Y.1731 SOAM PM PDU format version;
    - mefSoamLmCfgEnabled: PM session enable;
    - mefSoamLmCfgMeasurementEnable: Specific PM session measurement enables;
    - mefSoamLmCfgSessionType: PM session duration selection of proactive or on-demand;
    - mefSoamLmCfgTcaNextIndex: PM session next available TCA index number.
- LM session PDU transmission frequency and measurement interval size:
    - mefSoamLmCfgMessagePeriod: Interval between transmission of SOAM PM PDUs;
    - mefSoamLmCfgMeasurementInterval: PM session measurement interval. Calculations within a measurement interval are

15.2 Information Modeling of Carrier Ethernet Services for EMS and NMS

based upon a small time intervals Δt (delta_t) as configured by mefSoamLmCfgAvailabilityNumConsecutiveMeasPdus.
- mefSoamLmCfgNumIntervalsStored: Number of completed PM session measurement intervals stored.

- LM session PDU composition and length:
  - mefSoamLmCfgPriority: SOAM PM PDU frame priority;
  - mefSoamLmCfgCosType: SOAM PM PDU CoS type selection of VLAN ID, VLAN ID plus priority, or VLAN ID plus priority and DEI;
  - mefSoamLmCfgFrameSize – mefSoamLmCfgCosType: SOAM PM PDU frame size;
  - mefSoamLmCfgDataPattern: SOAM PDU data TLV fill pattern;
  - mefSoamLmCfgTestTlvIncluded: Selection between test TLV or data TLV fill;
  - mefSoamLmCfgTestTlvPattern: SOAM PDU test TLV fill pattern;
  - mefSoamLmCfgDei: Drop eligible indicator;
  - mefSoamLmTestId: SLM/1SL Test ID.

- LM session peer partner selection:
  - mefSoamLmCfgDestMacAddress: Target or destination MAC address field to be transmitted;
  - mefSoamLmCfgDestMepId: Target MEP ID of the target MEP;
  - mefSoamLmCfgDestIsMepId: Selection of the type of target MEP association: MEP or destination MAC address;
  - mefSoamLmCfgSourceMacAddress: Selection of the source MAC address field for the sink MEP.

- LM session start, stop, and repetition selection:
  - mefSoamLmCfgAlignMeasurementIntervals: PM session time of day and hour alignment;
  - mefSoamLmCfgAlignMeasurementOffset: PM session offset from time of day;
  - mefSoamLmCfgStartTimeType: PM session start time type (i.e., immediate, fixed, or relative);
  - mefSoamLmCfgFixedStartDateAndTime: PM session fixed UTC start time;
  - mefSoamLmCfgRelativeStartTime: PM session relative start time;
  - mefSoamLmCfgStopTimeType: PM session stop time type (i.e., none, fixed, or relative);
  - mefSoamLmCfgFixedStopDateAndTime: PM session fixed UTC stop time;
  - mefSoamLmCfgRelativeStopTime: PM session relative stop time;
  - mefSoamLmCfgRepetitionTime: PM session time between starts of a measurement interval.

- LM session availability configuration:

- mefSoamLmCfgAvailabilityMeasurementInterval: PM session availability measurement interval calculations within an availability measurement interval are based upon a small time intervals Δt (delta_t) as configured by mefSoamLmCfgAvailabilityNumConsecutiveMeasPdus;
- mefSoamLmCfgAvailabilityNumConsecutiveMeasPdus: Number of consecutive SOAM PM PDUs to be used in evaluating availability or FLR over a small time interval or Δt (delta_t);
- mefSoamLmCfgAvailabilityFlrThreshold: Availability threshold to be used in evaluating availability and unavailability status;
- mefSoamLmCfgAvailabilityNumConsecutiveIntervals: Number of consecutive Δt intervals to be used to determine the change in the availability status for each Δt interval, in other words, a sliding window of width n Δt is used to determine availability;
- mefSoamLmCfgAvailabilityNumConsecutiveHighFlr: Configurable number of consecutive p Δt intervals to be used in assessing CHLI in the sliding window to qualify a Δt interval as a CHLI.
- LM Session parameters for status:
  - mefSoamLmCfgSessionStatus: Current status of the PM session.
- LM session history statistic table clear:
  - mefSoamLmCfgHistoryClear: Object when written clears the PM session history table.

Alignment with MEF 10.3 is supported via the LM session availability configuration parameters where:

- mefSoamLmCfgAvailabilityNumConsecutiveIntervals is equivalent to MEF 10.3 [27] parameter "n."
- mefSoamLmCfgAvailabilityFlrThreshold is equivalent to MEF 10.3 [27] parameter C.
- mefSoamLmCfgAvailabilityNumConsecutiveHighFlr is equivalent to "p."
- mefSoamLmCfgAvailabilityNumConsecutiveMeasPdus times mefSoamLmCfgMessagePeriod for SLM/LMM/CCM is equivalent to MEF 10.3 [27] parameter Δt.

For the LM measured statistic, the LM measured statistic table, *mefSoamLmMeasuredStatsTable*, is created automatically when a LM session is created and contains the loss measurement statistic information from the last received SOAM PDU. It uses the same indices as the *mefSoamLmCfgTable* table.

- mefSoamLmMeasuredStatsForwardFlr: Last PM session forward FLR;
- mefSoamLmMeasuredStatsBackwardFlr: Last PM session backward FLR;
- mefSoamLmMeasuredStatsAvailForwardStatus: Last PM session forward availability status;

15.2 Information Modeling of Carrier Ethernet Services for EMS and NMS

- mefSoamLmMeasuredStatsAvailBackwardStatus: Last PM session backward availability status;
- mefSoamLmMeasuredStatsAvailForwardLastTransitionTime: Last PM session forward availability status transition time;
- mefSoamLmMeasuredStatsAvailBackwardLastTransitionTime: Last PM session backward availability status transition time.

For the LM current availability statistic, the LM current availability statistic table, *mefSoamLmCurrentAvailStatsTable*, is created automatically when an LM session is created and contains the availability statistics for the current availability measurement interval. It uses the same indices as the *mefSoamLmCfgTable* table, but the measurement interval (*mefSoamLmCfgAvailabilityMeasurementInterval*) is independent of the interval used for the *mefSoamLmCurrentStatsTable* (*mefSoamLmCfgMeasurementInterval*).

At the beginning of each availability measurement interval, the values in the current availability statistic table are copied to a new row in the *mefSoamLmHistoryAvailStatsTable* and the current availability statistic table counter and status values are reset to zero and the start time is updated to the new measurement interval start time.

The LM current availability statistic table objects are organized into four categories.

- LM availability interval start time and elapsed time:
  - mefSoamLmCurrentAvailStatsIndex: Current availability measurement interval for this PM session;
  - mefSoamLmCurrentAvailStatsStartTime: Current availability measurement interval start time;
  - mefSoamLmCurrentAvailStatsElapsedTime: Current availability measurement interval elapsed time.
- LM availability interval suspect status:
  - mefSoamLmCurrentAvailStatsSuspect: Current availability measurement interval suspect indicator.
- LM availability high loss and consecutive high loss:
  - mefSoamLmCurrentAvailStatsForwardHighLoss: Current availability measurement interval forward HLI;
  - mefSoamLmCurrentAvailStatsBackwardHighLoss: Current availability measurement interval backward HLI;
  - mefSoamLmCurrentAvailStatsForwardConsecutiveHighLoss: Current availability measurement interval forward CHLI;
  - mefSoamLmCurrentAvailStatsBackwardConsecutiveHighLoss: Current availability measurement interval backward CHLI.
- LM availability available and unavailable Δt intervals counters:
  - mefSoamLmCurrentAvailStatsForwardAvailable: Current availability/ measurement interval forward availability Δt intervals;

- mefSoamLmCurrentAvailStatsBackwardAvailable: Current availability measurement interval backward availability Δt intervals;
- mefSoamLmCurrentAvailStatsForwardUnavailable: Current availability measurement interval forward unavailability Δt intervals;
- mefSoamLmCurrentAvailStatsBackwardUnavailable: Current availability measurement interval backward unavailability Δt intervals.
- LM availability FLR counters:
  - mefSoamLmCurrentAvailForwardMinFlr: Current availability measurement interval forward minimum FLR;
  - mefSoamLmCurrentAvailForwardMaxFlr: Current availability measurement interval forward maximum FLR;
  - mefSoamLmCurrentAvailForwardAvgFlr: Current availability measurement interval forward average FLR;
  - mefSoamLmCurrentAvailBackwardMinFlr: Current availability measurement interval backward minimum FLR;
  - mefSoamLmCurrentAvailBackwardMaxFlr: Current availability measurement interval backward maximum FLR;
  - mefSoamLmCurrentAvailBackwardAvgFlr: Current availability measurement interval backward average FLR.

For the LM current statistic, the LM current statistic table, *mefSoamLmCurrentStatsTable*, is created automatically when an LM session is created and contains the currently enabled statistic counters and statuses for the current measurement interval. It uses the same indices as the *mefSoamLmCfgTable* table, but the measurement interval (*mefSoamLmCfgMeasurementInterval*) is independent of the interval used for the *mefSoamLmCurrentAvailStatsTable* (*mefSoamLmCfgAvailabilityMeasurementInterval*).

At the beginning of each measurement interval, the values in the current statistic table are copied to a new row in the *mefSoamLmHistoryStatsTable* and the current statistic table counter and status values are reset to zero and the start time is updated to the measurement interval start time.

The LM current statistic table objects are organized into five categories.

- LM interval start time and elapsed time:
  - mefSoamLmCurrentStatsIndex: Current measurement interval for this PM session;
  - mefSoamLmCurrentStatsStartTime: Current measurement interval start time;
  - mefSoamLmCurrentStatsElapsedTime: Current measurement interval elapsed time.
- LM interval suspect status:
  - mefSoamLmCurrentStatsSuspect: Current measurement interval suspect indicator.

15.2 Information Modeling of Carrier Ethernet Services for EMS and NMS

- LM forward transmitted/received frames, minimum, maximum and average (arithmetic mean) frame loss ratio, available and unavailable counters:
    - mefSoamLmCurrentStatsForwardTransmittedFrames: Current measurement interval forward transmitted frames;
    - mefSoamLmCurrentStatsForwardReceivedFrames: Current measurement interval forward received frames;
    - mefSoamLmCurrentStatsForwardMinFlr: Current measurement interval forward minimum FLR;
    - mefSoamLmCurrentStatsForwardMaxFlr: Current measurement interval forward maximum FLR;
    - mefSoamLmCurrentStatsForwardAvgFlr: Current measurement interval forward average FLR.
- LM backward transmitted/received frames, minimum, maximum and average (arithmetic mean) frame loss ratio, available and unavailable counters:
    - mefSoamLmCurrentStatsBackwardTransmittedFrames: Current measurement interval backward transmitted frames;
    - mefSoamLmCurrentStatsBackwardReceivedFrames: Current measurement interval backward received frames;
    - mefSoamLmCurrentStatsBackwardMinFlr: Current measurement interval backward minimum FLR;
    - mefSoamLmCurrentStatsBackwardMaxFlr: Current measurement interval backward maximum FLR;
    - mefSoamLmCurrentStatsBackwardAvgFlr: Current measurement interval backward average FLR.
- LM initiated and received measurement counts:
    - mefSoamLmCurrentStatsSoamPdusSent: Current measurement interval SOAM PM PDUs transmitted;
    - mefSoamLmCurrentStatsSoamPdusReceived: Current measurement interval SOAM PM PDUs received.

For LM availability history statistic, the LM availability history statistic table, *mefSoamLmHistoryAvailStatsTable*, is created automatically when the first availability measurement interval completes in a LM session. A new row is created as each availability measurement interval is completed with the information from the completed *mefSoamLmCurrentAvailStatsTable* entry. The duration of each availability measurement interval is determined by *mefSoamLmCfgAvailabilityMeasurementInterval* and is independent of the measurement interval used for the *mefSoamLmHistoryStatsTable*. The oldest row can be deleted after a period of time, but is mandatory to be persistent for 32 completed measurement intervals and recommended to be persistent for 96 completed measurement intervals.

The LM availability history statistic table contains the same four categories as the *mefSoamLmCurrentAvailStatsTable* table, except the first category is interval end time and elapsed time.

The objects are the same except they are listed as "history" instead of "current."

For the LM history statistic, the LM history statistic table, *mefSoamLmHistoryStatsTable*, is created automatically when the first measurement interval completes in a LM session. A new row is created as each measurement interval is completed with the information from the completed *mefSoamLmCurrentStatsTable* entry. The duration of each measurement interval is determined by *mefSoamLmCfgMeasurementInterval* and is independent of the availability measurement interval used for the *mefSoamLmHistoryAvailStatsTable*. The oldest row can be deleted after a period of time, but is mandatory to be persistent for 32 completed measurement intervals and recommended to be persistent for 96 completed measurement intervals.

The LM history statistic table uses the same indices as the *mefSoamLmCfgTable* table as well as the one additional index, the *mefSoamLmHistoryStatsIndex* number.

The LM History Statistic Table contains the same five categories as the *mefSoamLmCurrentStatsTable* table, except the first category is interval end time and elapsed time. The objects are the same except they are listed as "history" instead of "current."

15.2.13.2 Delay Measurement Objects

For PM MIB delay measurement (DM), the delay measurement objects are defined in three pairs of tables: mefSoamDmCfgTable and mefSoamDmCfgMeasBinTable, mefSoamDmCurrentStatsXTable and mefSoamDmCurrentStatsBinsTable, and mefSoamDmHistoryStatsXTable and mefSoamDmHistoryStatsBinsTable, and the mefDmMeasuredStatsTable.

The *mefSoamDmCfgTable* includes configuration objects for the delay measurement PM session. It uses the same indices that a MEP configuration does: *dot1agCfmMdIndex*, *dot1agCfmMaIndex*, and *dot1agCfmMepIdentifier*, as well as *mefSoamDmCfgIndex*, the specific DM PM session number on a MEP.

A delay measurement session is created on an existing MEP by first accessing the *mefSoamPmMepOperNextIndex* object and using this value as the *mefSoamDmCfgIndex* during row creation. A single delay measurement session tracks interframe delay, frame delay variation, and frame delay range.

Configuration/status options are organized into eight general categories.

- DM session type, session enable, and counter enables:
    - mefSoamDmCfgType: DM PM session type;
    - mefSoamDmCfgVersion: G.8013/Y.1731 SOAM PM PDU format version;
    - mefSoamDmCfgEnabled: PM session enable;
    - mefSoamDmCfgMeasurementEnable: Specific PM session measurement enables;
    - mefSoamDmCfgSessionType: PM session duration selection of proactive or on-demand;
    - mefSoamDmCfgTcaNextIndex: PM session next available TCA index number.

15.2 Information Modeling of Carrier Ethernet Services for EMS and NMS

- DM session PDU transmission frequency and measurement interval size:
  - mefSoamDmMessagePeriod: Interval between transmission of SOAM PM PDUs;
  - mefSoamDmCfgMeasurementInterval: PM session measurement interval;
  - mefSoamDmCfgNumIntervalsStored: Number of completed PM session measurement intervals stored.
- DM session PDU composition and length:
  - mefSoamDmCfgPriority: SOAM PM PDU frame priority;
  - mefSoamDmCfgCosType: SOAM PM PDU CoS type selection of VLAN ID, VLAN ID plus priority, or VLAN ID plus priority and DEI;
  - mefSoamDmCfgFrameSize: SOAM PM PDU frame size;
  - mefSoamDmCfgDataPattern: SOAM PDU data TLV fill pattern;
  - mefSoamDmCfgTestTlvIncluded: Selection between test TLV or data TLV fill;
  - mefSoamDmCfgTestTlvPattern: SOAM PDU test TLV fill pattern;
  - mefSoamDmCfgDei: Drop eligible indicator.
- DM session peer partner selection:
  - mefSoamDmCfgDestMacAddress: Target or destination MAC address field to be transmitted;
  - mefSoamDmCfgDestMepId: Target MEP ID of the target MEP;
  - mefSoamDmCfgDestIsMepId: Selection of the type of target MEP association: MEP or destination MAC address;
  - mefSoamDmCfgSourceMacAddress - Selection of the source MAC address field for the sink MEP.
- DM session start, stop, and repetition selection:
  - mefSoamDmCfgAlignMeasurementIntervals: PM session time of day hour alignment;
  - mefSoamDmCfgAlignMeasurementOffset: PM session offset from time of day;
  - mefSoamDmCfgStartTimeType: PM session start time type (i.e., immediate, fixed, or relative);
  - mefSoamDmCfgFixedStartDateAndTime: PM session fixed UTC start time;
  - mefSoamDmCfgRelativeStartTime: PM session relative start time;
  - mefSoamDmCfgStopTimeType: PM session stop time type (i.e., none, fixed, or relative);
  - mefSoamDmCfgFixedStopDateAndTime: PM session fixed UTC stop time;
  - mefSoamDmCfgRelativeStopTime: PM session relative stop time;
  - mefSoamDmCfgRepetitionTime: PM session time between starts of a measurement interval.

- DM session measurement bin configuration:
    - mefSoamDmCfgNumMeasBinsPerFrameDelayInterval - DM PM Session number of measurement bins per Frame Delay interval
    - mefSoamDmCfgNumMeasBinsPerInterFrameDelayVariationInterval: DM PM session number of measurement bins per interframe delay interval;
    - mefSoamDmCfgInterFrameDelayVariationSelectionOffset: DM PM session offset for interframe delay variation measurements;
    - mefSoamDmCfgNumMeasBinsPerFrameDelayRangeInterval: DM PM session number of measurement bins per frame delay range interval.
- DM session status:
    - mefSoamDmCfgSessionStatus: Current status of the PM session.
- DM session history statistic table clear:
    - mefSoamDmCfgHistoryClear: Object when written clears the PM session history table.

For the DM configuration bin, the *mefSoamDmCfgMeasBinTable* includes configuration objects for the delay measurement bin PM session. It uses the same indices as the *mefSoamDmCfgTable* as well as the *mefSoamDmCfgMeasBinType* and *mefSoamDmCfgMeasBinNumber*.

For each row the *mefSoamDmCfgMeasBinLowerBound* is selected, which defines the lower boundary of each bin. The set of bin boundaries indicates the time range for each of the defined bins.

For example, the selection of five bins via either the *mefSoamDmCfg-NumMeasBinsPerFrameDelayInterval* or *mefSoamDmCfgNumMeasBinsPerInter- FrameDelayVariationInterval* or *mefSoamDmCfgNumMeasBinsPerFrame- DelayRangeInterval* objects results in the set of default values for the *mefSoamDmCfgMeasBinLowerBound* of {0, 5,000, 10,000, 15,000, 20,000}. These values create bins with the following lower and upper boundaries, as shown in Table 15.9.

These default values can be updated based upon changing the individual *mefSoamDmCfgMeasBinLowerBound* object value in each row.

For DM measured statistics, *mefSoamDmMeasuredStatsXTable* is created automatically when a DM session is created and contains the delay measurement

Table 15.9 Delay Measurement Bin Default Boundaries

Bin Number	mefSoamDmCfg MeasBinLowerBound Default Values	Lower Boundary	Upper Boundary
1	0	≥0µs	<5,000µs
2	5,000	≥5,000µs	<10,000µs
3	10,000	≥10,000µs	<15,000µs
4	15,000	≥15,000µs	<20,000µs
5	20,000	≥20,000µs	<Infinity

15.2 Information Modeling of Carrier Ethernet Services for EMS and NMS

statistic information from the last received SOAM PDU. It uses the same indices as the *mefSoamDmCfgTable* table:

- mefSoamDmMeasuredStatsXFrameDelayTwoWay: Last PM session two-way frame delay;
- mefSoamDmMeasuredStatsXFrameDelayForward: Last PM session forward frame delay;
- mefSoamDmMeasuredStatsXFrameDelayBackward: Last PM session backward frame delay;
- mefSoamDmMeasuredStatsXIfdvTwoWay: Last PM session two-way interframe delay;
- mefSoamDmMeasuredStatsXIfdvForward: Last PM session forward interframe delay;
- mefSoamDmMeasuredStatsXIfdvBackward: Last PM session backward interframe delay.

For the DM current statistic, *mefSoamDmCurrentStatsXTable*, is created automatically when a DM session is created and contains the currently enabled statistic counters and statuses for the current measurement interval. It uses the same indices as the *mefSoamDmCfgTable* table.

At the beginning of each measurement interval, the values in the current statistic table are copied to a new row in the *mefSoamDmHistoryStatsXTable* and the current statistic table values and statuses are reset to zero and the start time is updated to the measurement interval start time.

The DM current statistic table objects are organized into six categories:

- DM interval start time and elapsed time:
  - mefSoamDmCurrentStatsXIndex: Current measurement interval for this PM session;
  - mefSoamDmCurrentStatsXStartTime: Current measurement interval start time;
  - mefSoamDmCurrentStatsXElapsedTime: Current measurement interval elapsed time.
- DM interval suspect status:
  - mefSoamDmCurrentStatsXSuspect: Current measurement interval suspect indicator.
- DM frame delay two-way, forward, and backward min, max, and average (arithmetic mean) counters:
  - mefSoamDmCurrentStatsXFrameDelayTwoWayMin: Current measurement interval frame delay two-way minimum;
  - mefSoamDmCurrentStatsXFrameDelayTwoWayMax: Current measurement interval frame delay two-way frame maximum;
  - mefSoamDmCurrentStatsXFrameDelayTwoWayAvg: Current measurement interval frame delay two-way average;

- mefSoamDmCurrentStatsXFrameDelayForwardMin: Current measurement interval frame delay forward minimum;
- mefSoamDmCurrentStatsXFrameDelayForwardMax: Current measurement interval frame delay forward maximum;
- mefSoamDmCurrentStatsXFrameDelayForwardAvg: Current measurement interval frame delay forward average;
- mefSoamDmCurrentStatsXFrameDelayBackwardMin: Current measurement interval frame delay backward minimum;
- mefSoamDmCurrentStatsXFrameDelayBackwardMax: Current measurement interval frame delay backward maximum;
- mefSoamDmCurrentStatsXFrameDelayBackwardAvg: Current measurement interval frame delay backward average.

• DM interframe delay variation two-way, forward, and backward min, max, and average (arithmetic mean) counters:
- mefSoamDmCurrentStatsXIfdvForwardMax: Current measurement interval interframe delay forward maximum;
- mefSoamDmCurrentStatsXIfdvForwardAvg: Current measurement interval interframe delay forward average;
- mefSoamDmCurrentStatsXIfdvBackwardMax: Current measurement interval interframe delay backward maximum;
- mefSoamDmCurrentStatsXIfdvBackwardAvg: Current measurement interval interframe delay backward average;
- mefSoamDmCurrentStatsXIfdvTwoWayMax: Current measurement interval inter-frame delay two-way maximum;
- mefSoamDmCurrentStatsXIfdvTwoWayAvg: Current measurement interval interframe delay two-way average.

• DM frame delay range two-way, forward, and backward, max, and average (arithmetic mean) counters:
- mefSoamDmCurrentStatsXFrameDelayRangeForwardMax: Current measurement interval frame delay range forward maximum;
- mefSoamDmCurrentStatsXFrameDelayRangeForwardAvg: Current measurement interval frame delay range forward maximum;
- mefSoamDmCurrentStatsXFrameDelayRangeBackwardMax: Current measurement interval frame delay range backward maximum;
- mefSoamDmCurrentStatsXFrameDelayRangeBackwardAvg: Current measurement interval frame delay range backward average;
- mefSoamDmCurrentStatsXFrameDelayRangeTwoWayMax: Current measurement interval frame delay range two-way maximum;
- mefSoamDmCurrentStatsXFrameDelayRangeTwoWayAvg: Current measurement interval frame delay range two-way average.

• DM initiated and received measurement counts:
- mefSoamDmCurrentStatsXSoamPdusSent: Current measurement interval SOAM PM PDUs transmitted;

- mefSoamDmCurrentStatsXSoamPdusReceived: Current measurement interval SOAM PM PDUs received.

For the DM current statistic bins, *mefSoamDmCurrentStatsBinsTable* is created automatically when a DM session is created and contains the currently enabled statistic bin counters for the current measurement interval. It uses the same indices as the *mefSoamDmCfgMeasBinTable*.

At the beginning of each measurement interval, the values in the current bin statistic table are copied to a new row, one for each bin number, in the *mefSoamDmHistoryStatsXBinsTable* and the current statistic bins table values are reset to zero.

The DM current bin statistic table contains one object per row per bin, *mefSoamDmCurrentStatsBinsCounter*, which indicates a count for the specific bin.

For the DM history statistic, *mefSoamDmHistoryStatsXTable*, is created automatically when the first measurement interval completes in a DM session. A new row is created as each measurement interval is completed with the information from the completed *mefSoamDmCurrentStatsXTable* entry. The oldest row can be deleted after a period of time, but is mandatory to be persistent for 32 completed measurement intervals and recommended to be persistent for 96 completed measurement intervals.

The DM history statistic table uses the same indices as the *mefSoamDmCfgTable* table as well as the one additional index, the *mefSoamDmHistoryStatsXIndex* number.

The DM history statistic table contains the same five categories as the *mefSoamDmCurrentStatsXTable* table, except the first category is the interval end time and elapsed time. The objects are the same except they are listed as "history" instead of "current."

For the DM history bin statistic, *mefSoamDmHistoryStatsBinTable* is created automatically when the first measurement interval completes in a DM session. One row for each bin is created as each measurement interval is completed with the information from the completed *mefSoamDmCurrentStatsBinsTable* entry. The oldest rows can be deleted after a period of time, but it is mandatory to be persistent for 32 completed measurement intervals and recommended to be persistent for 96 completed measurement intervals.

The DM history bin statistic table contains the same object as the *mefSoamDmCurrentStatsBinsTable* table, except it is listed as "history" instead of "current."

15.2.13.3 Threshold Crossing Alerts

For threshold crossing alert (TCA) configuration, there are two groups of tables: the mefSoamDmTcaCfgTable, used for DM thresholds, and the mefSoamLmTcaCfgTable, used for the LM thresholds. Each table configures specific thresholds for either the DM or LM PM session.

Two types of TCAs are supported: (1) stateless, which is generated when the measured value is above the threshold during a measurement interval; and (2) stateful, which is generated when a threshold is exceeded (SET) and again when the values falls below (CLEAR) the threshold in a measurement interval.

When a measurement value is above the threshold for a specific performance metric for a specific PM session within a measurement interval and the specific measurement counter is enabled and the specific threshold is enabled and the TCA stateless notification is enabled and a PM MIB stateless TCA notification has not already been generated during this measurement interval, a PM MIB TCA notification is generated.

When a measurement value exceeds the threshold for a specific performance metric for a specific PM session within a measurement interval and the specific measurement counter is enabled and the specific threshold is enabled and the TCA stateful notification is enabled and the previous measurement value did not exceed the threshold a PM MIB set TCA stateful notification is generated.

When a measurement value does not exceed the clear threshold for a specific performance metric for a specific PM session at the end of a measurement interval and the specific measurement counter is enabled and the specific threshold is enabled and the TCA stateful notification is enabled and the previous measurement value was exceeded at some point during (or at the end of) the previous measurement interval a PM MIB clear TCA notification is generated.

For LM threshold crossing alerts, the *mefSoamLmTcaCfgTable* is configured after the LM PM session is configured. Rows are not automatically created. Each threshold is individually enabled via the *mefSoamLmTcaCfgEnable* object after it has been created. One or more TCAs can be created per PM session. The same metric can be used to create multiple threshold levels that can be acted upon in different ways.

The *mefSoamLmTcaCfgTable* includes configuration objects for the LM PM session TCA. It uses the same indices that the LM PM session does: *dot1agCfmMdIndex*, *dot1agCfmMaIndex*, and *dot1agCfmMepIdentifier*, and *mefSoamLmCfgIndex*, as well as *mefSoamLmTcaCfgIndex*, the specific LM PM session TCA, and *mefSoamLmTcaCfgType*, the specific type of LM PM session TCA.

A LM PM session TCA is created on a specific PM session by first accessing the next available index number, *mefSoamLmCfgTcaNextIndex* object and using this value as the *mefSoamLmTcaCfgIndex* during row creation.

The following TCA configuration options are supported:

- mefSoamLmTcaCfgEnable: Specific TCA enable;
- mefSoamLmTcaCfgType: TCA PM metric selection (used as an index for the mefSoamLmTcaCfgTable);
- mefSoamLmTcaCfgAlarmType: Selection of TCA to be either stateless or stateful;
- mefSoamLmTcaCfgThresholdValue: TCA threshold value for stateless or stateful TCA SET;
- mefSoamLmTcaCfgClearValue: TCA threshold value for stateful TCA CLEAR;
- mefSoamLmTcaCfgAlarmCurrentState: Current state of the TCA notification.

The LM thresholds supported via the *mefSoamLmTcaCfgType* object are:

15.2 Information Modeling of Carrier Ethernet Services for EMS and NMS

- High loss interval (HLI) forward direction;
- Consecutive high loss interval (CHLI) forward direction;
- HLI backward direction;
- CHLI backward direction.

For DM threshold crossing alerts, the *mefSoamDmTcaCfgTable* is configured after the DM PM session is configured. Rows are not automatically created. Each threshold is individually enabled via the *mefSoamDmTcaCfgEnable* object after it has been created. One or more TCAs can be created per PM session. The same metric can be used to create multiple threshold levels that can acted upon in different ways.

The *mefSoamDmTcaCfgTable* includes configuration objects for the DM PM session TCA. It uses the same indices that the DM PM session does: *dot1agCfmMdIndex*, *dot1agCfmMaIndex*, and *dot1agCfmMepIdentifier*, and *mefSoamDmCfgIndex*, as well as *mefSoamDmTcaCfgIndex*, the specific DM PM session TCA, and *mefSoamDmTcaCfgType*, the specific type of DM PM session TCA.

A DM PM session TCA is created on a specific PM session by first accessing the next available index number, *mefSoamDmCfgTcaNextIndex* object and using this value as the *mefSoamDmTcaCfgIndex* during row creation.

The following TCA configuration options are supported:

- mefSoamDmTcaCfgEnable: Specific TCA enable;
- mefSoamDmTcaCfgType: TCA PM metric selection: forward/backward/two-way FD/FDR/IFDV bin value and maximum value (used as an index for the mefSoamDmTcaCfgTable);
- mefSoamDmTcaCfgAlarmType: Selection of TCA to be either stateless or stateful;
- mefSoamDmTcaCfgBinNumber: Bin value k needed for bin types of TCAs, UBC(k);
- mefSoamDmTcaCfgThresholdValue: TCA threshold value for stateless or stateful TCA SET;
- mefSoamDmTcaCfgClearValue: TCA threshold value for stateful TCA CLEAR;
- mefSoamDmTcaCfgAlarmCurrentState - Current state of the TCA notification.

The DM thresholds supported via the *mefSoamDmTcaCfgType* object are:

- Frame delay forward bin;
- Frame delay forward max;
- Frame delay range forward bin;
- Frame delay range forward max;
- Interframe delay variation forward bin;
- Interframe delay variation forward max;

- Frame delay backward bin;
- Frame delay backward max;
- Frame delay range backward bin;
- Frame delay range backward max;
- Interframe delay variation backward bin;
- Interframe delay variation backward max;
- Frame delay two-way bin;
- Frame delay two-way max;
- Frame delay two-way forward bin;
- Frame delay two-way forward max;
- Interframe delay variation two-way bin;
- Interframe delay variation two-way max.

15.2.13.4 SOAM PM Notifications

The following objects are specified for SOAM PM notifications:

- mefSoamPmNotificationObjDateAndTime: Contains the UTC time and date at the time that the notification event is detected;
- mefSoamPmNotificationObjThresholdId: Contains the object identifier of the object that caused the generation of the threshold notification;
- mefSoamPmNotificationObjThresholdConfig: Contains the configured threshold value of the object that caused the generation of the threshold notification;
- mefSoamPmNotificationObjThresholdValue: Contains the measured value of the object at the time of generation of the notification;
- mefSoamPmNotificationObjSuspect: Contains the suspect flag for the current measurement interval in which the notification was generated;
- mefSoamPmNotificationObjCrossingType: Contains the type of notification crossing;
- mefSoamPmNotificationObjDestinationMep: Contains the MAC address of the destination MEP associated with the notification event;
- mefSoamPmNotificationObjDestinationMepId: Contains the MEP identifier of the destination MEP associated with the notification event;
- mefSoamPmNotificationObjPriority: Contains the CoS priority associated with the notification event;
- mefSoamPmNotificationObjMeasurementInterval: Contains the time at the start of the measurement interval associated with the notification event;
- mefSoamPmNotificationObjSeverity: Contains the severity of the notification event;
- mefSoamPmNotificationObjAvailabilityStatus: Contains the availability status change for the notification event.

The following objects configure notifications:

- mefSoamPmNotificationCfgAlarmInterval: Contains the shortest time interval in seconds between the generation of the same notification type per PM session;
- mefSoamPmNotificationCfgAlarmEnable: Enables/disables specific types of notification.

The following SOAM PM notifications can be generated:

- mefSoamAvailabilityChangeAlarm: Is sent when the state of the availability of the indicated service changes;
- mefSoamLmSessionStartStopAlarm: Is sent when the state of the LM session changes;
- mefSoamDmSessionStartStopAlarm: Is sent when the state of the DM session changes;
- mefSoamPmThresholdCrossingAlarm: Is sent when the value of the threshold crossing object from mefSoamLmThresholdCfgTable, mefSoamLmTcaCfgTable, mefSoamDmThresholdCfgTable, or mefSoamDmTcaCfgTable as indicated by the mefSoamPmNotificationThresholdId is crossed.

For a notification to be sent the applicable measurement counter needs to be enabled and for TCA notifications a threshold needs to be configured and crossed during a measurement interval.

15.3 Service-Level Information Modeling

Proliferation of SDN and development of life-cycle service orchestration management reference architecture (LSO RA) by MEF [28], as depicted in Figure 15.31, required the development of service level information model for Carrier Ethernet services. MEF 7.3 [5] defines a common set of managed objects for the Carrier Ethernet services based on the LSO RA. This Carrier Ethernet services management information model serves as the base model for interface profiles and data modeling.

The service information model consists of a set of object classes, their attributes and the relationships among them. Figures 15.32 and 15.33 illustrate the overviews of object classes and their relationships for EVC and OVC services, respectively. To simplify the overview, some of the conditional packages are not shown in those figures.

Object classes are used to convey a static representation of an entity, including properties and attributes (i.e., data model, the static part of the model). In other words, defining object classes is about defining entities and not the operations acting on the entities.

There are three top-level classes, EVC, EvcEndPoint, and UNI for EVC service. An EVC is associated with two or more EvcEndPoint(s), which resides in UNI(s). All other classes are supporting classes for the top-level classes.

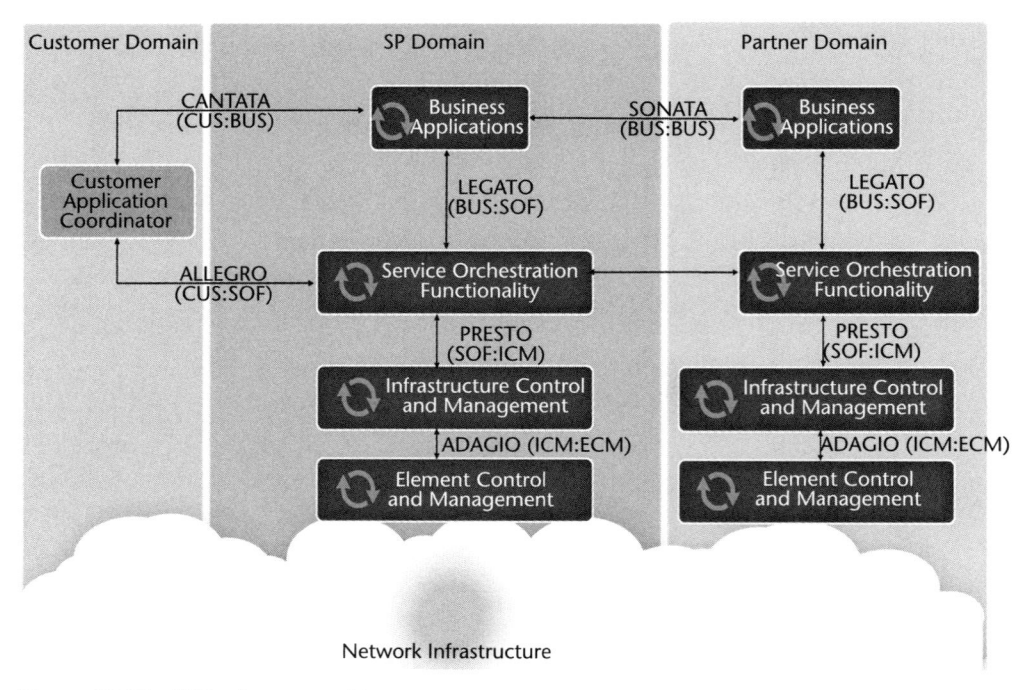

Figure 15.31 LSO reference architecture.

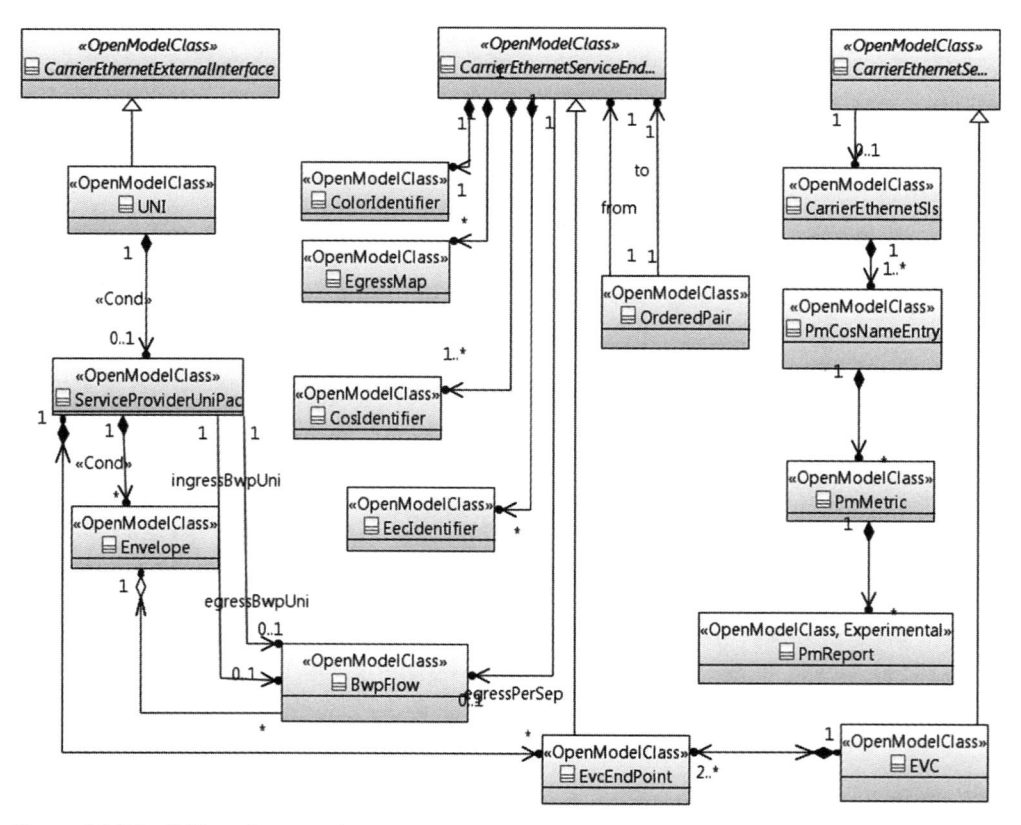

Figure 15.32 EVC service overview.

15.3 Service-Level Information Modeling

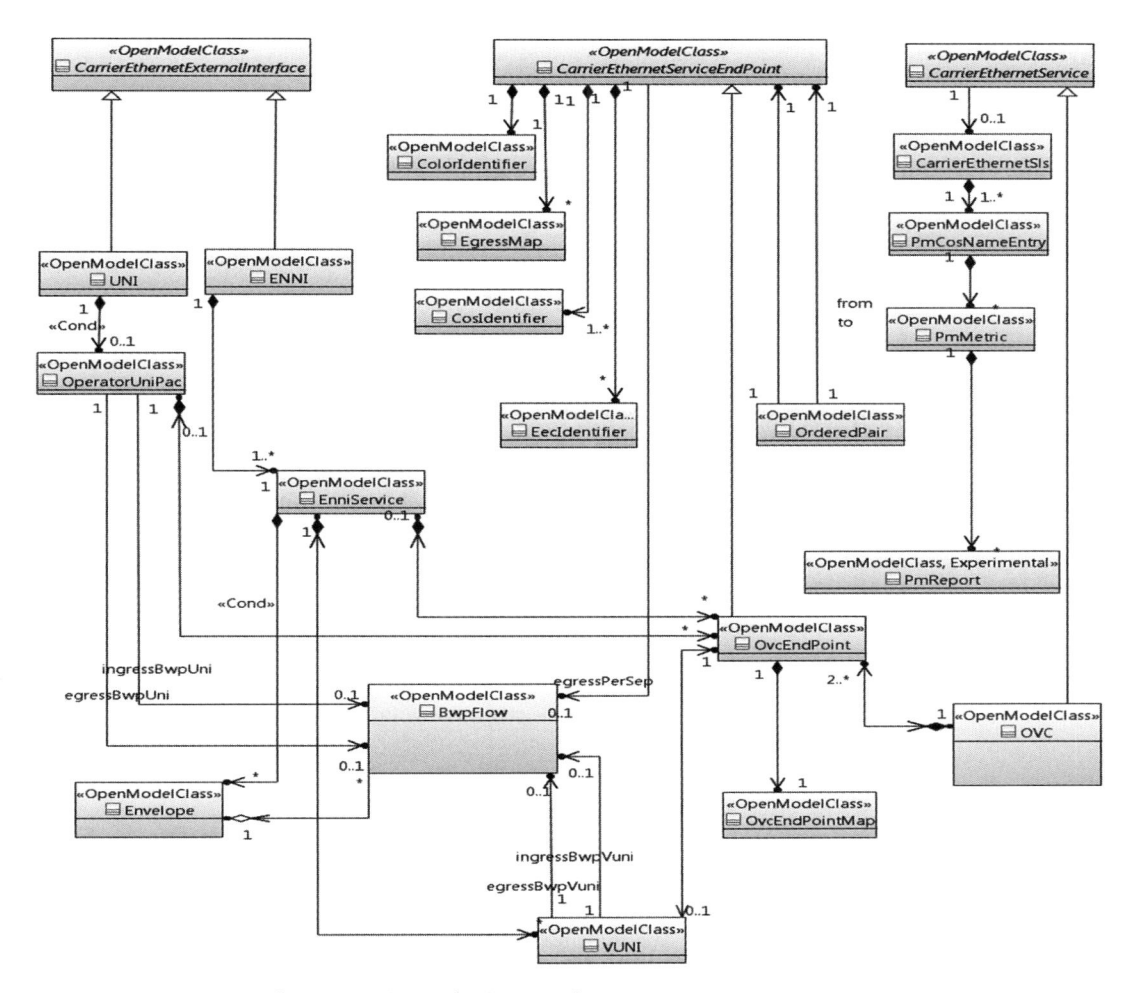

Figure 15.33 Carrier Ethernet service endpoint overview.

- EVC: A subclass of abstract class CarrierEthernetService representing the EVC service. The CarrierEthernetSls, representing Carrier Ethernet service specification, is the supporting class for EVC that are associated with all PM related classes.
- EvcEndPoint: A subclass of abstract class CarrierEthernetServiceEndPoint, representing the EVC end point. The supporting classes for EvcEndPoint include CosIdentifier representing class of service identifier), EecIdentifier representing egress equivalence class identifier), ColorIdentifier representing Color Identifier, EgressMap representing egress map, and BwpFlow representing the bandwidth profile flow.
- UNI: A subclass of abstract class CarrierEthernetExternalInterface representing UNI. The bandwidth profile-related classes are associated via the ServiceProviderUni_Pac representing the attribute set for UNI in service provider view.

There are three top-level classes for OVC service: OVC, OvcEndPoint, and ENNI (or UNI). An OVC is associated with 2 or more OvcEndPoint(s), which resides either in UNI(s) or in ENNI(s). All other classes are supporting classes for the top-level classes.

- OVC: A subclass of abstract class CarrierEthernetService representing the OVC service. The CarrierEthernetSls, representing Carrier Ethernet service specification is the supporting class for OVC that is associated with all PM-related classes.
- OvcEndPoint: A subclass of abstract class CarrierEthernetServiceEndPoint representing the OVC endpoint. The supporting classes for OvcEndPoint include CosIdentifier representing class of service identifier, EecIdentifier representing egress equivalence class identifier, ColorIdentifier representing color identifier, EgressMap representing egress map, and BwpFlow representing the bandwidth profile flow.
- ENNI: A subclass of abstract class CarrierEthernetExternalInterface representing ENNI. The bandwidth profile-related classes are associated via the EnniService_Pac representing ENI service attributes.
- UNI: A subclass of abstract class CarrierEthernetExternalInterface, representing UNI. The bandwidth profile related classes are associated via the OperatorUni_Pac representing the attribute set for UNI in the operator view. In addition, the class VUNI associated to ENNI via EnniService_Pac, OvcEndPoint, and BwpFlow

15.3.1 EVC

Figure 15.34 illustrates the object class of EVC with its attributes, including inherited attributes, and associations with the other object classes, including the inherited association.

Figure 15.35 illustrates the object class of OVC with its attributes, including inherited attributes, and associations with the other object classes, including the inherited associations.

The *CarrierEthernetServiceEndPoint* represents the EVC endpoint or the OVC endpoint. This is an abstract class and the super class of *EvcEndPoint* and *OvcEndPoint*. It contains the common attributes of *EvcEndPoint* and *OvcEndPoint*, as well as all common associations with the other object classes, such as a CarrierEthernetExternalInterface (i.e., UNI or ENNI), CosIdentifier(s), EecIdentifier(s), a ColorIdentifier, EgressMap(s), a CarrierEthernetService (i.e., EVC or OVC), and so forth.

Figure 15.36 illustrates the object class of *EvcEndPoint* representing EVC endpoint with its attributes and associations, including inherited attributes and associations.

EvcEndPoint is a subclass of *CarrierEthernetServiceEndPoint*. It represents the EVC endpoint which is EVC per UNI, provides all EVC endpoint service attributes, as well as all associations with the other object classes, such as an EVC, a UNI, CosIdentifier(s), EecIdentifier(s), a ColorIdentifier, an EgressMap(s), and so forth.

15.3 Service-Level Information Modeling

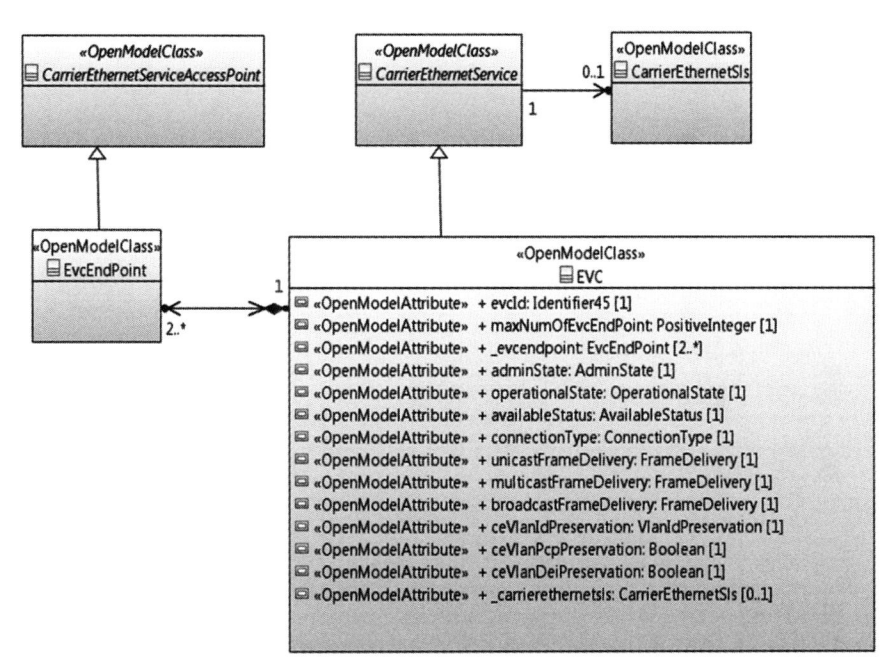

Figure 15.34 EVC class diagram.

An applied stereotype is OpenModelClass. [A stereotype is one of three types of extensibility mechanisms in the Unified Modeling Language (UML), the other two being tags and constraints. They allow designers to extend the vocabulary of UML in order to create new model elements, derived from existing ones.]:

15.3.2 OVC

Figure 15.37 illustrates the object class of OVC with its attributes, including inherited attributes, and associations with the other object classes, including the inherited associations.

The OVC is a subclass of *CarrierEthernetService* object class. It represents the MEF-defined OVC service with all OVC service attributes, as well as its associations with other object classes. It is associated with two or more *OvcEndPoint(s)*, a *CarrierEthernetSls* representing SLS. The applied stereotypes are OpenModelClass and support: MANDATORY.

OvcEndPoint provides all OVC endpoint service attributes, as well as all associations with the other object classes, such as an OVC, a CarrierEthernetExternalInterface (UNI or ENNI), CosIdentier(s), EecIdentifier(s), EgressMap(s), an OvcEndPointMap, a ColorIdentifer, VUNI(s), and so forth.

CarrierEthernetService object class representing the EVC service and the OVC service. It is an abstract class and the super class of EVC and OVC, containing all common attributes of EVC and OVC, including the common association with CarrierEthernetSls (representing SLS).

Applied stereotypes are OpenModelClass and support: MANDATORY.

The *CarrierEthernetServiceEndPoint* represents the EVC endpoint or the OVC endpoint. This is an abstract class and the super class of EvcEndPoint and

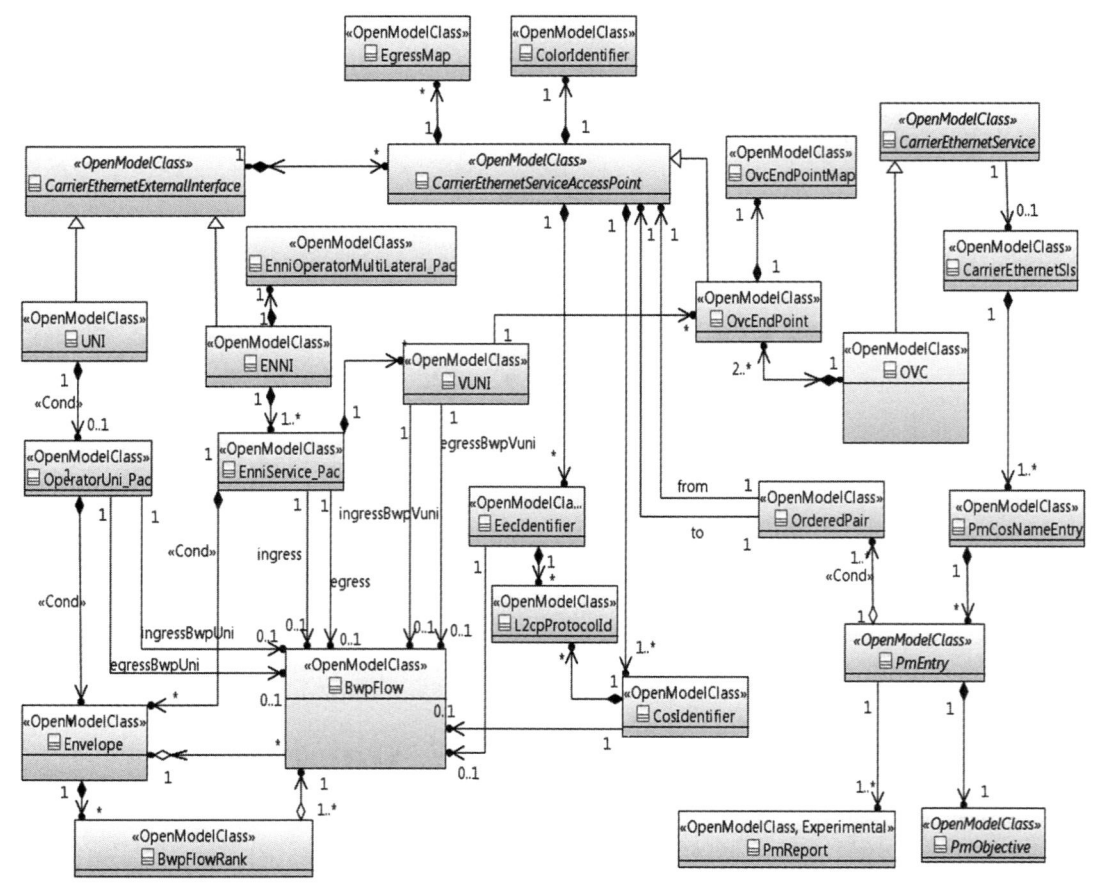

Figure 15.35 OVC class diagram.

OvcEndPoint. It contains the common attributes of *EvcEndPoint* and *OvcEndPoint*, as well as all common associations with the other object classes, such as a *CarrierEthernetExternalInterface* (i.e., UNI or ENNI), CosIdentifier(s), EecIdentifier(s), a ColorIdentifier, EgressMap(s), a CarrierEthernetService (i.e., EVC or OVC), and so forth.

Applied stereotypes are OpenModelClass and support: MANDATORY.

15.3.3 Carrier Ethernet External Interface

CarrierEthernetExternalInterface object class represents the Carrier Ethernet external interface (i.e., UNI and ENNI). It is an abstract class and the super class of UNI and ENNI, providing all common service attributes of UNI and ENNI, as well as all common associations with the other object classes, such as CarrierEthernetServiceEndPoint(s) (EvcEndPoint or OvcEndPoint), Envelope(s), BwpFlow(s), and so forth.

15.3.4 UNI

The UNI object class is a subclass of CarrierEthernetExternalInterface object class (see Figure 15.38). UNI may be managed by the service provider or by the operator,

15.3 Service-Level Information Modeling

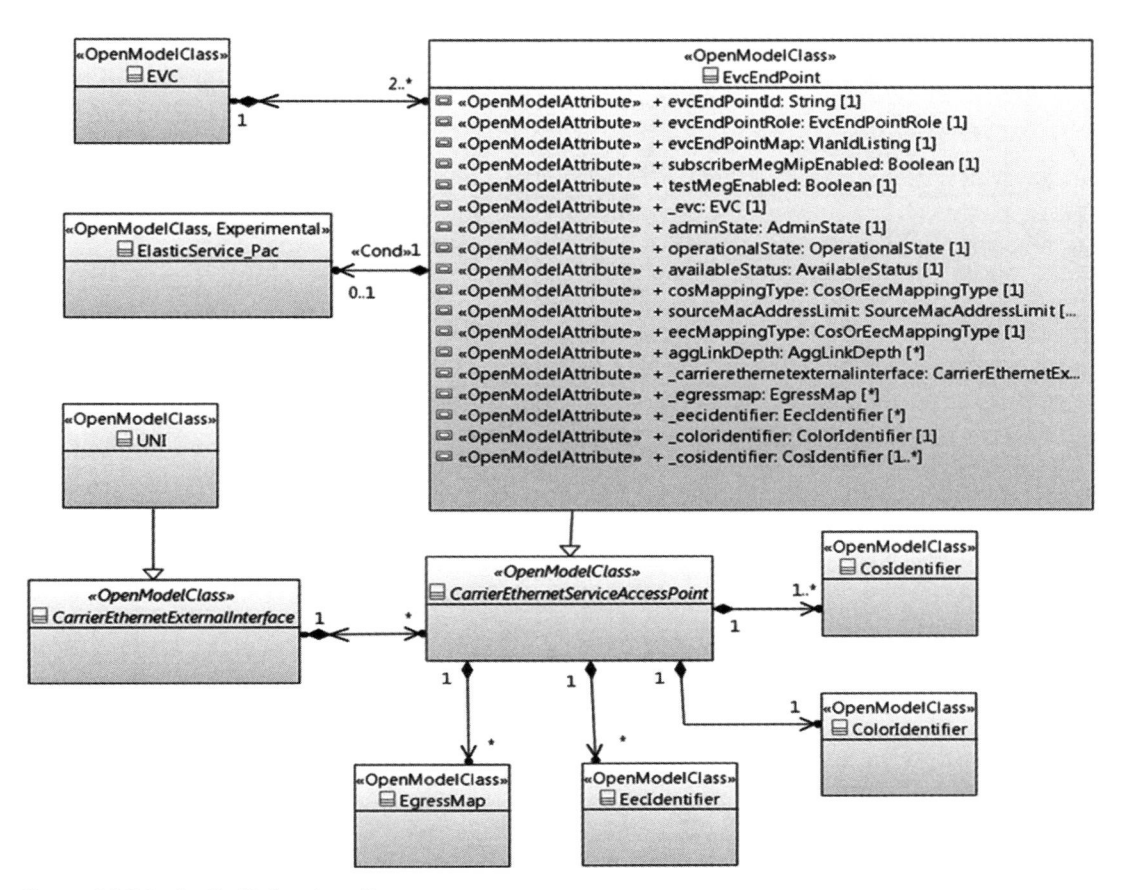

Figure 15.36 EvcEndPoint class diagram.

for VC services or for OVC services. Therefore, an UNI instance may contain ServiceProviderUniPac and/or OperatorUniPac, which consists of different attribute set, or different values for some service attributes. The bandwidth profile parameters [Envelope(s), BwpFlow(s)] are associated with the UNI indirectly via the ServiceProviderUniPac and the OperatorUniPac.

15.3.5 ENNI

Figure 15.39 illustrates the object class of ENNI, along with EnniService representing the ENNI service attributes, with their attributes, and associations with the other object classes, including inherited attributes and associations.

The ENNI object class is a subclass of *CarrierEthernetExternalInterface* object class. The ENNI service attributes consists of a set of common ENNI attributes, and multiple sets of ENNI service attributes. Via EnniService, the ENNI is associated with Envelope(s), BwpFlow(s), OvcEndPoint(s), and VUNI(s).

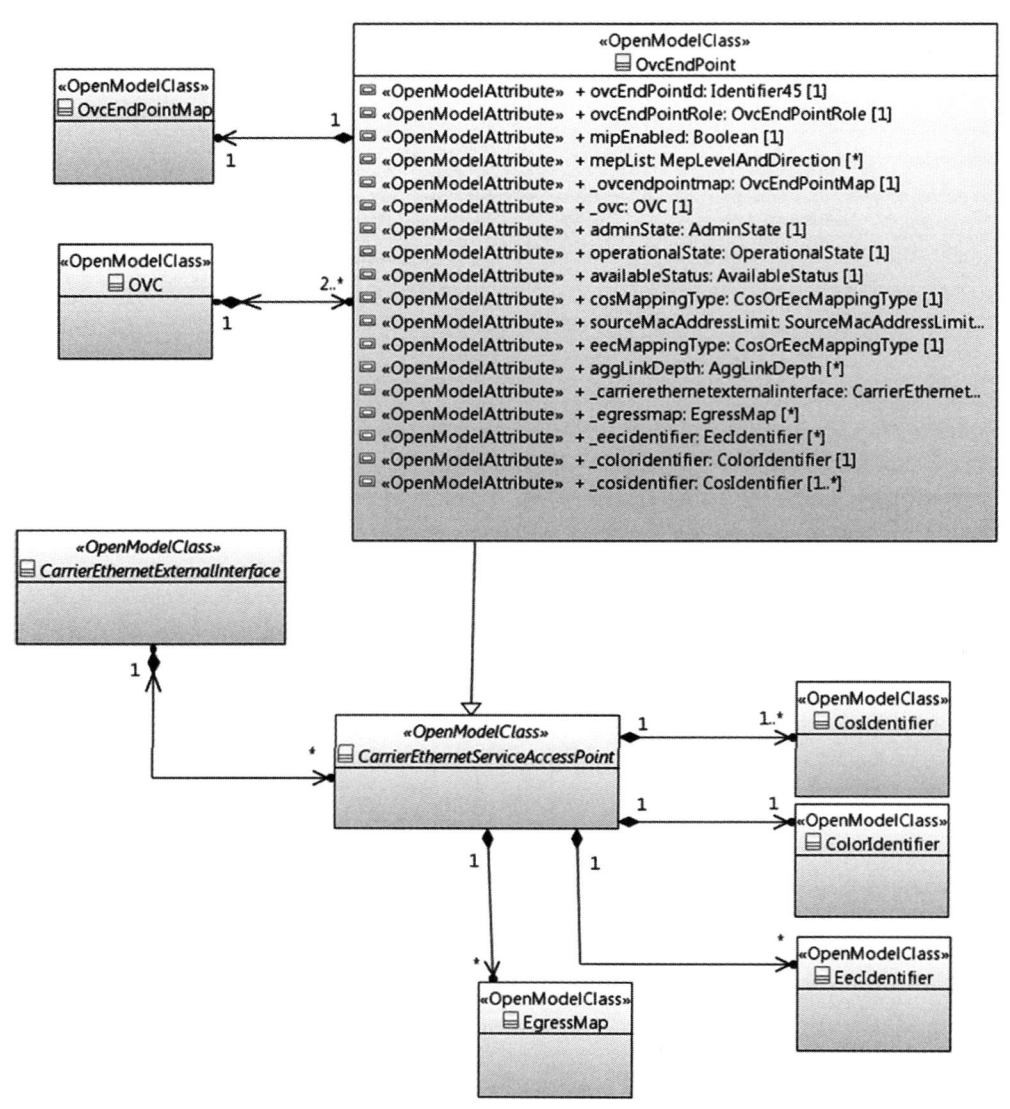

Figure 15.37 OvcEndPoint class diagram.

15.3.6 ENNI Service

This EnniService object class represents a set of ENNI service attributes. For each instance of an ENNI, there may be multiple sets of ENNI service attributes. The value for each ENNI service attribute in a set for an operator CEN is specific to a SP/SO [service provider/super operator (SO is the operator providing the end-to-end EVC that crosses multiple operators where some of the operators may have service provider-operator relationship)] that is using the ENNI. Each such value is agreed to by the SP/SO and the operator. For a given SP/SO that is using the ENNI, a given ENNI service attribute can have an identical value for each operator while another ENNI service attribute can have a different value for each operator. It is associated with Envelope(s), BwpFlow(s), OvcEndPoint(s), and VUNI(s).

15.3 Service-Level Information Modeling 443

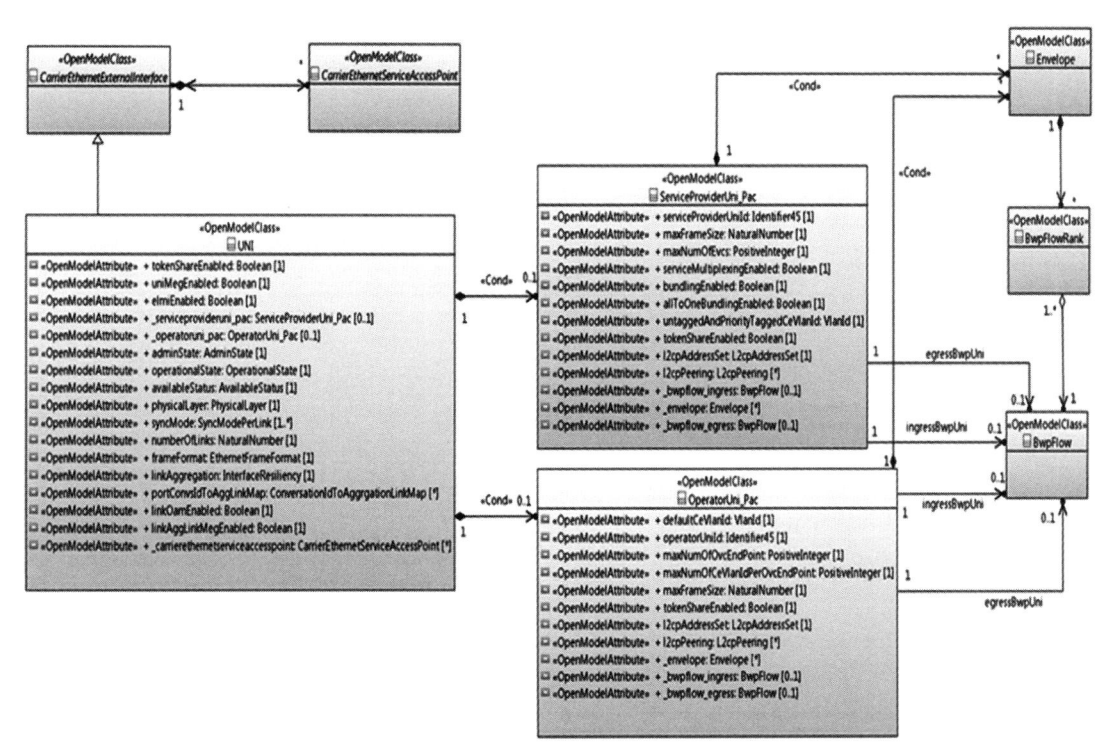

Figure 15.38 UNI class diagram.

The VUNI object class represents the VUNI (see Figure 15.40). It provides the MEF VUNI attributes, as well as the associations with the ENNI, the OVC endpoint and the bandwidth profile parameters.

Figure 15.41 illustrates the object class of CoSIdentifier, along with the conditional packages SepCosIdPac, PcpCosIdPac, and DscpCosIdPac, and their attributes and associations with the other object classes.

The CoSIdentifier object class represents the CoS identifier. Each ingress EI frame mapped to the given EVC/OVC endpoint has a single CoS. The CoS can be determined from inspection of the content of the ingress EI frame. It is associated with the SepCosIdPac, or the PcpCosIdPac, or the DscpCosIdPac (when the CoS identifier mapping type is service endpoint, or PCP values, or DSCP values, respectively). EI frames of L2CP protocols may be identified by a CoS identifier, mapping to specific CoS name.

15.3.7 Egress Equivalence Class Identifier

Figure 15.42 illustrates the object class of EecIdentifier, along with the conditional packages SepEecIdPac, PcpEecIdPac, and DscpEecIdPac and their attributes and associations with the other object classes.

The EecIdentifier object class represents the egress equivalence class identifier. Each egress EI frame mapped to the given EVC/OVC endpoint has a single egress equivalence class. The egress equivalence class can be determined from inspection of the content of the egress EI frame. It is associated with the SepCosIdPac, or the

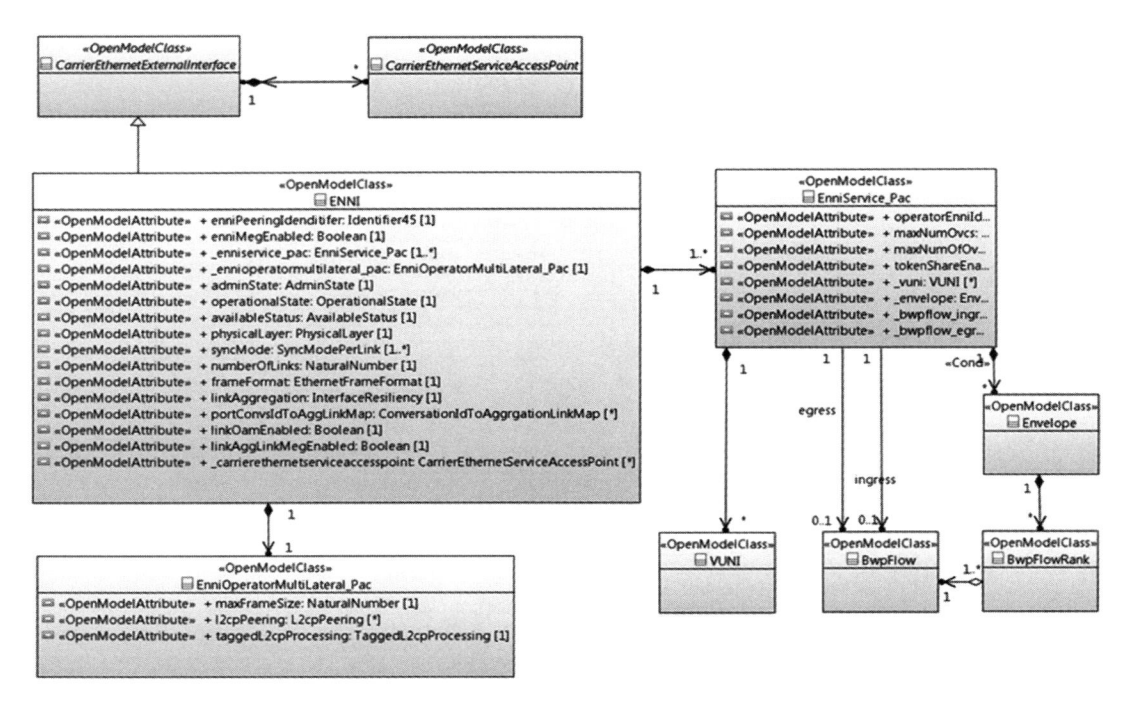

Figure 15.39 ENNI class diagram.

PcpCosIdPac, or the DscpCosIdPac (representing mapping to EVC/OVC endpoint, or PCP, or DSCP, respectively). EI frames of L2CP protocols may be identified by an egress equivalence class identifier, mapping to specific egress equivalence class name.

15.4 YANG Models for Carrier Ethernet Services

The key feature of SDN is autoprovisioning of telecom system and services. In order to accomplish that, open application interfaces for NEs, SDN controllers, and orchestrators are necessary. This will not only allow interoperability between various vendor equipment but also greatly minimizes introduction of equipment from various vendors into the network of a service provider.

Application program interface (APIs) based on NETCONF YANG models are becoming a de facto standards for managing NEs. The YANG modules for metro Ethernet services are depicted in Figure 15.43. NETCONF is an IETF network management protocol defining a simple mechanism through which a device can be managed, configuration data information can be retrieved, and new configuration data can be uploaded and manipulated.

YANG is a data modeling language to define the structure, syntax, and semantics of data that can be used for NETCONF operations, including configuration, state data, RPCs, and notifications. This allows the device to expose a full, formal API. Applications can use this API to send and receive full and partial configuration

15.4 YANG Models for Carrier Ethernet Services

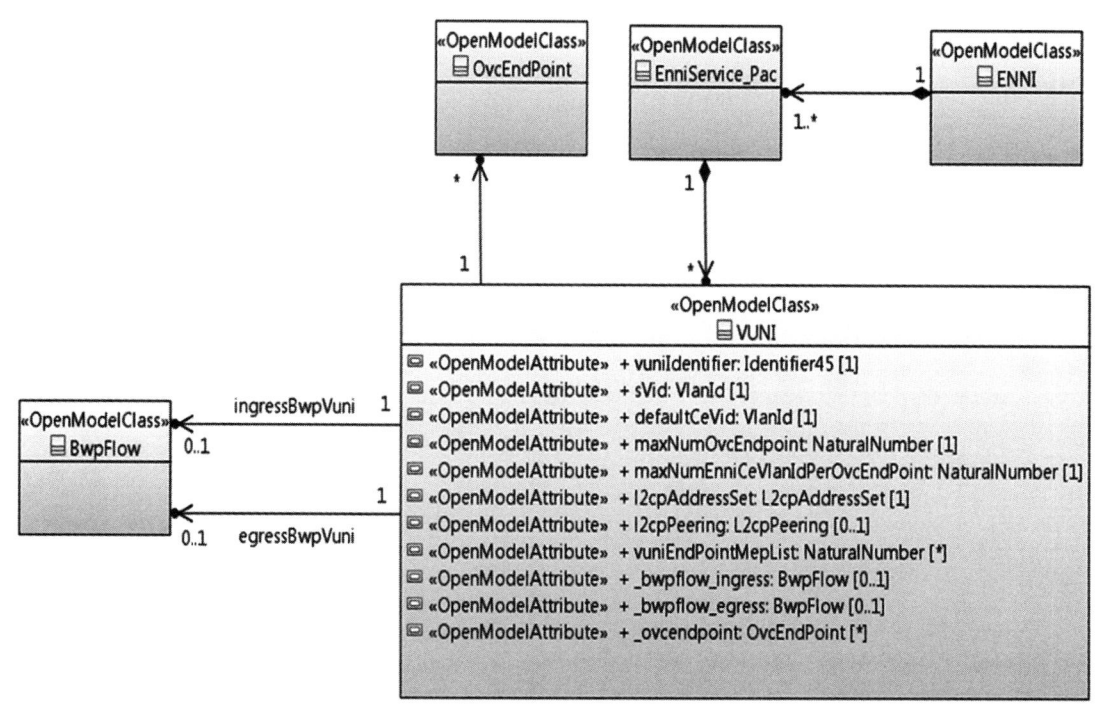

Figure 15.40 VUNI class diagram.

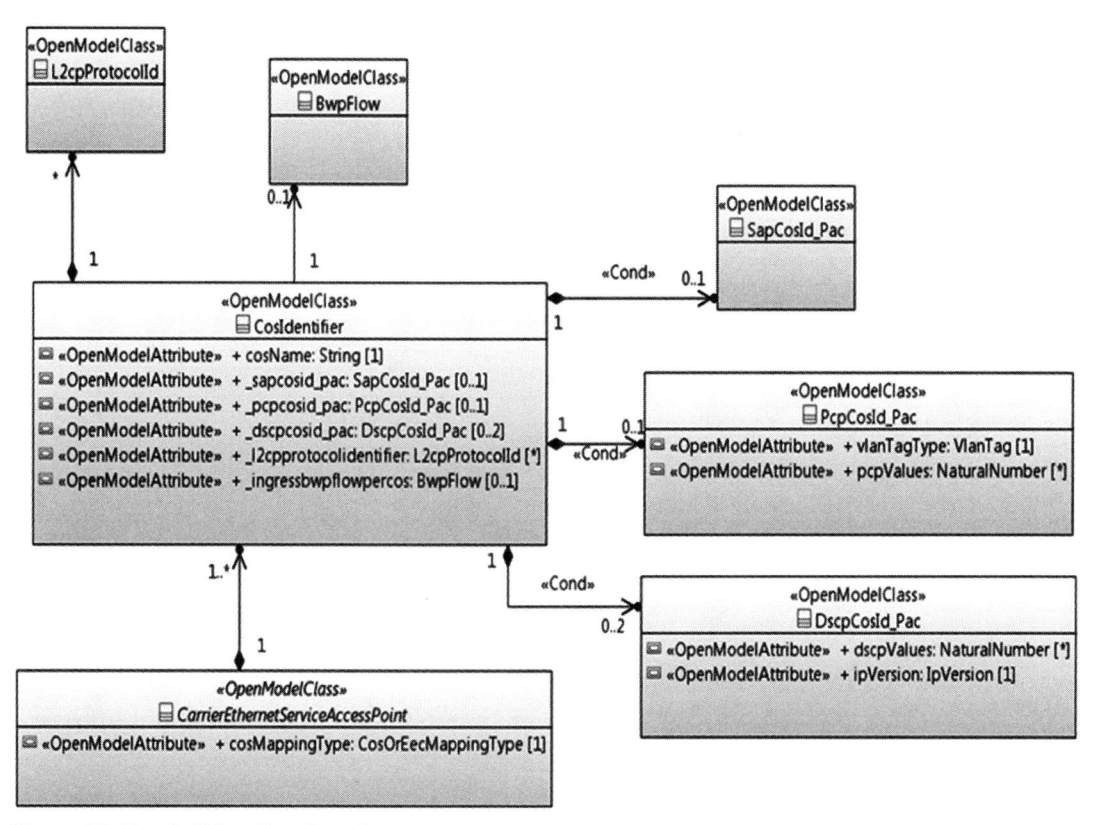

Figure 15.41 CoSIdentifier class diagram.

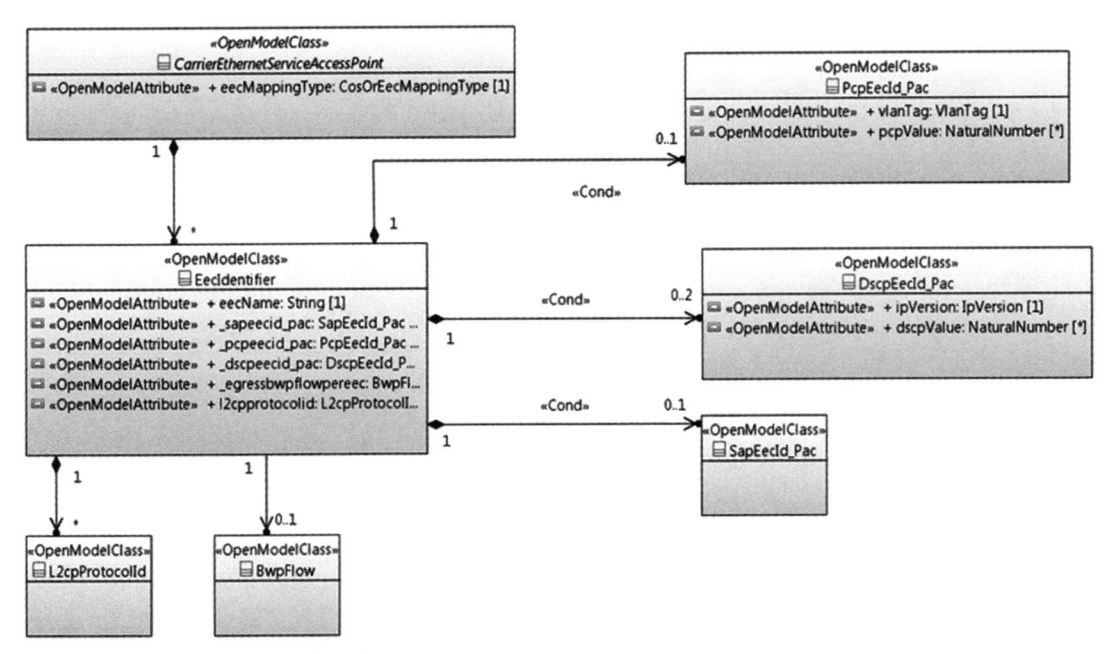

Figure 15.42 EecIdentifier class diagram.

data sets. NETCONF uses an Extensible Markup Language (XML)-based data encoding for the configuration data as well as the protocol messages.

The NETCONF protocol is defined in a set of RFCs. The base protocol is defined in RFC 6241 [41] and the mandatory transport mapping is defined in RFC 6242 [29]. A list of additional transport mappings and extensions to the protocol is maintained by the NETCONF Working Group in the IETF.

The YANG language standard is defined by the IETF in a set of RFCs, primary of which are RFC 6020 [8] and RFC 6087 [30].

15.4.1 SOAM YANG CFM Module [20]

The YANG CFM module implements the objects and functions found in IEEE 802.1Q [31]. IEEE 802.1ap [23] and the corresponding objects found in IEEE 8021-CFM-MIB [32], IEEE 8021-CFM-V2-MIB [33], and IEEE 8021-TC-MIB [34].

This module provides the top-level structure of Ethernet CFM including MD, MA/MEG and MEPs. These structures are then augmented with the SOAM FM data definitions in the SOAM YANG FM module.

The major content deviations from the IEEE 8021-CFM-MIB, IEEE 8021-CFM-V2-MIB, and IEEE 8021-TC-MIB are that there is no equivalent of *ieee8021Cfm-StackTable*. As the retrieval of information about maintenance points configured on any given interface, the CFM stack table (*ieee8021CfmStackTable*) can be implemented using the basic query features of NETCONF.

15.4 YANG Models for Carrier Ethernet Services

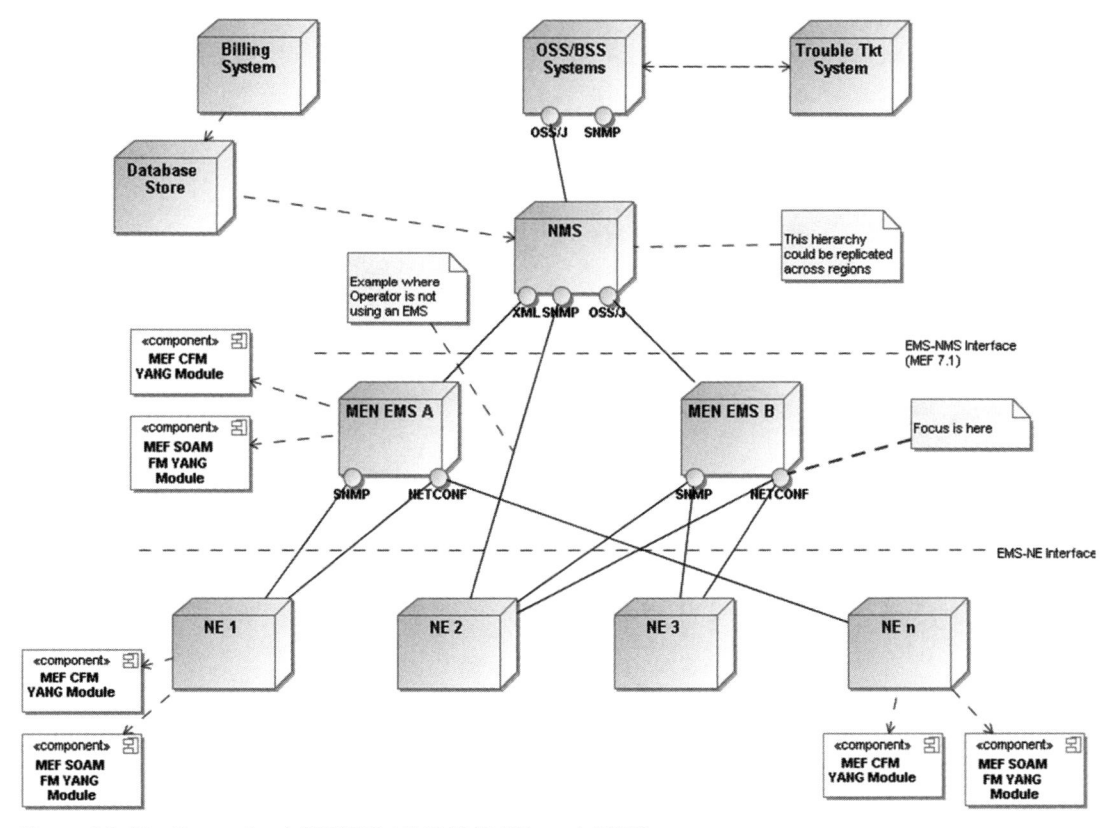

Figure 15.43 Generalized OSS/BSS-NMS-EMS-NE model [20].

The SOAM CFM YANG module defines the managed objects necessary to support SOAM CFM functionality. Its primary points of reference are ITU 802.1Q [31] and IEEE 802.1ap [23], and Y.1731 [6] with corresponding SNMP MIBs [32–34].

15.4.1.1 Type Definitions

Type definitions define derived types from base types using the typedef statement [8]. A base type can be either a built-in type or a derived type, allowing a hierarchy of derived types.

The following type definitions are defined within the SOAM CFM YANG module:

- error-conditions-type: A list of errors that may occur on creation or deletion of a MEP;
- fault-alarm-defect-bits-type: A set of bits indicating active defects;
- fault-alarm-defect-type: An enumerated value indicating the highest priority defect;
- id-permission-type: An enumerated value indicating what, if anything, is to be included in a sender ID TLV;

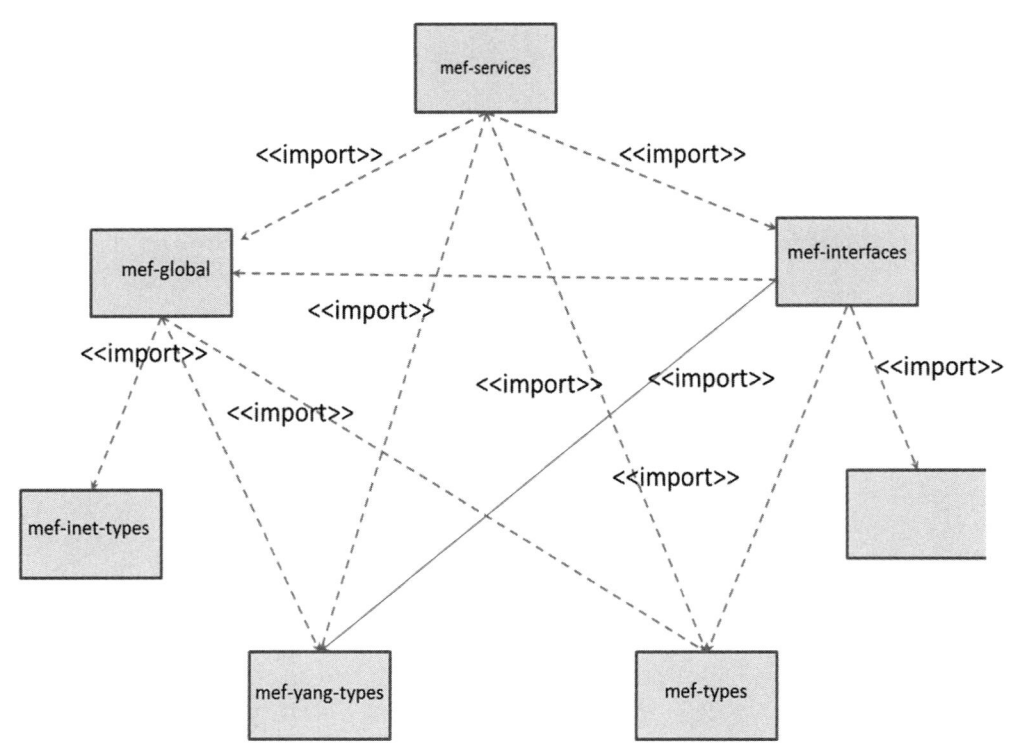

Figure 15.44 MEF YANG model architecture.

- interface-status-type: A value obtained from the interface status TLV of a CCM;
- lbm-transaction-id-type: The value to place in the loopback transaction identifier field in the next LBM frame;
- ltm-transaction-id-type: The value to place in the LTM transaction identifier field in the next LTM frame;
- md-level-type: A value indicating the MD level;
- mep-id-type: A MEP identifier, unique over a given MA;
- mhf-creation-type: A set of enumerated values to control the creation of a MIP for a VID on a bridge port;
- port-status-type: A value obtained from the port status TLV of a CCM;
- priority-type: A 3 bit priority value to be used in the VLAN tag;
- remote-mep-state-type: An enumerated value indicating the operational state of a remote MEP state machine;
- vlan-id-type: A 12-bit VLAN-ID used in the VLAN Tag header that uniquely identifies a VLAN;
- mac-address-and-uint-type: A MAC address and a two-octet unsigned integer.
- component-id-type: A component identifier used to distinguish between multiple virtual bridge instances.

15.4.1.2 Groupings

Groupings are reusable collections of nodes using the grouping statement. Groupings of definitions are instantiated using the uses statement. The grouping and uses statements are defined in RFC 6020 [8].

The following groupings are defined for CFM YANG:

- target-address-group: A group of data definitions identifying a target MEP or MAC address.
- loopback-parameters-group: A group of data definitions associated with specific loopback sessions
- linktrace-parameters-group: A group of data definitions associated with specific linktrace sessions.
- md-level-group: A group of data definitions related to the default MD level objects.
- port-id-tlv-group: A group of data definitions related to 802.1AB [35], Port ID TLV.
- sender-id-tlv-group: A group of data definitions related to IEEE 802.1Q [31] sender ID TLV.
- maintenance-domain-reference: A group of data definitions describing a reference to a specific maintenance domain.
- maintenance-association-reference: A group of data definitions describing a reference to a specific maintenance association.
- maintenance-association-end-point-reference: A group of data definitions describing a reference to a specific maintenance association endpoint.

15.4.1.3 Data Definitions

Data definitions define new data nodes in the data tree. Figure 15.45 describes the general structure of the YANG module. It closely corresponds to the definition of the CFM entities in IEEE 802.1Q [31] and is used as the basic structure to be extended by SOAM FM and PM.

The following list entries describe the relative location of the data definition in the SOAM CFM module tree, the informal name of all entries in bold text, for example, default MD level, and the formal YANG definition statement in bold italics enclosed in parenthesis, for example, (*list default-md-level*).

- Default MD levels (container default-md-levels): This container contains data definitions related to the global default MD level and the default MD levels associated with specific VLAN IDs (VIDs). The MD levels controls the MIP half function (MHF) creation for VIDs that are not contained in the list of VIDs attached to any specific maintenance association managed object. It also controls the transmission of the sender ID TLV by those MHFs. The parameters related to the global default MD level are located directly in this

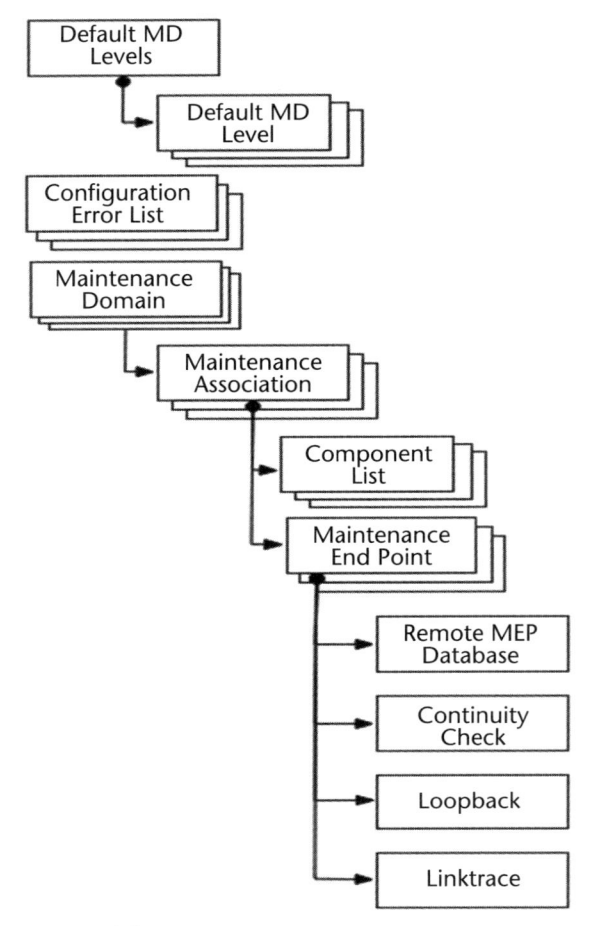

Figure 15.45 CFM YANG module structure.

container while the default MD levels associated with VIDs are located in the default-md-level list. This list is located on the top of the module.

- Default MD level (list default-md-level): This list contains the data definitions for the default MD level associated with specific VIDs. This list is contained in the default-md-levels container.
- Configuration error list (list configuration-error-list): This list contains bridge ports, aggregated ports, and VIDs that are incorrectly configured. This list is located on the top of the module.
- Maintenance domain (list maintenance-domain): This list contains all maintenance domains (MDs) on the bridge. An MD is required in order to create an MA with a MAID. From this maintenance domain-managed object, all maintenance association-managed objects associated with that maintenance domain-managed object can be accessed, and thus controlled. This list is located on the top of the module.
- Maintenance association (list maintenance-association): This list contains all maintenance association (MAs) on the bridge, one for each service instance for which an MP is defined on that bridge. From this maintenance

15.4 YANG Models for Carrier Ethernet Services

association-managed object, all maintenance association endpoint-managed objects associated with that maintenance association-managed object can be accessed, and thus controlled.

- Component list (list component-list): This list contains all bridge components on the bridge with associated configuration. This list is a member of the maintenance-association list.
- Maintenance association endpoint (list maintenance-association-end-point): This list contains all association MEPs on the bridge, one for each MEP defined within that bridge. From this association MEP-managed object, all management objects related to that MEP can be controlled. This list is a member of the maintenance-association list.
- The maintenance-association-end-point list may contain the following containers:

    - Continuity check (container continuity-check): This container contains data definitions related to continuity check (CC).
    - Loopback (container loopback): This container contains data definitions related to loopback (LB).
    - Linktrace (container link-trace): This container contains data definitions related to link trace (LT). This container contains a database container (container linktrace-database) that provides data definitions for historical linktrace results. The database container uses the linktrace-parameters-group grouping.
    - Remote MEP database (container remote-mep-database): This container contains data definitions related to the MEP database. The MEP database maintains received information about other MEPs in the maintenance association.

15.4.1.4 Remote Procedure Calls (RPCs)

Custom NETCONF RPCs are defined using the rpc statement. An RPC has a name and defines an optional set of input data using the input statement, and a set of return values using the output statement. The following RPCs are defined within the SOAM CFM YANG and are used by clients to perform specific tasks:

- transmit-loopback: This RPC creates a new loopback session associated with a specific MEP.
- abort-loopback: This RPC aborts a current loopback session associated with a specific MEP.
- transmit-linktrace: This RPC creates a new linktrace session associated with a specific MEP. The return value is the transaction identifier for the newly created linktrace session.

15.4.1.5 Notifications

Custom NETCONF notifications are defined using the notification statement [8]. A notification can contain arbitrary data defined in the YANG module.

The NETCONF notifications feature provides filtering capabilities allowing subscribers to filter out events based on event content. This feature makes the configuration to suppress specific notifications as defined in the *mefSoamAlarmEnable* MIB object in the MEF-SOAM-FM-MIB redundant.

The following notification is defined within the SOAM CFM YANG.

- fault-alarm: This notification signals an alarm condition associated with a specific MEP. This notification carries the following data definitions:
    - The ID of the affected MEP;
    - The current active defects.

15.4.2 SOAM FM YANG Module

The SOAM FM YANG module defines the managed objects necessary to support SOAM FM functionalities defined in the SOAM-FM Implementation Agreement [36].

Type definitions, groupings, data definitions, RPCs, and notifications in the SOAM FM YANG module are mandatory while data definitions and RPCs related to the augmented continuity check (CC) function, data definitions and RPCs related to alarm indication signal (AIS), data definitions and RPCs related to locking signal (ETH-LCK), and data definitions and RPCs related to test signal (ETH-Test) are optional to support.

The following type definitions are defined within the SOAM FM YANG module:

- interval-type-ais-lck-type: This type defines the AIS/LCK interval (transmission period for a frame).
- operational-state-type: This type defines the operational state (current capability) of the MEP.
- test-pattern-type: This type indicates the type of test pattern to be sent in an OAM PDU data TLV.

The following groupings are defined for the SOAM FM YANG module:

- locked-signal-parameters-group: Data definitions related to the administrative locking state of the MEP.
- test-signal-parameters-group: Data definitions related to generating and receiving test signals (ETH-Test) frames.
- test-signal-stats-in-group: Data definitions for statistics gathered while receiving test signals (ETH-Test) frames.
- test-signal-stats-out-group: Data definitions for statistics gathered while sending of test signals (ETH-Test) frames.

15.4.2.1 Data Definitions

Figure 15.46 describes the extensions to the CFM module provided by SOAM FM. The content of the FM module is depicted as boxes with dotted borders, extending certain existing CFM data definitions, and adding new data definitions to the tree.

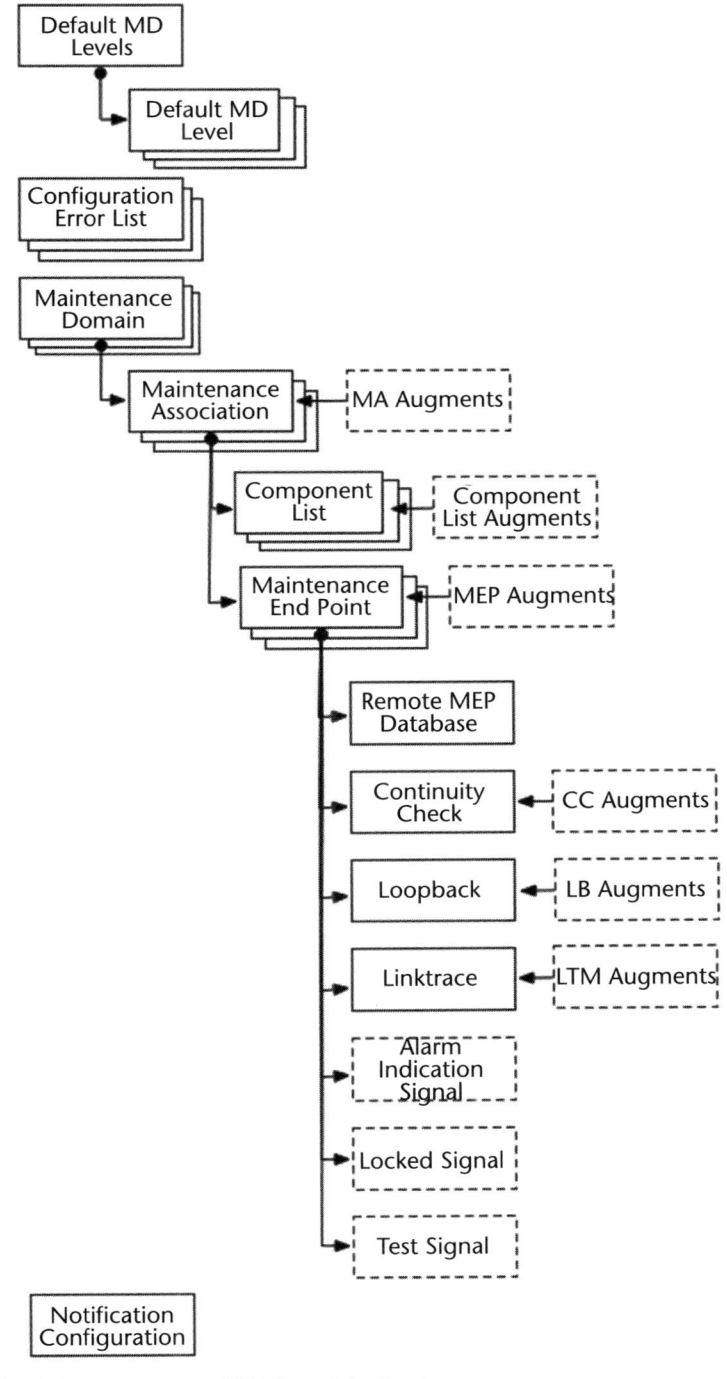

Figure 15.46 Fault management YANG module structure.

The list entries in the SOAM FM module tree are:

- Augment of CFM maintenance association (MA/MEG): A set of data definitions augmenting the CFM MA/MEG with parameters related to SOAM FM.
- Augment of CFM component list: A set of data definitions augmenting the CFM per-MA/MEG component list with parameters related to SOAM FM.
- Augment of CFM maintenance association endpoint (MEP): A set of data definitions augmenting the CFM MEP with parameters related to SOAM FM.
- Augment of CFM continuity check (CC): A set of data definitions augmenting the CFM continuity check with parameters related to SOAM FM.
- Augment of CFM link trace (LTM): A set of data definitions augmenting the CFM link trace with parameters related to SOAM FM.
- Container alarm indication signal (AIS): A set of data definitions augmenting the CFM MEP with parameters related to AIS.
- Locked signal (ETH-LCK): A set of data definitions extending the CFM MEP with parameters related to ETH-LCK. This container uses the locked-signal-parameters grouping.
- Test signal (ETH-Test): A set of data definitions extending the CFM MEP with parameters related to ETH-Test. This container uses the test-signal-parameters grouping.
- Notification configuration (container notification-configuration): A set of data definitions related to FM event notifications.

15.4.2.2 RPCs

Custom NETCONF RPCs are defined using the rpc statement. An RPC has a name and defines an optional set of input data using the input statement, and a set of optional return values using the output statement.

The following RPCs are defined within the SOAM FM YANG module and are used by clients to perform specific tasks.

- create-test-signal: This RPC creates a new test signal session associated with a specific MEP. It can be scheduled to execute immediately or at a later time.
- abort-test-signal: This RPC removes a scheduled test signal session that is scheduled to start at a later point in time or aborts an existing test signal session associated with a specific MEP.

15.4.2.3 Notifications

Custom NETCONF notifications are defined using the notification statement [8]. A notification can contain arbitrary data defined in the YANG module.

The NETCONF notifications feature provides filtering capabilities allowing subscribers to filter out events based on event content. This feature makes the configuration to suppress specific notifications as defined in, for example, the *mefSoamAlarmEnable* object in the MEF-SOAM-FM-MIB redundant.

The notifications of the SOAM FM YANG module are:

- mep-defect-alarm: This notification is generated when the value of mep-defects change. It indicates a persistent defect in the associated MEP. This notification carries the following data definitions:
    - The ID of the affected MEP;
    - The most recently sent defect;
    - The value of the remote MEP state at the time of notification.
- configuration-error-assert: This notification is generated when an entry is added to the cfm:configuration-error-list list in the CFM module. It indicates a configuration error during the setup of a SOAM FM entity.
- configuration-error-clear: This notification is generated when an entry is removed from the cfm:configuration-error-list list. It indicates that a configuration error has been removed during the setup of a SOAM FM entity. Its data definition is: The removed entry in the configuration-error leafref.
- mep-operational-state-changed: This notification is sent when the value of an operational-state associated with the MEP changes. Its data definitions are:
    - The ID of the affected MEP;
    - The operational state at the time of notification.
- lock-alarm: This notification is sent when either LCK transmit-status or receive-status changes. Its data definitions are:
    - The ID of the affected MEP;
    - The operational state at the time of notification.
- ais-alarm: This notification is sent when the state of either AIS transmit-status or receive-status changes. Its data definitions are:
    - A set of IDs identifying the affected MEP
    - The receive status at the time of notification
    - The transmit status at the time of notification

15.4.3 PM YANG Module [39]

The YANG PM module implements the objects and functions found in MEF SOAM PM IA [38] and various objects and tables from MEF-SOAM-PM-MIB [24] and MEF-SOAM-TC-MIB [39].

Type definitions, groupings, data definitions, RPCs, and notifications in the SOAM PM YANG module are mandatory, while data definitions and RPCs related to a single-ended responder, data definitions and RPCs related to loss measurement test TLV, availability, alignment and number consecutive high FLR, data definitions and RPCs related to availability and availability history statistics and data definitions and RPCs related to loss measurement current and history statistics are optional to support.

Data definitions and RPCs related to delay measurement test TLV and source MAC address align measurement offset and interframe delay variation offset; data definitions and RPCs related to delay measurement current and history statistics and thresholds for forward delay, two-way delay, and frame delay range are also optional to support.

15.4.3.1 Type Definitions

Type definitions define derived types from base types using the typedef statement [8]. A base type can be either a built-in type or a derived type, allowing a hierarchy of derived types.

Type definitions defined within the SOAM PM YANG module and used locally in the SOAM PM YANG module are:

- suspect-status-type: This Boolean data type indicates whether the measurement interval has been marked as suspect.
- performance-monitoring-interval-type: This integer data type indicates the transmission time between the SOAM PM frames for session.
- test-pattern-type: This enumeration data type indicates the type of test pattern to be sent in an OAM PDU data TLV.
- session-status-type: Indicates that the measurement instance is active.
- availability-type: Indicates that the availability is not known, for instance, because insufficient time has passed to make an availability calculation, the time has been excluded because of a maintenance interval, or because availability measurement is not enabled.
- measurement-bin-type: This enumeration data type indicates whether the bin number is for frame delay, interframe delay variation, or frame delay range.

15.4.3.2 Groupings

Groupings are reusable collections of nodes using the grouping statement. Groupings of definitions are instantiated using the uses statement.

The *grouping* and *uses* statements are defined in RFC 6020 [8] sections 7.11, and 7.12, respectively. The following groupings are defined for the SOAM PM YANG module and used throughout the PM module.

- remote-mep-group: This grouping includes objects that identify a remote MEP.
- measurement-timing-group: This grouping includes objects used for proactive and on-demand scheduling of PM measurement sessions.
- loss-measurement-configuration-group: This grouping includes configuration objects for the frame loss measurement functions.

- loss-availability-stats-group: This grouping includes availability statistics objects for a SOAM loss measurement session.
- loss-measurement-stats-group: This grouping includes statistics objects for a SOAM loss measurement session.
- delay-measurement-configuration-group: This grouping includes configuration objects for the delay measurement functions.
- delay-measurement-stats-group: This grouping includes statistics objects for a SOAM delay measurement session.
- delay-measurement-bins-content-group: This grouping contains result measurement bin objects for a SOAM delay measurement session.
- delay-measurement-bins-group: This grouping contains the specific bin objects for a SOAM delay measurement session.

15.4.3.3 Data Definitions

Figure 15.47 describes the extensions to the CFM module provided by SOAM PM. The content of the PM module is depicted as boxes with dotted borders, extending certain existing CFM data definitions and adding new data definitions to the tree.

The list entries describing the informal name of all grouped definitions and their relative location of the data definition in the SOAM FM module tree are:

- Augment of CFM MEP: A set of data definitions augmenting the CFM MEP with parameters related to SOAM PM, specifically with regards to the role of the MEP in PM tests.
- Augment of CFM MEP: A set of data definitions augmenting the CFM MEP with parameters related to SOAM PM, specifically with regards to loss measurement.
- Augment of CFM MEP: A set of data definitions augmenting the CFM MEP with parameters related to SOAM PM, specifically with regards to delay measurements.

15.4.3.4 RPCs

RPCs defined within the SOAM PM YANG module and used by clients to perform specific tasks are:

- create-loss-measurement: This RPC creates a new loss measurement session associated with a specific MEP. The output parameter provides the ID of the newly created loss measurement session.
- abort-loss-measurement: This RPC aborts a current loss measurement session with a specific session ID on a specific MEP.
- clear-loss-history-stats: This RPC clears the loss measurement history data associated with a specific session ID on a specific MEP.

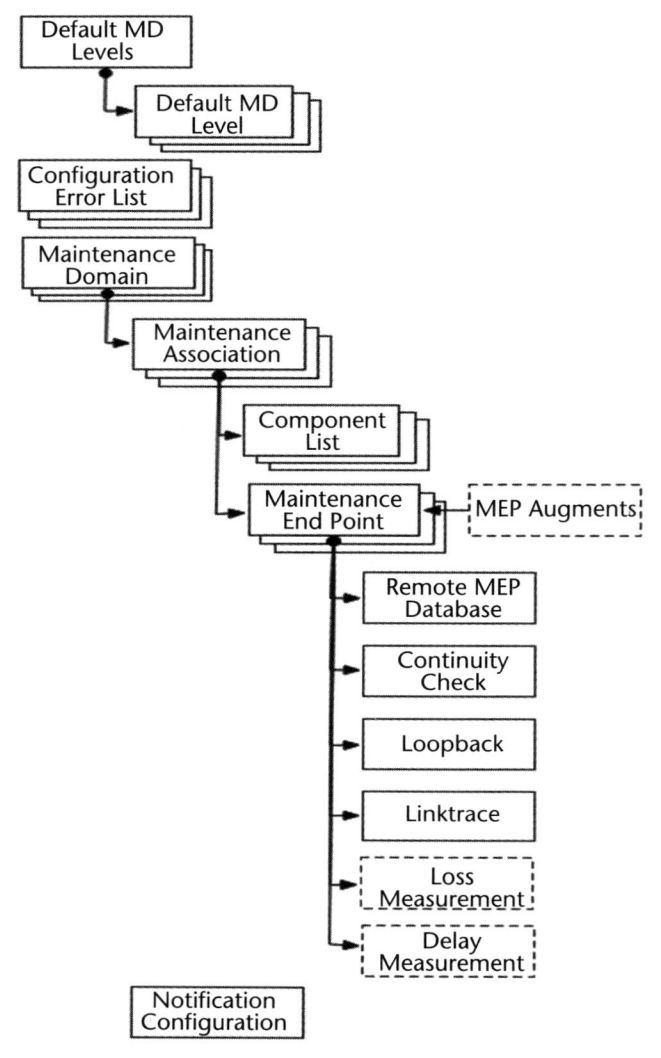

Figure 15.47 Extensions to the CFM module.

- create-delay-measurement: This RPC creates a new delay measurement session associated with a specific MEP. The output parameter provides the ID of the newly created delay measurement session.
- abort-delay-measurement: This RPC aborts a current delay measurement session with a specific session ID on a specific MEP.
- clear-delay-history-stats: This RPC clears the delay measurement history data associated with a specific session ID on a specific MEP.

15.4.3.5 Notifications

The following notifications are defined within the SOAM PM YANG module:

- availability-change-alarm: This notification is generated when the state of measured-availability-forward-status or measured-availability-backward-status associated with a specific loss-measurement session changes on a MEP. This notification carries the following data definitions:
    - The ID of the affected MEP;
    - The ID of the loss measurement session;
    - Availability forward and backward;
    - Last transition time for forward and backward availability;
    - Forward and backward unavailable and available status.
- loss-session-start-stop-alarm: This notification is generated when the state of session-status changes for a specific loss-measurement session. This notification carries the following data definitions:
    - The ID of the affected MEP;
    - The ID of the loss measurement session;
    - Session status.
- delay-session-start-stop-alarm: This notification is generated when the state of session-status changes for a specific delay-measurement session. This notification carries the following data definitions:
    - The ID of the affected MEP;
    - The ID of the delay measurement session;
    - Session status.
- threshold-crossing-alarm: This notification is generated if any of the criteria for above-alarm, set-alarm, or clear-alarm are met. This notification carries the following data definitions:
    - The ID of the affected MEP;
    - The ID of the loss or delay measurement session;
    - The notification crossing type;
    - The threshold ID and associated configured and measure value;
    - The current value of the suspect flag.

15.5 Network Resource Model for Carrier Ethernet Services

MEF initiated the Network Resource Provisioning (NRP) Project to define an API to support the design and development of network resource configuration and activation defined at the LSO PRESTO interface which is the between the service orchestrator and domain controller [8]. The API leverages with the ONF Core Information Model (CIM) [30] and is intended to support current physical network function (PNF) implementations and emerging virtual network function (VNF) implementations. The MEF NRP API for the LSO PRESTO interface can operate between a service orchestrator and one or more domain controllers. The APIs being worked by MEF are indicated in Figure 15.48.

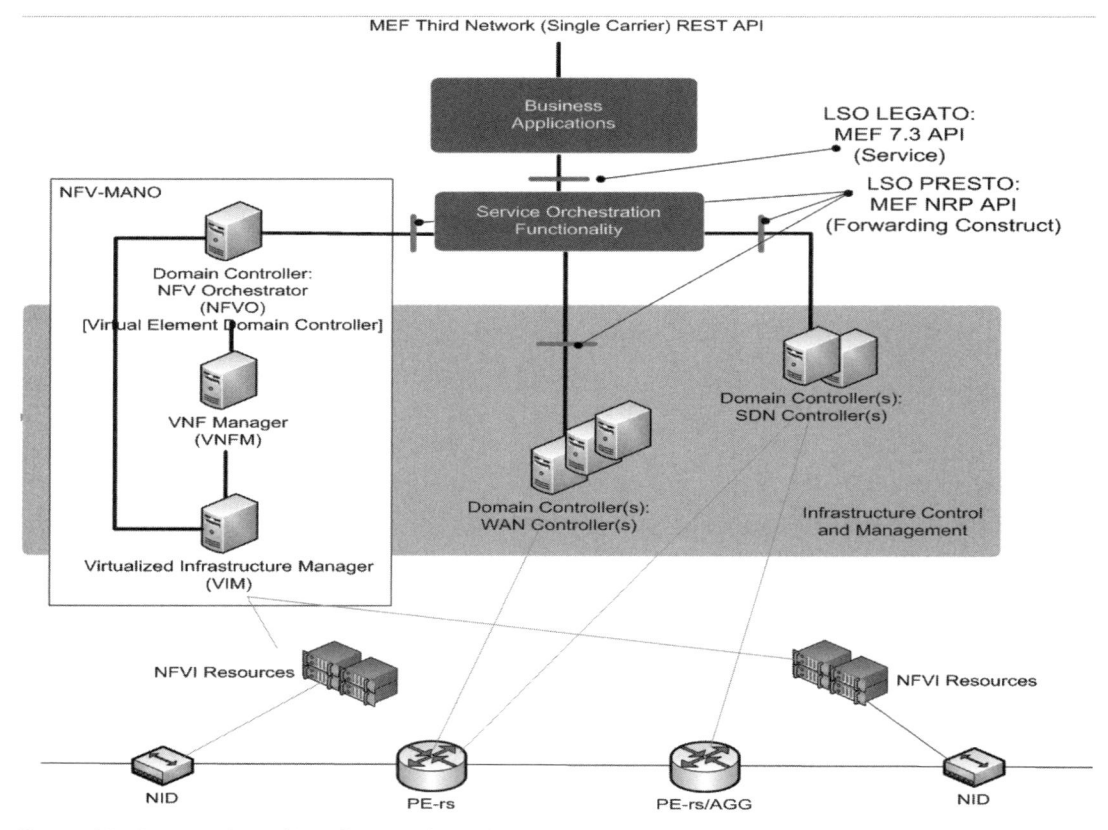

Figure 15.48 Mapping of interfaces and APIs being worked by MEF [6].

ONF defined the core network module [30] encompassing topology, termination, and forwarding. Some of the object classes are:

- LogicalTerminationPoint (LTP) object class encapsulates the termination, adaptation and OAM functions of one or more transport layers. The structure of LTP supports all transport protocols including circuit and packet forms. Each transport layer is represented by a LayerProtocol (LP) instance. The LayerProtocol instances of the LTP can be used for controlling termination and OAM functionality of that layer. It can also be used for controlling the adaptation (i.e., encapsulation and/or multiplexing of client signal).
- ForwardingDomain (FD) object class models the topological component which represents the opportunity to enable forwarding between points represented by the LTP in the model. The FD object can hold zero or more instances of ForwardingConstruct (FC) of one or more layer networks (e.g., OCh, ODU, ETH, and MPLS).
- ForwardingConstruct (FC) object class is used to effect forwarding of transport characteristic (layer protocol) information and offers the potential to enable forwarding models enabled potential for forwarding between two or more LTPs, and like the LTP, supports any transport protocol including all circuit and packet forms. The association of the FC to LTPs is made via

endpoints (essentially the ports of the FC) where each endpoint (EP) of the FC has a role in the context of the FC. The traffic forwarding between the associated EPs of the FC depends upon the type of FC and may be associated with FcSwitch object instances.

- FcRoute object class models the individual routes of an FC.
- EndPoint (EP) object class models the access to the FC function. Each EP instance has a role (e.g., working, protection, protected, hub, spoke, leaf, root) with respect to the FC function.
- FcSwitch object class models the switched forwarding of traffic (traffic flow) between EPs and is present where there is protection functionality in the FC.

Link object class models effective adjacency between two or more Forwarding-Domains (FD). In its basic form (i.e., point-to-point link) it associates a set of LTP clients on one FD with an equivalent set of LTP clients on another FD. Like the FC, the link has endpoints (LinkEnd) that take roles relevant to the constraints on flows offered by the link (e.g., root role or leaf role for a link that has a constrained tree configuration). A link may offer parameters such as capacity and delay.

MEF is in the process of building APIs using object classes defined by ONF. Figures 15.49, 15.50, and 15.51 depict examples of using ONF objects in representing Carrier Ethernet services.

15.6 Conclusion

In this chapter, we have summarized the approaches for information modeling of Carrier Ethernet services. Although MEF SNMP MIBs for NEs supporting Carrier

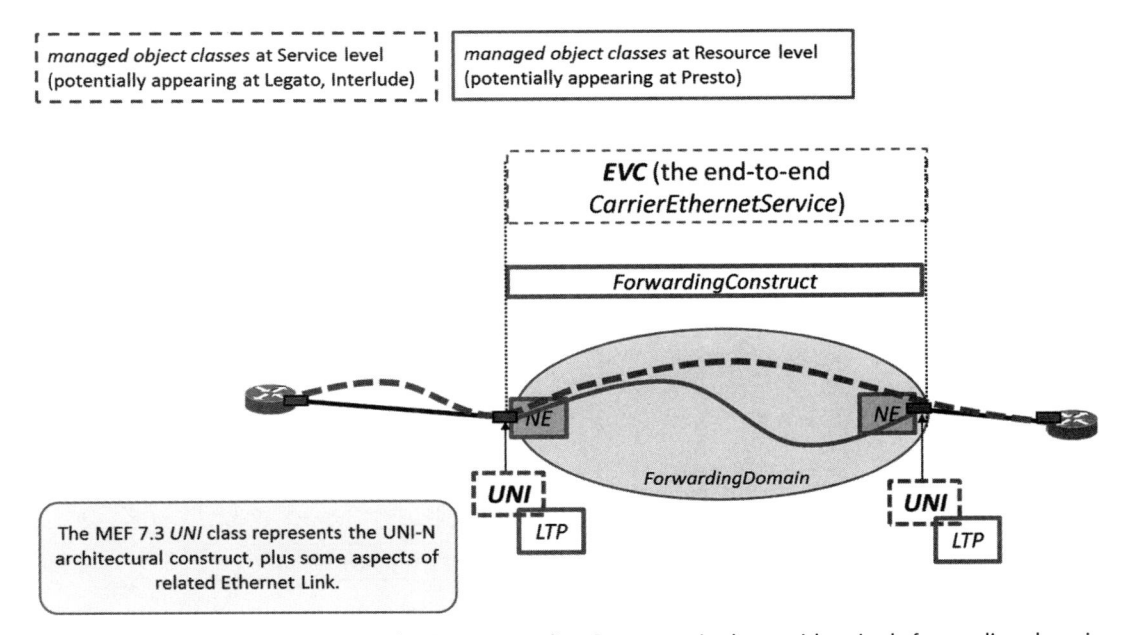

Figure 15.49 Mapping from service/EVC to ForwardingConstruct single-provider, single forwarding domain [40].

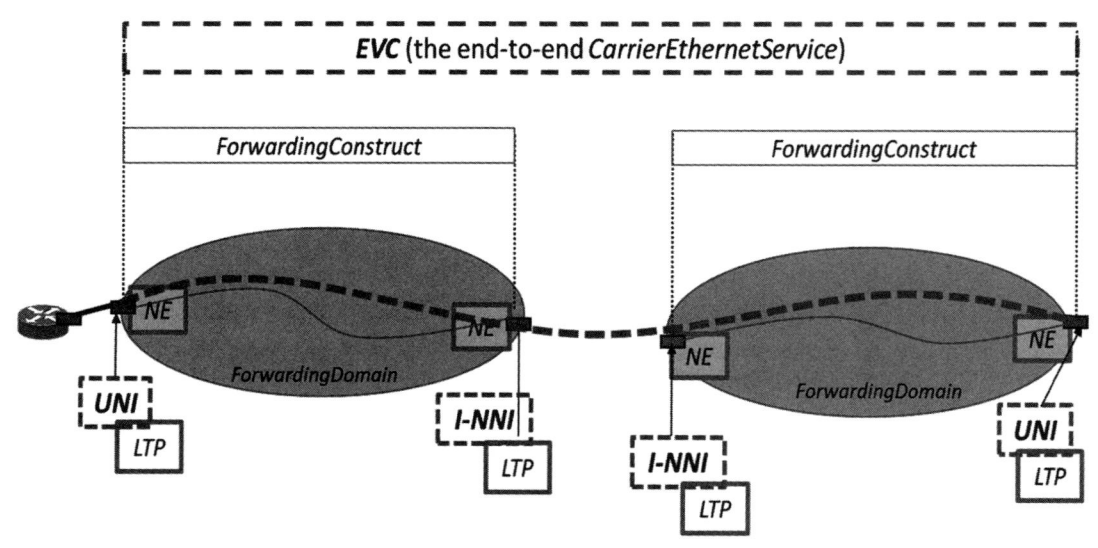

Figure 15.50 Mapping from service/EVC to ForwardingConstruct single provider, two FDs separately managed [25].

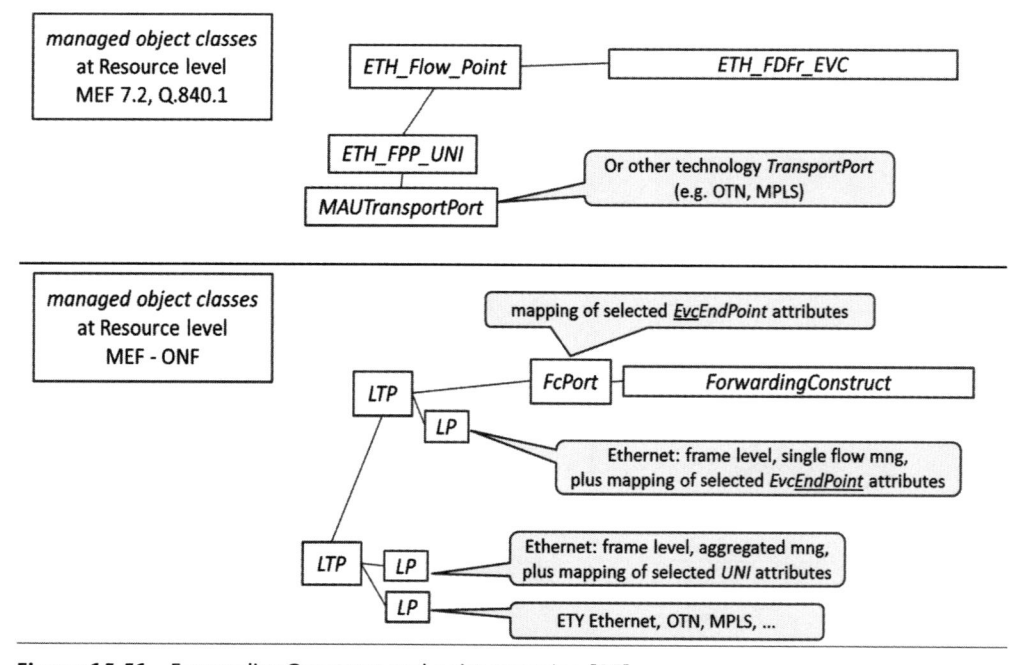

Figure 15.51 ForwardingConstruct and point mapping [25].

Ethernet services have been available for some time, their usage in the industry is not common. Due to the proliferation of SDN, the APIs developed by MEF in managing Carrier Ethernet services are likely to be used.

References

[1] ITU M.3100, "Generic Network Information Model," November 2005.
[2] ITU-T G.8010, "Architecture of Ethernet Layer Networks," February 2004.
[3] MEF 7.2, "Carrier Ethernet Management Information Model," April 2013.
[4] IEEE 802.3-2012, "Standard for Ethernet," December 2012.
[5] Draft MEF 7.3, "Carrier Ethernet Services Management Information Model," April 2016.
[6] MEF x.y, "Interface Profile Specification Network Resource Provisioning," August 1, 2016.
[7] https://wiki.mef.net/display/MTA/Network+Resource+Model.
[8] RFC 6020, "YANG: A Data Modeling Language for the Network Configuration Protocol (NETCONF)," October 2010.
[9] ITU-T Q.827.1, "Requirements and Analysis for the Common Management Functions of NMS-EMS Interfaces Amendment 1: Addition of a Common Managed Entity EMS," March 2007.
[10] ITU-T Q.840.1, "Requirements and Analysis for NMS-EMS Management Interface of Ethernet over Transport and Metro Ethernet Network (EoT/MEN)," March 2007.
[11] X.721 (02/92), "Information Technology - Open Systems Interconnection - Structure of Management Information: Definition of Management Information," February 1992.
[12] Q.822, "Generic Transport Performance Management," March 2003.
[13] Q.827.1, "Requirements and Analysis for the Common Management Functions of NMS-EMS Interfaces," October 2004.
[14] MEF 42, "ENNI and OVC Definition of Managed Objects," October 2013.
[15] MEF 46, "Latching Loopback Protocol and Functionality," 2014.
[16] MEF 31, "Service OAM Fault Management Definition of Managed Objects," January 2011.
[17] MEF 31.0.1, "Amendment to Service OAM SNMP MIB for Fault Management," January 2012.
[18] MEF 26.1, "External Network-Network Interface (ENNI) – Phase 2," January 2012.
[19] MEF 40, "UNI and EVC Definition of Managed Objects," April 2013.
[20] MEF 38, "Service OAM Fault Management YANG Modules," April 2012.
[21] MEF 28, "External Network-Network Interface (ENNI) Support for UNI Tunnel Access and Virtual UNI," October 2010.
[22] IEEE 802.1ag, "Connectivity Fault Management," December 2007.
[23] IEEE 802.1ap, "Management Information Base (MIB) Definitions for VLAN Bridges," March 2009.
[24] MEF 36.1, "Service OAM SNMP MIB for Performance Monitoring," April 2015.
[25] Mazzini, A., "Excerpt_Resource_Model_Mapping_Items_V1.2," MEF Contribution, December 2016.
[26] MEF 36, "Service OAM SNMP MIB for Performance Monitoring," January 2012.
[27] MEF 10.3, "Ethernet Services Attributes Phase 3," October 2013.
[28] MEF 55, "Lifecycle Service Orchestration (LSO): Reference Architecture and Framework," March 2016.
[29] RFC 6242, "Using the NETCONF Protocol over Secure Shell (SSH)," June 2011.
[30] RFC 6087, "Guidelines for Authors and Reviewers of YANG Data Model Documents," January 2011.

[31] IEEE 802.1Q-2014, "Bridges and Bridged Networks," 2014.
[32] IEEE 8021-CFM-MIB Definitions, October 15, 2008.
[33] IEEE 8021-CFM-V2-MIB Definitions, October 15, 2008.
[34] IEEE 8021-TC-MIB Definitions, October 15, 2008.
[35] IEEE 802.1AB, "Station and Media Access Control Connectivity Discovery," April 2005.
[36] MEF 30.1, "Service OAM Fault Management Implementation Agreement: Phase 2," April, 2013.
[37] MEF 39, "Service OAM Performance Monitoring YANG Module," April 2012.
[38] MEF 35.1, "Service OAM Performance Monitoring Implementation Agreement," May 2015.
[39] MEF-SOAM-TC-MIB Definitions, January 2012.
[40] ONF TR-512, "Core Information Model (CoreModel) Version 1.0," March 30, 2015.
[41] RFC 6241, "Network Configuration Protocol (NETCONF)," June 2011.

CHAPTER 16

Third Network

16.1 Introduction

The Third Network is initially defined as an agile, assured, and orchestrated network supporting CE 2.0 and employing SDN and NFV technologies [1]. The "agile "meant the delivery of new, dynamic, and on-demand services. The "assured" meant the delivery of performance and security guarantees. The "orchestrated" meant the delivery of automated service across service providers. The service was the connectivity services, which were initially Layer-2 Carrier Ethernet services, and later both Layer-2 Carrier Ethernet services and Layer-3 IP services, in addition to management services under the life-cycle service orchestration (LSO) umbrella. These connectivity services were considered to be used accessing cloud applications such as those provided by Amazon, Google, and so forth.

As Open Cloud Connect (OCC) merged MEF in 2016, the Third Network was expanded to include cloud services. Therefore, the Third Network became an umbrella concept that includes connectivity services, cloud applications, cloud services, and LSO for all.

The MEF is defining LSO capabilities and supporting APIs to streamline and automate the entire service life cycle in a sustainable fashion for coordinated management and control across all network domains responsible for delivering end-to-end services (i.e., CE 2.0, IP VPN, MPLS, SDN/NFV).

Figure 16.1 depicts a Third Network service supported by two operator networks employing SDN, NFV, and legacy technologies to an enterprise customer with multiple locations.

In the following section, we will describe LSO, virtualized Carrier Ethernet services and commercial cloud services.

16.2 Life-Cycle Service Orchestration

The life-cycle service orchestration (LSO) reference architecture is depicted in Figure 16.2 [2]. The descriptions of the components and interfaces are:

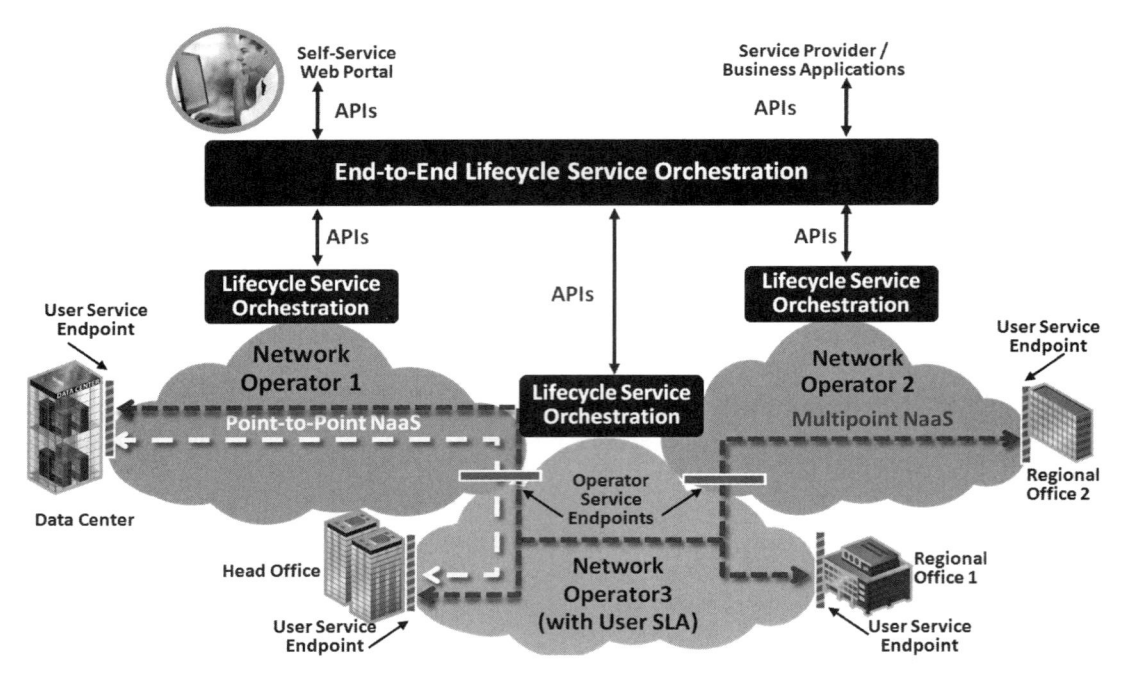

Figure 16.1 Third Network service for an enterprise customer.

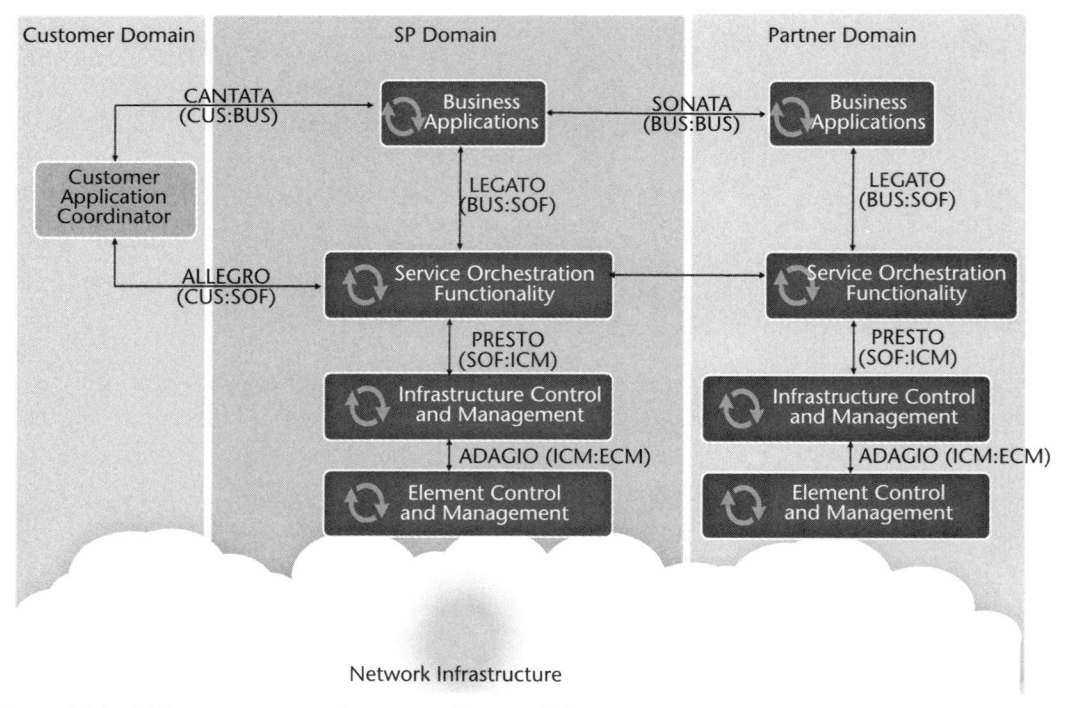

Figure 16.2 LSO management reference architecture [2]

- LSO components:
  - Service portal (SvP), which supports interactions with the customer or cloud application coordinator to request, modify, manage, control, and terminate their services.
  - Business applications (BUS), which are the provider functionality supporting the business management layer functionality such as product catalog, ordering, billing, and relationship management.
  - Service orchestration functionality, which is the set of service management layer functionality supporting an agile framework to streamlining and automating the service life cycle in a sustainable fashion for coordinated management supporting fulfillment, control, performance, assurance, usage, security, analytics, and policy capabilities encompassing all network domains that require coordinated end-to-end management and control to deliver connectivity services.
  - Infrastructure control and management (ICM), which is the set of functionality providing domain-specific resource management layer capabilities including configuration, control, and supervision of the network infrastructure.
  - Element control and management (ECM), which is the set of functionality supporting element management-layer management capabilities for groups of individual network elements.
  - Service gateway (SGW), which supports interactions with the partner service provider (e.g., access provider) to request, modify, manage, control, and terminate aspects of the connectivity services provided by the partner service provider.
- Interfaces:
  - Cantata, which is the management interface reference point that provides a customer (including enterprise customers) with capabilities via the service portal to support the operations interactions (e.g., ordering, billing, trouble management) for a portion of the service provider service capabilities related to the customer's services (e.g., customer service management interface).
  - Allegro, which is the management interface reference point that allows customer supervision, via the service portal, of the LSO service capabilities under its purview through interactions with the LSO orchestrator
  - Legato, which is the management interface reference point between the business applications and LSO needed to allow management and operations interactions supporting LSO services.
  - Sonata, which is the management interface reference point supporting the management and operations interactions (e.g., ordering, billing, trouble management) between two network providers (e.g., service provider and partner domain) (analogous to TMN X reference point).
  - Interlude, which is the management interface reference point that provides for the supervision of a portion of LSO services within the part-

ner domain that are coordinated by a service provider LSO within the bounds and policies defined for the service.

- Presto, which is the resource management interface reference point needed to manage the network infrastructure, including network view management functions.
- Animato, which is the management interface reference point between the application coordinator and the service portal. Animato allows cloud application coordination, via the service portal, of the LSO service capabilities under its purview.
- Adagio, which is the element management interface reference point needed to manage the network resources, including element view management functions.

16.3 LSO Management Abstractions and Constructs

LSO management abstractions and constructs define a common technology independent representation of connectivity, topology, and infrastructure, while providing the means to extend the model with technology-specific details.

There are three main abstracted management views in the LSO environment, as depicted in Figure 16.3:

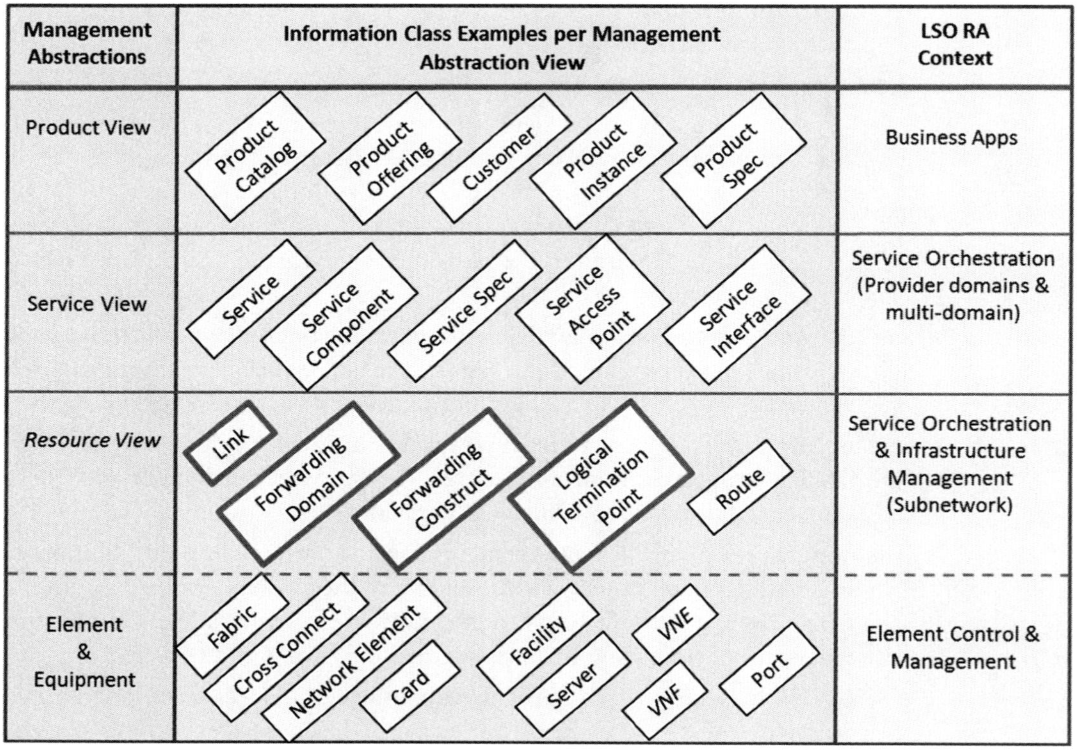

Management Abstractions	Information Class Examples per Management Abstraction View	LSO RA Context
Product View	Product Catalog, Product Offering, Customer, Product Instance, Product Spec	Business Apps
Service View	Service, Service Component, Service Spec, Service Access Point, Service Interface	Service Orchestration (Provider domains & multi-domain)
Resource View	Link, Forwarding Domain, Forwarding Construct, Logical Termination Point, Route	Service Orchestration & Infrastructure Management (Subnetwork)
Element & Equipment	Fabric, Cross Connect, Network Element, Card, Facility, Server, VNE, VNF, Port	Element Control & Management

Figure 16.3 Management view abstractions.

16.3 LSO Management Abstractions and Constructs

- Product view: The product view, which is commonly implemented in business support systems in the business management domain, appears at the interaction between the customer and the product offerings of a service provider. The product instance involves the purchasing or procurement of product offerings from a product catalog by a customer, and all other commercial aspects related to the customer's product instance, such as billing. Product specifications define the individual product characteristics that are used to create differentiated product offerings.

  In the product view, customers need to be able to express their needs in order to determine which product offerings can support their requirements. Similarly, service providers need to be able to match the customer requirements to technical specifications to realize the product offering. A product offering represents what is externally presented to the market that can be assembled from product specifications. A product catalog contains a list of product offerings for sale, with prices and illustrations. A product instance represents the subscription of a product offering by a customer, who normally is the purchaser of the product instance.

  Service providers create differentiated product offerings using their product specifications that define the characteristics of UNI/ENNI service interfaces, the EVC/OVC as connectivity services, and the associated service access points, or end points of the connection.

- Service view: A product instance is realized as one or more services and associated resources; therefore, services are tightly bound to product instances. A service is visible and directly usable by the customer, but may be divided within the service provider's infrastructure into one or more service components. Service components are not visible to the customer. Systems implementing service-related functionality have traditionally been operational support systems in the service management domain or service management systems.

  The service represents the intent of the service provider to deliver the features as specified in the customer's product instance. For example, the service may be a UNI-to-UNI EVC-based service offered by a service provider, or a UNI-to-ENNI, ENNI-to-ENNI OVC-based service offered by an operator.

  For example, if a service provider buys an OVC from an operator in order to provide an end-to-end EVPL service to the customer, the service provider and the operator have different views of the OVC. Within the service provider's management system, the OVC is viewed as a service component of the end-to-end EVPL service, whereas within the operator's management system, the OVC is viewed as the service.

- Resource view: Services are delivered via physical and logical resources in the network. Physical resources are actual hardware, and logical resources can be viewed as functionality provided by specific pieces of hardware. The resource view can be divided into the network and topology view and the element and equipment view. The network and topology view encompasses all the functions across network elements. The element and equipment view pertains to the management of a specific set of devices. Network and topology view functionality are usually implemented network management systems. The element and equipment view focuses on the physical and logical resources within a single network element or group of similar network elements that are implemented in element management systems.

The resource view comprises the network and topology view and element and equipment view abstractions:

- Network and topology view abstractions: The network control domain represents the scope of control that a particular network domain controller or WAN controller has with respect to a particular network. The topology of the network may be defined based on forwarding domains (FDs) and links, which represent adjacency between FDs. The FD is the topological component representing the opportunity to enable forwarding between points represented by logical termination points (LTPs) that encapsulate the termination, adaptation, and OAM functions of one or more transport layers.

  The FD contains instances of forwarding constructs (FCs) of one or more layer networks such as ETH and MPLS, defining the transport of the given service. The FD provides the context for instructing the formation, adjustment and removal of FCs and supports recursive aggregation such that the internal construction of an FD can be exposed as multiple lower-level FDs and associated links (partitioning). The FC effects protocol layer information between two or more LTPs where the association of the FC to LTPs is made via end points.

  The FC can represent many different structures including point-to-point (P2P), point-to-multipoint (P2MP), rooted-multipoint (RMP), and multipoint-to-multipoint (MP2MP) bridge and selector structure for linear, ring, or mesh protection schemes.

- Element and equipment view abstractions: The network element (NE) represents a network device in the data plane or a virtual network element visible in the interface. In the interface to an SDN controller, the NE defines the scope of control for the resources within the network element such as transferring user information between ports, encapsulation, multiplexing/demultiplexing, and OAM functions. Furthermore, the NE provides the scope of the naming space for identifying objects representing the resources within the NE.

  Where virtualization is employed, the NE represents a virtual network element (VNE). The mapping of the VNE to the NEs is performed by the network domain controller. NE instances can be created (or deleted) for providing (or removing) virtual views of the combination of slices of network elements in the data plane.

16.4 Virtualization

Virtualization is one of the most significant technologies that have greatly impacted the IT industry development in the past decade. With roots that can be traced back to the age of mainframes, virtualization recently regained its popularity in computing and has been widely employed especially in data centers, networks, and cloud infrastructures.

Network function virtualization (NFV) was initially proposed by ETSI in 2012 to address some of the fundamental challenges of networking by leveraging standard IT virtualization technologies. The basic idea of NFV lies in separating network functions from physical infrastructures by running virtual functions implemented by software upon standard server hardware. The ETSI NFV Industry Specialist Group (NFV-ISG) has published a set of documents that provide guidelines for realizing the NFV notion, including the requirements, architectural framework, infrastructure and software architecture, and various use cases for NFV [3].

Decoupling network functions and services from the underlying network infrastructures splits the traditional role of network service providers to multiple independent entities, including infrastructure providers, virtual network function providers, and virtual network providers and operators. Therefore, effective and flexible interactions among these entities play a crucial role in virtualization-based networks for service provisioning and service life cycle management.

Virtualization broadly describes the separation of a service or application from the underlying physical delivery of that service or application. Essentially, it provides a layer of abstraction between physical resources (including computing, storage, and networking hardware) and the applications running on them. Virtualization abstracts physical resources as virtual instances and enables independent multitenant access to a shared infrastructure substrate.

Some representative application scenarios of virtualization in computing include virtualization of servers, networking devices, and services. At the server level, virtualization abstracts platform hardware, such as processor, memory and hard drive, and I/O interfaces, into virtual resources for hosting various application instances. Virtualization may also be applied to networking devices, such as network interface cards (NICs), switches, and transmission links to form virtual networks (e.g., VLANs).

A key objective of server virtualization is to enable multiple independent virtual machines (VMs) to share the physical resources of a server. Figure 16.4 depicts general architecture of server virtualization, in which a VM manager, often referred to as a hypervisor, manages a group of VMs sharing server hardware. A VM is a software emulation of a physical machine. Each VM can run its own operating system, called the guest OS, and various applications upon the guest OS. Each VM is isolated from other VMs and behaves as if it is running on an individual physical machine. VMs are essentially isolated from one another in the same way that two physical machines would be on the same network. The guest OS on a VM has no knowledge of other VMs running on the same physical machine.

Hypervisor is the software providing an abstraction of server hardware resources and determines how such resources should be virtualized, allocated, and presented to VMs. There are two types of hypervisors. A type-I hypervisor runs directly on top of bare-metal hardware without using any operating system. A type-II hypervisor operates upon the operating system of the host server (host OS). Type-I hypervisors, with direct access to hardware resources, are typically more efficient than Type-II hypervisors and achieve greater scalability, robustness, and performance. However, Type II hypervisors, utilizing a standard host OS, may support a broader range of hardware configurations on hosting servers.

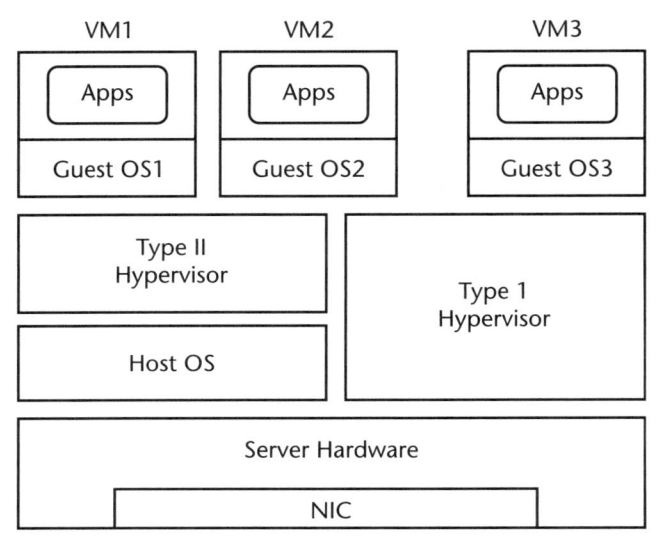

Figure 16.4 General architecture of server virtualization [3].

Container-based virtualization is another approach to server virtualization. In this approach, the operating system kernel runs on server hardware and allows multiple isolated user-space instances installed on top of it. Such isolated instances are called containers, which may look and feel like real servers from the point of view of their owners and users.

Unlike hypervisor-based server virtualization where each VM runs a complete guest OS, container-based virtualization allows all the virtual instances (containers) to share a common operating system. Therefore, container-based virtualization removes the overheads associated with VM guest OS and improves virtualization performance due to its lightweight implementation. However, the container-based virtualization approach requires each virtual instance on a host server to use the same operating system that the host is running, therefore limiting the flexibility that can be achieved by server virtualization.

In addition to standard computing resources such as CPU and storage, networking devices can be virtualized. The multiple VMs on the same server share the NIC of the server and expect their communication sessions to be isolated from each other. Consequently, the physical NIC of the server needs to be virtualized to a set of virtual NIC instances, one for each VM. NIC virtualization is typically a function provided by the hypervisor.

16.5 Network Virtualization

The main challenges to service provisioning in networks are the wide spectrum of services must be provisioned for meeting highly diverse application requirements and the various hetoergenous networking technologies that may be employed in different autonomous network domains for implementing the services.

The end-to-end design principle of IP networks require global agreement of coordination to deploy any fundamental change in network architecture. The significant capital investment in the Internet infrastructure and competing interests of its major stakeholders have created a barrier to introducing disruptive technologies in the Internet [4]. As a result, although the research community has attempted to take a clean-slate approach to developing new network architectures and protocols for facing the challenges of future networking, deployments of innovative technologies are basically limited to overlay-based approaches with little ability to introduce fundamental changes in network core architecture. As pointed by Anderson et al. in [5], the existing overlay approaches cannot provide an effective deployment path for disruptive networking technolgoies mainly due to two reasons: first, they are mostly used to deploy incremental solutions to specific problems without a holistic view of the interactions between coexisting overlays. Second, overlays are often designed and deployed in the application layer on top of IP, thus cannot support radically different network architecture.

A key to overcoming difficulties with the current network archtiecture and breaking the impasse to network innovations lies in decoupling the network functions for service provisioning and the network infrastructures for data transporation and processing [3]. Such decoupling allows alternative network architectures and new protocols to be developed and deployed for meeting various service requirements withtou being constrained by infrastructure implementations.

Virtualization technologies have already been employed in networking in various scenarios. For example, a virtual local area network (VLAN) forms a single broadcast domain by logically interconneting a group hosts and a virtual private network (VPN) enables private connections among multiple sites using secured tunnels over public networks. With the popularity of server virtualization, the demand of communications between VMs on the same physical server leads to virtual switches (e.g., Open vSwitch) that are typically implemented as part of a hypervisor. Virtualization has also been explored as a means to construct experimental platforms for conducting.

Therefore, the key objective of NV is to enable a networking environment that allows multiple independent virtual networks (VN) to share a common network infrastructue. Each VN may have its own network archtiecture, including packet format, addressing scheme, forwarding mechanism, and routing protocols, designed to provision various network services for meeting diverse application requirements. VNs are implemented on top of an infrastructure substrate comprising physical network resoruces.

16.6 Virtualized Carrier Ethernet Services

Carrier Ethernet services can be virtualized by dividing external interfaces (i.e., UNI and ENNI) and connection termination points (i.e., EVC at UNI or EVC termination point, OVC termination point) into virtualized and infrastructure components.

Figure 16.5 depicts virtualized and infrastructure components of an Ethernet private line (EPL) service provided by one operator. UNI and EVC termination have both infrastructure and virtualized components [6].

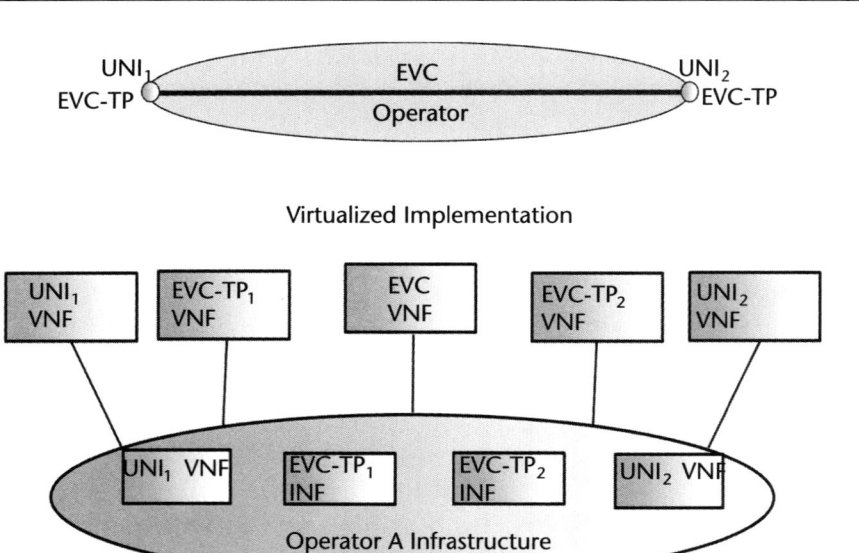

Figure 16.5 Virtualized components of an EVC between two UNIs.

Figure 16.6 depicts virtualized and infrastructure components of an EPL crossing multiple operators. UNI, ENNI, EVC termination point, and OVC termination point have both infrastructure and virtualized components.

16.6.1 Components of Virtualized Carrier Ethernet Services

We previously divided components of Carrier Ethernet services as virtualized and infrastructure components. How do we determine what function is virtual

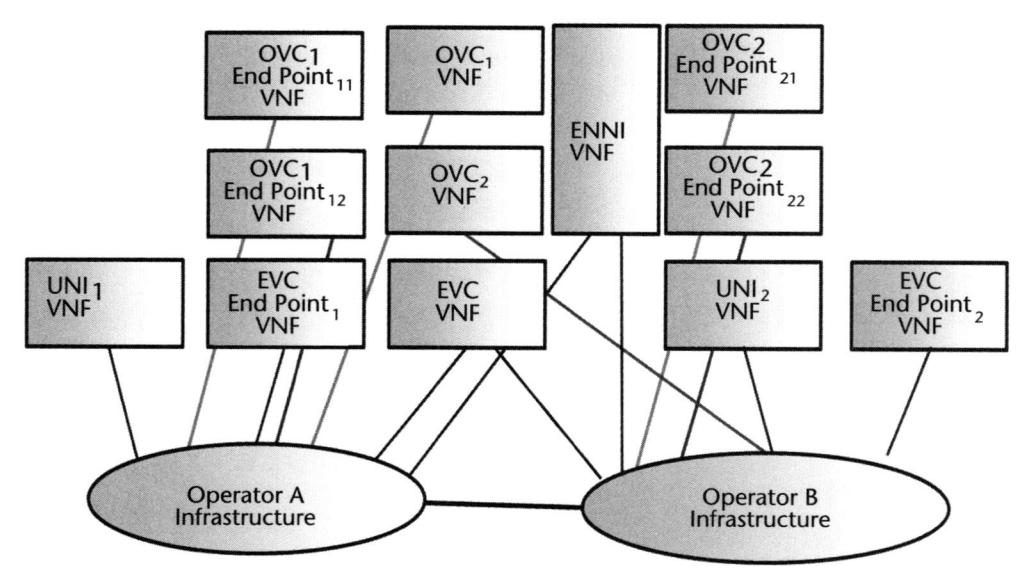

Figure 16.6 Virtualized components of an EPL crossing ENNI.

16.6 Virtualized Carrier Ethernet Services

and what function is infrastructure? Clearly this categorization depends on the implementation. In the following, we have categorized UNI attributes:

1. VNFUNI-Prov is a virtual function consisting of attributes that can be configured and supported by software.
2. INFUNI-Prov is an infrastructure function consisting of attributes that can be configured and supported by physical hardware.
3. VNFUNI-prot is a virtual function consisting of UNI protection attributes that can be configured and supported by software.
4. INFUNI-prot is an infrastructure function consisting of UNI protection attributes that can be configured and supported by hardware.
5. VNFUNI-loam is a virtual function consisting of UNI link OAM attributes that can be configured and supported by software.
6. INFUNI-loam is an infrastructure function consisting of UNI link OAM attributes that can be configured and supported by hardware.
7. VNFUNI-soam is a virtual function consisting of UNI service OAM attributes that can be configured and supported by software.
8. INFUNI-soam is an infrastructure function consisting of UNI service OAM attributes that can be configured and supported by hardware.
9. VNFUNI-sync is a virtual function consisting of UNI synchronization attributes that can be configured and supported by software.
10. INFUNI-sync is an infrastructure function consisting of UNI synchronization attributes that can be configured and supported by hardware.
11. VNFUNI-tsh is a virtual function consisting of UNI token sharing attributes that can be configured and supported by software.
12. INFUNI-tsh is an infrastructure function consisting of UNI token sharing attributes that can be configured and supported by hardware.
13. VNFUNI-env is a virtual function consisting of UNI envelope attributes for multiple bandwidth profile flows that can be configured and supported by software.
14. INFUNI-env is an infrastructure function consisting of UNI envelope attributes for multiple bandwidth profile flows that can be configured and supported by hardware.

From Table 16.1, we can identify basic and enhanced virtual capabilities, as well as infrastructure capabilities, of UNI. Possible implementation of UNI can be as shown in Table 16.2.

In Table 16.3, we have categorized the attributes of EVC termination point (i.e., EVC per UNI according to MEF specifications) as follows:

1. VNFEVC-Prov is a virtual function consisting of attributes that can be configured and supported by software.
2. INFEVC-Prov is an infrastructure function consisting of attributes that can be configured and supported by physical hardware.
3. VNFEVC-soam is a virtual function consisting of UNI service OAM attributes that can be configured and supported by software.

Table 16.1 UNI Service Attributes and Parameter Values for all Service Types

UNI Service Attribute [8]	Component of VNF or Infrastructure or Both	Categories of VNF and Infrastructure
UNI ID	VNF	$VNF_{UNI-Prov}$
Physical layer	Infrastructure	$INF_{UNI-Prov}$
Synchronous mode	Both	$INF_{UNI-sync} + VNF_{UNI-sync}$
Number of links	Both	$INF_{UNI-prot} + VNF_{UNI-prot}$
UNI resiliency	Both	$INF_{UNI-prot} + VNF_{UNI-prot}$
Service frame format	Infrastructure	$INF_{UNI-Prov}$
UNI maximum service frame size	Both	$INF_{UNI-Prov} + VNF_{UNI-Prov}$
Service multiplexing	Both	$INF_{UNI-Prov} + VNF_{UNI-Prov}$
CE-VLAN ID for untagged and priority tagged service frames	VNF	$VNF_{UNI-Prov}$
CE-VLAN ID/EVC Map	VNF	$VNF_{UNI-Prov}$
Maximum number of EVCs	Both	$INF_{UNI-Prov} + VNF_{UNI-Prov}$
Bundling	VNF	$VNF_{UNI-Prov}$
All to one bundling	VNF	$VNF_{UNI-Prov}$
Token share	Both	$INF_{UNI-tsh} + VNF_{UNI-tsh}$
Envelopes	VNF	$VNF_{UNI-env} + INF_{UNI-env}$
Ingress bandwidth profile per UNI	VNF	$VNF_{UNI-Prov} + INF_{UNI-Prov}$
Egress bandwidth profile per UNI	VNF	$VNF_{UNI-Prov} + INF_{UNI-Prov}$
Link OAM	Both	$VNF_{UNI-loam} + INF_{UNI-loam}$
UNI MEG	Both	$VNF_{UNI-soam} + INF_{UNI-soam}$
E-LMI	Both	$VNF_{UNI-elmi} + INF_{UNI-elmi}$
UNI L2CP address set	VNF	$VNF_{UNI-Prov}$
UNI L2CP peering	VNF	$VNF_{UNI-Prov}$
Test probes for ITU Y.1564 testing, RFC 6349 TCP testing	Both	$INF_{UNI-test} + VNF_{UNI-test}$

4. VNFUNI-eqc is a virtual function consisting of EVC equivalence attributes for multiple bandwidth profile flows that can be configured and supported by software.
5. INFUNI-env is an infrastructure function consisting of UNI envelope attributes for multiple bandwidth profile flows that can be configured and supported by hardware.

16.6 Virtualized Carrier Ethernet Services

Table 16.2 UNI Configurations

UNI Functionalities	VNF and INF Components Required (i.e., Service Functional Chaining Components)
Basic UNI provisioning	$VNF_{UNI\text{-}Prov} + INF_{UNI\text{-}Prov}$
Basic UNI provisioning + link OAM	$VNF_{UNI\text{-}Prov} + INF_{UNI\text{-}Prov} + VNF_{UNI\text{-}loam} + INF_{UNI\text{-}loam}$
Basic UNI Provisioning + Link Protection	$VNF_{UNI\text{-}Prov} + INF_{UNI\text{-}Prov} + VNF_{UNI\text{-}prot} + INF_{UNI\text{-}prot}$
Basic UNI provisioning + token sharing	$VNF_{UNI\text{-}Prov} + INF_{UNI\text{-}Prov} + VNF_{UNI\text{-}tsh} + INF_{UNI\text{-}tsh}$
Basic UNI provisioning + envelopes	$VNF_{UNI\text{-}Prov} + INF_{UNI\text{-}Prov} + VNF_{UNI\text{-}env} + INF_{UNI\text{-}env}$
Basic UNI provisioning + service OAM	$VNF_{UNI\text{-}Prov} + INF_{UNI\text{-}Prov} + VNF_{UNI\text{-}soam} + INF_{UNI\text{-}soam}$
Basic UNI provisioning + ELMI	$VNF_{UNI\text{-}Prov} + INF_{UNI\text{-}Prov} + VNF_{UNI\text{-}elmi} + INF_{UNI\text{-}elmi}$
Basic UNI provisioning + service OAM+ link OAM	$VNF_{UNI\text{-}Prov} + INF_{UNI\text{-}Prov} + VNF_{UNI\text{-}soam} + INF_{UNI\text{-}soam} + VNF_{UNI\text{-}loam} + INF_{UNI\text{-}Prov} + VNF_{UNI\text{-}loam} + INF_{UNI\text{-}loam}$

Table 16.3 VNFs and Infrastructure Components of EVC per UNI Service Attributes and Parameter Values for all Service Types

EVC per UNI Service Attribute [8]	Categories of VNF and Infrastructure
UNI EVC ID	$VNF_{EVC\text{-}Prov}$
Class of service identifier for data service frame	$VNF_{EVC\text{-}Prov}$
Class of service identifier for L2CP service frame	$VNF_{EVC\text{-}Prov}$
Class of service identifier for SOAM service frame	$VNF_{EVC\text{-}soam}$
Color identifier for service frame	$VNF_{EVC\text{-}Prov}$
Egress equivalence class identifier for data service frames	$VNF_{EVC\text{-}eqc}$
Egress equivalence class identifier for L2CP service frames	$VNF_{EVC\text{-}eqc}$
Egress equivalence class identifier for SOAM service frames	$VNF_{EVC\text{-}soam}$
Ingress bandwidth profile per EVC	$VNF_{EVC\text{-}Prov} + INF_{EVC\text{-}Prov}$
Egress bandwidth profile per EVC	$VNF_{EVC\text{-}Prov} + INF_{EVC\text{-}Prov}$
Ingress bandwidth profile per class of service identifier	$VNF_{EVC\text{-}Prov}$
Egress bandwidth profile per egress equivalence class	$VNF_{EVC\text{-}eqc}$
Source MAC address limit	$VNF_{EVC\text{-}Prov}$
Test MEG	$VNF_{EVC\text{-}soam} + INF_{EVC\text{-}soam}$
Subscriber MEG MIP	$VNF_{EVC\text{-}soam} + INF_{EVC\text{-}soam}$

In Table 16.4, we have mapped additional EVC attributes to VNF and INF categories of an EVC.

From these tables, we can identify virtual capabilities and infrastructure capabilities of EVC. Possible implementation of EVC can be as shown in Table 16.5.

16.6.2 Service Chaining for EPL

Let's assume that a user requests an EPL service between UNI1 and UNI2, from the E-line category in the product catalog as depicted in Figure 16.7. The main orchestrator needs to talk to the controller associated with UNIs and the NFV orchestrator associated with both UNIs and EVC. Let's assume that both UNIs belong to one vendor and are in the same subnetwork (domain); therefore, the same controller can configure both UNIs. Per request from the main orchestrator, the controller configures INFUNI-Prov for both UNIs. The VNFs, VNFUNI-Prov for both UNIs, can be configured independently from INFUNI-Prov . For a basic

Table 16.4 Categorization of EVC Attributes as VNF and Infrastructure Components

EVC Service Attribute [8]	Categories of VNF and Infrastructure
EVC type	$VNF_{EVC\text{-}Prov}$
EVC ID	$VNF_{EVC\text{-}Prov}$
UNI List	$VNF_{EVC\text{-}Prov}$
Maximum number of UNIs	$VNF_{EVC\text{-}Prov} + INF_{EVC\text{-}Prov}$
Unicast service frame delivery	$VNF_{EVC\text{-}Prov}$
Multicast service frame delivery	$VNF_{EVC\text{-}Prov}$
Broadcast service frame delivery	$VNF_{EVC\text{-}Prov}$
CE-VLAN ID preservation	$VNF_{EVC\text{-}Prov}$
CE-VLAN CoS preservation	$VNF_{EVC\text{-}Prov}$
EVC performance	$VNF_{EVC\text{-}Prov} + INF_{EVC\text{-}Prov}$
EVC maximum service frame size	$VNF_{EVC\text{-}Prov}$

Table 16.5 VNFs and Infrastructure Components of EVC

EVC Functionalities	VNF and INF Components Required (i.e., Service Function Chaining Components)
Basic EVC provisioning*	$VNF_{EVC\text{-}Prov} + INF_{EVC\text{-}Prov}$
Basic EVC provisioning + service OAM	$VNF_{EVC\text{-}Prov} + INF_{EVC\text{-}Prov} + VNF_{EVC\text{-}soam} + INF_{EVC\text{-}soam}$
Basic EVC provisioning + equivalence class support	$VNF_{EVC\text{-}Prov} + INF_{EVC\text{-}Prov} + VNF_{EVC\text{-}eqc}$

*EVC provisioning here describes only the provisioning of cross-connect between UNIs. In MEF, this term describes include provisioning of UNIs and EVC termination points (i.e., EVC per UNI).

16.7 Cloud Services Architectures

```
{User selects E-Line → EPL from}  →  Main Orchestrator  —INF_UNI-Prov→  Controller
 Product Catalog GUI                  Requests          INF_EVC-Prov    |INF_UNI-Prov|
                                                                         |INF_EVC-Prov|
                          VNF_UNI-Prov | VNF_EVC-Prov                          ↓
                                       ↓                                  CPE_A  CPE_B
                                      NFV
                                   Orchestrator
                          VNF_UNI-Prov | VNF_EVC-Prov
                                       ↓
                                   NFV Manager
                          VNF_UNI-Prov | VNF_EVC-Prov
                                       ↓
                              Virtual Infrastructure
                                    Manager
                          VNF_UNI-Prov | VNF_EVC-Prov
                                       ↓
                                  X.86 for CPE
                                      VNFs
```

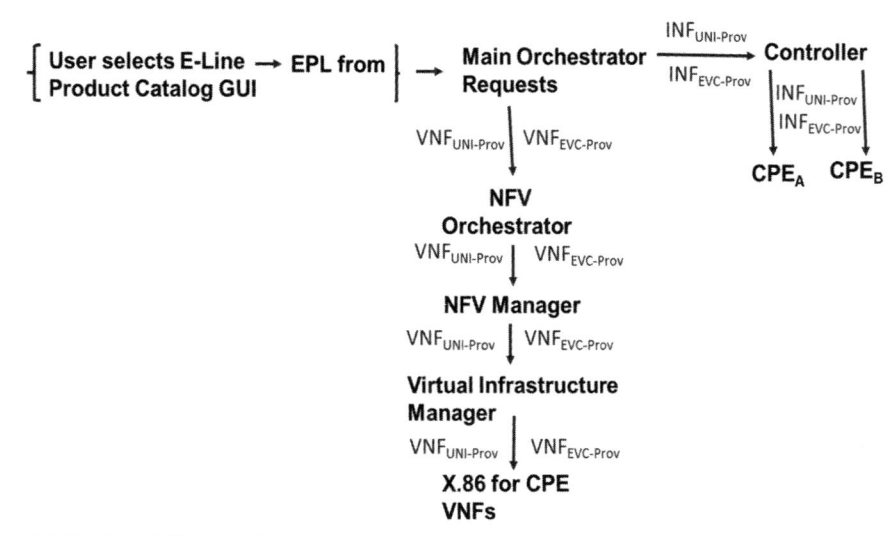

Figure 16.7 Basic EPL provisioning.

EPL provisioning, the process flows (i.e., service chains) are depicted in Figure 16.7 [3]. In this example, we considered each IUNI to be supported by a different CPE. Furthermore, we assumed that each CPE consists of virtualized and infrastructure components that may or may not be located at the same place.

The provisioning components in Figure 16.7 are quite different than objects defined in MEF [9]. We believe that what we define here constitutes a layer that is below a provisioning layer formed of service layer and resource layer objects defined by MEF.

16.7 Cloud Services Architectures

Cloud services architectures are defined by Open Cloud Connect [7]. The commercial cloud services include not only applications and their supporting infrastructure but also the networking path end-to-end between the user and application.

The key actors of cloud services are depicted in Figure 16.8 [5] where a cloud service provider (cSP) is responsible for providing an end-to-end cloud service to a cloud service user using cloud carrier(s) and cloud provider(s). The cSP may or may not own cloud carrier (cC) and cloud provider (cP) facilities, but provides a single bill to the cloud service user.

A user interfaces to a cSP facility via a standards interface called a cloud service user interface (cSUI), which is a demarcation point between the cSP and the cloud consumer (Figure 16.9). From this interface, the consumer establishes a connection, the cloud service connection (cSC), with a cP entity providing the application where the cP entity can be a VM with cloud service interface (cSI) or a physical resource such as storage with a cSUI. In addition, a cSC can be between two cP entities or between two cloud consumers.

The cSI can represent an interface of a vNIC, an interface of a virtual switch, an interface of a container, or an interface of a virtualized network function (VNF) (Figure 16.10), in addition to an interface of a VM.

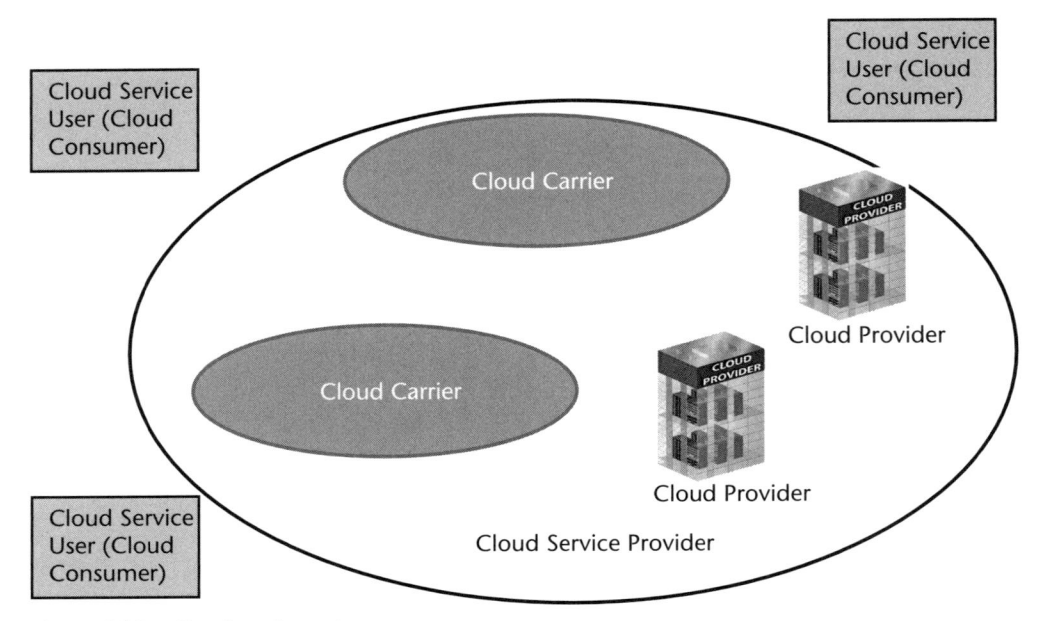

Figure 16.8 Cloud service actors.

When a cSC is between a cloud user and a cP physical or virtual resource, the cSC is established between two cloud service connection termination points (cSCTPs) residing at the user interface (i.e., cSUI) and the cP interface (i.e., cSUI or cSI).

The cSP may own the cP and cC facilities. When the cP and the cC are two independent entities belonging to two different operators, the standards interface between them is called cCcPI (cloud carrier cloud provider interface).

It is also possible for two or more cSPs to be involved in providing a cloud service to a cloud service user as depicted in Figure 16.11 where two cSPs interface to each other via a standards interface called cloud service provider-cloud service provider interface (cSPcSPI). In this scenario, only one of the cSPs needs to interface to the end user, coordinate resources, and provide a bill. The cSP that does not interface to the end user is called a cloud service operator (cSO).

The cSPs may employ a gateway to connect to each other, cloud service gateway (cSGW). The cSGW might provide connection multiplexing among other features that are required by cSPcSPI.

Further details of the architecture can be found in [3, 5].

16.7.1 Protocol Stacks and Interfaces

The previous section described interfaces between user and cSP, between cSPs, between cP and cC, between Network as a Service (NaaS) and cloud service application supporting entity. The protocol stack at each interface that can be supported is depicted in Figure 16.12. Each of the protocol layers may be further decomposed into their data, control and management plane components.

16.7 Cloud Services Architectures

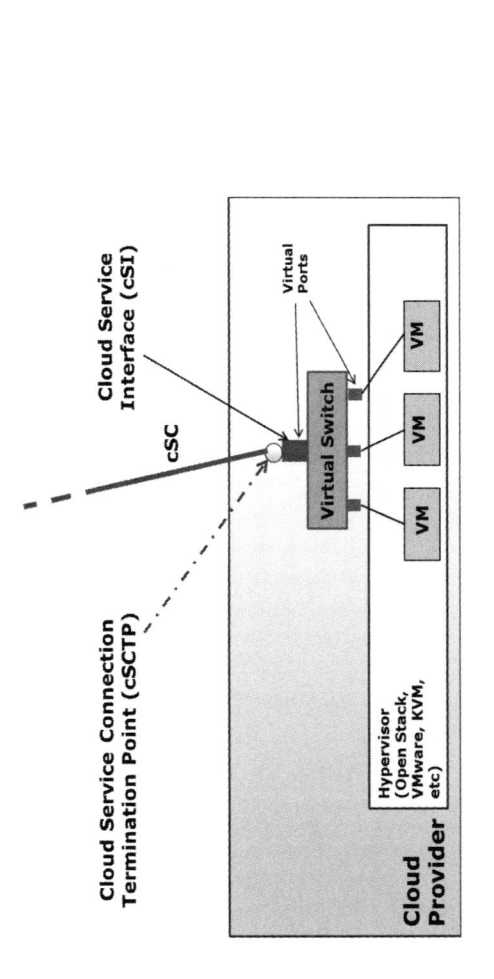

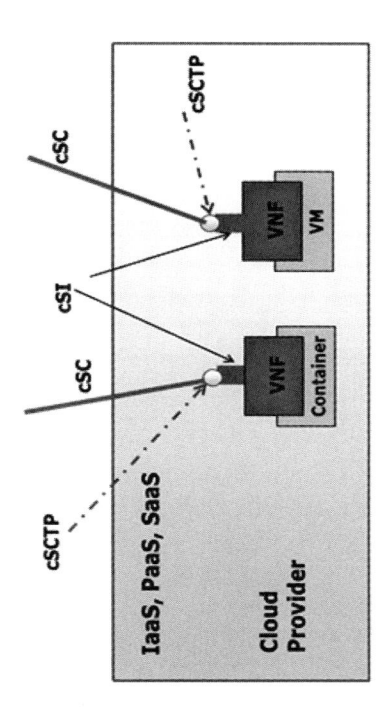

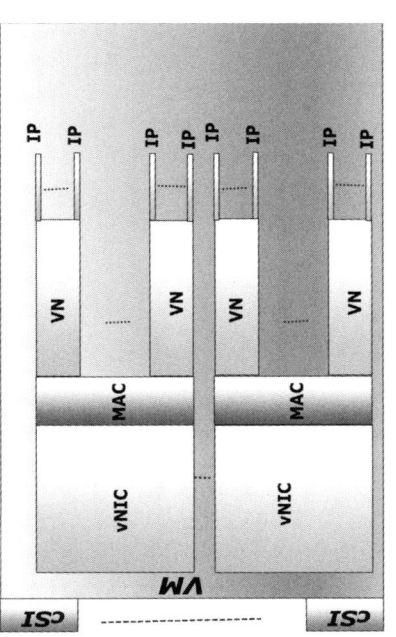

Figure 16.9 Virtual resources (i.e., VMs) and physical resources (i.e., computing and storage resources) that belong to one operator, providing cloud applications.

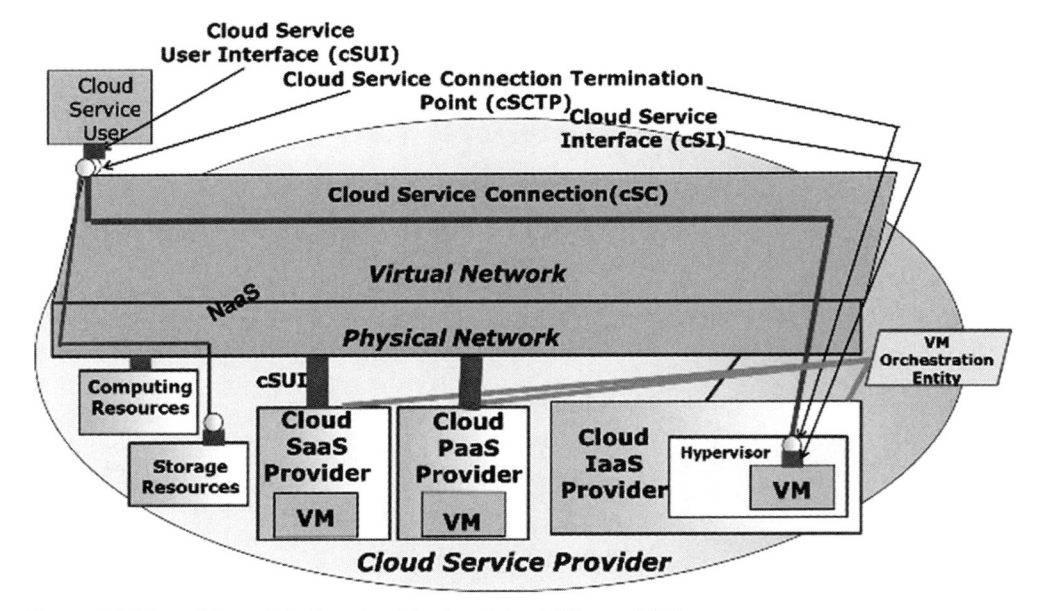

Figure 16.10 cSI is an interface to virtual switch, vNIC, or a VNF.

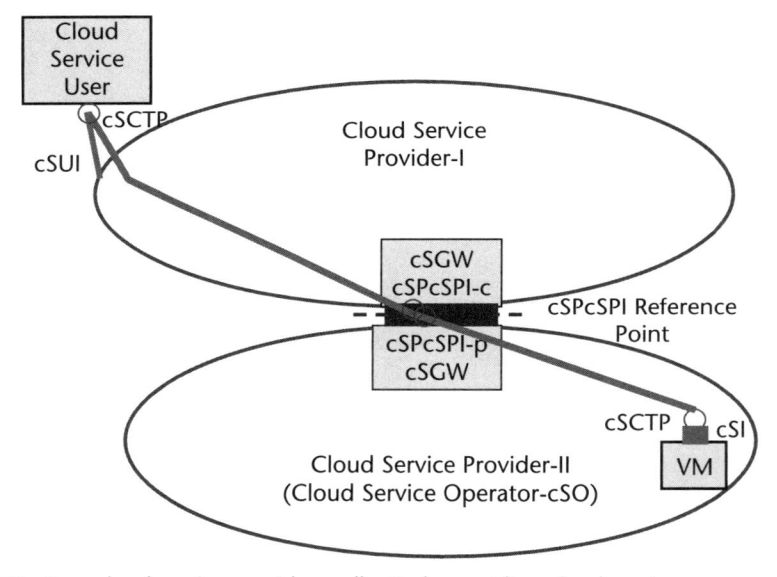

Figure 16.11 Two cloud service providers collectively providing cloud services.

cSUI may support a protocol stack from Layer 1 up to Layer 7. cSPcSPI and cCcPI may support a protocol stack from Layer 1 up to Layer 3. Given that TCP is used by routing protocols such as BGP, cCcPI and cSPcSPI might go up to L4. However, cSI may support a protocol stack from L2 up to Layer 7 without Layer 1.

The protocol stack for each interface is designed to support various network interfaces and applications.

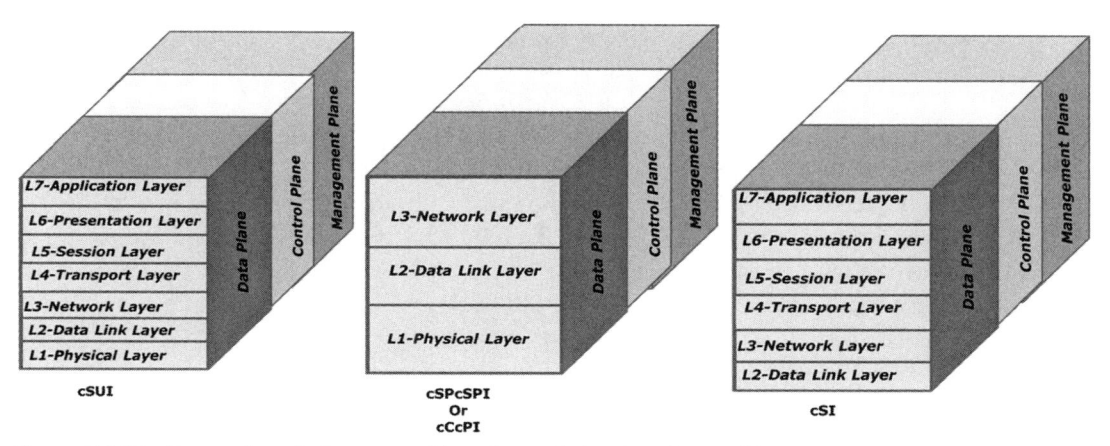

Figure 16.12 Protocol stacks for external interfaces of cloud services architecture.

16.7.2 Cloud Services

So far, we have described entities and interfaces between them and connection and its termination points for the transportation of cloud services. This section describes cloud services and their possible attributes.

A cloud service may include just cloud resources or cloud and noncloud resources. For example, a cloud service can include entities such as applications based on cloud resources, but the access network to the cloud applications may be based on noncloud resources. However, they form a cloud service end-to-end.

Software as a Service (SaaS), Platform as a Service (PaaS), and Infrastructure as a Service (IaaS) are among the well-known cloud services in the industry. They are the cloud applications provided by well-known cloud providers such as Amazon.

SaaS is an application running on a cloud infrastructure where the consumer does not manage or control the underlying cloud infrastructure including network, servers, operating systems, storage, or even individual application capabilities, with the possible exception of limited user-specific application configuration settings. SaaS examples include Gmail from Google, Microsoft "live" offerings, and salesforce.com.

PaaS is deploying onto the cloud infrastructure consumer-created or acquired applications created using programming languages and tools supported by the provider. The consumer does not manage or control the underlying cloud infrastructure including network, servers, operating systems, or storage, but has control over the deployed applications and possibly application hosting environment configurations.

PaaS provides the capability to build or deploy applications on top of IaaS. Typically, a cloud computing provider offers multiple application components that align with specific development models and programming tools. For the most part, PaaS offerings are built upon either a Microsoft-based stack such as Windows or an open source-based stack such as Apache.

IaaS is to provision processing, storage, networks, and other fundamental computing resources where the consumer is able to deploy and run arbitrary software. The software can include operating systems and applications. The consumer does not manage or control the underlying cloud infrastructure, but has control over operating systems, storage, deployed applications, and possibly limited control of selected networking components with firewalls.

The Open Cloud Connect (OCC) grouped services under Network as a Service (NaaS), IaaS, PaaS, SaaS, Communications as a Service (CaaS), and Security as a Service (SECaaS) for now. There is no hierarchy in these service offerings.

As depicted in Figure 16.13, it is possible to build cloud services in a hierarchical fashion starting with NaaS where each builds on the previous and provides services for the next in the hierarchy. The hierarchy from the bottom to the top would be NaaS, PaaS, IaaS, SaaS, CaaS, and SECaaS.

16.7.3 Network as a Service

Network as a Service (NaaS) delivers assured, dynamic connectivity services via virtual or physical and virtual service end points orchestrated over multiple operators' networks. Such services will enable users, applications, and systems to create, modify, suspend/resume, and terminate connectivity services through standardized Application Programming Interfaces (APIs). These services are assured from both performance and security perspectives.

NaaS is expected to support on-demand network configuration, secure and QoS guaranteed connectivity, and compatibility with heterogeneous networks. It is the responsibility of the NaaS provider, cSP, to maintain and manage the network resources. It is possible that cSP may not own NaaS, but provides coordination. NaaS offers network as a utility.

Possible NaaS services are:

- Load balancing among servers in the same location or over a geographical region consisting of multiple locations where servers are added and removed in real time. Load balancers can be with or without failover protection and automatic fallback. Dynamic load balancing automatically distributes incoming application traffic across multiple cSP service instances. As users can preallocate dynamic IP addresses, they can preallocate a dynamic load

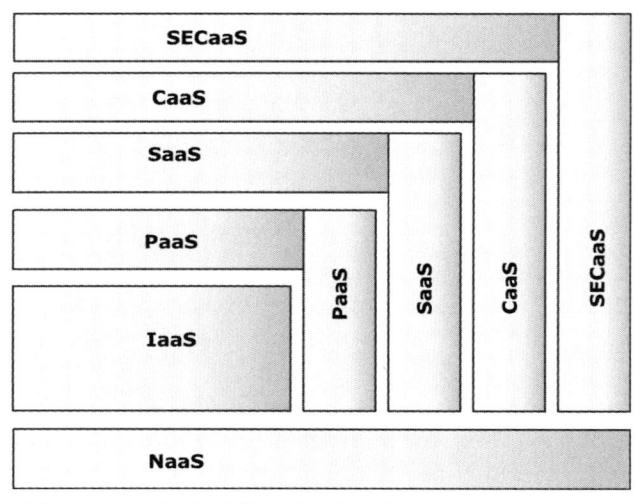

Figure 16.13 Possible hierarchy for building cloud services.

balancer so that its DNS name is already known, which can simplify the execution of protection.
- Domain registration services such as registering or transferring a domain name, full domain name system (DNS) control, geographically redundant DNS, and managed DNS.
- Hardware and software solutions to serve as routers, firewalls, VPN devices, and load balancers.
- IPv4 and IPv6-capable dual stack.
- Security.

NaaS can provide methods for users to provision cSP resources in a cloud virtual network that the user defines. The users have complete control over their virtual networking environments, including selection of user-owned IP address ranges, creation of subnets, and configuration of route tables and network gateways. This would enable users to create a VPN connection between the users' corporate data center and their cloud virtual network and leverage the cSP as an extension of the corporate data center. In the context of disaster recovery, users can use this virtual network to extend their existing network topology to the cloud.

16.7.4 Infrastructure as a Service

The capability provided to the consumer via IaaS is to provision processing, storage, networks, and other fundamental computing resources where the consumer is able to deploy and run arbitrary software, which can include operating systems and applications. The consumer does not manage or control the underlying cloud infrastructure but has control over operating systems, storage, deployed applications, and possibly limited control of select networking components such as firewalls.

IaaS cP configures, deploys, and maintains computing, storage, and networking resources to user. Also, IaaS cP provides the capability for users to use and monitor computing, storage, and networking resources so that they are able to deploy and run arbitrary software.

A customer portal could be provided to access the infrastructure. An API is needed to reduce human intervention for system management and total cost of operation.

In cloud computing IaaS, computing and storage resources can be provisioned on-demand. IT can access them on demand and create virtual datacenters from commodity servers, enabling IT to stitch together memory, I/O, storage, and computational capacity as a virtualized resource pool available over the network.

Servers can be:

- Bare metal servers with single processor, dual processors, or quad processors;
- Mass storage servers storing large amounts of data in solid state disks, hard disks, optical disks, or tapes;
- Virtual servers deployed on multitenant or single-tenant hosts as local or SAN storage. Portable storage can be added. Payment could be by the hour or month. Integration and migration between bare metal and virtual can

be performed. Users can customize their server configuration of computing cores, RAM, and storage on host servers.

The core of cloud computing services is flexible computer, storage, and network capacity, which can be adjusted up or down based on user demand. Within minutes, a user can create computing instances, which are VMs over which the user has complete control.

Machine images (MIs) can be preconfigured with operating systems and some application stacks. A user can also configure his or her own MIs. For disaster recovery, a user should own his or her MIs configured and identified so that they can be launched as part of the recovery procedure.

Application instances can be launched in separate availability zones to protect applications from the failure of a single location. Regions consist of one or more availability zones.

Each customer may be limited to a number of VMs, for example, 100 VMs, where VMs may be grouped into one or more virtual data centers (VDCs), each with an individual firewall policy. The user can scale infrastructure on demand by adding more resources where and when needed. When the flood of activity is over, the user can reduce capacity using a Web portal.

Storage IaaS can be:

- Simple storage service providing highly durable storage infrastructure designed for mission-critical and primary data storage.
- Dynamic block store (DBS) service providing the ability to create point-in-time snapshots of data volumes. Such snapshots can be used as the starting point for new DBS volumes, and to protect data for long-term durability.
- Import/export service for moving of large amounts of data into and out of a cP using portable storage devices for transport. The cP transfers user data directly onto and from storage devices by using NaaS.

The storage options can be:

- Memory to provide rapid access to data such as file caches, object caches, in-memory databases, and RAM disks.
- Message queues to provide temporary durable storage for data sent asynchronously between computer systems or application components.
- Storage area network (SAN) to provide block devices with the highest level of disk performance and durability for both business-critical file data and database storage. It can be used like a physical hard drive, typically by formatting it with the file system of user choice and using the file I/O interface provided by the instance operating system.
- Direct-attached storage (DAS) with local hard disk drives or arrays residing in each server providing higher performance than a SAN, but lower durability for temporary and persistent files, database storage, and operating system (OS) boot storage than a SAN.

16.7 Cloud Services Architectures

- Network attached storage (NAS) providing a file-level interface to storage that can be shared across multiple systems. NAS tends to be slower than either SAN or DAS.
- Databases such as a traditional SQL relational database, a NoSQL non-relational database, or a data warehouse where the underlying database storage typically resides on SAN or DAS devices or, in some cases, in memory.
- Backup and archive for data retained for backup and archival purposes which are typically stored on nondisk media such as tapes or optical media, which are usually stored off-site in remote secure locations for disaster recovery. There could be a limit on single archive and total amount of data in gigabytes, terabytes or petabytes.
- Durable reduced availability (DRA) storage buckets that can be introduced to have lower costs and lower availability, but to have the same durability as simple storage buckets. (Durability measures the length of a product's life. When the product can be repaired, estimating durability is more complicated. The item will be used until it is no longer economical to operate it. This happens when the repair rate and the associated costs increase significantly.)

Cloud storage allows users to enable DRA at the bucket level. The user can specify DRA storage at the time of bucket creation.

If a user wants to move data from a simple storage to a durable reduced availability storage bucket, the user needs to download the data from the simple storage bucket to his or her computer and then upload it to the durable reduced availability bucket.

A cP can provide a highly durable storage infrastructure designed for mission-critical and primary data storage where objects are redundantly stored on multiple devices across multiple facilities within a region.

A database IaaS can be set up, operated, and scaled on a relational database service (RDS) in the cloud. RDS can be used either in the preparation phase for disaster recovery to hold critical data in a running database already, and/or in the recovery phase to run the production database.

A simple database can be a highly available, flexible, nonrelational data store that offloads the work of database administration. It can also be used in the preparation and the recovery phase of disaster recovery. Users can also install and run their choice of database software on cP and can choose from a variety of leading database systems.

A database IaaS can be dedicated database instances with a cP database software. Users may have full administrative access via Secure Socket Shell (SSH), Structured Query Language (SQL) Developer, and other tools. The database could be a simple database with no administrative control. The choice of storage can be from GB to TB. RESTful Web Services can be used for data access.

The database IaaS service level may vary from customer managed to the cP managed providing full customer access. Resources can be dynamic such that the user can add or remove compute resources, memory, or storage as needed. Life-cycle management and security for database services can be also provided.

Disaster recovery is another IaaS providing recovery from hardware or software failure, a network outage, a power outage, physical damage to a building like fire or flooding, human error, or some other significant disaster.

Recovery time objective (RTO) and recovery point objective (RPO) are two parameters for disaster recovery IaaS. RTO is the duration of time and the service level to which a business process must be restored after a disaster (or disruption) to avoid unacceptable consequences associated with a break in business continuity. RPO describes the acceptable amount of data loss measured in time. For example, if the RPO was 1 hour, after the system was recovered, it would contain all data up to a point in time that is prior to 3:00 AM because the disaster occurred at 2:00 AM.

When reacting to a disaster, it is important to either quickly commission compute resources to run the user system in the cP domain or to orchestrate the failover to already running resources in cP domain.

The cloud user can choose the most appropriate location for the selected disaster recovery site, in addition to the site where the user system is fully deployed. A cP may have multiple regions where the selected recovery site can be chosen to be different.

A possible architecture for disaster recovery IaaS is given in Figure 16.14. When the server in Zone 1 failed:

- If the backup is 1:1 (i.e., active and standby configuration) and VM is already available in Zone 2, only the application is moved to Zone 2 from Zone 1.
- If backup is 1 + 1 (i.e., active and active configuration) and the application in Zone 2 is current, then from active connection cSC1 to backup connection cSC2 will be switched.
- If the only backup server is available in Zone 2, VM along with applications can be moved to Zone 2. In this case TCP/IP is likely to be used in moving VMs. The propagation time between Zone 1 and Zone 2 and the rate of connectivity between zones must be such that these factors will not be

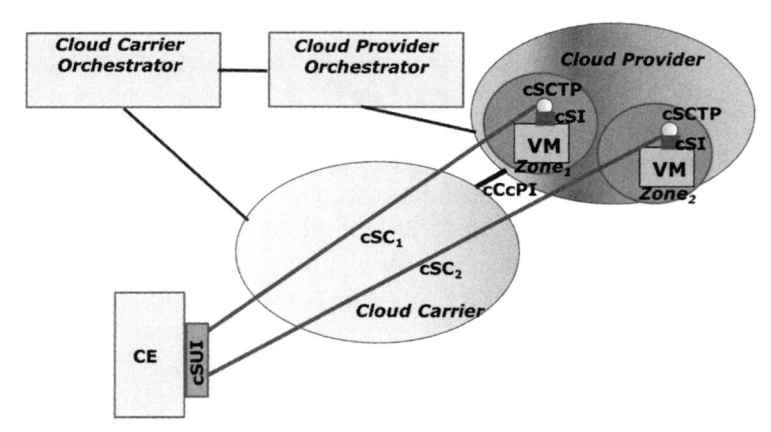

Figure 16.14 Protection coordination via orchestrators of cP and cC.

the dominant contributor of a TCP timeout. Therefore, VM moves between zones connected with high speed transport are more likely.

In order to support the configurations in Figure 16.14 in a timely manner, cloud carrier and cloud provider orchestrators must be in coordination.

16.7.5 Security as a Service

Security services such as connectivity security, application security, or content security can be provided by a cSP to cloud consumers. Such services are referred as Security as a Service (SECaaS).

With SECaaS, a consumer does not manage or control the underlying security transport negotiation, encryption, detection algorithms, threat intelligence or network inspection, but has control over the selection of security solutions and scope with respect to their data and network.

SECaaS can be:

- Security of storage services with managed authorized access and customized data leakage prevention technologies;
- NaaS security provided through network traffic data inspection and filtering, distributed denial-of-service (DDoS), and other intrusion attack vector protection;
- Threat intelligence where attack vectors are detected and propagated through cSP for mitigation;
- Traffic cleaning, where consumer network traffic that would not normally utilize the cSP is routed expressly for SECaaS.

Security around data storage services must allow consumer fine control of network access control list (ACL) for modification and accessibility of data stored in cSP. Additional security is provided by audit tracking of data access or modification, along with data leakage technologies applied to the network access between cloud users and cSP.

Network traffic between over a cSC is subject to protection from attack and intrusion vectors. cSP can provide the traffic inspection and intrusion/attack blocking via combination of traditional firewall/security appliances, alongside virtual security solutions. Both content inspection and packet inspection technologies should be utilized to provide high security.

The cSP may provide the service where security events and responses are utilized to gather threat intelligence and react in a manner to protect the consumer services. Should an attack or intrusion be detected, an automatic response to isolate the attack vector or continue to provide the service through alternate infrastructure can be taken.

SECaaS may provide network security functions through cSC set up for delivery of security functions by the cSP, regardless of whether the consumer traffic would normally access the cSP. Selection of routing or tunneling technologies to establish the cSC and security services is performed at cSUI.

16.7.6 Platform as a Service

By Platform as a Service (PaaS), the capability provided to the consumer is to deploy onto the cloud infrastructure consumer-created or acquired applications created using programming languages and tools supported by a cP. The consumer does not manage or control the underlying cloud infrastructure including network, servers, operating systems, or storage, but has control over the deployed applications and possibly application hosting environment configurations.

PaaS can be a stand-alone development environment that does not include technical, licensing, or financial dependencies on specific SaaS applications or Web services. These development environments are intended to provide a generalized development environment.

PaaS can be application delivery-only environments that do not include development, debugging, and test capabilities as part of the service, although they may be supplied offline. The services provided generally focus on security and on-demand scalability.

PaaS can be an open platform as a service that does not include hosting as such; rather it provides open source software to allow a PaaS provider to run applications. Some open platforms let the developer use any programming language, any database, any operating system, or any server to deploy their applications.

With PaaS, a scalable and high-performing network can be formed. As a fully managed application platform for running and consolidating software applications and databases in the cloud, PaaS includes:

- A virtualized, scalable infrastructure of application and database servers;
- Performance, reliability, and security of the network;
- Network, server, and storage infrastructure management;
- $24 \times 7 \times 365$ infrastructure monitoring and support;
- Built-in redundancy and security of data centers.

16.7.7 Software as a Service

The capability provided to the consumer via SaaS is to use the cloud provider's applications running on a cloud infrastructure. The applications are accessible from various client devices through a thin client interface such as a Web browser. The consumer does not manage or control the underlying cloud infrastructure including network, servers, operating systems, storage, or even individual application capabilities, with the possible exception of limited user-specific application configuration settings.

Software is installed on demand via customer portal, and licensed and billed monthly. Open-source and enterprise 32 and 64-bit operating system software options from various vendors are available. The following are a few examples of vendors and operating systems that could be installed: Microsoft, RedHat, CentOS, Debian, FreeBSD, Ubuntu, Vyatta Network, Cloud Linux, Parallels, cPanel, as well as:

- Server virtualization software such as VMWare ESX and ESXi, Citrix Xenserver, Citrix CloudPlatform, Parallels Virtuozzo, Microsoft Hyper-V;
- Security software such as McAfee Total Protection, McAfee Anti-Virus, Microsoft Windows Firewall, McAfee Host Intrusion Protection, Nimsoft Monitoring, APF Software Firewall;
- Database software such as Microsoft SQL Server (2000, 2005, 2008, 2012), MySQL, Cloudera Hadoop, MongoDB, Basho Riak;
- Control panel software such as cPanel/WHM with Fantastico, RVSkin and Softaculous, Parallels Plesk Panel.

In cloud content delivery network (CDN) SaaS, user content is distributed to a network of edge servers. Users can access the content from a server near them, ensuring faster load times. Large objects are delivered to many users with sustained high data transfer rates. If user traffic fluctuates, the service automatically adjusts as demand increases or decreases.

User content can be placed onto cloud object storage and then CDN enables the content. The user then visits a CDN site and requests files from the nearest edge server. The edge server delivers a local, cached copy or pulls one from cloud object storage. The object's time to live (TTL) will expire at intervals the user defines such as 24 hours. If the TTL has expired when the next request is made, the file is again retrieved from cloud object storage. The content is cached once again by the edge servers and the TTL restarts.

16.7.8 Communication as a Service

Real-time services such as Virtual PBX, voice and video conferencing systems, collaboration systems, and call centers can be considered as Communication as a Service (CaaS). CaaS capabilities can be:

- Business voice continuity avoiding missing a call even when disaster strikes;
- Unlimited inbound, local, and domestic long distance;
- Fixed mobile convergence, which removes the distinctions between fixed and mobile networks, providing a superior experience to customers by creating seamless services using a combination of fixed broadband and local access wireless technologies to meet their needs in homes, offices, other buildings, and on the go;
- Voicemail in user inbox or on user smartphone;
- Integrated business communications making calls from user desk or mobile phone and have it appear as user office number;
- Easy call management and feature editing through Microsoft Outlook, Internet Explorer, or Firefox;
- Fully managed and hosted;
- Point-to-point or multipoint video calling;
- Point-to-point or multipoint voice calling;
- Point-to-point or multipoint voice and video conferencing;

- Mobile application support allowing free download for both iOS and Android platforms;
- Professional voice recording service for user greetings and other messages recorded by an industry-leading voice talent;
- Bring your own device (BYOD) capabilities;
- SLAs including quality of service and availability such as next business day replacement of phones for equipment maintenance of virtual PBX service;
- Dynamic security policy including authentication, media encryption, and access control;
- Scalability.

16.8 Conclusion

We have described the Third Network and its components, LSO, virtualization, and cloud Services. As an example, we have described virtualization of Carrier Ethernet services.

The Third Network is still in its infancy. Development in this area will accelerate automation of service provisioning and maintenance as well as interoperability among operators' LSO systems.

References

[1] The MEF Carrier Ethernet Seminar, "The Third Network: LSO, SDN and NFV," May 2015.

[2] MEF 55, "Lifecycle Service Orchestration (LSO): Reference Architecture and Framework," March 2016.

[3] Duan, Q., and M. Toy, *Virtualized Software Defined Networks and Services*, Norwood, MA: Artech House, 2016.

[4] Turner, J. and D. E. Taylor, "Diversifying the Internet," Proceedings of the 2015 IEEE Global Telecommunications Conference (GLOBECOM '05), December, 2005.

[5] Anderson, T. et al., "Overcoming the Internet Impasse Through Virtualization," IEEE Computer Magazine, Vol. 38, No. 4, pp. 34–41, April 2005.

[6] Toy, M., "Framework for MEF Virtual Services," MEF Presentation, January 6, 2014.

[7] Toy, M., "OCC 1.0 Reference Architecture," December 2014.

[8] MEF 6.2, "EVC Ethernet Services Definitions Phase 3," August 2014.

[9] MEF 7.2, "Carrier Ethernet Management Information Model," April 2013.

About the Authors

Mehmet Toy received his Ph.D. in electrical and computer engineering from Stevens Institute of Technology, Hoboken, NJ. He is currently a Distinguished Member of Technical Staff in Verizon Communications and is involved in SDN, NFV, and cloud services architectures, testing, and standards. Prior to his current position, Dr. Toy held technical and management positions in well-known companies and startups including Comcast, Intel Corp., Verizon Wireless, Fujitsu Network Communications, AT&T Bell Labs, and Lucent Technologies. He also served as a tenure-track faculty and adjunct professor in various universities, including Stevens Institute of Technology, New Jersey Institute of Technology, Worchester Polytechnic Institute, and University of North Caroline at Charlotte.

Dr. Toy contributed to research and development of cloud, overlay, SDN, and virtualization-based commercial networks and services, Carrier Ethernet, IP multimedia systems (IMS), optical, IP/MPLS, and wireless and ATM technologies. He holds a patent, has five pending patent applications, and published numerous articles, seven books, and a video tutorial in these areas and in signal processing. Two of his books are being used as college textbooks and one has been translated into Turkish. Dr. Toy served on the Open Cloud Connect (OCC) board, IEEE Network Magazine editorial board, IEEE Communications Magazine as a guest editor, and for IEEE-USA and IEEE ComSoc in various capacities. He has received awards from Comcast, AT&T Bell Labs, and IEEE-USA. He is a senior member of IEEE, chairs the IEEE ComSoc Cable Networks and Services subcommittee, and serves in MEF and ETSI NFV.

Hakkı Candan Çankaya received his Ph.D. in computer engineering from Southern Methodist University, Dallas, TX. He is currently a Solutions Architect at Fujitsu Network Communications Inc. responsible for developing advanced networking solutions for a wide variety of customers and leveraging research and development activities into industry applications. He brings a long and distinguished network research and strategy career to his role. Prior to joining Fujitsu, Hakkı worked in various senior positions at Bell Laboratories and Alcatel-Lucent, and was a member of the Alcatel-Lucent Technical Academy for several years. He is also an adjunct professor of computer science and electrical engineering at Southern Methodist University. He has authored and co-authored a total of 19 patents in telecommunications and other communications networking technologies, as well as contributing numerous technical articles, lectures, and presentations to the technical community. He is actively involved in the Communications Society of Institute of Electrical and Electronics Engineers (IEEE), Open Networking Foundation (ONF), and is an MEF Carrier Ethernet Certified Professional (MEF-CECP).

Index

1 + 1 protection switching, 88–90
   bidirectional, 88, 89
      defined, 88
      unidirectional, 88–90
1:1 protection switching, 90–91
10-Gb Ethernet, 23–24
10-Mbps Ethernet, 22
40-Gb Ethernet, 24, 25
100-Gb Ethernet, 24, 25

A

Access links, 120–21, 122
Adaptation function headers, 274–76
Adaptation sink, 113
AIS/RDI, 203
Alarm propagation, 203, 204
ALOHA, 15
American National Standards Institute (ANSI), 35
Application-CoS-priority mapping, 178–81
   CoS identification, 178–81
   PCP and DSCP mapping, 181
ATM networks, 34, 35
Automatic protection switching (APS), 85
   detection of failures, 86
   entities, 86–87
   frames, 93
   PDU format, 91–95
   protection types, 94
   revertive and nonrevertive modes, 92–93
   timers, 94–95
   transmission and acceptance of, 93
Autonegotiation, 17
Availability
   calculation, 238–39
   defined, 238
   FLR-based, 240
   multipoint-to-multipoint EVC, 240
   service, 238–42
   for small time intervals, 239
   via sliding window, 239

B

Backbone core bridge (BCB), 8
Backbone edge bridge (BEB), 8, 308
   B-type (B-BEB), 8, 308
   IB-type (IB-BEB), 8, 308
   I-type (I-BEB), 8, 308
   provider (PBEB), 305, 308
   types of, 308
Bandwidth
   flow, 191
   incremental assignment, 108
   OAM, 267
Bandwidth profile
   algorithms, 183–86
   defined, 181
   for envelope with multiple flows, 186
   flags and switches, 184
   models and use cases for, 186–88
   no token recirculation use case, 189
   per CoS identifier, 184
   per EVC, 183
   per UNI, 183
   policer, 193, 194, 195
   recirculating tokens use case, 188
   for single flow per envelope, 187
   TCP throughput and, 191
   test, 252
   tokens, 185
   token sharing and, 186
BGP VPLS, 384–90
   auto-discovery, 384–85
   hierarchical, 389–90
   Layer-2 info extended community, 387
   multi-AS VPLS, 388–89
   NLRI, 386
   operation, 386

BGP VPLS (continued)
    operation illustration, 387
    route reflector, 385
    RT format, 384
    signaling, 385–86
Bidirectional Forward Detection (BFD), 352
Border gateway protocol (BGP)
    route target community, 12
    RT community, 384
    update message, 385
    VPLS and, 384–90
Bottleneck bandwidth, 195
Bridges, 18–19
B-type BEB (B-BEB), 8, 308
Business management layer (BML), 394

C

Canonical format indicator (CFI), 117
Carrier Ethernet
    architecture, 3–4, 109–22
    configuration management functions, 398
    E-LMI protocol, 283–96
    ENNI, 126–27
    ETH layer, 109–10
    ETH layer characteristic information, 114–18
    ETH layer functions, 118–20
    ETH links, 120–22
    EVC, 124–26
    EVC management, 399
    EVC services and attributes, 130–44
    fault management, 400, 403
    incremental bandwidth assignment, 108
    information modeling, 395–463
    interfaces, 4
    interfaces and types of connections, 122–30
    Layer-2, 1
    management entities (MEs), 402–3, 405, 409
    with multiple PBT domains and single domain, 315–16
    OAM, 199–267
    OVC, 127–28
    OVC management, 399
    OVC services and attributes, 144–53
    properties, 108
    protocol stack, 110
    security, 266
    services, 14
    synchronization of, 2
    testing, 244–63
    traffic management, 4–5, 163–97
    UNI, 123–24
    VUNI/RUNI, 128–30
Carrier Ethernet Management Information Model, 13–14
Carrier Ethernet networks (CENs), 101
Carrier Ethernet services
    common management objects, 402, 404
    information modeling of, 395–435
    model, 397
    network resource model, 459–61
    OAM configuration, 399, 401
    performance management, 400, 402
    virtualized, 473–79
    YANG models for, 444–59
CESoETH
    conclusion, 281
    control word, 274
    data frames, 278
    defects and alarms, 278–79
    defined, 269
    frames, 275, 279
    payload, 279
    performance monitoring of, 280
    RTP header, 276
    service configuration, 280–81
    TDM application signaling, 277–78
CE-VLAN preservation test, 253
Circuit emulation, 6–7
Circuit emulation over PSN (CESoPSN), 39
Circuit emulation services (CES), 269–81
    adaptation function headers, 274–76
    architecture, 270
    architecture with multiplexing, 270
    functional blocks, 271
    functional components and interface types, 273
    interworking function (CES-IWF), 271
    introduction to, 269–71
    IWF, 271–72
    as point to point, 270

synchronization, 276–77
Class of Service (CoS)
    bandwidth profile per, 184
    classes, 166
    flow classification, 4, 164
    flows, 164
    identification, 178–81
    multipoint performance objectives, 173–75
Client failure indication (CFI), 360
Client-server relationship, 112
Clock hierarchy, 45
Cloud services architectures, 479–92
    actors, 479, 480
    cloud services, 483–84
    Communication as a Service (CaaS), 491–92
    hierarchy, 484
    Infrastructure as a Service (IaaS), 483–84
    Network as a Service (NaaS), 484–85
    overview, 479–80
    Platform as a Service (PaaS), 483, 490
    protocol stacks and interfaces, 480–83
    Security as a Service (SECaaS), 489
    Software as a Service (SaaS), 483, 490–91
    virtual resources, 481
Color mode (CM), 164
Committed burst size (CBS)
    average TCP throughput, 192
    CIR versus, 190
    defined, 165
    determination, 189
    inversely proportional to CIR, 197
    large values, 197
    minimum size of, 188
    minimum value calculation, 193
    testing, 253
    tokens, 182
Committed information rate (CIR), 164, 165
    average TCP throughput and, 194
    CBS versus, 190
    TCP throughput equals, 193
Communication as a Service (CaaS), 14, 491–92
Complete service testing, 249
Composite performance metric (CPM), 240
Configurable availability threshold, 238
Conjoined Ethernet rings, 97

Connectivity Check, 352
Connectivity fault management (CFM), 206
Connectivity Verification, 352
Constant bit rate (CBR), 270
Container-based virtualization, 472
Continuity check messages (CCM), 5–6, 215–19
    attributes, 216–18
    defined, 215
    dual-ended loss measurement, 218
    format octet, 217
    frames, 215, 219
    illustrated, 216
    PDU, 217, 218
    PDU flags, 218
    transmission, 216, 217
    unintended connectivity detection, 216
Core Information Model (CIM), 459, 460–61
CoS performance objectives
    defined, 172
    for multipoint services, 173–75
    PT0.3, 173
    PT1, 173
    PT2, 174
    PT3, 174
    PT4, 175
Counters, 235–36
Coupling flag (CF), 164
CSMA/CD, 16–17
C-tag, 116
Customer UNI (UNI-C), 207
Cyclic redundancy check (CRC), 114

D

Data-link connection identifier (DLCI), 64
Decision point (DP) service attributes, 155
Delay measurement (DM) objects, 426–31
Delay measurements, 6, 234
Denial-of-service (DoS) attacks, 82
Device under test (DUT), 246–47, 249
Differential GPS (DGPS), 33
Differential mode delay (DMD), 21
DMM, 243, 244
DMR, 243, 244
Dual homing, 383
Dying gasp message propagation, 204

E

ECDX (emulated circuit/demultiplexing function), 271, 272
Egress equivalence class identifier, 443–44, 446
Element layer (EL), 394
Element management layer (EML), 394
Element management systems (EMSs), 2, 351, 393
    information modeling of services for, 395–435
    network management through, 200
E-LMI protocol, 200
ETH adaptation function (EAF), 119
ETH conditioning functions, 118
ETH connection function, 119
Ethernet
    1-Gbps interface types, 23
    10-Gb, 23–24
    10-Mbps, 22
    40-Gb, 24, 25
    100-Gb, 24, 25
    address, 29
    basic, 15–32
    bridges, 18–19
    circuit concept over, 1
    circuit emulation services (CES), 6–7
    conclusion, 32
    connection oriented, 108
    defined, 2, 15
    development of, 107
    elements, 20
    fast, 22
    frame types, 27–31
    full-duplex, 17
    gigabit, 22–23
    hubs, 17–18
    introduction to, 15
    LAN PHY, 24–25
    MAC sublayer, 25–26
    OAM, 5–6
    over fiber, 21
    PHY layer, 26
    physical layer, 20–26
    protection, 85–106
    repeaters, 17
    scalability, 108
    standards, 27
    switches, 19–20
    synchronous networks, 49–60
    temperature hardening, 27
Ethernet alarm indication signal (ETH-AIS), 227–29
    frames, 228, 229
    functionality, 227
    MEP configuration parameters, 229
    PDU, 227
    PDU format, 228
Ethernet automatic protection switching (EAPS)
    bidirectional, 87
    for Layer-2 loop protection, 86
    protected domain configuration, 87
    unidirectional, 87
Ethernet continuity check (ETH-CC), 215, 216, 217, 218, 228–29
Ethernet equipment clock (EEC), 58, 59–60
Ethernet flow termination function (EFTF), 272
Ethernet Frame, 114
Ethernet II (DIX), 29–30
Ethernet LAN (E-LAN) service, 131
Ethernet line (E-Line) service
    attributes and parameters, 140–41
    defined, 131
Ethernet Link Trace function (ETH-LT), 6, 224–26
Ethernet Local Management Interface (E-LMI), 7, 283–96
    bandwidth profile subinformation element, 291, 292
    conclusions, 296
    data instance information element, 291
    defined, 283
    EVC identifier subinformation element, 293
    EVC map entry subinformation element, 292
    EVC parameters subinformation element, 293
    EVC status coding, 290
    EVC status information element, 290
    EVC type coding, 293
    framing structure, 285
    general message organization example, 286
    illustrated, 284
    message elements, 288–92

messages, 286–88
message transfer, 284
message type coding, 288
parameters and procedures, 292–95
periodic polling process, 294–95
procedures, 283–84
report type coding, 288
sequence numbers information element, 289
status check example, 295
STATUS ENQUIRY message, 286–87
STATUS message, 286–87
termination in functional components, 284
UNI-C procedure, 295
UNI identifier subinformation element, 293
UNI-N procedure, 295–96
UNI status information element, 289
Ethernet locked signal (ETH-LCK), 230–31
Ethernet loopback function (ETH-LB), 6, 220–21
Ethernet loss measurement (ETH-LM), 234–36
Ethernet MD and MA relationship, 214
Ethernet private LAN (EP-LAN) service
  attributes and parameters, 142
  defined, 132–33
Ethernet private line (EPL) service
  attributes for EPL, 140–41
  defined, 131–32
  service chaining for, 478–79
  virtualized components, 473, 474
Ethernet private tree (EP-Tree) service
  attributes and parameters, 143–44
  defined, 134
Ethernet protection switched rings (EPSR), 86, 95, 96
Ethernet remote defect indication (ETH-RDI), 229–30
Ethernet ring protection (ERP), 86
Ethernet services layer, 110, 111
Ethernet Test Equipment (ETE), 250
Ethernet Test Equipment-Application (ETE-A), 250
Ethernet Test Equipment-Instrument (ETE-I), 250
Ethernet test support system (ETSS), 250
Ethernet tree (E-Tree) service, 131
Ethernet virtual connection (EVC)
  attributes, 137–38
  bandwidth profile per, 183
  class diagram, 439
  crossing ENNI, 128
  defined, 4, 124
  management, 397–98, 399
  multipoint-to-multipoint, 124–25
  multipoint-to-multipoint availability, 240
  performance metrics, 196–97
  per UNI attributes, 138–40
  point-to-multipoint, 125–26
  point-to-point, 131, 234, 270
  service frames and, 124
  service-level information modeling, 438–39
  services and attributes, 136
  types, 130
  UNI relationship, 125
  virtualized components, 474
  See also EVC services
Ethernet virtual private LAN (EVP-LAN) service
  attributes and parameters, 142–43
  defined, 133–34
Ethernet virtual private tree (EVP-Tree) service
  attributes and parameters, 144
  defined, 134–35
ETH EVC adaptation function (EEAF), 119
ETH EVC termination function (EETF), 119
ETH flow termination function (EFTF), 119
ETH layer, 3, 110
  characteristic information, 114–18
  functions, 118–20
  for instantiation, 110
  responsibilities, 110
  topological components, 109
  transport of adapted information, 110
  for unicast flow, 114
ETH-Test, 245
ETH to TRAN adaptation function (TAF), 119–20
EVC services
  applications, 135
  delivery, 135–36
  E-LAN, 131
  E-LAN attributes and parameters, 142–43
  E-Line, 131
  E-Line attributes and parameters, 140–41
  EPL, 131–32

EVC services (continued)
    EP-LAN, 132–33
    EP-LAN attributes and parameters, 143
    EPL attributes and parameters, 140–41
    EP-Tree, 134
    EP-Tree attributes and parameters, 143–44
    E-Tree, 131
    E-Tree attributes and parameters, 143–44
    EVC attributes, 136–40
    EVPL, 132
    EVP-LAN, 133–34
    EVP-LAN attributes and parameters, 142–43
    EVPL attributes and parameters, 141
    EVP-Tree, 134–35
    EVP-Tree attributes and parameters, 143–44
    types of, 130
    UNI attributes, 136–37
EVPL service
    attributes and parameters, 141
    illustrated, 132
    use of, 132
Excess burst size (EBS), 164, 181
    defined, 165
    testing, 253
    tokens, 182
Excess information rate (EIR), 164, 165
External network-to-network interface (ENNI), 101, 126–27, 150–51
    class diagram, 444
    DP flow chart, 159, 160
    L2CP peering service, 161
    MIBs, 408–10
    service, 442–43
    service attribute alignment, 416
    service-level information modeling, 441
    VPLS, 390–91
    VUNI related objects, 405–6, 412, 413

F

Facility data link (FDL), 280
Fast Ethernet, 22
Fault management
    Carrier Ethernet, 400, 403
    class diagram, 414
    connectivity (CFM), 206
    function set, 403
    objects, 406
    YANG module, 453
Flow control, 166
Flow point pool (FPP), 109
Flows, 113
Focused overload, 175–78
ForwardingConstruct (FC) class, 460, 461, 462
ForwardingDomain (FD) class, 460
Frame delay
    measurements, 242–44
    performance, 167–69
    range performance, 169–70
    variation, 170–71
Frame error ratio (FER), 280
Frame loss measurements, 234–38
    dual-ended, 236
    far-end frame loss, 235
    frame loss ratio (FLR) and, 234–35
    near-end frame loss, 235
    single-ended, 236
Frame loss ratio (FLR), 171–72, 234–35, 279
Frames
    defined, 21
    service, 118
    types of, 27–31
    See also specific types of frames

G

Generalized Multiprotocol Label Switching (GMPLS), 321
Generator test function (GTF), 250, 251
Generic associated channel (G-ACh), 348–49, 361–62
Grand master clock (GMC), 49
Guard timer, 94

H

Hierarchical VPLS (H-VPLS), 12, 174, 379–84
    architecture, 380
    BGP, 389–90
    defined, 379–80
    model using Ethernet access network, 383–84
    multitenant unit (MTU), 379–80
    protection via dual homing, 383

Index

spoke connectivity, 381
VPWS, 382–83
Hold-off timer, 94
Hot Standby Routing Protocols (HSRP), 204
Hubs, 17–18
Hypervisor, 471

I

IB-type BEB (IB-BEB), 8, 308
IEEE 802.1AB, 298–304
    architecture, 299–300
    frame format, 301–4
    See also Link Layer Discovery Protocol (LLDP)
IEEE 802.1ah frame format, 305–7
IEEE 802.2, 30–31
IEEE 802.3, 30
IEEE 1588
    boundary clocks, 49
    in CES application, 45
    defined, 43
    issues, 49
    messages, 49
    protocol stack, 44
    synchronous hierarchy, 48
    v2, 47–48
Information modeling, 13–14, 393–463
    common management objects, 402, 404
    conclusions, 461–63
    ENNI and OVC MIBs, 408–10
    Ethernet discovery functions, 400
    fault management, 400, 403
    fault management objects, 406, 414
    introduction to, 393–95
    management entities (MEs), 402–3, 405, 412, 413
    management information views, 393
    management of EVC, OVC, and ELMI, 397–99
    network resource model, 459–61
    OAM configuration, 399
    partitioning example, 396
    performance management, 400, 402
    performance monitoring objects, 408, 415
    service-level, 435–44

    of services for EMS and NMS, 395–435
    SOAM FM MIB, 410–13
    SOAM PM MIB, 413–35
    topological elements, 396
    YANG models, 444–59
Infrastructure as a Service (IaaS), 14, 483–84
    capability to consumer, 485
    cloud computing, 485
    database, 487
    disaster recovery, 488–89
    overview, 483–84
    servers, 485–86
    storage, 486–87
    See also Cloud services architectures
Interframe delay variation (IFDV), 170, 171, 241, 279
    measurements, 243–44
    round-trip, 243–44
Inverse multiplexing over ATM (IMA), 21
IP service classes, 176, 177
I-type BEB (I-BEB), 8, 308
IWF synchronization function, 56–57

J

Jitter buffer overrun condition, 279

L

Label Distribution Protocol (LDP), 10
    notification message, 80, 81
    SP-PE TLV, 78
    See also LDP-based VPLS
Label edge router (LER), 343
Label-switched paths (LSPs), 321, 323, 341–43
    monitoring, 357
    MPLS, 369
    tunnel, 366
Label switching router (LSR), 11, 343
Layer-2 Control Protocol (L2CP)
    address set service attribute, 157
    bridge reserved addresses, 154
    decision point, 155
    defined, 153
    destination MAC addresses, 153
    frames, 152–54
    in multi-CEN environment, 156–57
    peering service attribute, 158

Layer-2 Control Protocol (L2CP) (continued)
   peering service attribute list, 157
   port-based frames, 159
   in single CEN environment, 156
LBM, 220–24
   illustrated, 222
   multicast frames, 222
   PDU, 223
   PDU format, 224
   unicast frames, 220, 221
LBR, 220–24
   frames, 222–23
   illustrated, 222
   PDU, 223
   PDU format, 224
   validity, 221
LDP-based signaling, 376–78
LDP-based VPLS, 372–84
   auto-discovery and, 376
   data forwarding on Ethernet PW, 378–79
   discovery, 375–76
   flooding, forwarding, and address learning, 372–73
   hierarchical VPLS (H-VPLS), 379–84
   signaling, 376–78
   tunnel topology, 373–75
Leaky buckets, 182
Life-cycle service orchestration (LSO), 1, 395
   components, 467
   interfaces, 467–68
   management abstractions and constructs, 468–70
   management reference architecture, 466
Link aggregation, 100–106
   functional diagram, 104
   limitations, 106
   load balancing, 102
   operation, 103–5
   sublayer, 102
Link Aggregation Control Protocol (LACP)
   defined, 102
   keys, 105
   parameters, 105
   ports, 104
   priority, 105
Link aggregation groups (LAGs), 85
   ENNI with, 101

   formation of, 100
   objectives, 103
   in same STP groups, 105
   UNI with, 101
Link Layer Discovery Protocol (LLDP)
   agents, 299, 300, 303
   components, 300
   defined, 298
   frames, 300
   MIBs and, 303
   as one-way protocol, 301
   operation of, 299, 301
   principles of operation, 300–301
Link OAM, 202–5
LMM, 236–37
LMR, 236–38
Locked signal (LCK), 230–31
Lock instruct (LKI) command, 360
LogicalTerminationPoint (LTP) class, 460
Loss management (LM) objects, 420–26
Loss of connectivity (LOC), 91
LSP maintenance entity group (LMEG), 355
LSP Ping, 352
LSP SPME ME group (LSMEG), 355
LTM, 225–27
   egress identifier TLV format, 228
   flag form in PDU, 228
   frames, 225
   multicast, 226
   PDU format, 227
LTR, 225–27

M

MAC-in-MAC encapsulation, 304
Maintenance association (MA) endpoints, 208
Maintenance association endpoint identifier (MEPID), 209
Maintenance association identifier (MAID), 210
Maintenance endpoints (MEPs)
   counters, 235–36
   domain boundaries, 5, 206
   MA association, 211
   as maintenance functional entity, 207
   MPLS-TP OAM, 352, 354–56, 359
   provisioning, 209

Index

transit ETH flows and, 209
unexpected, 219
Maintenance entity groups (MEGs), 99–100, 207–8, 211
Maintenance intermediate points (MIPs), 5
Maintenance points (MPs)
   defined, 208
   functions, 209
Management entities (MEs), 207, 402–3, 405, 409
Management information bases (MIBs)
   defined, 299
   ENNI and OVC, 408–10
   SOAM FM, 410–13
   SOAM PM, 413–52
Maximum time interval error (MTIE), 59
Mean frame delay performance, 170
Media attachment unit (MAU), 231
MEG endpoints (MEPs), 208–9
MEG intermediate point (MIP), 209
Metcalf system, 15
Metro Ethernet Forum (MEF)
   APIs being worked by, 459, 460, 461
   Carrier Ethernet Management Information Model, 13
   Carrier Ethernet services, 107
   classes definition, 5
   defined, xvii
   Ethernet and, 1
   focus, 393
   management layering, 395
   Network Resource Provisioning (NRP) Project, 459
   Open Cloud Connect (OCC) and, 465
   protection specifications, 3
   YANG model architecture, 448
MPLS
   data plane, 340
   defined, 321
   fast reroute (FRR), 323
   internetworking, 318
   label-switched paths (LSPs), 321
   layers of network, 325
   LSPs, 369
   packet format, 338
   T-MPLS differences, 323–24
   See also Transport MPLS (T-MPLS)

MPLS-TP OAM
   client failure indication (CFI), 360
   data plane loopback, 359–60
   framework, 353
   functions, 352
   functions for proactive monitoring, 357–59
   hierarchy, 353–57
   hierarchy illustration, 354
   lock instruct (LKI) command, 360
   MEGs, 355, 356
   MEPs, 352, 354–56, 359
   MIPs, 352, 359
   packets, 352
   route tracing, 361
MPLS Traffic Engineering (MPLS-TE), 10
MPLS Transport Profile (MPLS-TP), 335–62
   architectural requirements, 339
   architecture, 339–52
   architecture for multisegment PW, 346
   architecture for network-layer clients, 346
   architecture for service LSP switching, 347
   architecture for single-segment PW, 345
   as client-agnostic, 336
   conclusions, 362
   control plane, 349–51
   control plane architecture context, 350
   data plane, 340–43
   defined, 10, 335
   ENNI interface, 344
   essential features of, 336–38
   extensions, 10
   forwarding, 339, 341
   frame format, 338–39
   generic associated channel, 348–49
   INNI interface, 344
   introduction to, 335–38
   IP transport service, 345–48
   label edge router (LER), 343
   label stack for IP and LSP clients, 347
   label stack using pseudowires, 346
   label switching router (LSR), 11, 343
   layer and client layer relationship, 341
   LSPs, 341–43
   monitoring in, 352
   network architecture with protected LSPs, 341
   network high level architecture, 340

MPLS Transport Profile (MPLS-TP) (continued)
  network layering, 340
  network management, 351–52
  network provisioning, 10
  NMS and, 10, 336
  OAM. *see* MPLS-TP OAM
  packet format, 338
  PE node, 11
  protection switching, 361
  protocol stack reference model, 350
  PWs and, 345
  router types, 343
  security considerations, 361–62
  service interfaces, 343–45
  standards, 335, 336
  static, 10, 336
Multi-AS VPLS, 388–89
Multiprotocol Label Switching. *See* MPLS
Multisegment PWs (MS-PWs)
  configuration with signaling mechanisms, 76
  defined, 73
  dynamic route selection, 77
  MPLS-TP, 343
  setup mechanisms, 76–78
  static configuration, 76
  switching alternatives, 77
  use circumstances, 75–76

N

Native service processing (NSP), 68
Network as a Service (NaaS), 14, 484–85
Network elements (NEs), 2, 393
Network errors, 297
Network flow, 113
Network function virtualization (NFV), 1, 471
Network layer reachability information (NLRI), 12
Network management layer (NML), 394
Network management systems (NMSs), 351, 393, 395–435
Network operations center (NOC), 395
Network Resource Provisioning (NRP) Project, 459
Network Time Protocol (NTP)
  clock reference, 40
  clock synchronization, 41, 42
  defined, 39
  implementations, 40
  scalability, 40
  version 4, 41
Network UNI (UNI-N), 207
Network virtualization, 472–73

O

OAM (operation, administration, and management), 5–6
  addressing, 212–15
  alarm propagation, 203, 204
  bandwidth profile, 267
  categories, 199–200
  conclusions, 267
  continuity check messages (CCM), 215–19
  discovery, 203
  dying gasp message propagation, 204
  elements of, 5, 200
  end-to-end, 313
  Ethernet alarm indication signal (ETH-AIS), 227–29
  Ethernet locked signal (ETH-LCK), 230–31
  Ethernet remote defect indication (ETH-RDI), 229–30
  frame delay measurements, 242–43
  frame format, 213
  frame loss measurements, 234–38
  frames, 209, 212
  interframe delay variation (IFDV) measurements, 243–44
  LBM, 220–24
  LBR, 220–24
  link, 202–5
  loopback testing, 204–5
  LTM, 224–27
  LTR, 224–27
  maintenance entity groups (MEPs), 207–8
  maintenance points (MPs), 208–11
  management entities (MEs), 207
  MPLS-TP, 352–61
  overview of, 5–6, 199–202
  PDUs, 202, 213
  PDUs, op codes, 215
  performance measurements, 231–33
  performance monitoring, 233–34, 263–66

Index 505

process operating boundaries, 201
security, 266
service, 205–7
service availability, 238–42
service configuration, 399, 401
standards summary, 200, 202
subscriber domain, 206
test frames, 234
testing, 244–63
O-LAN service
  defined, 145
  transit/access, 147
O-Line service
  defined, 145
  transit/access, 147
Open Cloud Connect (OCC), 14, 465, 479, 484
Operation, administration, and management. See OAM
Operation support systems (OSSs), 2, 13, 200, 250
Operator virtual connection (OVC), 127–28
  class diagram, 440
  defined, 4
  endpoint map example, 128
  EVC service over multiple-operator OVCs, 145
  functionalities, 148
  management, 398, 399
  MIBs, 408–10
  multiple OEP support, 127
  service attribute alignment, 417, 418
  service-level information modeling, 439–40
  See also OVC services
Organization, this book, 1–2
Organizational-Specific Slow Protocol (OSSP), 57
O-Tree service
  alternative combination, 147
  defined, 146–47
OVC services
  attributes, 151–52
  attributes per UNI, 152–53
  endpoint attributes per ENNI, 152
  ENNI service attributes, 150–51
  example delivery of, 149–50

existence of, 148–49
O-LAN, 145
O-Line, 145
O-Tree, 146–47
types of, 145–47
UNI service attributes, 151

P

Packet conditioning. See Traffic management
Packet switched network (PSN), 269
PAUSE, 17
PBEB (PB edge bridge), 305, 308
Performance management, 400, 402
Performance measurements, 6, 231–33
Performance monitoring, 233–34
  of CESoETH, 280
  class diagram, 415
  objects, 408
  SOAM, 402
Performance-monitoring (PM) solutions, 263–66
  dual-ended PM function, 265, 266
  layers, 264
  PM-1, 265
  PM-2, 265
  PM-3, 266
  PM-4, 266
  PM functions, 263–64
  single-ended PM function, 264, 265, 266
Periodic polling process, 294–95
Permanent virtual connection (PVC), 124
Platform as a Service (PaaS), 14, 483, 490
PM YANG module
  data definitions, 457
  groupings, 456–57
  notifications, 458–59
  overview, 455–56
  RPCs, 457–58
  type definitions, 456
  See also YANG models
Point-to-point service, 67
Policing, 164–65
Precision Time Protocol (PTP)
  data format, 44
  defined, 43
  frames, 48

Precision Time Protocol (PTP) (continued)
  network configuration/segmentation, 49
  synchronization computation, 46
Primary reference clock (PRC), 33, 34, 57–58
Primary reference source (PRS), 270
Priority code point (PCP)
  code point usage, 180
  defined, 117
  values for QoS and color, 180
Protection
  conclusion, 106
  Ethernet, 85–106
  linear, 87
  link aggregation, 100–106
  overview, 3
  ring, 95–100
  types, 94
  via dual homing, 383
Protection switching
  1 + 1, 88–90
  1:1, 90–91
  automatic (APS), 86, 91–95
  Ethernet automatic (EAPS), 87
  MPLS-TP, 361
  revertive/nonrevertive, 88
  T-MPLS, 328–33
  triggers, 91
  unidirectional, 87
Provider backbone bridge network (PBBN), 8, 9, 309, 310
Provider backbone bridges (PBBs), 1, 7
  802.1ah frame format, 305–7
  frame format, 306
  implementation, 304
  MAC address overlap, 311
  MAC-in-MAC encapsulation, 8, 304
  networks, 304, 308–11
  principles of operation, 307–8
Provider Backbone Bridging Traffic Engineering (PBBTE), 297–98
Provider backbone transport (PBT), 1, 7–9, 312–18
  802.1AB use, 313
  capabilities, 298
  combination of Ethernet extensions, 312
  defined, 297
  frame format, 9, 312

implementation, 314
MPLS internetworking, 318
multiple-domain network, 315
with non-PBT provider domain, 316
packets, 313
principles of operation, 313–15
single domain network, 316
spanning tree states and, 313
traffic-engineered (PBB-TE), 312, 313, 314, 316–18
underlying protocols, 8, 298
using, 312
Pseudowire emulation edge-to-edge. *See* PW3
Pseudowires (PWs), 2–3, 63–83
  architectures, 65–71
  conclusions, 82–83
  control plane, 71–73
  defined, 63
  demultiplexer layer and PSN, 70
  encapsulation, 69
  maintenance reference model, 70–71
  MPLS-TP and, 345
  multiple forwarding, 68
  multisegment architecture, 73–76, 343
  multisegment setup mechanisms, 76–78
  operations and maintenance, 79–81
  payload convergence layer, 68–70
  payload types, 64–65
  protocol layers, 63–64
  QoS and congestion control, 79
  resiliency, 78
  security, 81–82
  single segment (SS-PWs), 75, 342–43
  switching interprovider reference model, 75
  switching reference model, 74
  switching with AC reference model, 74
  in VPLS, 374
PW3
  control plane services, 71–72
  defined, 63
  encapsulation layer, 69
  maintenance reference model, 71
  network reference model, 66
  over IP PSN, 72
  over MPLS PSN, 73
  overview, 2–3
  preprocessing, 65–68

Index 507

preprocessing within network reference model, 66
protocol stack reference model, 64
signaling and encapsulation techniques, 75
PW maintenance entity group (PMEG), 355
PW SPME ME group (PSMEG), 355

Q

Quality-of-Service (QoS), 19, 79
Queuing, 165–66

R

Rapid Spanning Tree Protocol (RSTP), 85
Real-Time Protocol (RTP), 42, 276, 280–81
Real-Time Transport Control Protocol (RTCP), 42
Remote procedure calls (RPCs), 451, 454, 457–58
Remote UNI (RUNI), 128
Repeaters, 17
Retransmission timeout (RTO), 191, 192
RFC 2544, 246–49
Ring protection, 95–100
    architecture, 96
    conjoined rings, 97
    CSMA/CD, 16–17
    manual switching, 98
    MEPs, 99–100
    ring-fault operation, 97, 98
    switching, 97–100
    switching architecture for ring link failure, 99
Round-trip delay (RTD), 242
Round-trip interframe delay variation (IFDV), 243–44
Round-trip time (RTT), 191
Route tracing, 361

S

Section maintenance entity group (SMEG), 355
Security
    MPLS-TP, 361–62
    OAM, 266
    pseudowires (PWs), 81–82
    VPLS, 390
Security as a Service (SECaaS), 14, 489

Selective acknowledgement (SACK), 191
Service access points (SAPs), 31
Service activation measurement point (SAMP), 251
Service activation testing (SAT), 249–54
    appliances, 250
    Control Protocol, 254–55
    service activation criteria, 250
    service configuration test process, 252
    service configuration verification, 251
    Testing completion, 256–57
    Test sessions, 254, 256–63
Service availability, 238–42
Service classes
    3G traffic, 178
    defined by ETSI, 179
    IP, 176, 177
Service-level agreements (SLAs)
    defined, 5, 167, 397
    frame delay performance, 167–69
    frame delay range performance, 169–70
    frame delay variation, 170–71
    frame loss ratio, 171–72
    mean frame delay performance, 170
    performance measurements and, 231
    pseudowires (PWs), 79
Service-level information modeling, 435–44
    Carrier Ethernet endpoint overview, 437
    Carrier Ethernet external interface, 440
    egress equivalence class identifier, 443–44, 446
    ENNI, 440, 444
    ENNI service, 442–43, 445
    EVC, 438–39
    EVC service, 436
    LSO reference architecture, 436
    OVC, 439–40
    UNI, 440–41, 443
    See also Information modeling
Service-level specifications (SLSs)
    defined, 167, 397
    focused overload, 175–78
    multipoint CoS performance objectives, 173–75
    traffic management, 172–78
    See also Service-level agreements (SLAs)
Service management layer (SML), 394

Service OAM (SOAM), 205–7
Service performance objectives (CPOs), 252
Shaping, 188–97
Simple Network Time Protocol (SNTP), 39–41
Single-rate, three-color marking (SRTCM), 164
Single segment PWs (SS-PWs), 75, 342–43
SOAM FM MIB, 410–13
SOAM FM YANG module
    data definitions, 453–54
    fault management structure, 453
    groupings, 452
    list entries, 454
    notifications, 454–55
    overview, 452
    RPCs, 454
    type definitions, 452
    See also YANG models
SOAM PM MIB
    data definitions, 449–51
    delay measurement (DM) objects, 426–31
    groupings, 449
    loss management (LM) objects, 420–26
    notifications, 434–35, 452
    object groupings, 420
    objects, 413
    PM sessions, 414–15, 417–19
    remote procedure calls (RPCs), 451
    structure illustration, 450
    threshold crossing alerts (TCAs), 431–34
    See also YANG models
SOAM YANG CFM module
    overview, 446–47
    type definitions, 447–48
Software as a Service (SaaS), 14, 483, 490–91
Software-defined networking (SDN), 1, 2
Spanning Tree Protocol (STP), 85, 227
SP-PE TLV, 78
Stabilization period, 56
S-tag, 116
Standards
    Ethernet, 27
    MPLS-TP, 335, 336
    OAM, 200, 202
    synchronization, 38–39
Static MPLS-TP, 10, 336
Stratum clocking, 36, 40

Structure-agnostic TDM over packet (SAToP), 39, 269
Subnetwork Access Protocol (SNAP), 26, 31
S-VLANs, 309, 310
Switches, 19–20
Synchronization, 2, 33–60
    application requirements, 35–38
    applications, 37
    circuit emulation services (CES), 276–77
    clocking methods for, 52–54
    impact of packet network impairments on, 55–56
    introduction to, 33–35
    IWF function, 56–57
    NTP/SNTP, 39–42
    Precision Time Protocol (PTP), 43–49
    standards, 38–39
    synchronous Ethernet networks, 49–60
    types of, 36
Synchronous Ethernet networks
    clocking distribution methods, 49–50
    clocking methods for synchronization, 52–54
    EEC, 59–60
    frequency accuracy of slave clock, 59
    hierarchical timing distribution, 50
    IWF synchronization function, 56–57
    master-slave synchronization, 51
    operation modes, 58
    packet network impairments and, 55–56
    PRC, 57–58
    stabilization period, 56
    timing distribution via time stamps, 52
    timing recovery, 51, 53
Synchronous residual time stamp (SRTS), 52
Synchronous status messaging (SSM), 57–58

T

Tag Protocol Identifier (TPID), 117
Tandem connection monitoring (TCM), 353
TDM Pseudowire over Ethernet, 269
Temperature hardening, 27
Testing
    CBS, 253
    complete service, 249
    defined, 244

Index

device under test (DUT), 246–47, 249
EBS, 253
equipment, 244–45
importance of, 244
RFC 2544, 246–49
service activation, 249–54
Y.1731, 245–46
Test sessions, 254, 256–63
   backward, 261, 263
   forward, 257, 259
Third Network, 465–92
   cloud services architectures, 479–92
   conclusion, 492
   defined, 1, 14, 465
   introduction to, 465
   life-cycle service orchestration (LSO), 465–70
   network virtualization, 472–73
   service for enterprise customer, 466
   virtualization, 470–72
   virtualized Carrier Ethernet services, 473–79
Three-CoS model, 166–67, 168
Threshold crossing alerts (TCAs), 199, 431–34
Time deviation (TDEV), 59
Time division multiplexing (TDM)
   adaptation of signals, 35
   application signaling, 277–78
   local failure modification bits, 275
   network synchronization, 33–34
   over Ethernet for mobile backhaul, 272
   service emulation, 37
   service interface, 269
   virtual private line configurations, 271
Time domain reflectometer (TDR), 204–5
Time-phase requirements, 38
Timers, 94
TLVs (Type, Length, and Value)
   defined, 214
   generic format, 215
   interface status, 218
   organizationally specific format, 303
   sender ID, 218
   type field values, 214
Traffic-engineered PBB (PBB-TE), 312, 313, 314
   network, 316–18
   services, 317

topology, 317
trunks, 317, 318
Traffic management, 4–5
   application-CoS-priority mapping, 178–81
   bandwidth profile, 181–88
   CBS values, TCP, and shaping, 188–97
   conclusions, 197
   conditioning actions, 16
   defined, 4
   introduction to, 163–64
   policing, 164–65
   queuing, scheduling, and flow control, 165–66
   service-level agreements (SLAs), 167–72
   service-level specifications (SLSs), 172–78
   three-CoS model, 166–67, 168
Transport Control Protocol (TCP)
   average throughput, 192
   behavior with/without shaper, 195
   bottleneck, 195
   congestion control algorithm, 197
   flow bandwidth, 191
   flows, 190, 195, 196
   multiple flows, 196
   performance analysis, 190
   RTO and, 192
   segments received over time, 193, 194
   throughput limits, 192
   throughput through bandwidth profile policer, 191
   traffic interaction, 190
Transport MPLS (T-MPLS), 321–33
   APS channels field value description, 332
   APS payload structure, 331
   architectural principles, 321–22
   architecture, 324–28
   bidirectional LSPs, 323
   bidirectional switching, 332–33
   conclusion, 333
   data and control plane separation, 322
   defined, 9, 321
   forwarding behavior, 322
   frame structure, 328
   intended use, 322
   interfaces, 326–28
   introduction to, 321–23
   layers of network, 325

Transport MPLS (T-MPLS) (continued)
  mapping of frame structure, 329
  MPLS differences, 323–24
  MPLS versus, 9–10
  multi-operator network, 326
  network alternatives, 324
  networks, 10, 328–33
  NNI planes, 326, 327
  OAM methodology for, 325
  protection, 328–33
  as separate layer network, 9
  sublayer trail termination, 332
  subnetwork connection, 332
  target switching time, 333
  tunnels, services over, 330
  typical network, 329
Trunk links, 120, 121–22
TST frame, 245–46

U

User-to-network interface (UNI)
  attributes, 136
  bandwidth profile per, 183
  class diagram, 443
  configurations, 477
  customer (UNI-C), 207
  decision points at, 159
  defined, 123–24
  DP process flowchart at, 158
  EVC relationship, 125
  network (UNI-N), 207
  OVC attributes per, 152–53
  service attributes, 136, 151
  service attributes and parameters, 476
  service attributes for EPL, 140
  service-level information modeling, 440–41
  tunnel access, 128

V

Virtual circuit connectivity verification (VCCV), 80
Virtualization
  container-based, 472
  defined, 471
  general architecture, 472

  impact of, 470
  network, 472–73
  network function virtualization (NFV), 471
  popularity, 471
Virtualized Carrier Ethernet services, 473–79
  components, 474–78
  Ethernet private line (EPL) service, 473, 474
  overview, 473–74
  service chaining for EPL, 478–79
Virtual LANs (VLANs), 95, 96, 309–10, 473
Virtual private LAN service (VPLS), 2, 11–13, 365–91
  advantages of, 12–13, 369–70
  BGP approach, 384–90
  classification, 370–71
  conclusions, 391
  control plane, 384
  data forwarding on Ethernet PW, 378–79
  data plane, 370–72
  defined, 11, 367
  discovery, 375–76
  domain connected via 802.1ad links, 391
  emulating a LAN segment, 366
  encapsulation, 370
  ENNI interface, 390–91
  example, 378
  flooding, forwarding, and address learning, 372–73
  hierarchical (H-VPLS), 12, 379–84
  implementation, 368
  instances, 368, 371
  introduction to, 365–70
  key components, 11
  LDP-based, 12, 372–84
  MAC address learning and aging, 371–72
  multi-AS, 388–89
  multiple services, 375
  as multipoint service, 11
  network components, 367
  PEs, 371
  PW usage in, 374
  RFC 4364, 375
  security, 390
  single-bridge domain, 367
  solutions, 11, 365–66
  tunnel LSPs between PEs for, 366

Index

tunnel topology, 373–75
Virtual private networks (VPNs)
   Layer-2 multipoint, 365
   Layer-2 point-to-point, 365
   Layer-3 multipoint, 365
   MPLS-based, 365
   virtualization and, 473
Virtual private wire service (VPWS), 382–83
Virtual resources, 481
Virtual UNI (VUNI)
   decision points at, 159
   defined, 128
   DP process flowchart at, 158
   ENNI related objects, 405–6, 412, 413
   EVC implementation using, 129
   multiple, 130
   multiple EVCs supported by, 129

service attribute alignment, 417
service attributes, 128
VLAN identifier (VID), 117, 313
VPLS edge (VE), 12

W

Wait-to-restore (WTR) timer, 94–95

Y

YANG models, 444–59
   MEF architecture, 448
   NETCONF, 444, 446
   overview, 444–46
   PM module, 455–59
   SOAM CFM module, 446
   SOAM FM module, 452–55

Recent Titles in the Artech House Communications and Network Engineering Series

James Sterbenz, Senior Series Editor

Access Networks: Technology and V5 Interfacing, Alex Gillespie

Achieving Global Information Networking, Eve L. Varma et al.

Advanced High-Frequency Radio Communications, Eric E. Johnson et al.

ATM Interworking in Broadband Wireless Applications, M. Sreetharan and S. Subramaniam

ATM Switches, Edwin R. Coover

ATM Switching Systems, Thomas M. Chen and Stephen S. Liu

Broadband Access Technology, Interfaces, and Management, Alex Gillespie

Broadband Local Loops for High-Speed Internet Access, Maurice Gagnaire

Broadband Networking: ATM, SDH, and SONET, Mike Sexton and Andy Reid

Broadband Telecommunications Technology, Second Edition, Byeong Lee, Minho Kang, and Jonghee Lee

The Business Case for Web-Based Training, Tammy Whalen and David Wright

The Business Privacy Law Handbook, Charles H. Kennedy

Centrex or PBX: The Impact of IP, John R. Abrahams and Mauro Lollo

Chinese Telecommunications Policy, Xu Yan and Douglas Pitt

Communication and Computing for Distributed Multimedia Systems, Guojun Lu

Communications Technology Guide for Business, Richard Downey, Seán Boland, and Phillip Walsh

Community Networks: Lessons from Blacksburg, Virginia, Second Edition, Andrew M. Cohill and Andrea Kavanaugh, editors

Component-Based Network System Engineering, Mark Norris, Rob Davis, and Alan Pengelly

Computer Telephony Integration, Second Edition, Rob Walters

Creating Value-Added Services and Applications for Converged Communications and Networks, Johan Zuidweg

Customer-Centered Telecommunications Services Marketing, Karen G. Strouse

Delay- and Disruption-Tolerant Networking, Stephen Farrell and Vinny Cahill

Deploying and Managing IP over WDM Networks, Joan Serrat and Alex Galis, editors

Desktop Encyclopedia of the Internet, Nathan J. Muller

Digital Clocks for Synchronization and Communications, Masami Kihara, Sadayasu Ono, and Pekka Eskelinen

Digital Modulation Techniques, Second Edition, Fuqin Xiong

Disaster Recovery Planning for Communications and Critical Infrastructure, Leo A. Wrobel and Sharon M. Wrobel

E-Commerce Systems Architecture and Applications, Wasim E. Rajput

EMI Protection for Communication Systems, Kresimir Malaric

Engineering Internet QoS, Sanjay Jha and Mahbub Hassan

Error-Control Block Codes for Communications Engineers, L. H. Charles Lee

Essentials of Modern Telecommunications Systems, Nihal Kularatna and Dileeka Dias

FAX: Facsimile Technology and Systems, Third Edition, Kenneth R. McConnell, Dennis Bodson, and Stephen Urban

Fundamentals of Network Security, John E. Canavan

Gigabit Ethernet Technology and Applications, Mark Norris

The Great Telecom Meltdown, Fred R. Goldstein

Guide to ATM Systems and Technology, Mohammad A. Rahman

A Guide to the TCP/IP Protocol Suite, Floyd Wilder

Home Networking Technologies and Standards, Theodore B. Zahariadis

Implementing Value-Added Telecom Services, Johan Zuidweg

Information Superhighways Revisited: The Economics of Multimedia, Bruce Egan

Installation and Maintenance of SDH/SONET, ATM, xDSL, and Synchronization Networks, José M. Caballero et al.

Integrated Broadband Networks: TCP/IP, ATM, SDH/SONET, and WDM/Optics, Byeong Gi Lee and Woojune Kim

Internet E-mail: Protocols, Standards, and Implementation, Lawrence Hughes

Internet Technologies for Fixed and Mobile Networks, Toni Janevski

Introduction to Communication Networks, Tarmo Anttalainen and Ville Jääskeläinen

Introduction to Telephones and Telephone Systems, Third Edition, A. Michael Noll

An Introduction to U.S. Telecommunications Law, Second Edition, Charles H. Kennedy

IP Convergence: The Next Revolution in Telecommunications, Nathan J. Muller

LANs to WANs: The Complete Management Guide, Nathan J. Muller

The Law and Regulation of Telecommunications Carriers, Henk Brands and Evan T. Leo

Litigating with Electronically Stored Information, Marian K. Riedy, Susman Beros and Kim Sperduto

Managing Internet-Driven Change in International Telecommunications, Rob Frieden

Marketing Telecommunications Services: New Approaches for a Changing Environment, Karen G. Strouse

Mission-Critical Network Planning, Matthew Liotine

Multimedia Communications Networks: Technologies and Services, Mallikarjun Tatipamula and Bhumip Khashnabish, editors

Next Generation Intelligent Networks, Johan Zuidweg

Open Source Software Law, Rod Dixon

Performance Evaluation of Communication Networks, Gary N. Higginbottom

Performance of TCP/IP over ATM Networks, Mahbub Hassan and Mohammed Atiquzzaman

The Physical Layer of Communications Systems, Richard A. Thompson, David Tipper, Prashant Krishnamurthy, and Joseph Kabara

Power Line Communications in Practice, Xavier Carcelle

Practical Guide for Implementing Secure Intranets and Extranets, Kaustubh M. Phaltankar

Practical Internet Law for Business, Kurt M. Saunders

Practical Multiservice LANs: ATM and RF Broadband, Ernest O. Tunmann

Principles of Modern Communications Technology, A. Michael Noll

A Professional's Guide to Data Communication in a TCP/IP World, E. Bryan Carne

Programmable Networks for IP Service Deployment, Alex Galis et al., editors

Protocol Management in Computer Networking, Philippe Byrnes

Pulse Code Modulation Systems Design, William N. Waggener

Reorganizing Data and Voice Networks: Communications Resourcing for Corporate Networks, Thomas R. Koehler

Security, Rights, and Liabilities in E-Commerce, Jeffrey H. Matsuura

Service Assurance for Voice over WiFi and 3G Networks, Richard Lau, Ram Khare, and William Y. Chang

Service Level Management for Enterprise Networks, Lundy Lewis

SIP: Understanding the Session Initiation Protocol, Fourth Edition, Alan B. Johnston

Smart Card Security and Applications, Second Edition, Mike Hendry

SNMP-Based ATM Network Management, Heng Pan

Spectrum Wars: The Policy and Technology Debate, Jennifer A. Manner

Strategic Management in Telecommunications, James K. Shaw

Strategies for Success in the New Telecommunications Marketplace, Karen G. Strouse

Successful Business Strategies Using Telecommunications Services, Martin F. Bartholomew

Telecommunications Cost Management, S. C. Strother

Telecommunications Department Management, Robert A. Gable

Telecommunications Deregulation and the Information Economy, Second Edition, James K. Shaw

Telecommunications Technology Handbook, Second Edition, Daniel Minoli

Telemetry Systems Engineering, Frank Carden, Russell Jedlicka, and Robert Henry

Telephone Switching Systems, Richard A. Thompson

3D and HD Broadband Video Networking, Benny Bing

Third Networks and Services, Mehmet Toy and Hakkı Candan Çankaya

Understanding Modern Telecommunications and the Information Superhighway, John G. Nellist and Elliott M. Gilbert

Understanding Networking Technology: Concepts, Terms, and Trends, Second Edition, Mark Norris

Understanding SIP Servlets 1.1, Chris Boulton and Kristoffer Gronowski

Understanding Voice over IP Security, Alan B. Johnston and David M. Piscitello

Videoconferencing and Videotelephony: Technology and Standards, Second Edition, Richard Schaphorst

Virtualized Software-Defined Networks and Services, Qiang Duan and Mehmet Toy

Visual Telephony, Edward A. Daly and Kathleen J. Hansell

Wide-Area Data Network Performance Engineering, Robert G. Cole and Ravi Ramaswamy

Winning Telco Customers Using Marketing Databases, Rob Mattison

WLANs and WPANs towards 4G Wireless, Ramjee Prasad and Luis Muñoz

World-Class Telecommunications Service Development, Ellen P. Ward

For further information on these and other Artech House titles, including previously considered out-of-print books now available through our In-Print-Forever® (IPF®) program, contact:

Artech House
685 Canton Street
Norwood, MA 02062
Phone: 781-769-9750
Fax: 781-769-6334
e-mail: artech@artechhouse.com

Artech House
16 Sussex Street
London SW1V HRW UK
Phone: +44 (0)20 7596-8750
Fax: +44 (0)20 7630-0166
e-mail: artech-uk@artechhouse.com

Find us on the World Wide Web at: www.artechhouse.com